BIOFLUID DYNAMICS

Principles and Selected Applications

BIOFLUID DYNAMICS
Principles and Selected Applications

Clement Kleinstreuer

Taylor & Francis
Taylor & Francis Group
Boca Raton London New York

CRC is an imprint of the Taylor & Francis Group,
an informa business

Cover design: Christin C. Kleinstreuer

Published in 2006 by
CRC Press
Taylor & Francis Group
6000 Broken Sound Parkway NW, Suite 300
Boca Raton, FL 33487-2742

Printed in the United States of America on acid-free paper
10 9 8 7 6 5 4 3 2 1

International Standard Book Number-10: 0-8493-2221-9 (Hardcover)
International Standard Book Number-13: 978-0-8493-2221-1 (Hardcover)

Library of Congress Cataloging-in-Publication Data

Catalog record is available from the Library of Congress

informa
Taylor & Francis Group
is the Academic Division of Informa plc.

Visit the Taylor & Francis Web site at
http://www.taylorandfrancis.com

and the CRC Press Web site at
http://www.crcpress.com

To my family,
Christin, Nicole, and Joshua

CONTENTS

PREFACE

Due to its inherent complexity and increasing industrial applications, "Biofluid Dynamics" has evolved as an important research and teaching area, often in conjunction with "Biomechanics" and "Biomaterials." The focus is on "computational fluid-particle mechanics plus heat & mass transfer" as an integral part of biomedical engineering and related fields. This senior/graduate level text provides a unified treatment of the fundamental principles of biofluid dynamics and shows how to apply the principles towards solutions of transfer problems in the human body and medical devices. Prerequisites include junior-level courses in fluid mechanics, solid mechanics and heat transfer, at least freshman-level biology (see Glossary), as well as the standard engineering math series (see Section 1.3.5 and Appendix A). Employing an inductive, variably repeating approach for maximum pedagogical effects, *the overall goals are to provide the reader with a thorough biofluids background and good problem-solving skills for medical, regulatory, and industrial applications*. In order to achieve these modern objectives, a balanced physical-mathematical approach has to be taken, where a basic biology background is assumed. Once the reader has acquired a fundamental knowledge base and sufficient computational skills, advanced themes not directly addressed in the text can be readily explored. Many such themes are emerging from interfacing the life sciences and physical sciences. Examples include genetic manipulations, in-depth monitoring of complex biological processes, non-invasive surgical interventions, automatic in-vivo drug delivery, and, of course, comprehensive computer simulations of the dynamics of molecules, formation of drug compounds, cell interactions, neural networks, information processing, etc. However, knowing the underlying mechanisms of singular events is not enough – an understanding of the multi-level interaction processes is also a requirement for the development of new therapies and medical devices.

Existing biofluids books follow a more traditional approach, focusing mainly on flow in large arteries, usually excluding fluid-particle dynamics, lung airflow dynamics, and/or medical/regulatory applications. The books by Pedley (1980), Nichols & O'Rourke (1990), Mazumdar (1992), Chandran (1992), Fung (1997), and Fournier (1999) are typical examples. The undergraduate text by Truskey et al. (2004) nicely stresses mass transfer processes related to single and multiple cells as well as tissue and entire organ systems; but, they omitted heat transfer altogether. Then there is a number of edited books on the market, i.e., collections of proceedings, lecture series, and invited chapters on specific topics, being updated every five years or so.

Considering the number of established medical schools, medical device companies, and biomedical engineering departments as well as the formation of new bioengineering add-on programs, graduate centers, and start-up companies, this text should serve important market sectors.

Chapter I introduces and reviews basic biological and physical transport phenomena as well as two-phase flow and fluid-structure interaction aspects, stressing concepts, derivations and applications of the conservation laws plus closure models in a step-by-step, insightful fashion. Clearly, selecting topics from Chapter I is a "pick-and-choose" affair, depending upon the subject-reviews needed for: (a) the learning students and (b) the subsequent chapter/section selections by the instructor. Section 1.6 outlines a study guide for the serious student.

Advancing and applying the fundamentals discussed in Chapter I, Chapter II provides analyses and basic applications of exemplary biofluid flows in lumped-parameter compartments, tissues, knee joints, and microvascular beds (see Sect. 2.1) as well as in compartments of the cardiovascular system (see Sect. 2.2). *Chapters I and II form the "basic concept part," while Chapters III to V constitute the "applied part."*

Thus, the book may also serve as a text for a two-course sequence in "Biofluid Dynamics." Chapter III deals with physical factors of two critical arterial disease processes as well as surgical remedies based on computational analyses. Chapter IV focuses on biofluid dynamic aspects of the human lungs, kidneys, and liver. Chapter V then, relying on the material presented in the previous chapters, discusses modeling/design aspects of case studies ranging from drug delivery via microchannels and lung airways to stent-graft implantation to protect aortic aneurysms.

Four appendices provide: (i) review material of the engineering mathematics needed; (ii) a list of necessary partial differential equations; (iii) an overview of suitable numerical equation solvers, i.e., commercial software packages; and (iv) tables of material property data.

This well-balanced introductory text will fit nicely into existing and newly developed BME programs as well as engineering departments with bioengineering minors. Specifically, the book will be of interest to medical students, seniors, and first-year graduate students in the engineering sciences, as well as in applied mathematics and the physical/life sciences. It will also be a useful reference for practicing engineers, scientists, and research MDs.

Although this is the first rather comprehensive textbook in "biofluid dynamics," as all books, it relied on numerous sources, e.g., related books, journal articles and reviews, as well as the author's colleagues, research associates, and former graduate students. Special thanks go to Dr P. Worth Longest, Dr Junemo Koo, Dr Zhonghua Li, and Dr Zhe Zhang for their contributions to Chapter V. Mrs Joyce Sorensen and Mrs Joanne Self expertly typed the first draft of the manuscript. The critical comments and helpful suggestions provided by the expert reviewers Stan Berger (UC-Berkely, CA), Chris Bertram (UNSW, Australia), Danny Bluestein (SUNYSB, NY), Ashim Datta (Cornell Univ., NY), Samir Ghadiali (Lehigh Univ., PA), Jimmy Moore (Texas A & M, TX), Stanley Rittgers (UoA, OH), Geert Schmid-Schönbein (UC-SD, CA), Dalin Tang (WPI, MA), and Zahir Warsi (MSU, MS) are gratefully acknowledged as well. Also very valuable was the page-by-page critique from a student perspective by Robert Richter, a BME Senior at NCSU. Many thanks for their support go also to the editorial staff at CRC Press, especially Mr. Michael Slaughter, Ms Helena Redshaw and Ms Suzanne Lassandro, as well as to the professionals in the Mechanical Engineering Department and several Libraries at Stanford University.

The text could not have been produced without the expertise and help of Research Associate Professor Dr Zhe Zhang who reformatted the manuscript, generated the figures and graphs, and contributed to several problem solutions. Dr Zhang is the author of the Solutions Manual accompanying this book.

For technical correspondence, especially new homework assignments and future text changes based on teaching experience, please contact the author via email (ck@eos.ncsu.edu) or fax (919.515.7968).

Clement Kleinstreuer
Raleigh 2005

Glossary

A

absorption Transport of molecules across (epithelial) membranes into the body fluids.

ADH (Antidiuretic Hormone): It acts on the kidneys to promote water reabsorption, thereby decreasing the urine volume.

adrenals Glands near the kidneys secreting adrenaline.

arteriole Small terminal branch of an artery that typically connects with a capillary.

arteriosclerosis Any group of diseases characterized by thickening and hardening of the artery wall and narrowing of its lumen.

arteriovenous anastomosis Direct connection between an artery and a vein that bypasses the capillary bed.

atherosclerosis A common type of arteriosclerosis found in medium and large arteries in which raised areas, or "plaques," within the tunica intima are formed from smooth muscle cells, cholesterol, and other lipids. These plaques occlude arteries and serve as sites for the formation of thrombi.

ATP (Adenosine triphosphate): "The universal energy carrier of the cell."

autoregulation The ability of an organ to intrinsically modify the degree of constriction or dilation of its small arteries and arterioles and to thus regulate the rate of its own blood flow. Autoregulation may occur through myogenic or metabolic mechanisms.

B

bile Fluid produced by the liver and stored in the gallbladder that contains bile salts, bile pigments, cholesterol, and other molecules. The bile is secreted into the small intestine.

blood-brain barrier The structures and cells that selectively prevent particular molecules in the plasma from entering the central nervous system.

bronchiole The smallest of the air passages in the lungs, containing smooth muscle and cuboidal epithelial cells.

buffer A molecule that serves to prevent large changes in pH by either combining with H^+ or by releasing H^+ into solution.

C

calyx A cup-shaped cavity in an organ (pl. calices).

capillary The smallest vessel in the vascular system. Capillary walls are only one cell thick, and all exchanges of molecules between the blood and tissue fluid occur across the capillary wall.

carbohydrate An organic molecule containing carbon, hydrogen, and oxygen in a ratio of 1:2:1. The carbohydrate class of molecules is subdivided into monosaccharides, disaccharides, and polysaccharides.

cardiac muscle Muscle of the heart, consisting of striated muscle cells. These cells are interconnected into a mass called the myocardium.

cardiac output The volume of blood pumped by either the right or the left ventricle each minute.

carrier-mediated transport The transport of molecules or ions across a cell membrane by means of specific protein carriers. It includes both facilitated diffusion and active transport.

catalyst A substance that increases the rate of a chemical reaction without changing the nature of the reaction or being changed by the reaction.

cellular respiration The energy-releasing metabolic pathways in a cell that oxidize organic molecules such as glucose and fatty acids.

chemoreceptor A neural receptor that is sensitive to chemical changes in blood and other body fluids.

chemotaxis The movement of an organism or a cell, such as a leukocyte, toward a chemical stimulus.

cholesterol A twenty-seven-carbon steroid that serves as the precursor of steroid hormones.

chromosome A structure in the cell nucleus, containing DNA and associated proteins, as well as RNA, that is made according to the genetic instructions in the DNA.

cilia Tiny hairlike processes extending from the cell surface that beat in a coordinated fashion.

cirrhosis Liver disease characterized by the loss of normal microscopic structure, which is replaced by fibrosis and nodular regeneration.

clone **1.** A group of cells derived from a single parent cell by mitotic cell division; since reproduction is asexual, the descendants of the parent cell are genetically identical. **2.** A term used to refer to cells as separate individuals (as in white blood cells) rather than as part of a growing organ.

CNS (Central nervous system): That part of the nervous system consisting of the brain and spinal cord.

cochlea The organ of hearing in the inner ear where nerve impulses are generated in response to sound waves.

conducting zone The structures and airways that transmit inspired air into the respiratory zone transmit inspired air into the respiratory zone of the lungs, where gas exchange occurs. The conducting zone includes such structures as the trachea, bronchi, and larger bronchioles.

congestive heart failure The inability of the heart to deliver an adequate blood flow due to heart disease or hypertension. It is associated with breathlessness, salt and water retention, and edema.

connective tissue One of the four primary tissues, characterized by an abundance of extracellular material.

cortex **1.** The outer layer of an internal organ or body structure, as of the kidney or adrenal gland. **2.** The convoluted layer of gray matter that covers the surface of the cerebral hemispheres.

cotransport Also called *coupled transport or secondary active transport.* Carrier-mediated transport in which a single carrier transports an ion (e.g., NA^+) down its concentration gradient while transporting a specific molecule (e.g., glucose) against its concentration gradient. The hydrolysis of ATP is indirectly required for cotransport because it is needed to maintain the steep concentration gradient of the ion.

countercurrent exchange The process that occurs in the vasa recta of the renal medulla in which blood flows in U-shaped loops. This allows sodium chloride to be trapped in the interstitial fluid while water is carried away from the kidneys.

countercurrent multiplier system The interaction that occurs between the descending limb and the ascending limb of the loop of Henle in the kidney. This interaction results in the multiplication of the solute concentration in the interstitial fluid of the renal medulla.

crenation A notched or scalloped appearance of the red blood cell membrane caused by the osmotic loss of water from these cells.

cyto- Cell.

cytoplasm The semifluid part of the cell between the cell membrane and the nucleus, exclusive of membrane-bound organelles. It contains many enzymes and structural proteins.

cytoskeleton A latticework of structural proteins in the cytoplasm arranged in the form of microfilaments and microtubules.

D

Dalton's law The statement that the total pressure of a gas mixture is equal to the sum that each individual gas in the mixture would exert independently. The part contributed by each gas is known as the partial pressure of the gas.

deductive approach Explaining topics or solving problems starting with general theories, concepts and/or equations to arrive at a specific statement or problem solution. Opposite: **inductive approach**

dialysis A method of removing unwanted elements from the blood by selective diffusion through a semipermeable membrane.

diastole The phase of relaxation in which the heart fills with blood. Unless accompanied by the modifier *atrial*, diastole refers to the resting phase of the ventricles.

diastolic blood pressure The minimum pressure in the arteries that is produced during the phase of diastole of the heart. It is indicated by the last sound of Korotkoff when taking a blood pressure measurement.

diffusion The net movement of molecules or ions from regions of higher to regions of lower concentration.

dopamine A type of neurotrans-mitter in the central nervous system; also is the precursor of norepinephrine, another neurotransmitter molecule.

E

ECG (Electrocardiogram) (also abbreviated EKG): A recording of electrical currents produced by the heart.

edema Swelling due to an increase in tissue fluid.

efferent Conveying or transporting something away from a central location. Efferent nerve fibers conduct impulses away from the central nervous system, for example, and efferent arterioles transport blood away from the glomerulus.

electrolyte An ion or molecule that is able to ionize and thus carry an electric current. The most common electrolytes in the plasma are Na^+, HCO_3^-, and K^+.

emphysema A lung disease in which alveoli are destroyed and the remaining alveoli become larger. It results in decreased vital capacity and increased airway resistance.

endo- Within, inner.

endocrine glands Glands that secrete hormones into the circulation rather than into a duct; also called *ductless glands*.

endocytosis The cellular uptake of particles that are too large to cross the cell membrane. This occurs by invagination of the cell membrane until a membrane-enclosed vesicle is pinched off within the cytoplasm.

endothelium The simple squamous epithelium that lines blood vessels and the heart.

entropy A measure of the degree of disorder in a system, entropy increases whenever energy is transformed.

enzyme A protein catalyst that increases the rate of specific chemical reactions.

epi- Upon, over, outer.

epithelium One of the four primary tissue types; the type of tissue that covers and lines the body surfaces and forms exocrine and endocrine glands.

equal signs $:\hat{=}$ means "equivalent to" $:=$ "equal to computed value" or "equal to derived expression."

erythrocyte A red blood cell. Erythrocytes are the formed elements of blood that contain hemoglobin and transport oxygen.

exocrine gland A gland that discharges its secretion through a duct to the outside of an epithelial membrane.

F

facilitated diffusion The carrier-mediated transport of molecules through the cell membrane along the direction of their concentration gradients. It does not require the expenditure of metabolic energy.

fibrillation A condition of cardiac muscle characterized electrically by random and continuously changing patterns of electrical activity and resulting in the inability of the myocardium to contract as a unit and pump blood. It can be fatal if it occurs in the ventricles.

fibrin The insoluble protein formed from fibrinogen by the enzymatic action of thrombin during the process of blood clot formation.

Frank-Starling law of the heart The statement describing the relationship between end-diastolic volume and stroke volume of the heart. Basically, it is the principle that within physiology limits, the heart will pump all the blood which returns to it.

G

gas exchange The diffusion of oxygen and carbon dioxide down their concentration gradients that occurs between pulmonary capillaries and alveoli, and between systemic capillaries and the surrounding tissue cells.

gates A term used to describe structures within the cell membrane that regulate the passage of ions through membrane channels. Gates may be chemically regulated (by neurotransmitters) or voltage regulated (in which case they open in response to a threshold level of depolarization).

glomerular ultrafiltrate Fluid filtered through the glomerular capillaries into glomerular (Bowman's) capsule of the kidney tubules.

glomeruli The tufts of capillaries in the kidneys that filter fluid into the kidney tubules.

glycocalyx Glycoproteins on the surface of cells.

glycogen A polysaccharide of glucose – also called *animal starch* – produced primarily in the liver and skeletal muscles. Similar to plant starch in composition, glycogen contains more highly branched chains of glucose subunits than does plant starch.

gonads Sexual glands, e.g., ovary or testis

H

heart murmur An abnormal heart sound caused by an abnormal flow of blood in the heart due to structural defects, usually of the valves or septum.

heart sounds The sounds produced by closing of the AV valves of the heart during systole (the first sound) and by closing of the semilunar valves of the aorta and pulmonary trunk during diastole (the second sound).

hematocrit The ratio of packed red blood cells to total blood volume in a centrifuged sample of blood, expressed as a percentage.

heme The iron-containing red pigment that, together with the protein globin, forms hemoglobin.

hemoglobin The combination of heme pigment and protein within red blood cells that acts to transport oxygen and (to a lesser degree) carbon dioxide. Hemoglobin also serves as a weak buffer within red blood cells.

Henry's law The statement that the concentration of gas dissolved in a fluid is directly proportional to the partial pressure of that gas.

heparin A mucopolysaccharide found in many tissues, but in greatest abundance in the lungs and liver, that is used medically as an anticoagulant.

hepatitis Inflammation of the liver.

high-density lipoproteins (HDLs) Combinations of lipids and proteins that migrate rapidly to the bottom of a test tube during centrifugation. HDLs are carrier proteins that are believed to transport cholesterol away from blood vessels to the liver, and thus to offer some protection from atherosclerosis.

histamine A compound secreted by tissue mast cells and other connective tissue cells that stimulates vasodilation and increases capillary permeability. It is responsible for many of the symptoms of inflammation and allergy.

homeo Same.

homeostasis The dynamic constancy of the internal environment, the maintenance of which is the principal function of physiological regulatory mechanisms. The concept of homeostasis provides a framework for understanding most physiological processes.

hormone A regulatory chemical produced in an endocrine gland that is secreted into the blood and carried to target cells that respond to the hormone by an alteration in their metabolism.

hydrophilic Denoting a substance that readily absorbs water; literally, "water loving."

hydrophobic Denoting a substance that repels, and that is repelled by, water; literally, "water fearing."

hyperemia Excessive blood flow to a part of the body.

hyperplasia An increase in organ size due to an increase in cell numbers as a result of mitotic cell division.

hypertension High blood pressure. Classified as either primary, or essential, hypertension of unknown cause or secondary hypertension that develops as a result of other, known disease processes.

hypertonic A solution with a greater solute concentration and thus a greater osmotic pressure than plasma.

hyperventilation A high rate and depth of breathing that results in a decrease in the blood carbon dioxide concentration to below normal.

hypo- Under, below, less.

hypotension Abnormally low blood pressure.

hypothermia A low body temperature. This is a condition that is defended against by shivering and other physiological mechanisms that generate body heat.

hypovolemic A rapid fall in blood pressure as a result of diminished blood volume.

hypoxemia A low oxygen concentration of the arterial blood.

I

immunization The process of increasing one's resistance to pathogens. In active immunity a person is injected with antigens that stimulate the development of clones of specific B or T lymphocytes; in passive immunity a person is injected with antibodies made by another organism.

immunoassay Any of a number of laboratory or clinical techniques that employ specific bonding between an antigen and its homologous antibody in order to identify and quantify a substance in a sample.

infarct An area of necrotic (dead) tissue as a result of inadequate blood flow (ischemia).

insulin A polypeptide hormone secreted by the beta cells of the islets of Langerhans in the pancreas that promotes the anabolism of carbohydrates, fat, and protein. Insulin acts to promote the cellular uptake of blood glucose and, therefore, to lower the blood glucose concentration; insulin deficiency produces hyperglycemia and diabetes mellitus.

inter- Between, among.

interferons Small proteins that inhibit the multiplication of viruses inside host cells and that also have antitumor properties.

intra- Within, inside.

intrapulmonary The space within the air sacs and airways of the lungs.

in vitro Occurring outside the body, in a test tube or other artificial environment.

in vivo Occurring within the body.

ionization The dissociation of a solute to form ions.

ischemia A rate of blood flow to an organ that is inadequate to supply sufficient oxygen and maintain aerobic respiration in that organ.

iso- Equal, same.

isotonic solution A solution having the same total solute concentration, osmolality, and osmotic pressure as the solution with which it is compared; a solution with the same solute concentration and osmotic pressure as plasma.

L

Laplace, law of The statement that the pressure within an alveolus is directly proportional to its surface tension and inversely proportional to its radius.

larynx A structure consisting of epithelial tissue, muscle, and cartilage that serves as a sphincter guarding the entrance of the trachea. It is the organ responsible for voice production.

lesion A wounded or damaged area.

leukocyte A white blood cell.

ligament A tough cord or fibrous band of dense regular connective tissue that contains many parallel arrangements of collagen fibers. It connects bones or cartilages and serves to strengthen joints.

ligand A smaller molecule that chemically binds to a larger molecule, which is usually a protein. Oxygen, for example, is the ligand for the heme in hemoglobin, and hormones or neurotransmitters can be the ligands for specific membrane proteins.

lipid An organic molecule that is nonpolar and thus insoluble in water. Lipids include triglycerides, steroids, and phospholipids.

low-density lipoproteins (LDLs) Plasma proteins that transport triglycerides and cholesterol to the arteries and that are believed to contribute to ateriosclerosis.

lumen The cavity of a tube or hollow organ.

lung surfactant A mixture of lipoproteins (containing phospholipids) secreted by type II alveolar cells into the alveoli of the lungs. It lowers surface tension and prevents collapse of the lungs, as occurs in hyaline membrane disease when surfactant is absent.

lymph The fluid in lymphatic vessels that is derived from tissue fluid.

lymphatic system The lymphatic vessels and lymph nodes.

lymphocyte A type of mononuclear leukocyte, the cell responsible for humoral and cell-mediated immunity.

lymphokine Any of a group of chemicals released from T cells that contribute to cell-mediated immunity.

M

macromolecule A large molecule; a term commonly used to refer to protein, RNA, and DNA.

macrophage A large phagocytic cell in connective tissue that contributes to both specific and nonspecific immunity.

malignant A structure or process that is life threatening. Of a tumor, tending to metastasize.

mean arterial pressure An adjusted average of the systolic and diastolic blood pressures. It averages about 100 mmHg in the systemic circulation and 10 mmHg in the pulmonary circulation.

medulla oblongata A part of the brain stem that contains neural centers for the control of breathing and for regulation of the cardiovascular system via autonomic nerves.

membrane potential The potential difference of voltage that exists between the inner and outer sides of a cell membrane. It exists in all cells, but is capable of being changed by excitable cells (neurons and muscle).

meta- Change.

metabolism All of the chemical reactions in the body; it includes those that result in energy storage (anabolism) and those that result in the liberation of energy (catabolism).

metastasis A process whereby cells of a malignant tumor separate from the tumor, travel to a different site, and divide to produce a new tumor.

microvilli Tiny fingerlike projections of a cell membrane. They occur on the apical (luminal) surface of the cells of the small intestine and in the renal tubules.

mitosis Cell division in which the two daughter cells receive the same number of chromosomes as the parent cell (both daughters and parent are diploid).

molar Pertaining to the number of moles of solute per liter of solution.

mole The number of grams of a chemical that is equal to its formula weight (atomic weight for an element or molecular weight for a compound).

monoclonal antibodies Identical antibodies derived from a clone of genetically identical plasma cells.

monocyte A mononuclear, nongranular leukocyte that is phagocytic and that can be transformed into a macrophage.

mucous membrane The layers of visceral organs that include the lining epithelium, submucosal connective tissue, and (in some cases) a thin layer of smooth muscle (the muscularis mucosa).

myocardial infraction An area of necrotic tissue in the myocardium that is filled in by scar (connective) tissue.

myogenic Originating within muscle cells; this term is used to describe self-excitation by cardiac and smooth muscle cells.

N

necrosis Cellular death within tissues and organs as a result of pathological conditions. This differs histologically from the physiological cell death of apoptosis.

negative feedback A mechanism of response that serves to maintain a state of internal constancy, or homeostasis. Effectors are activated by changes in the internal environment, and the actions of the effectors serve to counteract these changes and maintain a state of balance.

neoplasm A new, abnormal growth of tissue, as in a tumor.

nephron The functional unit of the kidneys, consisting of a system of renal tubules and a vascular component that includes capillaries of the glomerulus and the peritubular capillaries.

Nernst equation The equation used to calculate the equilibrium membrane potential for given ions when the concentrations of those ions on each side of the membrane are known.

nerve A collection of motor axons and sensory dendrites in the peripheral nervous system.

neuron A nerve cell, consisting of a cell body that contains the nucleus, short branching processes, called dendrites, that carry electrical charges to the cell body, and a single fiber, or axon, that conducts nerve impulses away from the cell body.

neurotransmitter A chemical contained in synaptic vesicles in nerve endings that is released into the synaptic cleft, where it stimulates the production of either excitatory or inhibitory postsynaptic potentials.

nexus A bond between members of a group; the type of intercellular connection found in single-unit smooth muscles.

nucleolus A dark-staining area within a cell nucleus; the site where ribosomal RNA is produced.

nucleoplasm The protoplasm of a nucleus.

nucleotide The subunit of DNA and RNA macromolecules. Each nucleotide is composed of a nitrogenous base (adenine, guanine, cytosine, and thymine or uracil); a sugar (deoxyribose or ribose); and a phosphate group.

nucleus brain An aggregation of neuron cell bodies within the brain. Nuclei within the brain are surrounded by white matter and are deep to the cerebral cortex.

nucleus, cell The organelle, surrounded by a double saclike membrane called the nuclear envelope, that contains the DNA and genetic information of the cell.

O

organ A structure in the body composed of two or more primary tissues that performs a specific function.

organelle A structure within cells that performs specialized tasks. The term includes microchondria, Golgi apparatus, endoplasmic reticulum, nuclei, and lysosomes;

it is also used for some structures not enclosed by a membrane, such as ribosomes and centrioles.

osmolality A measure of the total concentration of a solution; the number of moles of solute per kilogram of solvent.

osmosis The passage of solvent (water) from a more dilute to a more concentrated solution through a membrane that is more permeable to water than to the solute.

osmotic pressure A measure of the tendency for a solution to gain water by osmosis when separated by a membrane from pure water. Directly related to the osmolality of the solution, it is the pressure required to just prevent osmosis.

osteo- Pertaining to bone.

osteoporosis Demineralization of bone, seen most commonly in postmenopausal women and patients who are inactive or paralyzed. It may be accompanied by pain, loss of stature, and other deformities and fractures.

oxidizing agent An atom that accepts electrons in an oxidation-reduction reaction.

P

pacemaker of the heart The fastest depolarizing cell, usually in the sinoatrial node (SA node). These autorythmic cells in the right atrium serve as the main pacemaker of the heart.

pathogen Any disease-producing microorganism or substance.

peptic ulcer An injury to the mucosa of the esophagus, stomach, or small intestine caused by acidic gastric juice.

perfusion The flow of blood through an organ.

peri- Around, surrounding.

perilymph The fluid between the membranous and bony labyrinths of the inner ear.

peripheral resistance The resistance to blood flow through the arterial system. Peripheral resistance is largely a function of the radius of small arteries and arterioles. The resistance to blood flow is proportional to the fourth power of the radius of the vessel.

peristalsis Waves of smooth muscle contraction in smooth muscles of the tubular digestive tract that involve circular and longitudinal muscle fibers at successive locations along the tract. It serves to propel the contents of the tract in one direction.

PET, PTFE, ePTFE Synthetic graft materials, either Dacron or Teflon based.

pH The symbol (short for potential of hydrogen) used to describe the hydrogen ion (H^+) concentration of a solution. The pH scale in common use ranges from 0 to 14. Solutions with a pH of 7 are neutral; those with a pH lower than 7 are acidic; and those with a higher pH are basic.

pituitary gland Also called the *hypophysis*. A small endocrine gland joined to the hypothalamus at the base of the brain. The pituitary gland is functionally divided into anterior and posterior portions. The anterior pituitary secretes ACTH, TSH, FSH, LH, growth hormone, and prolactin. The posterior pituitary secretes oxytocin and antidiuretic hormone (ADH).

plasma The fluid portion of the blood. Unlike serum (which lacks fibrinogen), plasma is capable of forming insoluble fibrin threads when in contact with test tubes.

plasma cells Cells derived from B lymphocytes that produce and secrete large amounts of antibodies. They are responsible for humoral immunity.

platelet A disc-shaped structure, 2 to 4 micrometers in diameter, that is derived from bone marrow cells called megakaryocytes. Platelets circulate in the blood and participate (together with fibrin) in forming blood clots.

PNS The peripheral nervous system, including nerves and ganglia.

polymer A large molecule formed by the combination of smaller subunits, or monomers.

portal system A system of vessels consisting of two capillary beds in series, where blood from the first is drained by veins into a second capillary bed, which in turn is drained by veins that return blood to the heart. The two major portal systems in the body are the hepatic portal system and the hypothalamo-hypophyseal portal system.

positive feedback A mechanism of response that results in the amplification of an initial change. Positive feedback results in avalanche-like effects, as occur in the formation of a blood clot or in the production of the LH surge by the stimulatory effect of estrogen.

posterior At or toward the back of an organism, organ, or part; the dorsal surface.

potential difference In biology, the difference in charge between two solutions separated by a membrane. The potential difference is measured in voltage.

prophylaxis Prevention of protection.

protein The class of organic molecules composed of large polypeptides, in which over a hundred amino acids are bonded together by pepticle bonds.

protoplasm A general term for the colloidal complex of protein that includes cytoplasm and nucleoplasm.

R

reabsorption The transport of a substance from the lumen of the renal nephron into the peritubular capillaries.

receptive field An area of the body that, when stimulated by a sensory stimulus, activates a particular sensory receptor.

renal Pertaining to the kidneys.

renal plasma clearance The milliliters of plasma that are cleared of a particular solute each minute by the excretion of that solute in the urine. If there is no reabsorption or secretion of that solute by the nephron tubules, the renal plasma clearance is equal to the glomerular filtration rate.

repolarization The reestablishment of the resting membrane potential after depolarization has occurred.

respiratory zone The region of the lungs including the respiratory bronchioles, in which individual alveoli are found, and the terminal alveoli. Gas exchange between the inspired air and pulmonary blood occurs in the respiratory zone of the lungs.

REV (Representative Elementary Volume): Open system (e.g., a cube) of differential size for the derivation of balance equations.

S

semilunar valves The valve flaps of the aorta and pulmonary artery at their juncture with the ventricles.

semipermeable membrane A membrane with pores of a size that permits the passage of solvent and some solute molecules but restricts the passage of other solute molecules.

sensory neuron An afferent neuron that conducts impulses from peripheral sensory organs into the central nervous system.

serum The fluid squeezed out of a clot as it retracts; the supernatant when a sample of blood clots in a test tube and is centrifuged. Serum is plasma from which fibrinogen and other clotting proteins have been removed as a result of clotting.

shock As it relates to the cardiovascular system, a rapid, uncontrolled fall in blood pressure, which in some cases becomes irreversible and leads to death.

sinus A cavity.

sinusoid A modified capillary with a relatively large diameter that connects the arterioles and venules in the liver, bone marrow, lymphoid tissues, and some endocrine organs. In the liver, sinusoids are partially lined by phagocytic cells of the reticuloendothelial system.

skeletal muscle pump A term used with reference to the effect of skeletal muscle contraction on the flow of blood in veins. As the muscles contract, they squeeze the veins, and in this way help move the blood toward the heart.

smooth muscle A specialized type of nonstriated muscle tissue composed of fusiform, single-nucleated fibers. It contracts in an involuntary, rhythmic fashion in the walls of visceral organs.

sodium/potassium pump An active transport carrier, with ATPase enzymatic activity that acts to accumulate K^* within cells and extrude Na^+ from cells, thus maintaining gradients for these ions across the cell membrane.

Starling forces The hydrostatic pressures and the colloid osmotic pressures of the blood and tissue fluid. The balance of the Starling forces determines the net movement of fluid out of or into blood capillaries.

steroid A lipid, derived from cholesterol, that has three six-sided carbon rings and one five-sided carbon ring. These form the steroid hormones of the adrenal cortex and gonads.

surfactant In the lungs, a mixture of phospholipids and proteins produced by alveolar cells that reduces the surface tension of the alveoli and contributes to the lungs' elastic properties.

synergistic Pertaining to regulatory processes or molecules (such as hormones) that have complementary or additive effects.

systemic circulation The circulation that carries oxygenated blood from the left ventricle via arteries to the tissue cells and that carries blood depleted in oxygen via veins to the right atrium; the general circulation, as compared to the pulmonary circulation.

systole The phase of contraction in the cardiac cycle. Used alone, this term refers to contraction of the ventricles; the term atrial systole refers to contraction of the atria.

T

T cell A type of lymphocyte that provides cell-mediated immunity (in contrast to B lymphocytes, which provide humoral immunity through the secretion of antibodies). There are three subpopulations of T cells: cytotoxic (killer), helper, and suppressor.

tendon The dense regular connective tissue that attaches a muscle to the bones of its origin and insertion.

thorax The part of the body cavity above the diaphragm; the chest.

threshold The minimum stimulus that just produces a response.

thrombin A protein formed in blood plasma during clotting that enzymatically converts the soluble protein fibrinogen into insoluble fibrin.

thrombosis The development or presence of a thrombus.

thrombus A blood clot produced by the formation of fibrin threads around a platelet plug.

thymus A lymphoid organ located in the superior portion of the anterior mediastrinum. It processes T lymphocytes and secretes hormones that regulate the immune system.

total minute volume The product of tidal volume (ml per breath) and ventilation rate (breaths per minute).

transplantation The grafting of tissue from one part of the body to another part, or from a donor to a recipient.

transpulmonary pressure The pressure difference across the wall of the lung, equal to the difference between intrapulmonary pressure and intrapleural pressure.

V

vasa-, vaso- Pertaining to blood vessels.

vasa vasora Blood vessels that supply blood to the walls of large blood vessels.

vasoconstriction A narrowing of the lumen of blood vessels as a result of contraction of the smooth muscles in their walls.

ventilation Breathing; the process of moving air into and out of the lungs.

villi Fingerlike folds of the mucosa of the small intestine.

vital capacity The maximum amount of air that can be forcibly expired after a maximal inspiration.

Chapter I

ELEMENTS OF CONTINUUM MECHANICS

A sound understanding of the physics of fluid flow, heat/mass transfer, two-phase flow and stress-strain theory as well as a mastery of basic solution techniques is an important prerequisite for studying and applying biofluid dynamics systems. As always, the *objective* is to learn to develop mathematical models; here, approximate representations of actual biofluid flow phenomena in terms of differential or integral equations. The (numerical) solutions to the describing equations should produce testable predictions and allow for the analysis of biofluid system variations leading to a deeper *understanding* and possibly to new or improved medical procedures or devices *by design*. Fortunately, most biofluid dynamics systems are governed by continuum mechanics laws with notable micro- and nano-scale exceptions, requiring molecular models solved via statistical mechanics or molecular dynamics simulations.

Traditionally, the answer to a given (flow) problem is obtained by copying suitable equations, submodels, and boundary conditions with their appropriate solution techniques from available sources. This is called "matching" and may result in a good first-step learning experience; however, it should be augmented later on by more independent work, e.g., deriving governing equations, plotting and visualizing results, improving basic submodels, finding new, interdisciplinary applications, exploring new concepts, interpreting observations in a more generalized form, or even pushing the envelope of existing solution techniques. In any case, the triple pedagogical goals of *understanding*, *skills*, and *design* can be achieved only via *independent* practice, hard work, and creative thinking. To reach these lofty goals, a "bottom-up," i.e., from-the-simple-to-the-complex, approach is adopted throughout this text, where more general biofluid dynamics principles are recognized and assembled from special cases. For example, in this chapter basic transport phenomena needed in biofluid dynamics modeling (see Chapter II) are reviewed, while in Chapters III and IV major systems are introduced in increasing order of complexity. In Chapter V the case studies are presented again in an inductive fashion, i.e., building on simple examples to arrive at more general methods or conclusions. Repetition of some key concepts or topics from different angles is another pedagogical device used throughout the text. In general, as for all engineering subjects, the reader should strive for an equal balance between *physical insight* and *mathematical application skills* (see App. A).

1.1 BIOLOGICAL TRANSPORT PROCESSES

After sufficient uptake of air, water, minerals, and nutrients, complex life forms require well-regulated *transport processes*, e.g., ions, molecules, proteins, and organelles moving in fluids and interacting/binding with membranes, cells, tissue, and organs. A nice description of diffusive mass transport within a cell, across a cell, between cells (i.e., inside tissues), and in organs may be found in the book by Truskey et al. (2004). Some biomechanical processes are even more complex than convection/diffusion. For example, monolayers of cells lining blood vessels or lung airways are exposed to fluid mechanical loads, such as lumen pressure, wall shear, and cell-cell contact forces, which in turn change the shape of the cells and trigger the production of different molecules as well as changes in gene expression. In general, endothelial cell shape and metabolic activity is greatly influenced by wall shear stress magnitude and temporal/spatial changes, while intramural cells (smooth muscle cells and fibroblasts) are controlled by the pressure-induced circumferential (or hoop) stress. Biomedical devices, say, for kidney dialysis, drug delivery, or blood-vessel stenting, rely on such fluid-particle transport (and fluid-structure interactions) as well. As mentioned, bio-transport processes go beyond conventional *convection* and *diffusion*. Especially on the micro-scale level, biological transport can also occur via blood vessel contraction/dilation, ion channeling, binding interaction, cell signaling, solute carrying, protein motoring, or endocytosis. Thus, biochemical transport and reaction models have to be modified or new ones developed to describe, at least qualitatively, such exotic events. For more quantitative results, rigorous nano- to pico-scale analyses will be necessary, employing (future) computational fluid-particle dynamics or fluid-particle-structure interaction simulations. After all, biomedical engineering and the life sciences are becoming more and more subsets of physics, which is like a tree trunk where its branches are all the other sciences, as already noted by René Descartes (1596-1650).

1.1.1 Micro- to Macro-scale Systems

Of the many micro-size *particles* in any biological system, proteins are the most fascinating ones. Their multiple functions include binding of molecules from the extracellaur fluid, nutrient transport across cells, regulation of cell attachment, and signal transmission, to name a few. Of the many micro-size particles most important for biofluid dynamics considerations are platelets (for blood clotting after injury), red blood cells (as oxygen carriers), monocytes (members of the white blood cell group), endothelial cells (regulators lining all blood vessels), epithelial cells (regulators lining inner organ walls, such as the lungs, stomach, intestines, kidneys, etc.) and, of course, bacteria and viruses. Figure 1.1.1 provides a sketch of a cell with key elements and their functions (see Alberts et al., 2002).

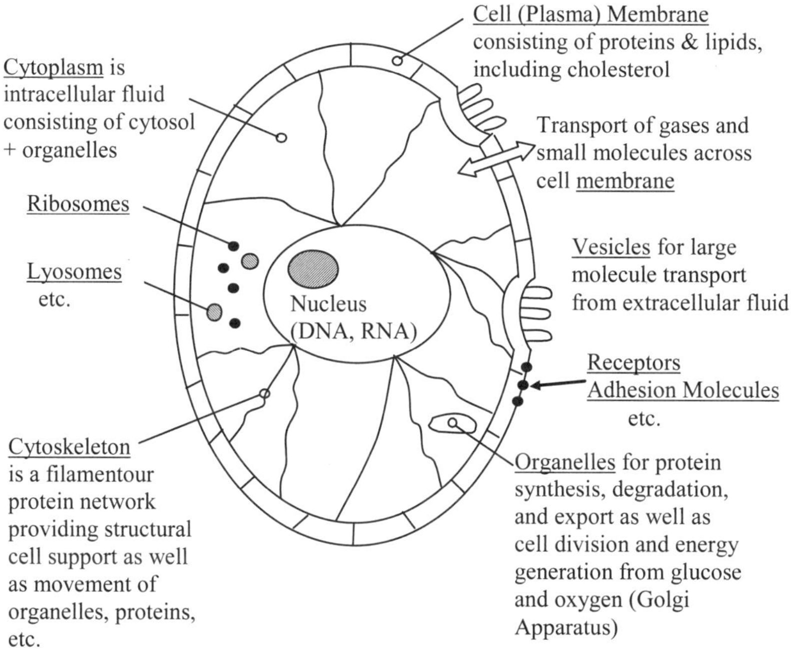

Fig. 1.1.1 Schematic of a mammalian (eukaryotic) cell with basic elements.

Tissue, such as muscle, nerve or connective tissue, is always composed of different cell types. As a first approximation, fluid and particle transport through tissue, i.e., cell junctions, could be modeled as "flow through porous media" (see Sect. 1.4.4). The microstructure, i.e., the capillary network of tissue is labeled the extracellular matrix (ECM). It gives tissue strength, shape and form, regulates cells, including metabolic activities, and it provides an aqueous environment for cell migration, macromolecule diffusion, and protein/cell anchoring. Tissues form organs and organ groups which execute special functions, such as cardio-vascular, digestive, pulmonary, reproductive, or cleansing ones. For example, the cardiovascular system (heart, blood plus vessels) distributes oxygen, nutrients, and hormones throughout the body. In addition, waste products

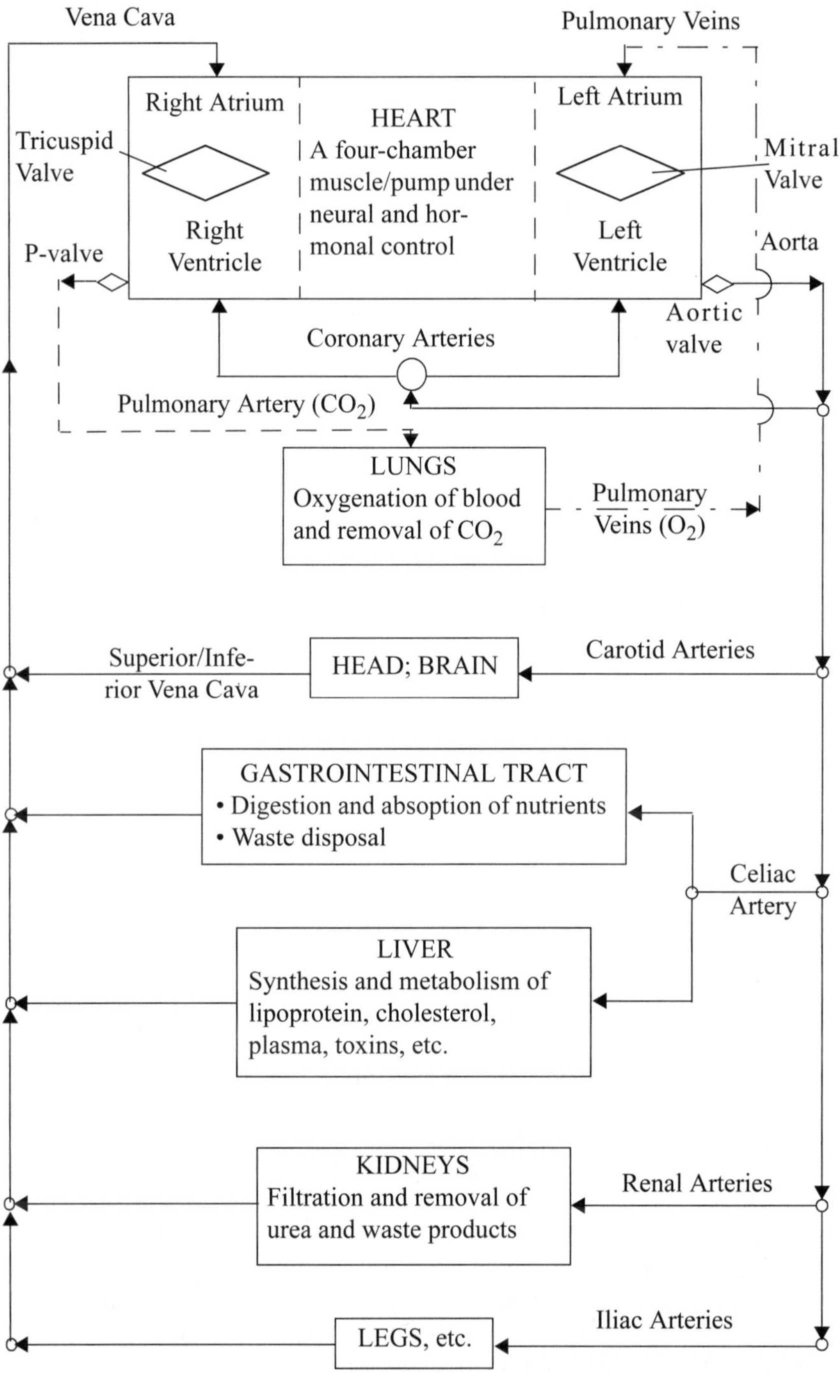

Fig. 1.1.2 Schematic of key organs with arterial and venous systems.

are carried to the liver and kidneys for synthesis and removal; when needed, blood clots are produced and antibodies as well as white blood cells are provided at sites of injury and infection, respectively. Figure 1.1.2 depicts major organ groups and associated blood flow conduits.

Blood vessels are large arteries, small arteries (arterioles), capillaries and venules, as well as large and small veins (Fig.1.1.2). Arteries, $1mm \leq D \leq 8mm$, are three-layer composites, made of the intima (i.e., layers of endothelial cells and collagen matrix), the media (i.e., extracellular matrix, elastic lamina, and smooth muscle cells), and the adventitia (i.e., connective tissue, smooth muscle cells, and fibroblast with capillaries and lymphatic vessels embedded). Capillaries, $1\mu m \leq D \leq 12\mu m$, consist of an endothelium layer plus a basement membrane. Venules are reservoirs receiving blood from where it flows via the heart to the lungs for oxygenation (see Fig. 1.1.2). Blood itself consists mainly of plasma (i.e., water, proteins, ions, sugars, and amino acids), red blood cells (with hemoglobin, an oxygen-carrying protein), white blood cells (fighting pathogens, etc.), and platelets (to mend injuries via clotting). In contrast to the *circulatory* blood flow, tissue fluid (i.e., lymph) drains *unidiectionally* into lymphatic vessels (i.e., lymphatics) that deliver lymph into the venous system.

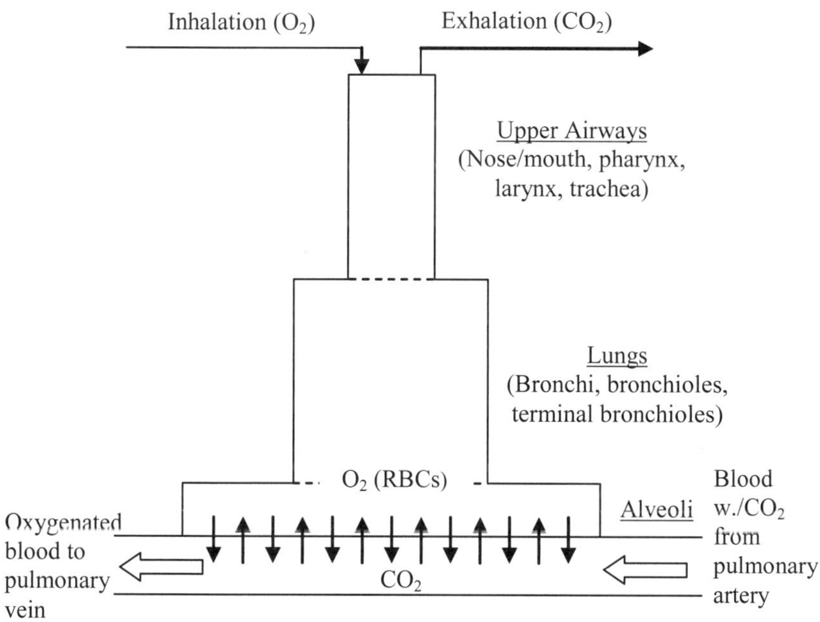

Fig. 1.1.3 Respiratory system with O_2-CO_2 exchange.

While the *heart* (see Fig.1.1.2) is a finely tuned pump, the *respiratory system* is a very complex, highly efficient O_2-CO_2 gas exchange apparatus (see Fig. 1.1.3). Specifically, air inhalation convects oxygen into the alveolar region where the millions of alveoli (i.e., tiny sacs with a surface area of 130m^2) are surrounded by pulmonary capillaries (with 115m^2 in surface area). Thus, oxygen diffuses rapidly from the alveoli through several surface barriers into the capillary bed. The reverse occurs with the CO_2 in the blood arriving in the pulmonary artery (see Fig. 1.1.2). Surfactants and associated surface tension forces play important roles in governing lung mechanics.

The *liver* and *kidneys* are highly perfused because of their multitude of vital functions. The liver is responsible for detoxification, metabolism of cholesterol, lipoprotein and various ions, synthesis of bile, vitamins and retinoids, as well as energy metabolism and the processing of amino acids, fatty acids, and sugars. The kidneys are responsible for waste removal via blood plasma filtration and water reabsorption as well as blood pressure regulation.

Additional biofluid dynamics applications, e.g., fluid motion and traveling waves in the cochlea for hearing or fluid motion in the ear's vestibular system for motion sensing, are discussed in more specialized texts.

In summary, of interest in this book are:

- fluids, i.e., air, O_2, and CO_2 as well as water, solvents, solutions, suspensions, serum, lymph, and blood;
- particles, including molecules, proteins, cells, clots, toxins, and drugs in the nanometer and micrometer ranges, assuming in most cases a quasi-spherical shape; and
- structures, such as membranes, blood vessels, organ walls, and tissue as well as implants, grafts, and medical devices.

Material transport inside and across cells is largely driven by concentration gradients (i.e., diffusion), while fluid flow and particle transport in lymphatic vessels, blood conduits, and airways is mainly due to pressure gradients. Under nonpathological conditions fluid flow is laminar, i.e., $0.1 < Re_D \leq 1800$, throughout the body. However, local obstructions, e.g., in the upper airways or in large arteries, may cause transition to low-level turbulence. Thus, at such low speeds and pressure differentials, all fluids are incompressible. Most molecular mass and energy transfer across natural barriers (e.g., membranes, endothelium, epithelium, etc.) are greatly aided by very large surface areas; as, for example, in the gastrointestinal tract, the alveolar region of the lung, the liver, and the kidneys. Thus, large near-wall gradients and surface areas generate high transfer rates of fluids, molecules, particles, and energy.

1.1.2 Solute Transport

Clearly, the proper transport of dissolved species ranging from gases (e.g., CO_2 and O_2) and liquids (e.g., lymphate) to electrolytes (i.e., charged biological molecules), nutrients, and waste products is most important for *homeostasis*. Such solute transport occurs across membranes, e.g., walls of capillaries, driven by fluid and/or osmotic pressure differentials and concentration gradients, aided by large membrane surface areas and selective membrane pores or channels. While these transport processes are very slow when compared to convection, the micro-distances such molecules have to travel and the huge (membrane) surface areas available make diffusion, osmosis and ultrafiltration locally very effective. Membranes are semi-permeable, i.e., only certain species or fluids may permeate. For example, cells need proteins, nucleotides, and other molecules for proper cell structure and function; hence, those species cannot cross the plasma membranes.

Membrane processes can be divided into passive and active transport. Passive transport, such as diffusion and osmosis, is driven by concentration and/or pressure gradients. In contrast, active transport goes against gradients and hence requires expenditure of metabolic energy from adenosine tri-phosphate (ATP), involving specific carrier proteins.

Diffusion across a membrane. Known as Brownian motion, molecules in gases as well as dissolved molecules and ions in a liquid (solvent) are moving constantly around in a random fashion, as a result of their thermal energy. If any differences in solute concentration exist anywhere in the solution, there will be a net movement, i.e., net diffusion, along the concentration gradient until all differences are eliminated (see Fig. 1.1.4a). Thus, the unstable state of a species mass gradient, having a low entropy level because of the highly ordered molecule arrangement, resolves into a stable, disordered state (i.e., high entropy) with uniform species concentration throughout the solution.

Larger, polar molecules (e.g., glucose) and charged inorganic ions (e.g., sodium Na^+ and potassium K^+) cannot permeate the membrane and hence need carrier proteins and ion channels formed by integral proteins, respectively, to migrate across the cell membrane. Carrier-mediated transport, occurring against a concentration gradient and therefore requiring expenditure of cellular energy obtained from ATP, is a type of active transport. A basic example are the kidney tubules (see Sect. 4.2), where glucose is moved from the lumen (i.e., site of lower concentration) to the blood (i.e., site of higher concentration). Similarly, all cells extrude Ca^{++} into the calcium-ion-rich intracellular environment via a hinge-like motion of the integral protein subunit (see Fig. 1.1.4b). Such active transport carriers are referred to as "pumps"; the most important one is the sodium-potassium (Na^+/K^+) pump. Each cell has numerous Na^+/K^+ pumps; for example, several million per cell as part of the tubules of the kidneys.

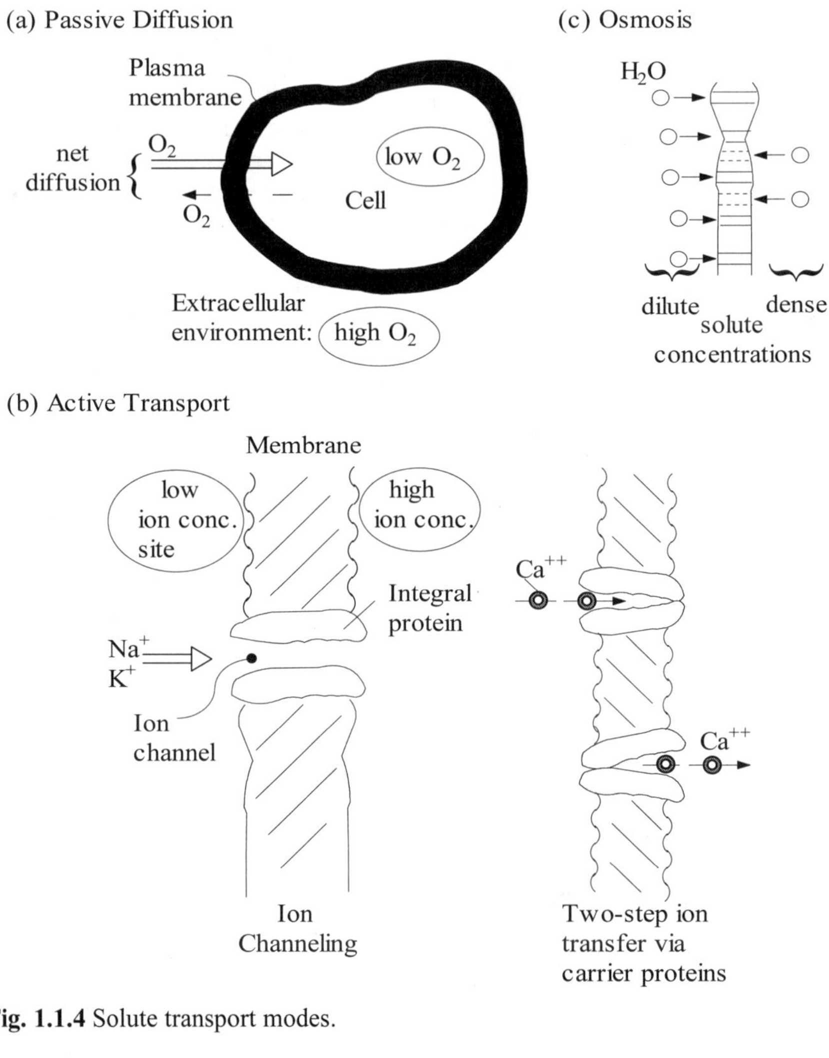

Fig. 1.1.4 Solute transport modes.

As a result of a membrane's semi-permeable function and the Na+/K+ pump actions, a cell inside is negatively charged compared to the outside. This transverse difference in voltage is known as the membrane potential. At equilibrium, the difference in electric charge across a membrane is balanced by a difference in ion concentrations. Thus, the equilibrium potential, say, ±90 milli-Volts (mV), counteracts ion diffusion, resulting in zero net transport. Thus, as expressed in the Nernst equation for a specific ion, a membrane's equilibrium potential depends on the ratio of the ion concentrations inside and outside a cell, i.e., in simple form:

$$\Delta\psi_k = \frac{61}{n}\log\left(\frac{C_o}{C_i}\right) \tag{1.1.1}$$

where $\Delta\psi_k$ is the equilibrium potential in mV for ion k, n is the ion valence, and $C_{o/i}$ are the outside/inside ion concentrations.

The gradient of the electrical potential, $\nabla\psi$, leads to a solute flux, where C_k is the k-ion concentration and $\vec{\mathbf{v}}_k$ is the ion migration velocity vector, i.e.,

$$\vec{\mathbf{j}}_k\Big|_{electric} = \vec{\mathbf{v}}_k C_k \tag{1.1.2a}$$

For dilute, i.e., non-interacting, suspensions,

$$\vec{\mathbf{j}}_k = -\frac{D_k C_k z_k F}{RT}\Delta\psi \tag{1.1.2b}$$

where D_k is the diffusion coefficient, z_k is the net charge of the molecule, F = 96,487 Coulombs/mole is Faraday's constant, R is the universal gas constant, and T is the temperature.

For example in 1-D, assuming equilibrium between the membrane potential and concentration gradient, Eq. (1.1.2b) can be rewritten with K=constant as

$$\frac{dc}{dx} = KC\frac{d\psi}{dx} \tag{1.1.2c}$$

which integrated yields Eq. (1.1.1).

Osmosis. When a membrane is more permeable to water molecules than to solute, there will be a net movement of water by diffusion across the membrane from the more dilute solution side, where more H_2O-molecules are concentrated, to the less dilute side (see Fig.1.1.4c). The driving force is the osmotic pressure difference, $\Delta\pi$, of the two solutions, where π depends on the molar concentration of the solute, i.e.,

$$\pi = C_M RT \tag{1.1.3a}$$

where $C_M = C/M$ (i.e., C $\hat{=}$ mass concentration and M $\hat{=}$ molecular weight of solute), R is the universal gas constant, and T is the temperature.

Clearly, for pure water $\pi \equiv 0$, which implies in view of Fig. 1.1.5 that an osmotic pressure results:

$$\pi = \gamma h \tag{1.1.3b}$$

where γ is the specific weight of the solution. This static pressure is needed to keep its solvent in equilibrium with the pure water at a given temperature. The fluid flow rate across a barrier, i.e., membrane or porous wall, can be estimated with *Starling's law of filtration*:

$$j_v = \frac{Q_v}{S} = L_p(\Delta p - \sigma_p \, \Delta \pi) \tag{1.1.4}$$

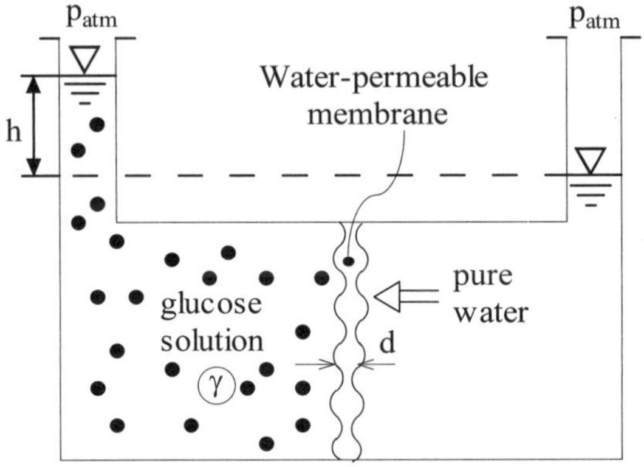

Fig. 1.1.5 Equilibrium between a solution and pure water due to osmotic pressure.

where j_v in $[L^3/(L^2T)]$ is the solution flux, S is the surface of the membrane and L_p its hydraulic conductivity, Δp is the hydrostatic pressure difference, Δπ is the osmotic pressure difference, and σ_s is the osmotic reflection coefficient, $0 \leq \sigma_s \leq 1.0$, i.e., $\sigma_s = 1.0$ implies that the membrane (or vessel wall) is impermeable to the solvent, say, glucose molecules.

For solute diffusion across a membrane (or microvessel wall), the Kedem & Katchalsky (1958) approach has been typically employed, i.e.,

$$j_s = j_v(1 - \sigma_f)\overline{C}_s + P_s(C_i - C_0) \tag{1.1.5}$$

where j_v is the solution flux of Eq. (1.1.4), σ_f is the filtration reflection coefficient, $0 \leq \sigma_f \leq 1$, $\overline{C}_s \approx (C_i - C_o)/2$ is the average solute concentration inside the membrane, $P_s = D_{eff}/d$ [L/T] is the membrane permeability, i.e., the solute effective diffusivity divided by the membrane thickness, and C_i, C_o are the inside/outside solute concentrations. An alternative approach, given by Patlek et al. (1964), is discussed by Truskey et al. (2002), p. 437.

Multi-component mixtures. Solute fluxes in mixtures are described by

$$j_i = \rho_i \, \vec{\mathbf{v}}_i \tag{1.1.6a}$$

where the velocity of each component is

$$\vec{\mathbf{v}}_i = \vec{\mathbf{v}}_D + \vec{\mathbf{v}} \tag{1.1.6b}$$

with $\vec{\mathrm{v}}_D$ being the species diffusive velocity and $\vec{\mathrm{v}}$ being the bulk, mixture or mass-averaged velocity, i.e.,

$$\vec{\mathbf{v}} = \frac{1}{\rho}\sum_{i=1}^{h} \rho_i \vec{\mathbf{v}}_i \tag{1.1.6c}$$

Then, the diffusive mass flux, $\vec{\mathbf{j}}$, can be defined as:

$$\vec{\mathbf{j}}_i^D = \rho_i(\vec{\mathbf{v}}_i - \vec{\mathbf{v}}) = \rho_i \vec{\mathbf{v}}_D \tag{1.1.7}$$

When additional forces act on the solute, the resulting total flux may be due to ordinary diffusion (see Eq. (1.1.9)), an applied electric field (see Eq. (1.1.2a), pressure differential (see Eq. (1.1.5), and thermal diffusion, i.e.,

$$\vec{\mathbf{j}}_i^{total} = \vec{\mathbf{j}}_i^{diff.} + \vec{\mathbf{j}}_i^{electr.} + \vec{\mathbf{j}}_i^{press} + \vec{\mathbf{j}}_i^{thermal} \tag{1.1.8}$$

As mentioned, most frequently encountered is ordinary diffusion, specifically in a dilute binary mixture, for which Fick's law holds:

$$\vec{\mathbf{j}}_i^D = -D_{ij}\nabla C_i \tag{1.1.9}$$

where D_{ij} is the diffusion coefficient of solute i in solvent j. Equation (1.1.9) plays a key role in the mass transfer equation derived and discussed in Section 1.3.

The diffusion coefficient. Key in ordinary diffusion is the binary diffusion coefficient, typically picked for a given solute, solvent, and temperature from a table in a biochemical engineering handbook. Here we consider a dilute binary mixture of spherical particles (i.e., solute). The approach is to combine a 1-D force balance for a representative, randomly moving particle with 1-D kinetic theory and Einstein's random-walk behavior of a diffusing particle (i.e., "translational Brownian motion") which is based on Fick's law of diffusion (see Eq. (1.1.9)) and probabilistic arguments (Probstein, 1994). Specifically,

$$\underbrace{\tfrac{1}{2} m_p \langle \mathrm{v}_x^2 \rangle}_{\text{kinetic energy}} = \underbrace{\tfrac{3}{2} k_B T}_{\text{thermal energy}} \tag{1.1.10}$$

and

$$\langle x^2 \rangle = 6 D_{ij} t \tag{1.1.11}$$

where m_p is the particle mass, v_x is its velocity in the x-direction, $k_B = 1.38 \times 10^{-23}\ JK^{-1}$ is the Boltzmann constant, T is the temperature, $<x>$ is the mean random particle displacement, and D is the binary diffusion coefficient of interest. Now, with a 1-D force balance for Stokes flow (Re≤1) we obtain:

$$m_p \frac{d^2x}{dt^2} = -F_D + F_C \tag{1.1.12}$$

where $F_D = fv_x$ is the drag force with f being the friction coefficient, $f = 6\pi\mu r_0$, and F_c is a random collision force which is negligible in dilute suspensions.

In order to generate an $m_p\langle v_x^2\rangle$-term and to use Eq. (1.1.10), we multiply Eq. (1.1.12) through by x and expand, i.e.,

$$x\frac{d^2x}{dt^2} := \frac{d}{dt}\left(x\frac{dx}{dt}\right) - \left(\frac{dx}{dt}\right)^2$$

which yields after ensemble-averaging:

$$m_p\langle v_x^2\rangle = \frac{f}{2}\frac{d\langle x^2\rangle}{dt} + \frac{m_p}{2}\frac{d^2\langle x^2\rangle}{dt^2} \tag{1.1.13}$$

Employing Eq. (1.1.10) and solving for $\langle x^2\rangle$ subject to $\langle x^2\rangle$ = $d\langle x^2\rangle / dt = 0$ at t=0 yields the mean-squared "random walk" displacement:

$$\langle x^2\rangle = \frac{6k_BT}{f}t - \frac{6m_pk_BT}{f^2}\left[1 - \exp\left(-\frac{f}{m_p}\right)t\right] \tag{1.1.14}$$

For relatively large observation times, say, $t >> \frac{m_p}{f} = o\,(10^{-7}\ s)$ for particles with 1 μm radius,

$$\langle x^2\rangle \approx \frac{6k_BT}{f}t \quad \text{or} \quad \langle x^2\rangle = 6D_{ij}t$$

based on Eqs. (1.2.19). Hence, for spherical particles of radius r ,

$$D_{ij} = \frac{k_BT}{f} := \frac{k_BT}{6\pi\mu r_0} \tag{1.1.15a, b}$$

Equation (1.1.15) is known as the Stokes-Einstein equation which also holds for random motion, i.e., solute diffusion in three dimensions. For nonspherical particles, $f \rightarrow \bar{f}$ (see Happel & Brenner, 1983). Examples 1.8 and 1.21 illustrate 1-D diffusive mass transfer while Sects.5.3.1 and 5.3.2.3 discuss the use of Eq. (1.1.15).

1.2 BASIC MOMENTUM, HEAT, AND MASS TRANSFER CONCEPTS

There are a number of basic fluid mechanics texts available on an introductory level (e.g., White, 2003; Fox et al., 2004; Middleman, 1998; or Smits, 2000), intermediate level (e.g., Pnueli & Gutfinger, 1992; Potter & Wiggert, 1997; or Wilkes, 1999), and graduate level (e.g., White, 1991; Panton, 1996; Kleinstreuer, 1997; Warsi, 1999; or Schlichting & Gersten, 2000). Supplementary information on heat and mass transfer may be found in Bird et al. (2002), Naterer (2003), Gebhart et al. (1988), Kays & Crawford (1993), Brenner & Edwards (1993), and Arpaci et al. (1999). Naturally, this review of transport phenomena relies mainly on the author's previous text (Kleinstreuer, 2003) as well as Bird et al. (2002). The fundamental principles of mechanics are the conservation laws of mass, linear/angular momentum, and energy, as well as the inequality of entropy (cf. Second Law of Thermodynamics). They can be readily defined for an identifiable constant mass (Lagrange's closed system) and also derived for an open system (Euler's control volume), which may be fixed or moving (see Sect. 1.2.2). The underlying hypothesis for the conservation equations is that all materials (or media) form a continuum (see Sect. 1.2.1). Actually, continuum mechanics encompasses both solid and fluid materials, which are treated in separate mechanics disciplines solely for historical (and feasibility) reasons.

By definition, *fluids* deform continuously under (shear) stress, ignoring molecular effects which may be important in microfluidics (cf. Gad-el-Hak, 1999) when the *continuum mechanics* assumption (see Sect. 1.2.1) is violated. The continuum hypothesis postulates that fluid elements, actually material points, are distributed continuously. It implies that the mean-free-path, λ, between fluid (gas) molecules is much less than an appropriate minimum length-scale L of the flow system, as expressed with the Knudsen number

$$Kn = \lambda/L \tag{1.2.1}$$

For example, $L=\rho/|\partial\rho/\partial y|$ to calculate the *local* Kn-number, or $L=l_{device}$ for the *global* Kn-number. The Knudsen number is appropriately used for gases because in a liquid the molecules are predominantly in vibrational modes rather than in random motion. Alternatively, for liquids the mean-free-path, λ, may be replaced by the intermolecular length, where $\lambda_{IM} \ll l$; for example, λ_{IM} for water is about 3 Å, i.e., 3.0×10^{-10} m. In any case, the continuum approach is valid as long as $Kn < 0.1$ as depicted in Fig. 1.2.1, which also lists modeling approaches for different flows. For example, a 1-mm cube of standard air contains 2.54×10^{7} molecules with a mean free path of λ=0.065 mm, which is the average distance traveled by (air) molecules between collisions, while the intermolecular spacing is only about 0.0034 mm. Clearly, for normal devices or systems where 0.001 m $\leq$

$L \leq 10$ m, $Kn << 1$. Thus, in the conventional sense, a fluid is a continuum when for all properties p convergence occurs for all points i, that is,

$$\lim_{V \to V'} p = p_i \tag{1.2.2}$$

where V is a volume around point i in the flow and V' is traditionally "a very small volume containing a sufficient number/mass of molecules or particles." Clearly, regions of matter are continua of properties that can be represented as differentiable functions (see Sect.1.2.1). For example, in Sect. 1.4, not only the fluid but also the dispersed phase, e.g., a collection of micro-particles, is considered to be a continuum, i.e., a "fluid" as well (cf. Sect. 1.4.3).

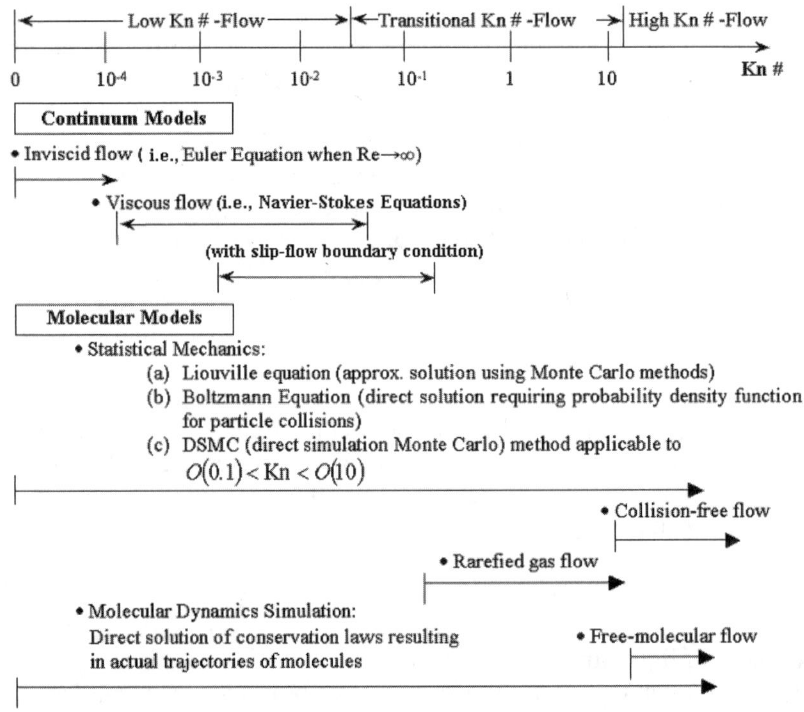

Fig. 1.2.1 Knudsen number regimes, modeling approaches, and application limits for gas flows (after Kleinstreuer, 2003).

Another potential restriction is inherent in the type of fluid and its limiting time scales. Specifically, constitutive equations, not containing time as a variable, have the implicit assumption that fluxes result instantaneously from the imposition of the corresponding gradients. There are two "fluid time scales" that characterize (instantaneous) signal propagation, i.e., the time for molecule displacement (e.g., $\Delta t_{gas} = 10^{-10}$ s and $\Delta t_{liquid} = 10^{-15}\,\text{s}$) and the elapsed time between collisions among molecules (e.g., $\Delta t_{gas} = 10^{-12}\,\text{s}$ and $\Delta t_{liquid} = 10^{-13}\,\text{s}$). These time intervals are so small that there is no need for basic and generalized Newtonian fluids to be concerned with relaxation effects in the constitutive equations (cf. Bird et al., 1987). Thus, for all practical processes, *single-phase* flow under the continuum mechanics assumption $10^{-4}<\text{Kn}<10^{-1}$(see Fig.1.2.1) can be grouped as shown in Fig. 1.2.2. Obviously, the left column represents the easier set of fluid flow problems. In order to connect

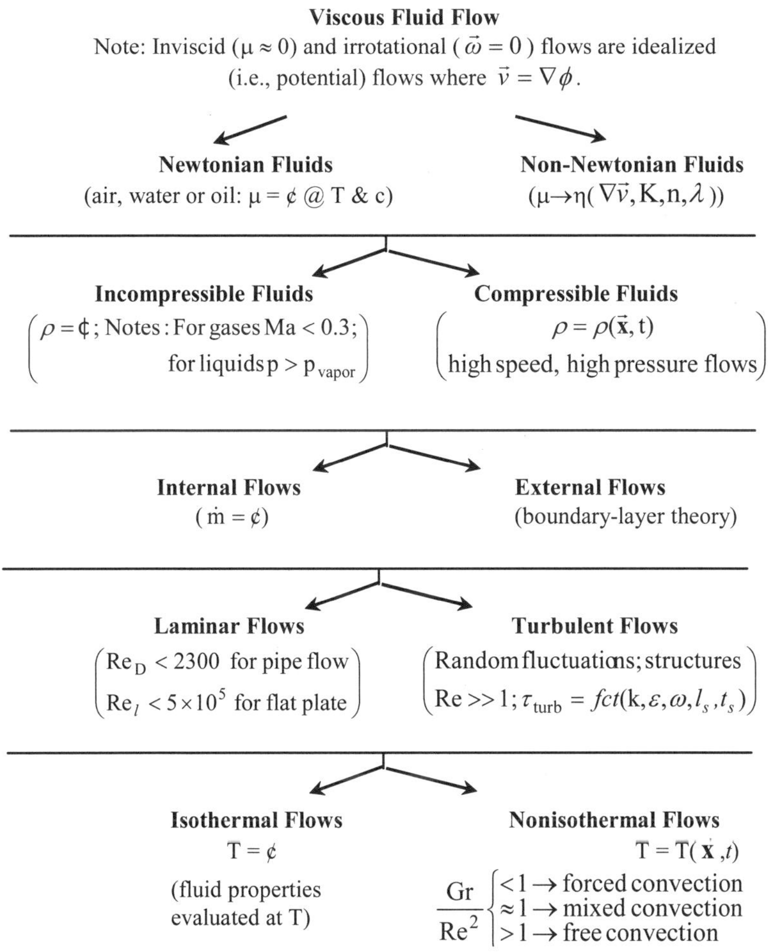

Fig. 1.2.2 Special cases of viscous fluid flows.

Fig. 1.2.1 with Fig. 1.2.2, simulation approaches for "flow through porous media" may serve as a self-explanatory example (cf. Table 1.2.1). It should be noted that classical *statistical mechanics* endeavors to obtain continuum transport laws from the knowledge of discrete molecular-dynamics laws (Chandler, 1987). For example, Brownian motion of submicron particles in a suspension was basic to the development of kinetic theory, represented by the Boltzmann equation. Thus micro-continuum transport laws could be derived from underlying sub-continuum molecular processes. In turn, statistical averaging schemes and volume- or time- or ensemble-averaging methods have been employed to capture and express some microtransport phenomena on the macrotransport level.

Table 1.2.1 Level of Resolution in Computer Simulations Illustrated for Flow through Porous Media.

Molecular Dynamics	**Microtransport**	**Macrotransport**
<u>Needed</u>: • Exact location and boundaries of porous material • Molecular dynamics equations for fluid molecules and particle trajectories	<u>Needed</u>: • Approximated BCs; for example, medium is a stack of spheres, or rods, etc. • Property values for ρ, ν, etc.	<u>Needed</u>: • Porosity, permeability, friction factors, etc., describing the porous medium
<u>Results</u>: • Fluid flow is the interactions of billions of molecules as well as with the porous material • Transport properties, boundary conditions, etc., are not required	<u>Results</u>: • Point-wise solution of the extended Navier-Stokes equations • Pressure and shear stresses as new concepts	<u>Results</u>: • Mean velocities from solving the extended Darcy equation • Hydraulic conductivity as a new concept

Now, back to the objective at hand: the overall goal is to find and analyze the interactions between *fluid forces*, e.g., pressure, gravity/buoyancy, drag/friction, inertia, etc., and *fluid motion*, i.e., the velocity vector field from which everything else can be directly obtained or derived (see Fig. 1.2.3):

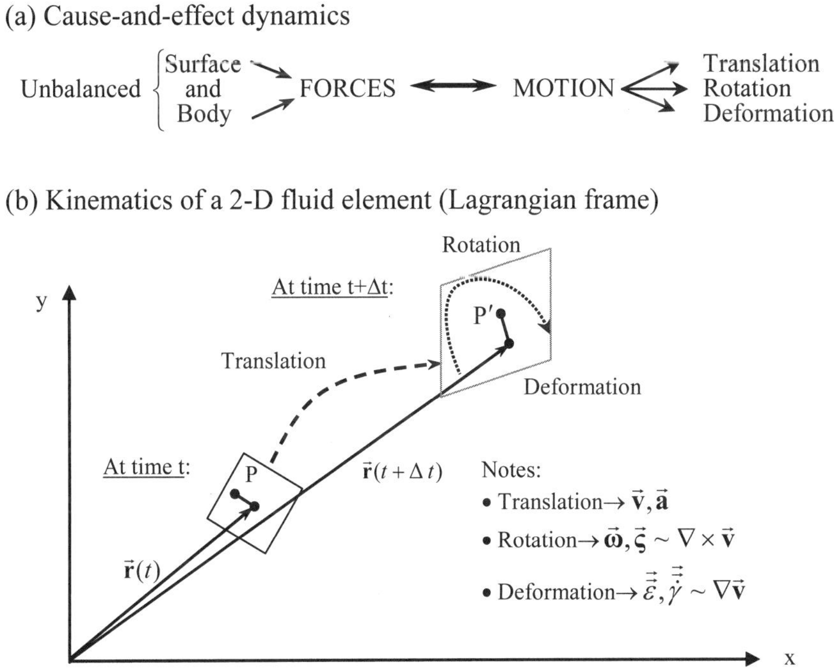

Fig. 1.2.3 Dynamics and kinematics of fluid flow: (a) force-motion interactions; and (b) 2-D fluid kinematics.

In turn, scalar transport equations can be solved based on the velocity field to obtain critical magnitudes and gradients (or fluxes) of mass and energy.

Exact flow problem *identification*, especially in industrial settings, is one of the more important and sometimes the most difficult first task. After obtaining some basic information and reliable data, it helps to think and speculate about the physics of the fluid flow, asking:

(i) What category (see Fig. 1.2.2) does the given flow system fall into, and how does it respond to normal as well as extreme changes in operating conditions?

(ii) What variables and system parameters play an important role in the observed transport phenomena?

(iii) What are the key dimensionless groups and what are their expected ranges?

Answers to these questions assist in grouping the flow problem at hand and determining the main assumptions listed below. Once a given fluid dynamics problem has been categorized (Fig. 1.2.2), some *justifiable assumptions* have to be considered in order to simplify the general transfer equations, as exemplified here:

Flow Assumption		*Consequence*
• Time-dependence	→	$\frac{\partial}{\partial t}=0$: steady; $\vec{\mathbf{v}}=\vec{\mathbf{v}}(t)$: transient
• Dimensionality	→	Required number of space coordinates
• Directionality	→	Required number of velocity components
• Development phase	→	$\frac{\partial}{\partial s}=0$: fully developed ($s \hat{=}$ axial coordinate)
• Symmetry	→	$\frac{\partial}{\partial n}=0$: midplane ($n \hat{=}$ normal coordinate)
		$\frac{\partial}{\partial \theta}=0$: axisymmetry

Information on a given flow problem, in terms of the viscous flow grouping in conjunction with a set of proper assumptions, allows for the selection of a suitable solution technique (see Sect. 1.3.5 and App. A). That decision, however, requires first a brief review of possible *flow field descriptions*, i.e., Lagrangian vs. Eulerian framework in continuum mechanics, and in case of probabilistic events, the type of probability density function in the framework of statistical mechanics.

1.2.1 Continuum Mechanics Axioms

Newton's second law of motion holds for both molecular dynamics, i.e., interacting molecules, and continua, like blood, air, vessel walls, tissues, bones, and implants. Such *solid and fluid* structures are assumed continua, because for a fluid $Kn \ll 1$ [see Eq.(1.2.1)] and, in general, it would be more than cumbersome to model them on a molecular (or multi-particle) dynamics level. Focusing on solids, a body (or structure) with a closed surface forms a continuum. When a piece breaks or tissue grows, the surfaces of the fracture or new cells form new (external) surfaces. Thus, the key continuum mechanics assumptions are:

(i) When a force is applied, any two neighboring body particles at one time remain neighbors at all times.

(ii) The applied tangential or normal forces result in body stresses, material deformations (strains or strain rates), and/or material element motion (fluid flow); these stresses, strains, pressures (part of the normal stress), and velocities can be defined at all points in the body/structure.

(iii) Relationships exist for stress, strain, and strain rates in terms of constitutive equations which may also take into account temperature and electrical charge effects.

Taking the lung (or the cardiovascular circulation) as an example, it is a matter of desired resolution and the availability of reliable data sets if such a system is represented as a single continuum or numerous interacting continua.

For fluids, when Kn>0.1 (see Fig.1.2.1), the behavior of representative molecules have to be described, using statistical means or molecular dynamics. Problems where the continuum hypothesis breaks down include rocket-air interactions in the lower atmosphere (continuum assumption) and later in the upper thin atmosphere (statistical description) as well as blood-particle flows in medium-to-large arteries (c.a.) and certain flows in micro-vessels (s.d.).

1.2.2 Flow Field Descriptions

Any flow field can be described at either the microscopic or the macroscopic level. The microscopic or molecular models consider the position, velocity, and state of every molecule of the single fluid or multiple 'fluids' at all times. In gas flow, deterministic or probabilistic molecular dynamics models (i.e., the three conservation laws or the Boltzmann equation plus suitable probability density functions) are employed for direct numerical simulation. The goal of the *statistical approach* is to compute the probability of finding a molecule at a particular position, velocity, and state. Occasionally, both approaches are employed for (two-phase) flow problem solutions. Averaging discrete-particle information (i.e., position, velocity, and state) over a local fluid volume yields macroscopic quantities, e.g., the velocity field $\vec{\mathbf{v}}(\vec{\mathbf{x}},t)$, at any location in the flow. The advantages of the molecular approach include general applicability (i.e., all Knudsen numbers, see Fig. 1.2.1), no need for submodels (e.g., for the stress tensor, heat flux, turbulence, wall conditions, etc.), and an absence of numerical instabilities (e.g., due to steep flow field gradients). However, considering myriads of molecules, atoms, and nanoparticles requires enormous computer resources, and hence only *simple* channel or stratified flows with a finite number of interacting molecules (i.e., solid spheres) can be presently evaluated. For example, in a 1-mm cube there are about 34 billion water molecules, which makes molecular dynamics simulation prohibitive, but on the other hand, intuitively validates the continuum assumption and hence the use of the Navier-Stokes equations. Further discussions of molecular models and solution procedures, especially the direct simulation Monte Carlo (DSMC) method for *dilute gases*, can be found in Bird (1994) as well as in the review by Oran et al. (1998).

Within the continuum mechanics framework, two basic flow field descriptions are of interest, i.e., the *Lagrangian* viewpoint and the *Eulerian* approach.

1.2.2.1 Lagrangian Description

Consider particle A moving on a pathline with respect to a fixed Cartesian coordinate system. Initially, the position of A is at $\vec{\mathbf{r}}_{\mathbf{o}} = \vec{\mathbf{r}}_{\mathbf{o}}(\vec{\mathbf{x}}_{\mathrm{o}}, t_{\mathrm{o}})$ and a moment later at $\vec{\mathbf{r}}_{\mathbf{A}} = \vec{\mathbf{r}}_{\mathbf{A}}(\vec{\mathbf{r}}_{\mathbf{o}}, t_{\mathrm{o}} + \Delta t)$ as depicted in Fig. 1.2.4, where $\vec{\mathbf{r}}_{\mathbf{A}} = \vec{\mathbf{r}}_{\mathbf{o}} + \Delta\vec{\mathbf{r}}$.

Considering all distinct points and following their motion for $t > t_{\mathrm{o}}$, fluid (particle) motion, steady or unsteady, can be described with the position vector

$$\vec{\mathbf{r}} = \vec{\mathbf{r}}(\vec{\mathbf{r}}_{\mathbf{o}}, t) \tag{1.2.3}$$

where in the limit,

$$\frac{d\vec{\mathbf{r}}}{dt} = \vec{\mathbf{v}} \tag{1.2.4}$$

and

$$\frac{d^2\vec{\mathbf{r}}}{dt^2} = \frac{d\vec{\mathbf{v}}}{dt} = \vec{\mathbf{a}} \tag{1.2.5}$$

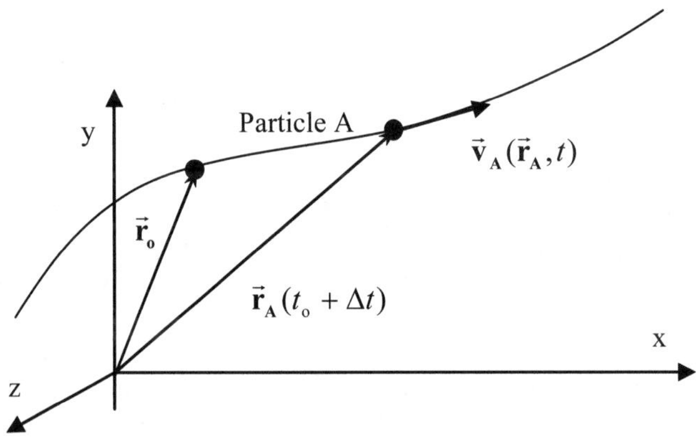

Fig. 1.2.4 Incremental fluid particle motion.

Now, the material-point concept is extended to a material volume with constant identifiable mass, forming a "*closed system*" that moves and deforms with the flow but no mass crosses the material volume surface ever. Again, the system is tracked through space and as time expires, it is of interest to know what the changes in system mass, momentum, and energy are. This can be expressed in terms of the system's extensive property N_s which is either mass m, momentum $m\vec{\mathbf{v}}$, or energy E. Thus, the key question in mathematical shorthand is: "How can we express the fate of N_s, i.e., $DN_s/Dt = ?$" Clearly, the material or Stokes derivative $D/Dt \equiv \partial/\partial t + \vec{\mathbf{v}} \cdot \nabla$ follows the closed system and records the total time-rate-of-change of whatever is being tracked (see Sect.1.3).

Obviously, we cannot use *d/dt* from "particle dynamics" for this task. For example, focusing on the fluid density, *dρ/dt* is ambivalent, implying either a density change with time of a specific fluid element at a point (*x,y,z*) in the flow field *or* a rate-of-change of density at the point because of new fluid elements traveling during *dt* past point (*x,y,z*). Thus, to avoid misconceptions, we briefly revisit the various *time derivatives*, i.e., $\partial/\partial t$ (local), d/dt (total of a material point or solid particle), and D/Dt (total of a fluid element). Their differences can be illustrated using acceleration:

- $a_{x,local} = \frac{\partial u}{\partial t}$, where u is the fluid element velocity;
- $\vec{\mathbf{a}}_{particle} = \frac{d\vec{\mathbf{v}}}{dt}$ as employed in solid particle dynamics; and
- $\vec{\mathbf{a}}_{\substack{fluid\\element}} = \frac{D\vec{\mathbf{v}}}{Dt} := \underbrace{\frac{\partial \vec{\mathbf{v}}}{\partial t}}_{\vec{a}_{local}} + \underbrace{(\vec{\mathbf{v}} \cdot \nabla)\vec{\mathbf{v}}}_{\vec{a}_{convective}}$

The derivation of $D\vec{\mathbf{v}}/Dt$ is left as a HW problem in Sect. 1.7.

1.2.2.2 Eulerian Description

In the Eulerian framework an "*open system*" is considered where mass, momentum, and energy may readily cross, i.e., being convected across the *control volume* surface and local fluid flow changes may occur within the control volume over time. The fixed or moving control volume may be a large system/device with inlet and outlet ports, it may be small finite volumes generated by a computational mesh, or it may be in the limit a "point" in the flow field. In general, the Eulerian observer fixed to an inertial reference frame records temporal and spatial changes of the flow field at all "points," or in case of a control volume, the transients inside and fluxes across its control surfaces.

In contrast, the Lagrangian observer stays with each fluid element or material volume and records its basic changes while moving through space. While Sect. 1.4.3.1 applies the Lagrangian approach directly to evaluate particle trajectories, the Sect. 1.3 employs both viewpoints to describe mass, momentum, and heat transfer in integral form, known as the Reynolds Transport Theorem (RTT). A surface-to-volume integral transformation then yields the conservation laws in differential form in the Eulerian framework.

1.2.3 Derivation Approaches

There are basically four ways of obtaining specific transport equations reflecting the conservation laws. The points of departure for each of the four methods are either given (e.g., Boltzmann equation or Newton's Second Law) or derived based on differential (REV) balances.

(i) *Molecular Dynamics Approach*: Fluid properties and transport equations can be obtained from kinetic theory and the Boltzmann equation, respectively, employing statistical means. Alternatively, $\sum \vec{\mathbf{F}} = m\vec{\mathbf{a}}$ is solved for each molecule using direct numerical simulation.

(ii) *Integral Approach*: Starting with the Reynolds Transport Theorem (RTT) for a fixed open control volume (Euler), specific transport equations in integral form can be obtained (see Sect.1.3).

(iii) *Differential Approach*: Starting with the Generalized Transport Equation (Sect. 1.3), the mass, momentum, and energy transfer equations in differential form can be formulated. Alternatively, the RTT is transformed where in the limit the field equations in differential form are obtained. Also shown in Sect.1.3 is a third approach where differential balance equations for a representative elementary volume (REV) are derived.

(iv) *Phenomenological Approach*: Starting with balance equations for an open system, transport phenomena in complex transitional, turbulent, or multiphase flows representing biomedical systems are derived, largely based on *empirical* correlations and dimensional analysis considerations.

These approaches were already alluded in Fig. 1.2.1. After a brief illustrative Lagrange vs. Euler example, approaches (ii) and (iii) are employed in Sect. 1.3.

<u>*Example 1.1*</u>: Lagrangian vs. Eulerian Flow Description

In the Eulerian fixed coordinate frame, river flow is approximated as steady 1-D, i.e.,

$$\mathrm{v}(x) = \mathrm{v}_0 + \Delta\mathrm{v}\left(1 - e^{-ax}\right)$$

which implies that at x=0, say, the water surface moves at v_0 and then accelerates downstream to $\mathrm{v}(x \to \infty) = \mathrm{v}_0 + \Delta\mathrm{v}$. Derive an expression for $\mathrm{v} = \mathrm{v}(\mathrm{v}_0, t)$ in the Lagrangian frame.

<u>*Recall:*</u> $\vec{\mathbf{v}} = d\vec{\mathbf{r}}/dt$ or in our 1-D case

$$\frac{dx}{dt} = v(x) = v_0 + \Delta v\left(1 - e^{-ax}\right)$$

<u>*Solution:*</u> Separation of variables and integration yield

$$\int_0^x \frac{dx}{(v_0 + \Delta v) - \Delta v e^{-ax}} = \int_0^t dt$$

so that

$$x + \frac{1}{a}\ln\left[1 + \frac{\Delta v}{v_0}\left(1 - e^{-ax}\right)\right] = (v_0 + \Delta v)t$$

Now, replacing the two x-terms with expressions from the v(x)-equation, i.e., $x = -\frac{1}{a}\ln\left(1 - \frac{v - v_0}{\Delta v}\right)$ and $e^{-ax} = 1 - \frac{v - v_0}{\Delta v}$, we can express the Lagrangian velocity as

- $$v(t) = \frac{v_0(v_0 + \Delta v)}{v_0 + \Delta v \exp[-a(v_0 + \Delta v)t]}$$

Given typical parameter values, the velocities v(x) and v(t) in the Eulerian and Lagrangian frames are graphed below.

<u>*Graphs:*</u>

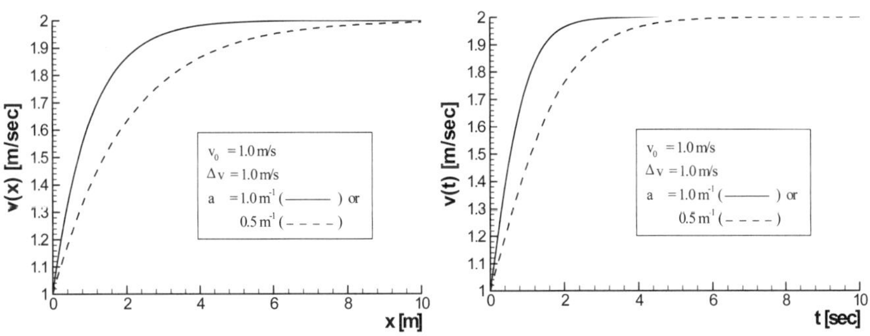

1.3 CONSERVATION LAWS

A Transport Theorem in Differential Form. Conservation of mass, momentum, and energy can be expressed as one equation in *differential form*, labeled by Kleinstreuer (1997) as the *generalized transport equation* (GTE) (see Sect. 1.7 HW problem for derivation):

- $$\frac{\partial(\rho\psi)}{\partial t}+\nabla\cdot(\rho\psi\,\vec{\mathbf{v}})+\nabla\cdot J-\rho\phi=0 \tag{1.3.1}$$

where ψ is the specific quantity being conserved. Specifically, $\psi \equiv 1$ (for mass), $\vec{\mathbf{v}}$ (for momentum), or $(\tilde{\mathrm{u}}+\mathrm{v}^2/2)$ (for internal plus kinetic energies); J is the (molecular) flux, i.e., $\mathrm{J}=0,\ -\vec{\vec{\mathrm{T}}},\text{ or }\vec{q}-\vec{\vec{\mathrm{T}}}\cdot\vec{\mathbf{v}}$, respectively; and ϕ is the source term, i.e., $\phi=0,\ \vec{\mathbf{g}},\text{ or }\vec{\mathbf{g}}\cdot\vec{\mathbf{v}}$, respectively. The total or Cauchy stress tensor $\vec{\vec{T}}$ and the heat flux vector $\vec{\mathbf{q}}$ are discussed below. An alternative derivation of the field equations in differential forms is outlined in Sects. 1.3.1-1.3.3.

A transport theorem in integral form. Consider N to be an arbitrary extensive quantity of a *closed system* or a moving material volume (see Sect.1.2.2.1). Examples include a piston-cylinder assemblage with constant gas mass or a (yellow) vapor cloud of fixed mass moving through the air; N could be the system's mass, momentum, or energy. The first task is to express in the Lagrangian sense the material derivative, i.e., the total time-rate-of-change of N. Clearly, with $m_{system} \equiv m = \text{constant}$, we can write:

- $N \equiv m_{system} \rightarrow \quad \dfrac{Dm}{Dt}=0$ (conservation of mass)
- $N \equiv (m\vec{\mathbf{v}})_{system} \rightarrow \quad m\dfrac{D\vec{\mathbf{v}}}{Dt}=m\vec{\mathbf{a}}=\sum\vec{\mathbf{F}}_{surface}+\sum\vec{\mathbf{F}}_{body}$

  (conservation of momentum or Newton's Second Law)
- $N \equiv E_{system} \rightarrow \quad \dfrac{DE}{Dt}=\dot{Q}-\dot{W}$

  (conservation of energy or First Law of Thermodynamics)

Now, the conservation laws in terms of *DN/Dt* are related to an *open system* (Eulerian frame) in terms of a fixed control volume (C.V.) with material streams flowing in and out, i.e., across the control surfaces (C.S.), and possibly accumulating inside. Descriptively,

$$\begin{Bmatrix}\text{Total time-rate-of}\\ \text{change of system}\\ \text{property } N\end{Bmatrix} = \begin{Bmatrix}\text{Local time-rate-of}\\ \text{change of N/V} \equiv \rho\eta\\ \text{within the C.V.}\end{Bmatrix} + \begin{Bmatrix}\text{Net efflux of } (\rho\eta)\text{, i.e.,}\\ \text{material convection across}\\ \text{control surfaces (C.S.)}\end{Bmatrix}$$

or mathematically,

- $$\left.\frac{DN}{Dt}\right|_{\substack{closed\\ system}} = \frac{\partial}{\partial t}\iiint_{C.V.} \rho\eta dV + \iint_{C.S.} \rho\eta\vec{\mathbf{v}}\cdot d\vec{\mathbf{A}} \tag{1.3.2}$$

which is the *Reynolds Transport Theorem (RTT)*, for a control volume at rest, where the specific property $\eta = \frac{N}{m} := \begin{cases}1\\ \vec{\mathbf{v}}\\ \mathrm{e}\end{cases}$

Notes:

(i) For a *moving control volume*, $\vec{\mathbf{v}}$ is replaced by $\vec{\mathbf{v}}_{\text{relative}} = \vec{\mathbf{v}}_{\text{fluid}} - \vec{\mathbf{v}}_{\text{C.V.}}$. The operator $\partial/\partial t$ acting on the first term on the R.H.S. has to be replaced by d/dt when the control volume is *deformable*.

(ii) For a *non-inertial* coordinate system, e.g., when tracking an accelerating system, e.g., a rocket, $\sum\vec{\mathbf{F}}$ is expressed as $m\vec{\mathbf{a}}_{abs} = m(d\vec{\mathbf{v}}/dt + \vec{\mathbf{a}}_{rel})$, where $m\vec{\mathbf{a}}_{rel}$ accounts for noninertial effects, so that $\sum\vec{\mathbf{F}} - m\vec{\mathbf{a}}_{\text{rel}} = m\,d\,\vec{\mathbf{v}}/dt$, and finally the extended momentum equation reads (i.e., $N \equiv (m\vec{\mathbf{v}})_{system}$)

$$\sum\vec{\mathbf{F}} - \int_{C.V.}\vec{\mathbf{a}}_{rel}dm = \frac{d}{dt}\left(\int_{C.V.}\vec{\mathbf{v}}\rho dV\right) + \int_{C.S.}\vec{\mathbf{v}}\rho\left(\vec{\mathbf{v}}_{rel}\cdot\hat{n}\right)dA \tag{1.3.3}$$

Applications of the RTT are given in Sects. 1.3.1 -1.3.3 and as Examples 1.2 and 1.3.

(iii) There are a few sequential steps necessary for *tailoring the general RTT* [cf. Eq. (1.3.2)] toward a specific flow system description and solving the resulting integral equations (cf. Examples 1.2 and 1.3 in Sect. 1.3.2).

(1) Identify the extensive quantity N_{system}, e.g., the mass of the identifiable material, linear or angular momentum, or total energy. As a result, the specific property of the closed system, $\eta = N/m_{system}$, is known.

(2) Determine $DN/Dt|_{system}$ for each conservation case, i.e., mass, momentum, or energy.

(3) Select a "smart" control volume and determine if:
- The control volume is fixed, or moving at $V_{C.V.}$=¢, or accelerating a_{rel}=¢, or accelerating and rotating;
- the flow problem is steady or transient;
- the fluid properties are constant or variable;
- the inflow/outflow velocity fields are constant, i.e., uniform, or a function of inlet/exit space variables; and
- the resulting integral balance equations are decoupled.

(4) Set up the momentum equation, i.e., *force balance*, for each coordinate direction.

(5) Solve the volume and/or surface integrals (cf. Integration Tables), watching the inflow/outflow sign convention, i.e., IN $\hat{=}$ "-" and OUT $\hat{=}$ "+", and check the results for correctness.

Numerous sample problem solutions further illustrating the use of the RTT may be found in Fox et al. (2004) and other undergraduate texts.

1.3.1 Mass Conservation

Taking $N \equiv m$ and hence $\eta = 1$, the RTT [Eq. (1.3.2)] reads

$$0 = \frac{\partial}{\partial t} \int_{C.V.} \rho dV + \int_{C.S.} \rho \vec{\mathbf{v}} \cdot d\vec{\mathbf{A}} \tag{1.3.4}$$

Employing the Gauss divergence theorem

$$\iint_{S_0} \rho \vec{\mathbf{v}} \cdot dS = \iiint_{V_0} \nabla \cdot (\rho \vec{\mathbf{v}}) dV \tag{1.3.5}$$

where $\nabla \cdot (\rho \vec{\mathbf{v}}) \equiv div(\rho \vec{\mathbf{v}})$ and $\vec{\mathbf{v}} \cdot d\vec{S} = \vec{\mathbf{v}} \cdot \hat{n} dS$, Eq. (1.3.4) can be rewritten as

$$0 = \iiint_{C.V.} \left[\frac{\partial \rho}{\partial t} + \nabla \cdot (\rho \vec{\mathbf{v}}) \right] dV \tag{1.3.6}$$

Since $dV \neq 0$, the *continuity equation* follows as

- $$\frac{\partial \rho}{\partial t} + \nabla \cdot (\rho \vec{\mathbf{v}}) = 0 \tag{1.3.7a}$$

Note, for steady flow,

$$\nabla \cdot (\rho \vec{\mathbf{v}}) = 0 \tag{1.3.7b}$$

and for incompressible flow,

- $$\nabla \cdot \vec{\mathbf{v}} = 0 \quad \text{or} \quad div\, \vec{\mathbf{v}} = 0 \tag{1.3.8a, b}$$

Mass balance derivation. Alternatively, Eq. (1.3.7a) can be derived with a more physical approach, based on a fluid mass balance over an open system, say, a cube (cf. Fig. 1.3.1).

- Global mass balance : $\sum \dot{m}_{in} - \sum \dot{m}_{out} = \left.\frac{\partial m}{\partial t}\right|_{\Delta V}$
- In a 1-D differential form:

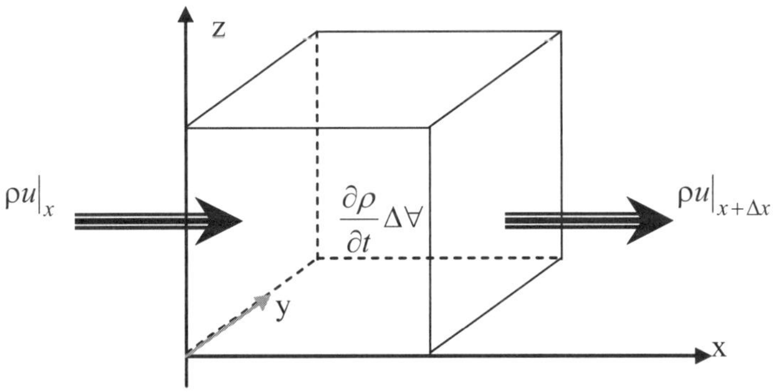

Fig. 1.3.1 One-dimensional fluid mass balance.

$$\left[(\rho u)|_x - (\rho u)|_{x+\Delta x}\right]\Delta y \Delta z = \frac{\partial \rho}{\partial t}\Delta V$$

- With Taylor's truncated series expansion, $f|_{x+\Delta x} = f|_x + \frac{\partial f}{\partial x}\Delta x$,

we obtain

$$-\frac{\partial(\rho u)}{\partial x}\Delta x \Delta y \Delta z = \frac{\partial \rho}{\partial t}\Delta x \Delta y \Delta z$$

- Adding the other two net fluxes (ρv) and (ρw) in the y- and z-direction, respectively, and dividing by the arbitrary volume ΔV, yields

  $$-\frac{\partial(\rho u)}{\partial x} - \frac{\partial(\rho v)}{\partial x} - \frac{\partial(\rho w)}{\partial x} = \frac{\partial \rho}{\partial t}$$

  or

- $\frac{\partial \rho}{\partial t} + \nabla \cdot (\rho \vec{v}) = 0$ (1.3.7a)

1.3.2 Momentum Conservation (Integral Approach)

Taking $N \equiv m\vec{v}$ and hence $\eta \equiv \vec{v}$, the RTT reads

$$m\frac{D\vec{\mathbf{v}}}{Dt} = m\vec{\mathbf{a}} = \vec{\mathbf{F}}_{body} + \vec{\mathbf{F}}_{surface} = \frac{\partial}{\partial t}\int_{C.V.}\rho\vec{\mathbf{v}}dV + \int_{C.S.}\vec{\mathbf{v}}\rho\vec{\mathbf{v}}\cdot d\vec{\mathbf{A}} \quad (1.3.9)$$

where $\vec{\mathbf{F}}_{body} \equiv \int_{C.V.}\rho\vec{\mathbf{f}}_b dV$ and $\vec{\mathbf{F}}_{surface} \equiv \int_{C.S.}\vec{\vec{T}}\cdot d\vec{\mathbf{A}}$. Thus, $\vec{\vec{T}}$ being the total stress tensor (see Eq.(1.3.13), we have

$$\frac{\partial}{\partial t}\int_{C.V.}\rho\vec{\mathbf{v}}dV + \int_{C.S.}\vec{\mathbf{v}}\rho\vec{\mathbf{v}}\cdot d\vec{\mathbf{A}} = \int_{C.S.}\vec{\vec{T}}\cdot d\vec{\mathbf{A}} + \int_{C.V.}\rho\vec{f}_b dV \quad (1.3.10)$$

Note: For a *rotating* material volume, the fluid angular momentum per unit volume $\eta \equiv \rho(\vec{\mathbf{r}}\times\vec{\mathbf{v}})$ has to be considered. The law of *conservation of angular momentum* states that the rate-of-change of angular momentum of a material volume is equal to the resultant moment on the volume.

Example 1.2: Force on Disk Moving Axially into an Axisymmetric Jet

A steady uniform round jet hits an approaching conical disk as shown. Find the force exerted on the disk, where v_{jet}, A_{jet}, v_{disk}, diameter D, angle θ, and fluid layer thickness t are given.

sketch:

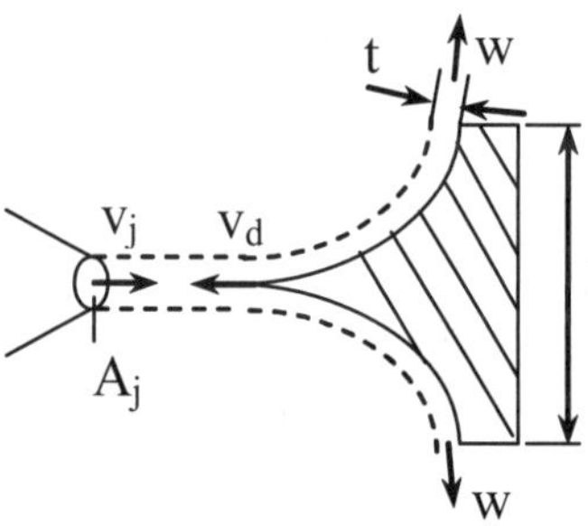

control volume:

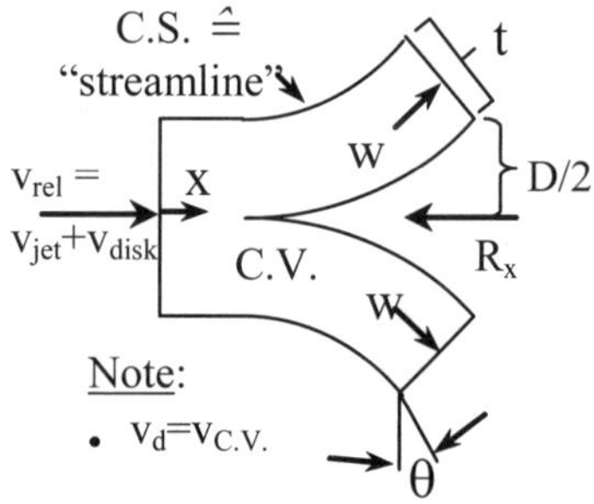

Assumptions: Constant averaged velocities and properties; moving C.V.

Approach: RTT (mass balance and 1-D force balance)

Solution:

(a) Mass Conservation: $0 = \int_{C.S.}\rho\vec{\mathbf{v}}\cdot d\vec{\mathbf{A}} = -\int_{A_{jet}}\rho v_{rel}dA + \int_{A_{exits}}\rho w dA$

where $v_{rel} \equiv v_{fluid} - v_{C.V.} = v_{jet} - (-v_{disk}) = v_j + v_d$ and $A_{exit} \approx \pi D\, t$.

$$\therefore \quad -\rho A_j\left(\mathrm{v}_j+\mathrm{v}_d\right)+\rho\left(\pi D\, t\right)\mathrm{w}=0 \quad \succ \quad \mathrm{w}=\frac{\left(\mathrm{v}_j+\mathrm{v}_d\right)A_j}{\pi D\, t}$$

(b) Momentum Conservation:

$N=\left(m\vec{\mathbf{v}}\right)_s$; $\eta=\vec{\mathbf{v}}$; $DN/Dt := \vec{\mathbf{F}}_{surf} = -R_x$ (resulting force)

$$\sum \vec{\mathbf{F}}_{\text{surf}} = \int_{\text{C.S.}} \vec{\mathbf{v}}\rho\vec{\mathbf{v}}_{\text{rel}}\cdot d\vec{\mathrm{A}} \xrightarrow{\text{x-component}} -R_x = \int_{C.S.} u\rho \mathrm{v}_{rel}\, dA$$

Thus, $-R_x = -\mathrm{v}_{rel}\rho \mathrm{v}_{rel} A_{\text{jet}} + \mathrm{w}^2 \rho A_{exit}\sin\theta$

where $\mathrm{w}=\dfrac{\mathrm{v}_{rel}A_{jet}}{A_{exit}}$ so that

- $R_x=\rho A_j\left(\mathrm{v}_j+\mathrm{v}_d\right)^2\left[1-\dfrac{A_j\, sin\,\theta}{\pi t D}\right]$

Comment: The resulting fluid-structure force $R_x=\left(\dot{m}\mathrm{v}_x\right)_{out}-\left(\dot{m}\mathrm{v}_x\right)_{in}$, i.e., R_x is the result of a change in fluid flow momentum. If the disk would move *away* with $\mathrm{v}_d=\mathrm{v}_j$, v_{rel}=0 and hence R_x=0.

<u>*Example 1.3*</u>: Force on a Submerged Body

Find the drag per unit length on a submerged elliptic rod of characteristic thickness *h*, based on a measured velocity profile downstream from the body, i.e.,

$$u(y)=\frac{u_\infty}{4}\left(1+\frac{y}{h}\right); \quad 0\le y\le 3h$$

Sketch:

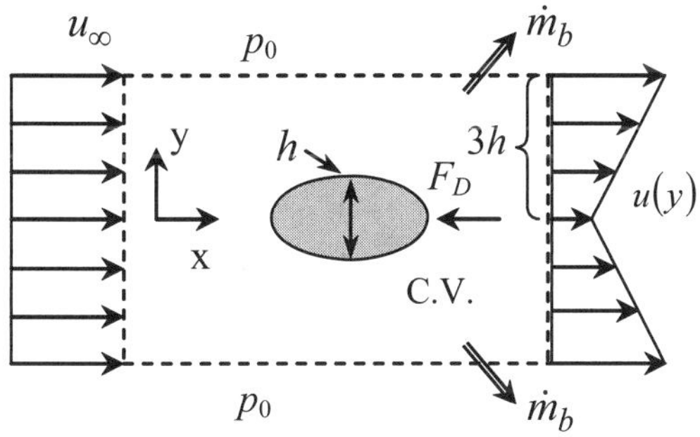

Approach:

- Integral mass & momentum balances (RTT)

Assumptions: steady state, 2-D, constant properties, fixed C.V., symmetry.

Solution:

(a) Mass Conservation: $\dot{m}_{in} = \dot{m}_b + \dot{m}_{out}$

$$0 = \int_{C.S.} \rho\vec{\mathrm{v}} \cdot d\vec{\mathbf{A}} = -\rho u_\infty 2(3h) + \dot{m}_b + 2\rho \frac{u_\infty}{4} \int_0^{3h} \left(1 + \frac{y}{h}\right) dy$$

$$\therefore \quad \dot{m}_b = 6\rho h u_\infty - \frac{15}{4}\rho h u_\infty = \frac{9}{4}\rho h u_\infty$$

(b) Momentum Conservation (x-momentum):

$$-F_D = \int_{C.S.} \mathrm{v}_x \rho\vec{\mathrm{v}} \cdot dA := \left(\dot{m}\mathrm{v}_x\big|_{exit} + \dot{m}\mathrm{v}_x\big|_b\right) - \dot{m}\mathrm{v}_x\big|_{inlet}$$

$$-F_D = -u_\infty \rho u_\infty (6h) + u_\infty \left(\frac{9}{4}\rho h u_\infty\right) + 2\rho \int_0^{3h} [u(y)]^2 dy$$

- $F_D = \frac{9}{8}\rho h u_\infty^2$ in [N/m]

Comment: The fluid flow structure inside the C.V., especially behind the submerged body, is very complex. The RTT treats it as a "black box" and elegantly obtains $F_D = F_{pressure} + F_{friction}$ via "velocity defect" measurements.

1.3.2.1 Stress Tensors and Stress Vectors

While in most cases $\vec{\mathbf{f}}_\mathrm{b}$ is simply gravity per unit volume, $(\rho\vec{\mathbf{g}})$, the Cauchy or total stress tensor $\vec{\vec{T}}$ in Eq. (1.3.10) is a new unknown that constitutes a *closure problem*, i.e., $\vec{\vec{T}}$ has to be related to the principal variable $\vec{\mathrm{v}}$ or its derivatives. In expanded form

$$\vec{\vec{T}} = -p\vec{\vec{I}} + \vec{\vec{\tau}} \tag{1.3.11a}$$

where p is the *thermodynamic pressure* (or the static pressure for $\Delta\vec{\mathbf{v}} = 0$), $\vec{\vec{I}}$ is the unit tensor, and $\vec{\vec{\tau}}$ is the stress tensor. For any coordinate system, the stress vector $\vec{\boldsymbol{\tau}}$ relates to the *symmetric* second-order tensor $\vec{\vec{T}}$ as

$$\vec{\boldsymbol{\tau}} = \hat{\mathbf{n}} \cdot \vec{\vec{T}} = \vec{\vec{T}} \cdot \hat{\mathbf{n}}$$

where $\hat{\mathbf{n}}$ is the normal (unit) vector. Without tensor symmetry, i.e., $T_{ij} = T_{ji}$, an infinitesimal fluid element ($\Delta\forall \to 0$) would spin at $|\vec{\omega} \to \infty|$. In tensor notation the total stress tensor is written for a Cartesian coordinate system as:

$$T_{ij} = -p\delta_{ij} + \tau_{ij} \tag{1.3.11b}$$

where $-p\delta_{ij}$ is the isotropic part (e.g., fluid statics and inviscid flow) and τ_{ij} is the deviatoric part for which a constitutive equation has to be found. Physically, $\tau_{ij} \mathrel{\hat{=}} \tau_{(i=force)(j=direction)}$ represents a force field per unit area as a result of the resistance to the rate of deformation of fluid elements, i.e., internal friction. This insight leads for fluids to the postulate

$$\tau_{ij} = fct(\varepsilon_{ij})$$

where $\varepsilon_{ij} = \frac{1}{2}(\mathrm{v}_{i,j} + \mathrm{v}_{j,i})$ is the rate-of-deformation tensor. The relation between τ_{ij} and ε_{ij} (plus vorticity tensor ζ_{ij}) can be more formally derived, starting with a fluid element displacement from point P (with $\vec{\mathbf{v}}$ at t) to point P′ ($\vec{\mathbf{v}} + d\vec{\mathbf{v}}$ at $t + dt$) a distance ds apart. Expanding the total derivative in Cartesian coordinates

$$d\vec{\mathbf{v}} = \begin{bmatrix} \frac{\partial u}{\partial x} & \frac{\partial u}{\partial y} & \frac{\partial u}{\partial z} \\ \frac{\partial \mathrm{v}}{\partial x} & \frac{\partial \mathrm{v}}{\partial y} & \frac{\partial \mathrm{v}}{\partial z} \\ \frac{\partial w}{\partial x} & \frac{\partial w}{\partial y} & \frac{\partial w}{\partial z} \end{bmatrix} d\vec{s}_i = \nabla\vec{\mathbf{v}}\, d\vec{s} \tag{1.3.12a}$$

the spatial changes, or deformations, the fluid element is experiencing during dt can be expressed as the "rate-of-deformation" tensor, i.e.,

$$\frac{d\vec{\mathbf{v}}}{d\vec{s}} = \nabla\vec{\mathbf{v}} := \frac{\partial \mathrm{v}_i}{\partial x_j} \tag{1.3.12b}$$

It can be decomposed into strain-rate tensor ε_{ij} <symmetrical part> and the vorticity (or rotation tensor) ζ_{ij}:

$$\frac{\partial \mathrm{v}_i}{\partial x_j} = \varepsilon_{ij} + \zeta_{ij} \tag{1.3.13}$$

It can be readily shown (see Sect. 1.7) that:

- $\zeta_{yx} = -\zeta_{xy} = \omega_z$, $\zeta_{xz} = -\zeta_{zx} = \omega_y$, and $\zeta_{zy} = -\zeta_{yz} = \omega_x$, where $\omega_z = \frac{1}{2}\left(\frac{\partial \mathrm{v}}{\partial x} - \frac{\partial u}{\partial y}\right)$, etc.; thus, $2\vec{\omega} = \nabla \times \vec{\mathbf{v}} = \vec{\zeta}$
- $\varepsilon_{ii} := \dot{\gamma}_{ii} \equiv \partial \mathrm{v}_i / \partial x_i$ indicate volume change (dilation)
- $\varepsilon_{ij} := \frac{1}{2}\dot{\gamma}_{ij}$, $i \neq j$, represent element distortion
- The shear-rate tensor $\dot{\gamma}_{ij} \equiv \frac{\partial \mathrm{v}_i}{\partial x_j} + \frac{\partial \mathrm{v}_j}{\partial x_i}$ and hence the $\frac{1}{2}$ in $\varepsilon_{ij} = \frac{1}{2}\dot{\gamma}_{ij}$ is mathematically necessary in order to match Eq. (1.3.13).

Now, Stokes suggested that $\vec{\vec{\tau}}$ is a *linear* function of $\vec{\vec{\varepsilon}}$, which is not the case for non-Newtonian fluids, rarefied gases, and some fluid flows in microscale devices, e.g., MEMS. Specifically, for Newtonian fluids

$$\vec{\vec{\tau}} = \lambda(\nabla \cdot \vec{\mathrm{v}})\vec{\vec{I}} + 2\mu\vec{\vec{\varepsilon}} \tag{1.3.14}$$

where the viscosity coefficients λ and μ depend only on the thermodynamic state of the fluid. For incompressible flow, $\nabla \cdot \vec{\mathbf{v}} = 0$ (see Eq.(1.3.8)) and the total stress tensor reduces to

$$T_{ij} = -p\delta_{ij} + 2\mu\varepsilon_{ij} \tag{1.3.15}$$

where

$$2\mu\varepsilon_{ij} \equiv \tau_{ij} = \mu\left(\frac{\partial \mathrm{v}_\mathrm{i}}{\partial \mathrm{x}_\mathrm{j}} + \frac{\partial \mathrm{v}_\mathrm{j}}{\partial \mathrm{x}_\mathrm{i}}\right) := \mu\dot{\gamma}_{ij} \tag{1.3.16}$$

and $\dot{\gamma}_{ij} \equiv 2\varepsilon_{ij}$ is called the shear-rate tensor. Section 1.5 provides further discussions of stresses, strains, and deformations in (solid) continua.

Example 1.4: Shear Stress in Simple Couette Flow

Provide a physical explanation and a mathematical derivation for τ_{xy} in unidirectional flow.

(a) Tangential force, F_{pull}, keeping a viscous fluid in motion:

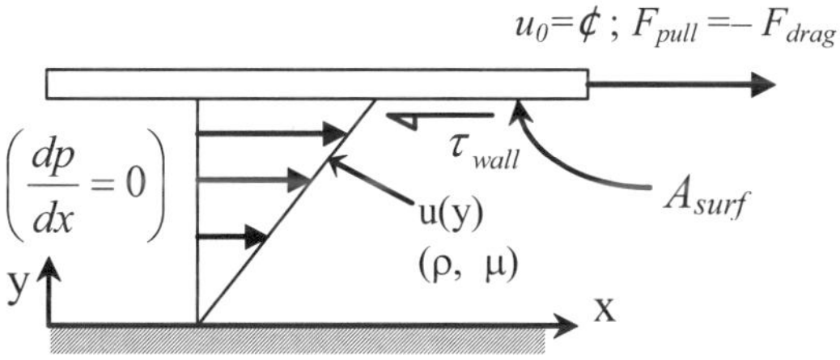

We observe:

$$\tau_{wall} = \frac{F_{drag}}{A_{surface}}$$

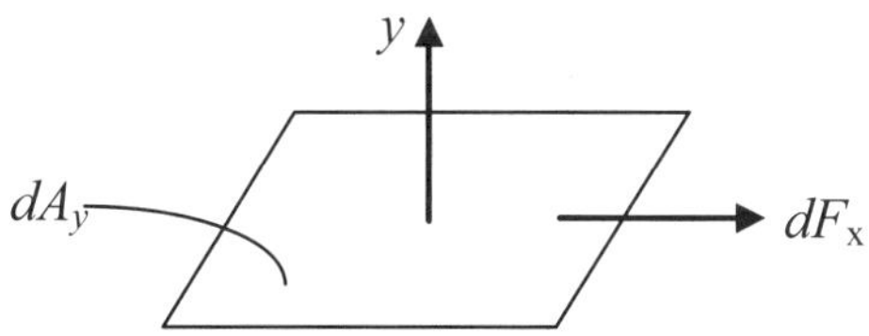

or anywhere in the fluid

$$\tau_{yx} = \frac{dF_x}{dA_y}$$

(b) Resistance to fluid element deformation:

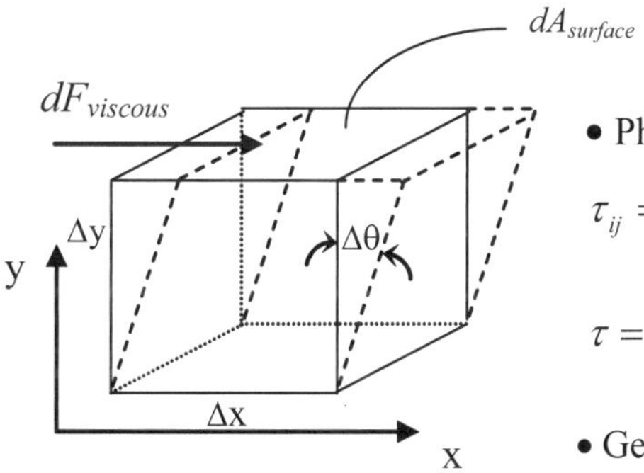

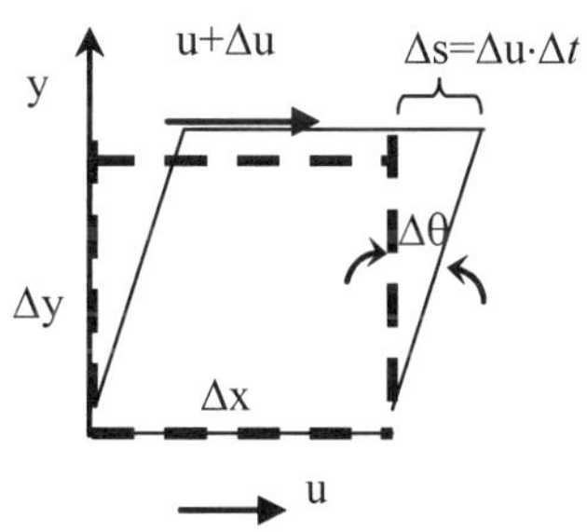

- Physics:

$$\tau_{ij} = \lim_{\delta A_i \to 0} \frac{\delta F_i}{\delta A_j} \hat{=} \frac{\textit{surface force}}{\textit{unit area}}$$

$$\tau = \frac{dF_v}{dA_s} \sim \frac{\Delta\theta}{\Delta t}$$

- Geometry:

$$\tan \Delta\theta \approx \Delta\theta = \frac{\Delta s}{\Delta y} = \frac{\Delta u \cdot \Delta t}{\Delta y}$$

- Combining both:

$$\tau \sim \frac{\Delta u}{\Delta y}$$

and in the limit with the proportionality factor μ,

- $\tau_{yx} = \mu \dfrac{du}{dy}$

in unidirectional flows.

Comments: • u(y) is linear as in simple Couette-type flows, $\tau_{yx} = \tau_{wall} = ¢$.

• This result for τ_{yx} is a very special case of Eq. (1.3.16).

1.3.2.2 Equation of Motion and its Special Cases

Returning now to Eq. (1.3.10) and including Eq. (1.3.11), the RTT for linear momentum transfer can be written as

$$\frac{\partial}{\partial t}\int_{C.V.}\rho\vec{v}dV + \int_{C.S.}\vec{v}\rho\vec{v}\cdot d\vec{A} = \int_{C.S.}\left(-p\vec{\vec{I}}+\vec{\vec{\tau}}\right)\cdot d\vec{A} + \int_{C.V.}\rho\vec{g}dV \qquad (1.3.17)$$

Transforming the surface integrals, using Eq. (1.3.5), leads to the *equation of motion*, which is valid for any fluid

• $$\frac{\partial(\rho\vec{v})}{\partial t} + \nabla\cdot(\rho\vec{v}\vec{v}) = -\nabla p + \nabla\cdot\vec{\vec{\tau}} + \rho\vec{g} \qquad (1.3.18)$$

(i) Navier-Stokes Equations:

For constant fluid properties and using Eq. (1.3.16):

• $$\frac{\partial\vec{v}}{\partial t} + (\vec{v}\cdot\nabla)\vec{v} = -\frac{1}{\rho}\nabla p + \nu\nabla^2\vec{v} + \vec{g} \qquad (1.3.19)$$

where $\frac{\partial\vec{v}}{\partial t} + (\vec{v}\cdot\nabla)\vec{v} \equiv \frac{D\vec{v}}{Dt} = \vec{a}_{total}$ and $\nu = \mu/\rho$ is the kinematic viscosity. Equation (1.3.19) together with the continuity equation, $\nabla\cdot\vec{v} = 0$, form the *Navier-Stokes (N-S) equations*. For steady two-dimensional (2-D) flows, the N-S equations read in rectangular coordinates:

$$\frac{\partial u}{\partial x} + \frac{\partial v}{\partial y} = 0$$

$$u\frac{\partial u}{\partial x} + v\frac{\partial u}{\partial y} = -\frac{1}{\rho}\frac{\partial p}{\partial x} + \nu\left(\frac{\partial^2 u}{\partial x^2} + \frac{\partial^2 u}{\partial y^2}\right) + g_x \qquad (1.3.20a\text{-}c)$$

$$u\frac{\partial v}{\partial x} + v\frac{\partial v}{\partial y} = -\frac{1}{\rho}\frac{\partial p}{\partial y} + \nu\left(\frac{\partial^2 v}{\partial x^2} + \frac{\partial^2 v}{\partial y^2}\right) + g_y$$

(ii) Prandtl's Boundary-Layer Equations:

For thin wall boundary layers where the ratio of boundary layer thickness to flat-plate length $\delta/l \ll 1$ and $v \ll u$ because $Re_l \gg 1$, the axial momentum diffusion, $\nu\,\partial^2 u/\partial x^2$, is negligible when compared to all other terms in the x-momentum equation (1.3.20b). In the y-momentum equation, only $\partial p/\partial y$ is relatively significant (see HWA in Sect. 1.7). Thus, the steady 2-D boundary-layer equations read (cf. Figs. 1.3.4 and 1.3.6 as well as Schlichting & Gersten, 2000 or Kleinstreuer, 1997):

- $$\frac{\partial u}{\partial x}+\frac{\partial \mathrm{v}}{\partial y}=0$$
- $$u\frac{\partial u}{\partial x}+\mathrm{v}\frac{\partial u}{\partial y}=-\frac{1}{\rho}\frac{\partial p}{\partial x}+\nu\frac{\partial^2 u}{\partial y^2} \qquad (1.3.21a\text{-}c)$$
- $$0=-\frac{1}{\rho}\frac{\partial p}{\partial y}$$

The last equation implies that the pressure normal to the wall is constant, i.e., $p(\mathrm{x})$ of the outer flow is imposed onto the thin shear layer.

(iii) Stokes Equation:

When the viscous forces are dominant, i.e., Re<1, the $(\vec{\mathbf{v}}\cdot\nabla)\vec{\mathbf{v}}$ term in Eq. (1.3.19) is negligible and the *Stokes equation* is obtained.

- $$\rho\frac{\partial \vec{\mathbf{v}}}{\partial t}=-\nabla p+\mu\nabla^2\vec{\mathbf{v}} \qquad (1.3.22)$$

(iv) Euler's Inviscid Flow Equation:

For frictionless flow, Eq. (1.3.19) reduces to

- $$\rho\frac{D\vec{\mathbf{v}}}{Dt}=-\nabla p+\rho\vec{\mathbf{g}} \qquad (1.3.23)$$

which is the *Euler equation*. Equation (1.3.23) applied in 2-D to a representative streamline along s yields

$$\frac{\partial \mathrm{v}_s}{\partial t}+\mathrm{v}_s\frac{\partial \mathrm{v}_s}{\partial s}+\frac{1}{\rho}\frac{\partial p}{\partial s}+g\underbrace{\sin\theta}_{=\partial z/\partial s}=0 \qquad (1.3.24)$$

Multiplying Eq. (1.3.24) through by ∂s and integrating yields for steady incompressible inviscid flow the *Bernoulli equation*:

- $$\frac{\mathrm{v}^2}{2}+\frac{p}{\rho}+gz=C \qquad (1.3.25)$$

where v and p are locally averaged quantities along a representative streamline and the z-coordinate extends against the direction of gravity. Thus, for two points on a streamline the total energy per unit mass is balanced, i.e.,

$$\frac{v_1^2}{2}+\frac{p_1}{\rho}+gz_1=\frac{v_2^2}{2}+\frac{p_2}{\rho}+gz_2 \tag{1.3.26}$$

For example, for a given system where $v_2=0$ (e.g., a wall) and $g\Delta z \approx 0$, we have

$$p_2 = p_1 + \frac{\rho v_1^2}{2} \tag{1.3.27}$$

where p_2 is the *total* or *stagnation point pressure*; p_1 is the *static pressure* at point 1; and $\rho v_1^2/2$ is the *dynamic pressure* at point 1. One application of Eq. (1.3.27) is the Pitot tube where $v_1 = \sqrt{2(p_2 - p_1)/\rho}$.

1.3.2.3 Force Balance Derivation

A more physical approach for deriving the (linear) momentum equation starts with a force balance for a representative elementary volume (REV). Employing rectangular coordinates and an incompressible fluid, external surface and body forces accelerate an REV of mass *m*, so that we can write per unit volume in the Lagrangian framework (cf. Fig. 1.3.2 and Eq.(1.3.9)):

$$\rho\frac{D\vec{\mathbf{v}}}{Dt}=\sum\vec{\mathbf{f}}_{surface}+\sum\vec{\mathbf{f}}_{body} \tag{1.3.28}$$

From the Eulerian point of view, the REV turns into a control volume and we record local and convective momentum changes due to *net* pressure, viscous, and gravitational forces, viz.:

$$\rho\left[\frac{\partial\vec{\mathbf{v}}}{\partial t}+\left(\vec{\mathbf{v}}\cdot\nabla\right)\vec{\mathbf{v}}\right]=\vec{\mathbf{f}}_{\substack{net\\pressure}}+\vec{\mathbf{f}}_{\substack{net\\viscous}}+\vec{\mathbf{f}}_{\substack{net\\buoyancy}} \tag{1.3.29a}$$

For example, considering a 1-D force balance (cf. Fig. 1.3.3), the net pressure force per unit volume $\Delta\forall$ in the *x*-direction is

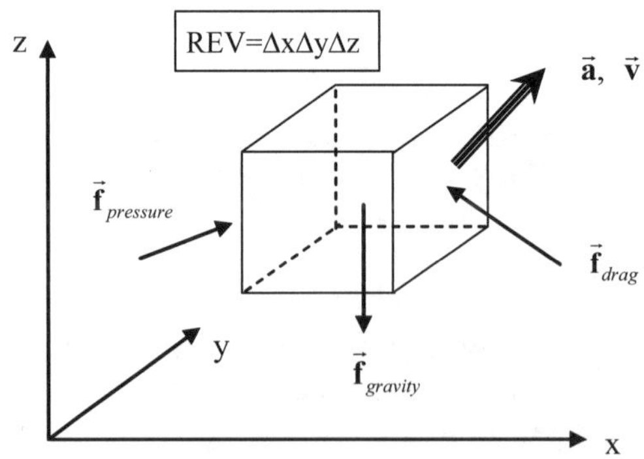

Figure 1.3.2 Closed system, i.e., accelerating material volume.

$$f_{\substack{net\\pressure}} = f_p\Big|_x - f_p\Big|_{x+\Delta x} = -\frac{\partial f_p}{\partial x}\Delta x$$

and with $f_p \equiv \dfrac{p\Delta A}{\Delta V}$, $\quad f_{\substack{net\\pressure}} = -\dfrac{\partial p}{\partial x}\dfrac{\Delta y \Delta z}{\Delta x \Delta y \Delta z}\Delta x = -\dfrac{\partial p}{\partial x}$,

or in 3-D

$$\vec{\mathbf{f}}_{\substack{net\\pressure}} = -\nabla p = -\left(\frac{\partial p}{\partial x}\vec{\mathbf{i}} + \frac{\partial p}{\partial y}\vec{\mathbf{j}} + \frac{\partial p}{\partial z}\vec{\mathbf{k}}\right) \tag{1.3.29b}$$

Similarly, the net viscous force per unit volume in the x-direction reads:

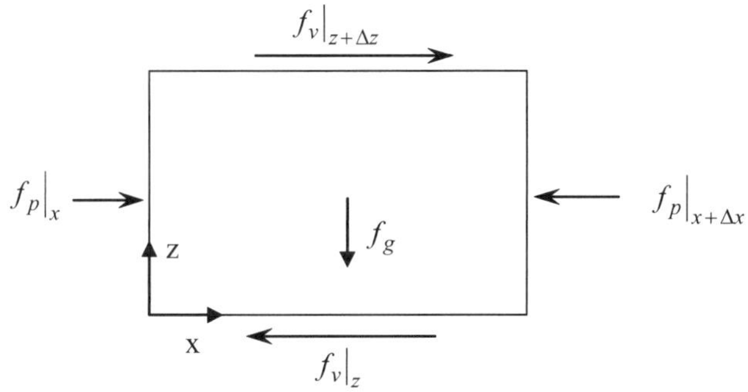

Fig. 1.3.3 Control volume for 1-D force balances.

$$f_{\substack{net\\viscous}} = f_v\Big|_z - f_v\Big|_{z+\Delta z} = -\frac{\partial f_v}{\partial z}\Delta z$$

and with $f_v \equiv \dfrac{\tau\Delta A}{\Delta V}$, $\quad f_{\substack{net\\viscous}} = -\dfrac{\partial \tau}{\partial z}\dfrac{\Delta x \Delta y}{\Delta x \Delta y \Delta z}\Delta z = -\dfrac{\partial \tau}{\partial z}$,

or in 3-D

$$\vec{\boldsymbol{f}}_{\substack{net\\viscous}} = \nabla\cdot\vec{\vec{\tau}} = \begin{pmatrix} \left(\dfrac{\partial \tau_{xx}}{\partial x} + \dfrac{\partial \tau_{yx}}{\partial y} + \dfrac{\partial \tau_{zx}}{\partial z}\right)\vec{\mathbf{i}} + \\ \left(\dfrac{\partial \tau_{xy}}{\partial x} + \dfrac{\partial \tau_{yy}}{\partial y} + \dfrac{\partial \tau_{zy}}{\partial z}\right)\vec{\mathbf{j}} + \\ \left(\dfrac{\partial \tau_{xz}}{\partial x} + \dfrac{\partial \tau_{yz}}{\partial y} + \dfrac{\partial \tau_{zz}}{\partial z}\right)\vec{\mathbf{k}} \end{pmatrix} \tag{1.3.29c}$$

Now, with Stokes' hypothesis for incompressible Newtonian fluids, Eq. (1.3.14) reduces to

• $\qquad \vec{\vec{\tau}} = \mu\left(\nabla\vec{\mathbf{v}} + \nabla\vec{\mathbf{v}}^{\mathrm{tr}}\right)$

so that Eq. (1.3.29a) reads (cf. N-S equation)

- $$\rho\left[\frac{\partial \vec{\mathbf{v}}}{\partial t}+(\vec{\mathbf{v}}\cdot\nabla)\vec{\mathbf{v}}\right]=-\nabla p+\mu\nabla^2\vec{\mathbf{v}}+\rho\vec{\mathbf{g}} \qquad (1.3.19)$$

Example 1.5: Parallel and Nearly Parallel Flows

The assumptions of steady laminar fully-developed or *unidirectional* flow lead to analytic solutions of the Navier-Stokes equations. Examples include Poiseuille flow, Couette flows, Stokes I and II flows, etc. (cf. Kleinstreuer, 1997). There is another group of analytical solutions where the wall flow is *nearly parallel*; for example, flow in a slightly tapered tube, flow in a slider bearing, levitated disk flows, coating flows, etc. (cf. Kleinstreuer, 1997; Middleman, 1998; Bird et al., 2001; among others). The trick for quasi-unidirectional flows is to solve the parallel flow equations and then incorporate the nonparallel wall effects via (the) boundary condition(s).

Problem: Load-Carrying Capacity of a Planar Bearing (l, W, θ)

Sketch:

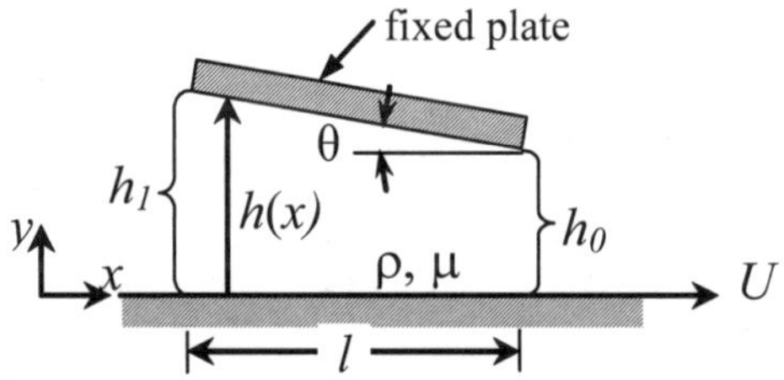

Assumptions:

- steady laminar 2-D flow
- $\mathrm{Re}(=Uh_1/\nu) << 1$ and $v << u$, i.e., $v \approx 0$
- $p = p(x)$ only
- constant fluid properties
- no end effects

Solution:

Based on the assumptions, the starting equations are Eqs. (1.3.21a-c). However, with $v \approx 0$

$$\frac{\partial u}{\partial x}+\frac{\partial v}{\partial y}=0 \quad \text{reduces to} \quad \frac{\partial u}{\partial x}\approx 0\text{, hence } 0=-\frac{\partial p}{\partial x}+\mu\frac{\partial^2 u}{\partial y^2}$$

Clearly, the viscous forces are dominant and axial diffusion is negligible. Also, $0=-\frac{\partial p}{\partial y}$ because the normal velocity component is small and its second derivatives are negligible. The boundary conditions are $u(y=0)=U$ and $u[y=h(x)]=0$.

With the nondimensionalizations

$$\hat{x} = x / h_1, \quad \hat{y} = y / h_1; \quad \hat{u} = u / U, \quad \hat{v} = v / U;$$
$$\hat{p} = p h_1 / (\mu U) \quad \text{and} \quad \text{with} \quad \eta = h(x) / h_1$$

where $h(x) = h_1 - \frac{(h_1 - h_0)}{l} x$, inserted into the simplified x-momentum equation, we have after double integration and invoking the boundary conditions $\hat{u}(\hat{y}=0)=1, \hat{u}(\hat{y}=\eta)=0$, $\hat{p}(\hat{x}=0)=p_0$, and $\hat{p}\left(\hat{x}=\frac{l}{h_1}\right)=p_0$:

- $$\hat{u}(\hat{x},\hat{y}) = \left(1 - \frac{\hat{y}}{\eta}\right) - \frac{\eta^2}{2}\left(\frac{d\hat{p}}{d\hat{x}}\right)\left[\frac{\hat{y}}{\eta} - \left(\frac{\hat{y}}{\eta}\right)^2\right]$$

The pressure gradient $d\hat{p} / d\hat{x}$ can be obtained from the mass balance

$$\dot{m}_{\text{lube}} = \rho W \int_0^{h(x)} u(x,y)dy = \rho W U h_1 \int_0^{\eta} \hat{u} d\hat{y} = \rho Q$$

where W is the width and Q is the volumetric flow rate, i.e.,

- $\frac{d\hat{p}}{d\hat{x}} = 12\left[\frac{1}{2\eta^2} - \frac{æ}{\eta^3}\right]$, where $æ \equiv \frac{Q}{UWh_1}$ is typically known.

Now various pressure distributions $p(x)$ based on design parameters h_1, h_0, l, and $\dot{m}$ can be postulated; in turn, $p(x)$ determines the lubricant velocity field and the load-carrying capacity

$$L = \int_0^l p(x) \cos\theta dx$$

within the stated restrictions.

Example 1.6: Transient Flow in a Tube

Most naturally occurring flows are unsteady, e.g., blood flow in arteries and air flow in the lung, or transient effects, such as start-up, shut-down, pressure waves, etc., have to be considered for industrial flows. Pulsatile flow in a tube has been analyzed by Womersley (cf. Nichols & O'Rourke, 1990 or Zamir, 2000); "start-up" due to a suddenly applied pressure gradient is discussed below.

Problem: Sudden Pressure Applied to Horizontal Flow in a Pipe

Sketch:

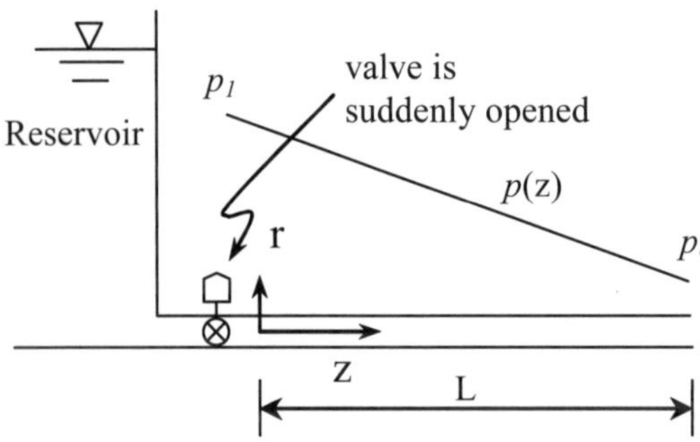

Assumptions:

- transient laminar 1-D flow
- constant fluid properties
- $\partial p/\partial z = ¢$ for $t > 0$

Solution:

With the postulates $\vec{\mathbf{v}} = [u(r,t),0,0]$, $-\partial p/\partial z = (p_1 - p_0)/L = ¢$ and $\partial/\partial\theta = 0$, continuity states

$$\frac{\partial u}{\partial z} = 0$$

Thus the *z*-momentum equation (see App. B) reduces to

$$\frac{\partial u}{\partial t} = -\frac{1}{\rho}\left(\frac{\partial p}{\partial z}\right) + \frac{\nu}{r}\frac{\partial}{\partial r}\left(r\frac{\partial u}{\partial r}\right)$$

subject to $u(r,t=0)=0$, $u(r=r_0,t)=0$ and $\partial u/\partial r|_{r=0} = 0$ for all t. Now there are various techniques for solving the PDE available: direct numerical solution, product solution leading to two ODEs, or superposition with the postulate

$$u(r,t) = u(r)\big|_{ss} + u(r,t)\big|_{transient}$$

With the assumption that $-\partial\, p/\partial\, z = \Delta\, p/L = ¢$ at all times,

$$\frac{\Delta p}{L} = 4\mu u_{max}\left[1 - \left(\frac{r}{r_0}\right)^2\right]$$

where $u_{max}\left[1-\left(\frac{r}{r_0}\right)^2\right] = u_{ss}$ is the Poiseuille solution.

Knowing $u(r)\big|_{ss}$, we obtain for the transient part with

$$\hat{u} = u_{tran} / u_{\max}, \ \hat{r} = r / r_0, \ \text{and} \ \hat{t} = \frac{\mu}{\rho r_0} t$$

$$\frac{\partial \hat{u}}{\partial \hat{t}} = \frac{1}{\hat{r}} \frac{\partial}{\partial \hat{r}} \left(\hat{r} \frac{\partial \hat{u}}{\partial \hat{r}} \right)$$

subject to $\hat{u}(\hat{t} = 0) = \hat{r} - 1, \ \hat{u}(\hat{r} = 1) = 0, \ \text{and} \ \left.\frac{\partial \hat{u}}{\partial \hat{r}}\right|_{\hat{r}=0} = 0.$

The solution to this Sturm-Liouville problem is (see Özisik, 1993)

$$\hat{u} = -8 \sum_{n=1}^{\infty} \frac{J_0(\lambda_n \hat{r})}{\lambda_n^3 J_1(\lambda_n)} e^{-\lambda_n^2 \hat{t}}$$

The eigenvalues λ_n=2.41, 5.52, 8.65 for n=1, 2, 3 and $\lambda_{n+1} = \lambda_n$+p for $n \geq 3$ are the zeros of the Bessel function J_0. The profiles shown illustrate the evolution of the axial velocity, $u(r,t) / u_{max}$, throughout the pipe.

Graph:

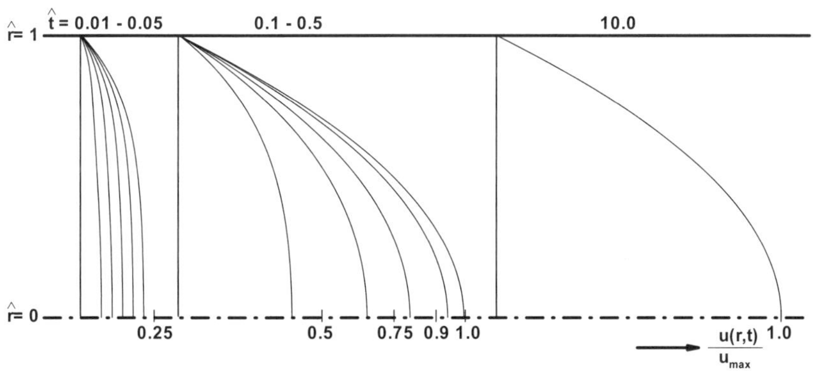

Additional exact and approximate solutions based on the reduced Navier-Stokes equations may be found in Kleinstreuer (1997), Nichols & O'Rourke (1998), Schlichting & Gersten (2000), Bird et al. (2002), and Panton (2005), among others.

1.3.3 Energy Conservation

Taking $N \equiv E_{total}$ and hence $\eta \equiv e_t$, the RTT reads [Recall: First Law of Thermodynamics]:

$$\left.\frac{DE_t}{Dt}\right|_{\substack{closed\\system}} \equiv \dot{Q} - \dot{W} = \frac{\partial}{\partial t}\int_{C.V.} \rho e_t dV + \int_{C.S.} \rho e_t \vec{\mathrm{v}} \cdot d\vec{\mathrm{A}} \tag{1.3.30}$$

Typically, $E_{total}=E_{kinetic}+E_{internal}:=\frac{m}{2}|\vec{\mathbf{v}}|^2 + m\tilde{u}$, i.e., the L.H.S. has to be separately expressed. Employing the momentum conservation law, Eq. (1.3.18) is scalarly multiplied through by $\vec{\mathbf{v}}$ (see App. A), i.e.,

$$\rho\vec{\mathbf{v}} \cdot \frac{D\vec{\mathbf{v}}}{Dt} \hat{=} \frac{D}{Dt}\left(\frac{\rho}{2}|\vec{\mathbf{v}}|^2\right) = \left(\nabla \cdot \vec{\vec{\mathrm{T}}}\right) \cdot \vec{\mathbf{v}} + \rho\vec{f}_b \cdot \vec{\mathbf{v}}$$

Integration over the material (or control) volume V yields

$$\frac{DE_{kin}}{Dt} = \int_V \left(\nabla \cdot \vec{\vec{\mathrm{T}}}\right) \cdot \vec{\mathbf{v}} dV + \int_V \rho\left(\vec{\mathbf{f}}_b \cdot \vec{\mathbf{v}}\right) dV$$

The integrand $\left(\nabla \cdot \vec{\vec{\mathrm{T}}}\right) \cdot \vec{\mathbf{v}} \equiv \nabla \cdot \left(\vec{\vec{\mathrm{T}}} \cdot \vec{\mathbf{v}}\right) - \vec{\vec{\mathrm{T}}} : \vec{\vec{\varepsilon}}$, so that with the Gaussian divergence theorem

$$\frac{DE_{kin}}{Dt} = \int_S \left(\vec{\vec{\mathrm{T}}} \cdot \vec{\mathbf{v}}\right) \cdot d\vec{\mathbf{S}} - \int_V \vec{\vec{\mathrm{T}}} : \vec{\vec{\varepsilon}} \mathrm{dV} + \int_V \rho\left(\vec{\mathbf{v}} \cdot \vec{\mathbf{f}}_b\right) dV$$

where the integral

$$\int_S \left(\vec{\vec{\mathrm{T}}} \cdot \vec{\mathbf{v}}\right) \cdot d\vec{\mathbf{S}} \equiv \int_S \left(\vec{\vec{\mathrm{T}}} \cdot \vec{\mathbf{v}}\right) \cdot \vec{n} dS = \int_S \left(\vec{\mathbf{v}} \cdot \vec{\tau}\right) dS$$

Hence,

$$\frac{D}{Dt}\left(E_{kin} + E_{int}\right) = \int_S \left(\vec{\mathbf{v}} \cdot \vec{\tau}\right) dS + \int_V \rho\left(\vec{\mathbf{v}} \cdot \vec{\mathbf{f}}_b\right) dV + \int_S \vec{\mathbf{q}}_t \cdot d\vec{\mathbf{S}} \tag{1.3.31}$$

$$\begin{Bmatrix}\text{Time-rate-of-change}\\ \text{of } E_{total} \text{ in material}\\ \text{volume moving with}\\ \text{the flow}\end{Bmatrix} = \begin{Bmatrix}\text{Rate of work done}\\ \text{on the volume by}\\ \text{surface plus body}\\ \text{forces}\end{Bmatrix} + \begin{Bmatrix}\text{Net heat flux}\\ \text{across the}\\ \text{material volume}\end{Bmatrix}$$

The frictional work term, i.e., $\int_V \vec{\vec{\mathrm{T}}} : \vec{\vec{\varepsilon}} \mathrm{dV}$, dropped out and the heat flux $\vec{\mathbf{q}}_t = \vec{\mathbf{q}}_{cond.} + \vec{\mathbf{q}}_{conv.} + \vec{\mathbf{q}}_{rad.}$ was added as part of the internal energy contribution. In order to complete the energy conservation law, a distributed internal heat source term, e.g., $\int_V \rho \hat{q}_{int} dV$, may have to be added. Finally, Eq. (1.3.31) reads for a stationary control volume:

$$\frac{DE_t}{Dt} = \frac{\partial}{\partial t} \int_{C.V.} \rho e_t dV + \int_{C.S.} \rho e_t \vec{\mathbf{v}} \cdot d\vec{\mathbf{A}} \equiv \int_{C.V.} \rho\left(\vec{\mathbf{v}} \cdot \vec{\mathbf{f}}_b\right) dV + \int_{C.S.} \left(\vec{\mathbf{v}} \cdot \vec{\vec{\tau}}\right) dA + \int_{C.S.} \vec{\mathbf{q}}_t \cdot d\vec{\mathbf{A}} + \int_{C.V.} \rho \hat{q}_{int} dV \tag{1.3.32}$$

Again, using the divergence theorem (1.3.5), Eq. (1.3.32) can be rewritten in differential form as

$$\rho \frac{De_t}{Dt} \equiv \rho\left[\frac{\partial e_t}{\partial t} + (\vec{\mathbf{v}} \cdot \nabla) e_t\right] = \rho\left(\vec{\mathbf{f}}_b \cdot \vec{\mathbf{v}}\right) + \nabla \cdot \left(\vec{\vec{\mathbf{T}}} \cdot \vec{\mathbf{v}}\right) + \nabla \cdot \vec{\mathbf{q}}_t \tag{1.3.33}$$

where $e_t = \frac{1}{2}|\vec{\mathbf{v}}|^2 + \tilde{u} \cong \frac{1}{2}|\vec{\mathbf{v}}|^2 + c_v \mathrm{T}$ and $\vec{q}_t = -k\nabla\mathrm{T} + \ldots$, where T is now the temperature.

1.3.3.1 Heat and Mass Transfer Equations

A simpler derivation of the energy equation resulting in a directly applicable form employs the GTE [Eq.(1.3.1)] as a point of departure. Setting $\Psi \equiv h = \tilde{u} + p/\rho$ <enthalpy per unit mass>, $J \equiv \vec{\mathbf{q}}$ <diffusive heat flux>, and $\phi \equiv \nu\Phi$ <energy dissipation due to viscous stress> yields

$$\frac{\partial}{\partial t}(\rho h) + \nabla \cdot (\rho \vec{\mathbf{v}} h) = -\nabla \cdot \vec{\mathbf{q}} + \mu\Phi \tag{1.3.34}$$

with $dh = c_p dT$, or $h = c_p T$ when c_p=constant, $\vec{\mathbf{q}} = -k\nabla T$ after Fourier and $\mu\Phi = \tau_{ij}\partial v_i / \partial x_j$, we obtain for thermal flow with constant fluid properties the heat transfer equation in the form:

$$\frac{\partial T}{\partial t} + (\vec{\mathbf{v}} \cdot \nabla) T = \alpha \nabla^2 T + \frac{\mu}{\rho c_p} \Phi \tag{1.3.35a}$$

where $\alpha = k/(\rho c_p)$ is the thermal diffusivity. This scalar equation

$$\frac{DT}{Dt} = \alpha \nabla^2 T + S_T \tag{1.3.35b}$$

has the same form as the species mass transfer equation

$$\frac{Dc}{Dt} = \mathcal{D} \nabla^2 c + S_c \tag{1.3.36}$$

where $\mathcal{D}$ is the binary diffusion coefficient and S_c possible species sinks or sources (Bird et al., 2002).

1.3.3.2 Basic Heat and Mass Transfer Applications

In order to highlight the use of Equations (1.3.35) and (1.3.36), two sample problems are provided employing analytical solutions. Additional examples may be found in Bejan (1995) and Bird et al. (2002).

Example 1.7: Convection Heat Transfer in a Tube

Considering hydrodynamically and thermally fully developed flow, i.e., $\partial u/\partial z = 0$ and Peclet number $Pe_l = Re_l\ Pr = \bar{u}l / \alpha > 100$, an analytic solution of the heat transfer equation and hence an expression for the Nusselt number can be obtained, provided that q_{wall} =¢ is the thermal boundary condition.

Sketch:

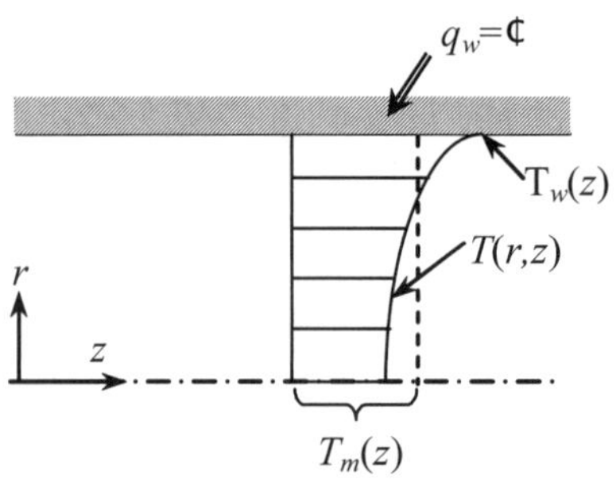

Assumptions:

- Poiseuille flow
- axial convection balanced by radial conduction (or radial thermal diffusion)
- constant fluid properties
- constant wall heat flux

Solution:

Based on the stated restrictions, Eq. (1.3.35b) reduces in cylindrical coordinates to

$$\bullet \qquad u(r)\frac{\partial T}{\partial z} = \frac{\alpha}{r}\frac{\partial}{\partial z}\left(r\frac{\partial T}{\partial r}\right) \tag{E1.7-1a}$$

subject to

$$k\frac{\partial T}{\partial r}\bigg|_{r=r_0} = q_w \quad [\text{or } T(r = r_0) = T_w(z)] \tag{E1.7-1b}$$

and

$$\frac{\partial T}{\partial r}\bigg|_{r=0} = 0 \qquad \langle symmetry \rangle \tag{E.1.7-1c}$$

In order for the flow to be "thermally fully developed" there has to be a generalized temperature profile $\Theta(r)$ for which

$$\frac{\partial\Theta}{\partial z}=0$$

With $\Theta=\dfrac{T_w-T}{T_w-T_m}$, where $T_m=\dfrac{1}{A\bar{u}}\int_A u(r)T(r,z)dA$, we obtain from the condition $\bar{\bar{f}}=\bar{f}$

$$\frac{\partial T}{\partial z}=\frac{\partial T_w}{\partial z}-\Theta\frac{\partial T_w}{\partial z}+\Theta\frac{\partial T_m}{\partial z}$$

i.e., with all gradients being constant for q_{wall}=¢ and $T_w-T_m=$¢ (cf. Newton's law of cooling), $dT_w/dz=dT_m/dz$, so that

$$\frac{\partial T}{\partial z}=\frac{\partial T_m}{\partial z}=¢$$

Now, the governing PDE can be rewritten as

- $$u_{max}\left[1-\left(\frac{r}{r_0}\right)^2\right]\left(\frac{dT_m}{dz}\right)=\frac{\alpha}{r}\frac{\partial}{\partial r}\left(r\frac{\partial T}{\partial r}\right) \quad \text{(E1.7-2)}$$

Direct integration yields

- $$T(r,z)=T_w(z)-\frac{u_{max}}{\alpha}\left(\frac{dT_m}{dz}\right)\left(\frac{3}{16}r_0^2+\frac{r^4}{16r_0^2}-\frac{r^2}{4}\right) \quad \text{(E1.7-3)}$$

The "mixing-cup" temperature T_m can now be computed from its definition and then $(T_m - T_w)$ can be formed, i.e.,

$$T_m-T_w=\frac{11}{96}\frac{u_{max}}{\alpha}\left(\frac{dT_m}{dz}\right)r_0^2$$

An energy balance for a cylindrical control volume, $\pi r_0^2\Delta z$, in the tube yields

- $$q_w=\frac{r_0}{4}\rho u_{max}C_p\left(\frac{dT_m}{dz}\right);\quad u_{max}=2\bar{u} \quad \text{(E1.7-4)}$$

Now, with $q_w=h(T_w-T_m)$=¢ and $Nu_D=\dfrac{hD}{k}$ where $\alpha=\dfrac{k}{\rho C_p}$ and D=$2r_0$,

- Nu_D=4.364

Example 1.8: Diffusive Mass Transfer across a 1-D Membrane or Cell

Consider two solutions of constant concentrations, where $c_1 > c_2$ at all times, separated at *t=0* by a thin flat membrane, $0 \le x \le h$, with initial solute concentration $c(x, 0) = 0$. Find: (a) *c(x,t)* as well as the mass flux *j(t;x=0)* and then (b) apply the mathematical solution to a cell of characteristic extension *h*, surrounded by a reservoir with solute concentration c_0. Calculate the ratio of diffusing solute mass *m(t)* to the solute mass at equilibrium, i.e., $m_0 = m(t \to \infty)$.

Sketch:

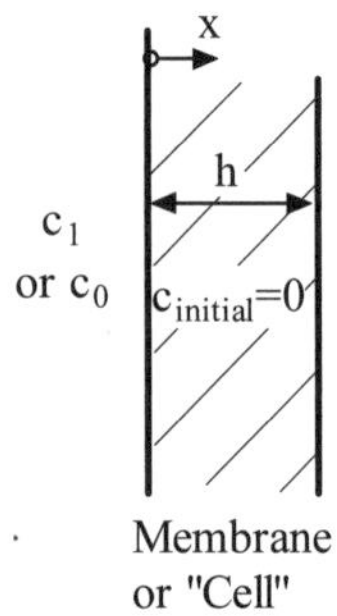

Assume:

- Transient 1-D binary diffusional mass transfer
- Constant diffusivity

Concept/Approach:

- Transient 1-D diffusion Equation
- Separation of Variables method
- Neglect higher-order term

Solution:

With the implicit assumptions of $\vec{\mathbf{v}} \equiv 0$ and c=c(x,t) only, Eq. (1.3.36) can be reduced to

$$\frac{\partial c}{\partial t} = D\frac{\partial^2 c}{\partial x^2} \tag{E1.8-1}$$

Subject to

$$c(x, 0) = 0, \quad c(0, t) = c_1 \text{ and } c(h, t) = c_2$$

Postulating the product solution (cf. Ozisik, 1993; among others)

$$c(x, t) = X(x) \cdot T(t) \tag{E1.8-2}$$

leads to

$$\frac{1}{DT}\frac{dT}{dt} = \frac{1}{X}\frac{d^2X}{dx^2} := -\lambda^2 = \text{constant}$$

so that

$$\frac{dT}{dt} = -\lambda^2 DT \Rightarrow T(t) = Ae^{-\lambda^2 t}$$

and

$$\frac{d^2X}{dx^2} = -\lambda^2 X \Rightarrow X(x) = \begin{cases} A_1 x + A_2 & \text{for } \lambda = 0 \\ B\sin\lambda x + C_1 \cos\lambda x & \text{for } \lambda \neq 0 \end{cases}$$

As a result, with the boundary conditions for the $\lambda = 0$ case,

$$c(x,t) = c_1 + \left(\frac{c_2 - c_1}{h}\right)x + (B\sin\lambda x + C_1\cos\lambda x)e^{-\lambda^2 Dt}$$

Invoking the B.C.s for $\lambda \neq 0$ yields

$$c(x,t) = c_1 + \left(\frac{c_2 - c_1}{h}\right)x + \sum_{n=1}^{\infty} B_n \sin(\lambda_n x)e^{-\lambda_n^2 Dt}$$

where

$$\lambda_n = \frac{n\pi}{h} \qquad \text{n=1,2,3,......}$$

The initial condition, using the Fourier series expansion, yields

$$A_n = \frac{2}{n\pi}[(-1)^n c_2 - c_1]$$

(a1) Finally,

- $$c(x,t) = c_1 + \left(\frac{c_2 - c_1}{h}\right)x + \frac{2}{\pi}\sum_{n=1}^{\infty}\frac{1}{n}[(-1)^n c_2 - c_1] \qquad \text{(E1.8-3a)}$$

$$\sin\left(\frac{n\pi x}{h}\right)\exp(-n^2\pi^2 Dt/h^2)$$

Retaining only the terms for n=1 because the others being proportional to n^2 vanish rapidly with time, we have

- $$c(x,t) \approx c_1 + \left(\frac{c_2 - c_1}{h}\right)x - \frac{2}{\pi}(c_1 + c_2)\sin\left(\frac{\pi x}{h}\right)\exp(-\pi^2 Dt/h^2)$$

(E1.8-3b)

(a2) The mass flux $j\left[\frac{M}{L^2T}\right]$ diffusing into the membrane at x=0 is

- $$j_0 = D\frac{\partial c}{\partial x}\bigg|_{x=0} \qquad \text{(E1.8-4)}$$

i.e., approximately

- $$j_0 \approx \frac{D}{h}(c_2 - c_1) - \frac{2D}{h}(c_1 + c_2)\exp(-\pi^2 Dt/h^2) \tag{E1.8-5}$$

(b1) The solute concentration in a 1-D cell exposed to a solution reservoir for which

$$c_1 = c_2 = c_0$$

we have from Eq. (E1.8-3a)

- $$C_c(x, t) = C_0\left[1 - \frac{4}{\pi}\sum_{n=1,3,5,\ldots}^{\infty}\frac{1}{n}\sin\left(\frac{n\pi x}{h}\right)\exp(-n^2\pi^2 Dt/h^2)\right] \tag{E1.8-6}$$

(b2) The solute mass ratio for the cell can be expressed as

- $$\frac{m(t)}{m_0} = \frac{\int_0^h C_c(x, t)dx}{\int_0^h C_c(x, \infty)dx} \tag{E1.8-7}$$

$$= 1 - \frac{8}{\pi^2}\sum_{n=1,3,5,\ldots}^{\infty}\frac{1}{n^2}\exp(-n^2\pi^2 Dt/h^2) \tag{E1.8-8a}$$

Again, except for very short intervals (n>1)-terms can be neglected, so that

- $$\frac{m(t)}{m_0} \approx 1 - \frac{8}{\pi^2}\exp(-\pi^2 Dt/h^2) \tag{E1.8-8b}$$

<u>*Notes:*</u>

- For 1-D *convection* and diffusion, Eq. (1.3.36) can be reduced to

$$\frac{\partial c}{\partial t} + u\frac{\partial c}{\partial x} = D\frac{\partial^2 c}{\partial x^2}$$

which can be transformed with

$$\xi = x - ut \quad \text{and} \quad \eta = t$$

to

$$\frac{\partial c}{\partial \eta} = D\frac{\partial^2 c}{\partial \xi^2} \tag{E1.8-1}$$

- Now, the same solution method (E1.8-2) can be applied when solving simplified forms of Eq. (1.3.35b) where $\alpha \triangleq D$.

1.3.4 Turbulent Flow Equations

The vast majority of natural and industrial flows are *turbulent*. In fact, blood flow in highly stenosed arteries, urine flow, and airflow in the upper respiratory system are frequently turbulent. Thus, the basic knowledge and mathematical skills gained from laminar flow studies have to be significantly extended in order to evaluate, at least approximately, *transition* from laminar to turbulent flow and to compute quantitatively *turbulence effects* in a variety of practical applications. Although turbulence is dissipative, i.e., requiring higher pressure gradients to overcome extra drag, energy losses, etc., turbulence is most desirable for high throughput and especially for enhanced mixing or dispersion.

After a brief discussion of laminar flow instabilities and turbulence characteristics, approaches to turbulent flow modeling are summarized. Examples then deal with turbulent flow applications.

1.3.4.1 Aspects of Turbulence

Transitional flows. Changes from laminar-to-turbulent and/or turbulent-to-laminar flows are inadvertently of importance in numerous applications including particle-hemodynamics in locally occluded blood vessels, air flow in the larynx and trachea, drag reduction with laminarized flows, and transition control of fighter jets. For all practical purposes, at least two problems have to be solved: (i) determination of the onset of transition and (ii) modeling of the influence of the transition region on the specific flow field. In general, numerous "irregularities and disturbing sources" exist a priori in free shear flows and on surfaces of wall-bounded flows. *At sufficiently large Reynolds numbers*, these flow perturbations (e.g., free-stream turbulence, flow instabilities in form of vortices, wall roughness effects, etc.) are not damped out; just the opposite, they amplify and start to interact with the entire flow field. Specifically, in two-dimensional flows, the natural transition begins with the amplification of Tollmien-Schlichting (T-S) waves, i.e., the linear and nonlinear development of initially small amplitude T-S waves into large amplitude disturbances that break down into a chaotic motion (cf. Fig. 1.3.4). Free stream fluctuations, Taylor- or Goertler-type vortices, body surface roughness, wall non-uniformities or vibration, and acoustic or particle environments all can contribute to the onset of transition and the development of (locally) turbulent flow fields. Linear stability theory allows the calculation of characteristics of these T-S waves as eigen-solutions of the linearized Navier-Stokes equations (cf. Warsi, 1999). A crude evaluation of the transition region may be obtained with an intermittency function, γ, applied to the turbulent shear stress $\tau_{total} = \tau_{laminar} + \gamma\tau_{turb}$; here, $\gamma=0$ in laminar flow and $\gamma \geq 1$ in turbulent flow, where $\gamma > 1$ near the end of the transition region.

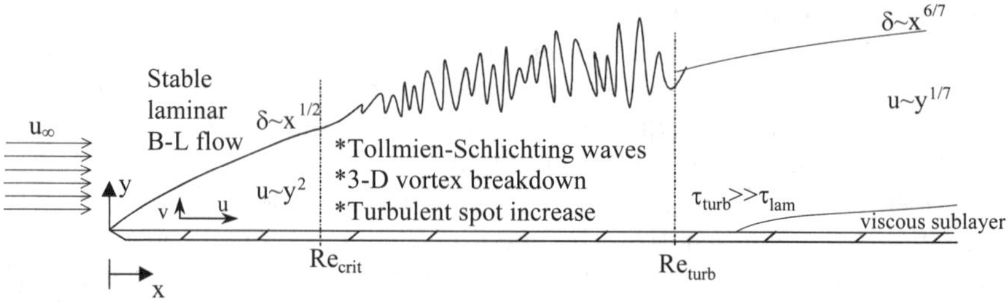

Fig. 1.3.4 Flat plate laminar/transitional/turbulent boundary-layer characteristics (after White, 1991).

Turbulent flows. The development of a general theory for *turbulent flow*, although treated as a continuum phenomenon, is hampered by the fact that turbulence is very complex. Turbulence exhibits *random velocity fluctuations*; thus, it is chaotic, nonlinear, three-dimensional, inhomogeneous, and nonstationary, and it is largely dependent upon its origin and the given flow configuration. However, the detection of various types of spatially *coherent structures* in important engineering flows has spawned a new class of turbulence research. Coherent structures, associated with the large-scale motion, are basically recognizable patterns with characteristic orientations that recur throughout the flow field. A (Von Karman) vortex street behind a cylinder at high Reynolds numbers may serve as an example. The coherent structures contain a significant fraction of the turbulence kinetic energy and they are associated with the transport of scalar quantities in the flow field. Thus, they are believed to be responsible for the *apparent stresses* or *Reynolds stresses* and for most of the energy-containing motion. On a smaller scale, coherent substructure phenomena exist such as longitudinal vortices, hairpin structures, and horseshoe vortices of random position and variable shape and size. In wall turbulence for example, there exists a hierarchy of eddy scales with the smaller one proportional to ν/u_τ, where $u_\tau \equiv (\tau_w/\rho)^{\frac{1}{2}}$ is the friction velocity. In contrast, there are also detached *incoherent* motions, e.g., eddies of the Kolmogorov length scale, $(\nu^3/\varepsilon)^{\frac{1}{4}}$, which are responsible for the *energy dissipation*, expressed in terms of the eddy dissipation rate ε, as discussed below.

Turbulence modeling. The traditional approach for modeling turbulence is the use of the Reynolds-averaged Navier-Stokes equations. Because of the nonlinear terms, the time-averaging or time-smoothing process of these equations produces dominant (apparent) stresses that make *turbulence closure models* necessary. Since averaging occurs over all the turbulence dynamics simultaneously, it is necessary to model all structures requiring flow system-specific rather than universal relationships. This closure problem, i.e., the necessary specification of some auxiliary equations in order to match the number of new unknowns, is usually solved with eddy-viscosity models or, now more frequently, with Reynolds-stress transport models. In any case, the classical point of departure is Reynolds' decomposition of the instantaneous, dependent variable, v, i.e.,

• $$\mathrm{v} = \overline{\mathrm{v}} + \mathrm{v}' \tag{1.3.37}$$

into a time-averaged $\overline{\mathrm{v}}$, and a randomly fluctuating component v', where v could be a tensor, a vector, or a scalar. Substituting the decomposition into the basic transport equations (cf. Sect. 1.3) yields after averaging, for example, time-averaging,

• $$\bar{f} = \frac{1}{\Delta t}\int_{t_1}^{t_2} f\, dt \tag{1.3.38}$$

the turbulent transport equations. Basic averaging axioms include $\overline{f+g} = \bar{f} + \bar{g}$, $\overline{cf} = c\bar{f}$, $\overline{\bar{f}\, g} = \bar{f}\,\bar{g}$, and assuming that $\bar{\bar{f}} = \bar{f}$, where f and g are random functions and c is a constant.

For example, the Navier-Stokes equations (Sect. 1.3.2) can be transformed as follows. Note, for the reason of simplicity, the arrow over velocity vectors is now omitted, i.e., $\vec{\mathbf{v}} \equiv \mathrm{v}$

(a) Continuity: $div\,\overline{\mathrm{v}} + div\,\mathrm{v}' = 0$

With $\overline{div\,\overline{\mathrm{v}}} = div\,\overline{\mathrm{v}}$, we obtain

$$div\,\overline{\mathrm{v}} = 0 \tag{1.3.39a}$$

which, of course, implies that $\overline{div\ \mathrm{v}'} = div\ \overline{\mathrm{v}'} = 0$ because $\overline{\mathrm{v}'} \equiv 0$.

(b) Momentum:

$$\frac{\partial\overline{\mathrm{v}}}{\partial t} + \frac{\partial \mathrm{v}'}{\partial t} + \nabla\cdot(\overline{\mathrm{v}}\overline{\mathrm{v}}) + \nabla\cdot(\overline{\mathrm{v}}\mathrm{v}') + \nabla\cdot(\mathrm{v}'\overline{\mathrm{v}}) + \nabla\cdot(\mathrm{v}'\mathrm{v}')$$
$$= -\nabla\left(\frac{\bar{p}}{\rho}\right) - \nabla\left(\frac{p'}{\rho}\right) + \nu\nabla^2\overline{\mathrm{v}} + \nu\nabla^2\mathrm{v}'$$

After each term is averaged or filtered (cf. Warsi 1999), the Reynolds-averaged Navier-Stokes equation appears:

• $$\frac{\partial\overline{\mathrm{v}}}{\partial t} + \nabla\cdot(\overline{\mathrm{v}}\overline{\mathrm{v}}) = -\nabla\left(\frac{\bar{p}}{\rho}\right) + \nu\nabla^2\overline{\mathrm{v}} - \nabla\cdot\left(\overline{\mathrm{v}'\mathrm{v}'}\right) \tag{1.3.39b}$$

Note the following:

- Subtracting the last two equations yields an equation for the flow perturbations.
- The dyad, $-\rho\overline{\mathrm{v}'\mathrm{v}'}$, is called the *Reynolds stress tensor*, i.e., it contains nine, actually, due to symmetry, six different, so-called apparent or turbulent stress components. For example, in rectangular coordinates

$$\rho\overline{\mathrm{v}'\mathrm{v}'} = \begin{vmatrix} \rho\overline{u'^2} & \rho\overline{u'\mathrm{v}'} & \rho\overline{u'w'} \\ \rho\overline{u'\mathrm{v}'} & \rho\overline{\mathrm{v}'^2} & \rho\overline{\mathrm{v}'w'} \\ \rho\overline{u'w'} & \rho\overline{\mathrm{v}'w'} & \rho\overline{w'^2} \end{vmatrix}$$

- The sum of the diagonal in the form:

$$\frac{1}{2}\left(\overline{u'^2} + \overline{\mathrm{v}'^2} + \overline{w'^2}\right) = k := 0.5\overline{(\mathrm{v}_i')^2} \quad (1.3.40)$$

  is the *turbulence kinetic energy,* an important measure of turbulent flows.
- In Cartesian coordinates, using the convention of summation over repeated indices, the turbulent incompressible flow equations read

- $$\frac{\partial \overline{\mathrm{v}}_i}{\partial x_i} = 0 \quad (1.3.41)$$

- $$\frac{\partial \overline{\mathrm{v}}_i}{\partial t} + \overline{\mathrm{v}}_j \frac{\partial \overline{\mathrm{v}}_i}{\partial x_j} = -\frac{1}{\rho}\frac{\partial \overline{p}}{\partial x_i} + \underbrace{\nu \frac{\partial^2 \overline{\mathrm{v}}_i}{\partial x_j \partial x_j}}_{\nabla\cdot\vec{\vec{\tau}}_{laminar}} - \underbrace{\frac{\partial}{\partial x_j}\left(\overline{\mathrm{v}_i'\mathrm{v}_j'}\right)}_{\nabla\cdot\vec{\vec{\tau}}_{turbulent}} \quad (1.3.42)$$

- While $\tau_{ij}^{total} = \tau_{ij}^{lam} + \tau_{ij}^{turb}$ throughout a flow field, we know that

  $\tau_{ij}^{turb} >> \tau_{ij}^{lam}$ everywhere in turbulent flow fields except in the near-wall region, where the local Reynolds number approaches zero and a "viscous sublayer" is formed (cf. Fig. 1.3.4).
- For two-dimensional incompressible turbulent *boundary-layer* flows

- $$\frac{\partial \overline{u}}{\partial x} + \frac{\partial \overline{\mathrm{v}}}{\partial y} = 0 \quad (1.3.43)$$

and

- $$\frac{\partial \overline{u}}{\partial t} + \overline{u}\frac{\partial \overline{u}}{\partial x} + \overline{\mathrm{v}}\frac{\partial \overline{u}}{\partial y} = -\frac{1}{\rho}\frac{\partial \overline{p}}{\partial x} + \frac{\partial}{\partial y}\left(\nu\frac{\partial \overline{u}}{\partial y} - \overline{u'\mathrm{v}'}\right) \quad (1.3.44)$$

Turbulence modeling, i.e., a solution to the closure problem, revolves around the turbulent stresses [cf. Eq. (1.3.42)]

$$\tau_{ij}^{turb} = -\rho\overline{\mathrm{v}'_i\,\mathrm{v}'_j} \quad (1.3.45a)$$

One prevailing idea is the *Boussinesq concept,* i.e., the eddy viscosity model (EVM), where in an extension to Stokes' hypothesis [cf. Eq. (1.3.16)]

- $$\frac{\tau_{ij}^{turb}}{\rho} = \nu_{turb} \frac{\partial \overline{v}_i}{\partial x_j} \tag{1.3.45b}$$

Here, the turbulent eddy viscosity, ν_{turb}, is a rather complex function. In general, $\nu_t = \nu_t$ *(eddy size or length scale, turbulence kinetic energy, turbulence energy dissipation, turbulence time scale, velocity field gradients, system geometry, wall roughness, fluid viscosity, etc.)*.

Depending upon the level of turbulence complexity, which is typically associated with the type of flow geometry, we may need several ensemble-averaged PDEs in addition to the mean flow equations, to model turbulence effects. For example, Prandtl's Mixing Length Hypothesis (MLH) is an *algebraic* equation for the eddy viscosity of the form $\nu_t[l(y)]$. It is called a "zero-equation" model because no additional PDE is required for the solution of a particular turbulent flow. On the other hand, if two extra PDEs for the *turbulence kinetic energy*

- $$k = \frac{1}{2}\overline{(v'_i)^2} \tag{1.3.46}$$

and the *dissipation function*

- $$\varepsilon_{ij} = 2\nu_t \overline{\frac{\partial v'_i}{\partial x_k}\frac{\partial v'_j}{\partial x_k}} \tag{1.3.47}$$

are solved to form, for example, $\nu_t = C(k^2/\varepsilon)$ the approach is called a "two-equation" model, and so on. The equations for k and ε, forming the k-ε turbulence model as well as the high-Reynolds number RNG k-ε and "low"-Reynolds-number k-ω models are discussed in the next subsection.

Alternatively, the cumbersome solution of *any* differential equation can be avoided when streamwise *velocity profiles for specific turbulent flows* are directly *postulated.* Examples include the log-profiles and power-law profiles. Other recent advances in turbulence modeling have not matured enough and/or are too costly to be considered in complex turbulent systems analyses. Examples include the theory of dynamic systems where chaotic yet deterministic behavior of solution trajectories, controlled by sets of strange attractors, is being investigated; or direct turbulence simulation on parallel supercomputers, where the spatial and temporal resolutions have to be fine enough to represent the smallest and fastest turbulent flow phenomena. Since the exact simulation of the turbulence dynamics for complex flows is still unattainable, traditional closure concepts are and will be used for today's and tomorrow's engineering computations. It is also evident that there is not one turbulence model that can compute a whole range of flows to acceptable engineering accuracy, and each particular model requires "fine tuning" with empirical data that depend on the flow type and system geometry.

1.3.4.2 Turbulence Scales

Before a specific turbulence model is discussed, it may be of interest to consider first characteristic time and length scales of turbulent flows. Such scales are usually constructed from representative flow system parameters, such as approach velocity, wall friction velocity, or mean velocity; mean vorticity; body length, shear layer thickness or duct radius; fluid properties (i.e., ν, ρ, k, and c_p); mean temperature, etc.

Time scales:

- convection time scale $t_c = \frac{L}{U}$ where L is a typical system dimension and U is a mean flow velocity;
- viscous diffusion time scale $t_\nu = \frac{L^2}{\nu}$;
- Kolmogorov time scale $t_k = \left(\frac{\nu}{\varepsilon}\right)^{\frac{1}{2}}$ where the eddy dissipation rate $\varepsilon \propto \frac{U^3}{L}$ or $\varepsilon \propto \nu\left(\frac{u'}{l'}\right)^2$ with u' and l' are eddy velocity and size scales; here, the viscous diffusion time is about equal to the eddy convection time l'/u';
- heat diffusion time scale $t_h = \frac{L^2}{\alpha}$ where $\alpha = k/(\rho c_p)$.

Length scales:

- viscous wall length $l_\tau = \nu/u_\tau = O(l')$ where $u_\tau \equiv \sqrt{\tau_w/\rho}$ is the friction velocity;
- mixing length $l_m = \kappa y$ or $l_m = \kappa y[1 - exp(-y/A)]$ where $\kappa = 0.41$ and A being the Van Driest damping length;
- Kolmogorov length scale $l_k = \left(\frac{\nu^3}{\varepsilon}\right)^{\frac{1}{4}}$ or with $Re_L = UL/\nu$,

  $\frac{l_k}{L} \sim Re_L^{-\frac{3}{4}}$

The brief discussion of the complexities of turbulence should imply that *turbulence modeling* is a very challenging, never-ending task. Numerous volumes have been written on this subject matter, and hence it is only appropriate to provide here a brief *summary* of some viable engineering approaches.

1.3.4.3 Summary of Turbulence Modeling

Despite some frustrating failures (in the past), computational fluid dynamics (CFD) modeling of turbulence is advantageous because of lower turnaround time, lower cost, more flexibility, and occasionally higher accuracy when compared to physical modeling. In a *decreasing order of complexity*, which implies reduced needs for computational resources and measured "constants," *deterministic* turbulence modeling approaches could be grouped as follows:

1. *Direct Numerical Simulation (DNS)* of turbulence, which requires time and length scales small enough to resolve turbulent fluctuations and tiny, near-wall eddies. In other words, DNS does not require any turbulence *model*.
2. *Large Eddy Simulation (LES)* with subgrid scale (SGS) modeling where the coherent, large-scale structures are directly computed with the filtered Navier-Stokes equations while the small-scale eddies are modeled based on the Smagorinsky eddy-viscosity concept. (cf. Ferziger & Peric, 1999).
3. *Reynolds Stress Modeling (RSM)* where transport equations (PDEs) for each of the important components of the turbulence stress tensor, $\rho\overline{u'_i u'_j} = \tau_{ij}^{turb}$, plus the turbulence energy, $k = \frac{1}{2}\overline{u_i'^2}$, have to be solved. Numerous empirical coefficients have to be tuned to match system-specific turbulent flow patterns (cf. Pope, 2000).
4. *Eddy Viscosity Modeling (EVM)*, based on the postulates, or constitutive equations, by Boussinesq (1877) and Kolmogorov (1942), i.e.,

   $$\overline{u'_i u'_j} = \frac{\tau_{ij}^t}{\rho} = \frac{2}{3}k\delta_{ij} - \nu_t\left(\frac{\partial \bar{u}_i}{\partial x_j} + \frac{\partial \bar{u}_j}{\partial x_i}\right) \tag{1.3.48}$$

   It requires the calculation of one or more turbulence quantities (e.g., length scale l, turbulence kinetic energy k, dissipation function ε, pseudovorticity ω; time scale τ, etc.), which are then combined and directly related to the "turbulent eddy viscosity"

   $$\nu_t = \frac{\mu_t}{\rho} = \text{fct}(l, k, \varepsilon, \omega, \tau, \text{etc.})$$

   Zero-equation models, such as Prandtl's Mixing Length Hypothesis (MLH) and the Van Driest extension for wall damping, require the solution of an *algebraic* equation, for example, an eddy length scale, where $\nu_t \sim l_m^2$. One-equation models require the solution of *one* transport equation (PDE), for example, $k = 1/2\overline{u_i'^2}$ where $\nu_t \sim \sqrt{k}$. Two-equation models require the solution of *two* transport equations (PDEs), for example, k and where $\nu_t \sim k^2/\varepsilon$ as mentioned earlier (cf. Wilcox, 1998).

5. *Empirical Correlations:* The use of *empiricism* for fully developed pipe flows and flat plate boundary-layer (B-L) flows allows bypassing the equations of motion and hence eliminates the need for basic turbulence modeling. For example, turbulent pipe flow problems are solved with the extended Bernoulli equation, which relates pressure drops to losses in terms of the friction factor *f*.

$$\Delta z + \frac{\Delta p}{\rho g} = h_f = f \frac{L}{d} \frac{v^2}{2g} \tag{1.3.49}$$

where $f = f\left(Re_d, \frac{e}{d}\right)$ from Moody Chart or formulas by Colebrook, Blasius, Von Karman, etc. Pipe sizing, flow rates, and pump requirements are directly related to the friction loss h_f, where

$$h_f = \frac{8Q^2 Lf}{\pi^2 g d^5} \tag{1.3.50}$$

with $f = 4c_f = \frac{8\tau_w}{\rho v^2}$; power $P = F_D \cdot \mathrm{v}$ with $F_D = \tau_w A_{surf}$.

Alternatively, on a more differential basis, semiempirical *turbulent velocity profiles* could be employed

$$\bar{u} \sim \begin{cases} \left(\frac{y}{r_o}\right)^{\frac{1}{7}} \text{ or } \left(\frac{y}{\delta}\right)^{\frac{1}{7}} & \text{power laws for pipe or B-L flows} \\ \ln\left(\frac{y}{r_o}\right) \text{ or } \ln y & \text{log laws for pipe or B-L flows} \end{cases} \tag{1.3.51a-d}$$

Most turbulent velocity profiles $\bar{u}(x, y)$ require knowledge of the friction velocity $u_\tau = (\tau_w/\rho)^{\frac{1}{2}}$ (Schlichting & Gersten, 2000).

<u>*Notes to 1 and 2 above:*</u> DNS is useful for studying the detailed physics of turbulent "building block" flows. This understanding may lead to new, or at least improved, turbulence models. Current DNS examples include wall-bounded flows such as channel and boundary-layer flows as well as free

shear-flows, e.g., mixing layers and plane jets. However, all flow examples are presently restricted to relatively *low* Reynolds numbers and *simple* geometries.

Incorporation of length and time *scales* in turbulent flow calculations is as follows:

- Kolmogorov length or dissipation scale

$$l_k = \left(\frac{\nu^3}{\varepsilon}\right)^{\frac{1}{4}} \quad \text{with} \quad \varepsilon \sim \frac{U^3}{L}$$

where ε is the energy dissipation function. Thus,

$$\frac{l_k}{L} \sim Re_L^{-\frac{3}{4}} \quad \text{with} \quad Re_L = \frac{UL}{\nu}$$

where L is a reference length (e.g., B-L thickness or largest eddy size) and U is a reference velocity (e.g., u_{max} or u). In DNS, the Navier-Stokes equations have to "cover" *all* length scales of the turbulence field, i.e., $\Delta x \ll l_k$.

- Kolmogorov time scale

$$t_k = \left(\frac{\nu}{\varepsilon}\right)^{\frac{1}{2}} \quad \text{where again} \quad \varepsilon \sim \frac{U^3}{L}$$

$$\therefore \quad \frac{t_k}{t_{ref}} \sim Re_L^{-\frac{1}{2}} \quad \text{with} \quad t_{ref} = \frac{L}{U}$$

Note: Time step $t \ll t_k, t_{ref}$.

In conclusion, DNS is presently too costly for advanced engineering flows. LES, although presently capable of depicting more complex flows than DNS, is still too cost-intensive and not yet suitable for complex engineering problem solutions. In any case, a good indicator of what is generably feasible are the types of turbulence models available in present-day CFD software.

Note to 3: RSM accounts for the history and nonlocal effects of the mean velocity gradients, including relaxational effects, wall curvature, nonparallel flows, countergradient transport, etc. Employing again

$$u_i = \bar{u}_i + u'_i \quad \text{where} \quad \bar{u}_i = \frac{1}{\Delta t}\int_{t_1}^{t_2} u_i dt$$

results after time-averaging in the mean flow equations:

$$\bar{u}_{i,i} = 0 \quad \text{and} \quad \frac{D\bar{u}_i}{Dt} = -\frac{1}{\rho}\bar{p}_{,i} - \left(\overline{u'_i u'_j}\right)_{,j} + \nu\bar{u}_{i,jj} \qquad (1.3.52a, b)$$

Obviously, the apparent stresses $-\rho\overline{u_i' u_j'}$ have to be modeled in order to gain closure. The Reynolds Stress Transport (RST) equations are of the differential or algebraic form.

(i) Differential closures:

$$\frac{D}{Dt}\left(\overline{u'_i\, u'_j}\right) = D_{ij} + P_{ij} + T_{ij} + \Pi_{ij} - \varepsilon_{ij} \tag{1.3.53}$$

$$D_{ij} = \nu\left(\overline{u'_i\, u'_j}\right)_{,kk} \quad \hat{=} \text{ diffusive transport}$$

(i.e., on the viscous laminar level)

$$P_{ij} = -\overline{u'_i\, u'_k}\, u_{j,k} - \overline{u'_j\, u'_k}\, u_{i,k} \quad \hat{=} \text{ stress production}$$

$$T_{ij} = -\left[\overline{u'_i\, u'_j\, u'_k} + \frac{1}{\rho}\left(\overline{u'_i\, p'}\delta_{ik} + \overline{u'_j\, p'}\delta_{ik}\right)\right]_{,k} \quad \hat{=}$$

turbulent diffusion transport (to be modeled)

$$\Pi_{ij} = \frac{1}{\rho}\overline{p'\left(u'_{i,j} + u'_{j,i}\right)} \quad \hat{=} \text{ pressure-strain correlation}$$

(to be modeled)

$$\varepsilon_{ij} = 2\nu\overline{u'_{i,k}\, u'_{j,k}} \quad \hat{=} \text{ viscous dissipation (to be modeled)}$$

(ii) Algebraic closures:

Assuming the transport of $\overline{u'_i\, u'_j}$ to be proportional to the transport of turbulence energy, $k = \frac{1}{2}\overline{u'^2_i}$, the L.H.S. of the RST equation is approximated as (Rodi, 1980)

$$\frac{D}{Dt}\left(\overline{u'_i\, u'_j}\right) \approx \frac{\overline{u'_i\, u'_j}}{k}\frac{Dk}{Dt} \tag{1.3.54}$$

resulting in algebraic stress equations for

$$\overline{u'_i\, u'_j} = \text{fct.}\left(k, \frac{Dk}{Dt}, \varepsilon, \text{etc.}\right)$$

where simultaneously solutions to auxiliary PDEs for k and ε have to be supplied.

In conclusion, although very computer- and data-intensive, RSM is now being incorporated into fluid dynamics software packages. However, *wall effects*, such as rapid near-wall variations especially in the Π_{ij} terms, may still cause substantial errors. Furthermore, RSM relies typically on a single turbulence time scale to characterize rate processes in turbulence. However, at least two independent time scales accounting for the different response mechanisms of the large-scale and the small-scale motions are needed in some applications.

Note to 4: EVM, typically based on zero-equation models or two-equation models, work reasonably well for nonseparating, *near-parallel* shear flows. However, model inclusions of correction factors and extension terms, representing effects due to the presence of walls, streamline curvature, flow separation, flow rotation, fluid compressibility, relatively low Reynolds numbers, and other factors, have kept EVM as the first choice for a large variety of engineering problems. Virtually all fluid dynamics software packages entertain MLH (i.e., zero-equation) and k-ε as well as RNG k-ε (i.e., two-equation) closures. Again, proper near-wall shear-layer modeling is crucial for successful turbulent flow simulations.

The underlying concept is Boussinesq's analogy with molecular transport of momentum. Thus, in simplified form

$$\tau_t = \rho \nu_t \frac{\partial \bar{u}}{\partial y} \quad \text{where} \quad \rho \nu_t \equiv \mu_t$$

Now, for wall-bounded shear layers, based on Prandtl's mixing length hypothesis (MLH):

$$\nu_t = l^2 \left| \frac{\partial \bar{u}}{\partial y} \right|, \qquad l = \kappa y \quad \text{(Prandtl \& Von Karman)}$$

or

$$\nu_t = l^2 \left| \frac{\partial \bar{u}}{\partial y} \right| \tag{1.3.55}$$

where

$$l = \begin{cases} \kappa y & \text{for} \quad 0 \le \frac{y}{\delta} \le \frac{\lambda}{\kappa} \\ \lambda \delta & \text{for} \quad \frac{\lambda}{\kappa} \le \frac{y}{\delta} \le 1 \end{cases} \qquad \text{(Cebeci \& Bradshaw)}$$

$$\kappa = 0.435 \quad \text{and} \quad \lambda = 0.09$$

or

$$\nu_t = \begin{cases} \nu_t^{inner} = fct\left(l^2, \left|\frac{\partial u}{\partial y}\right|, \gamma^i\right) & \text{for} \quad 0 \le y \le y_c \\ \nu_t^{outer} = fct\left(U_o, \delta_1, \gamma^o\right) & \text{for} \quad y_c \le y \le \delta \end{cases} \qquad \text{(Cebeci \& Smith)}$$

where $l = \kappa y\left[1-\exp\left(\frac{-y}{A}\right)\right]$, A being the Van Driest damping length, γ^{i} and γ^{o} are intermittency factors, U_0 is the outer flow, and δ is the B-L thickness; $\tau_t^{inner} = \tau_t^{outer}$ at $y = y_c$.

For a *complete* eddy-viscosity turbulence model, at least two turbulence quantities have to be specified, i.e., typically the turbulence kinetic energy k plus a length scale L, energy dissipation rate ε, or dissipation time scale τ. For example,

(i) the k-L model: $\nu_t = c\sqrt{k}\,L$ (Kolmogorov-Prandtl)

(ii) the k-ε model:

$$\nu_t = C_\mu \frac{k^2}{\varepsilon}, \quad C_\mu = 0.09; \quad \varepsilon \propto \frac{k^{\frac{3}{2}}}{L} \text{ (Jones \& Launder)}$$

where

$$\frac{Dk}{Dt} = P_{ij} - \varepsilon + \frac{\partial}{\partial x_j}\left(\frac{\nu_t}{\sigma_k}\frac{\partial k}{\partial x_j}\right) + \nu\nabla^2 k \tag{1.3.56}$$

and

$$\frac{D\varepsilon}{Dt} = G_{ij} - C_\varepsilon \frac{\varepsilon^2}{k} + \frac{\partial}{\partial x_i}\left(\frac{\nu_t}{\sigma_\varepsilon}\frac{\partial \varepsilon}{\partial x_i}\right) + \nu\nabla^2\varepsilon \tag{1.3.57}$$

with $\sigma_k \approx 1.0$, $\sigma_\varepsilon \approx 1.92$, $P_{ij} \triangleq$ k-production terms, and $G_{ij} \triangleq$ -generation terms.

(iii) the RNG k-ε model, capturing near-wall effects such as flow separation better than the basic k-ε turbulence model, i.e., again

$$\nu_t = C_\mu k^2 / \varepsilon \tag{1.3.58}$$

and the transport equations for k and ε are the same but the coefficients differ (cf. Yakhot & Orzag, 1986). For example, C_ε is not a constant anymore but

$$C_{\varepsilon_1} = 1.42 - \frac{\eta(1-\eta/\eta_\infty)}{1+\beta\eta^3}$$

where

$$\eta = S \cdot k/\varepsilon,\ \eta_\infty = 4.38,\ \beta = 0.015, \text{ and } S = \left(2\bar{S}_{ij}\bar{S}_{ij}\right)^{1/2}$$

with $\bar{S}_{ij}$ being the mean rate-of-strain tensor.

(iv) the *k*-ω model, which in the version of Wilcox (1998) is a low-Reynolds-number model covering laminar, transitional, and turbulent flow regions:

$$\nu_t = C_\mu f_\mu k / \omega \tag{1.3.59}$$

where $C_\mu \approx 0.09,\ f_\mu = exp\left[-3.4/(1+Re_t/50)^2\right]$ with $Re_t = \rho k/(\mu\omega)$, and the two PDEs for *k* and w as well as numerous coefficients/functions for sample applications are given in Wilcox (1998).

<u>Note to 5:</u> The turbulent boundary-layer velocity profile

$$u^+ \equiv \frac{\bar{u}}{u_\tau} = \frac{1}{\kappa} ln\, y^+ + B;\, y^+ = u_\tau \frac{y}{\nu} \tag{1.3.60}$$

can be *derived* based on Prandtl's near-wall shear stress hypotheses. The empirical constants are κ = 0.40 and B = 5.24. The friction velocity $u_\tau \equiv \sqrt{\tau_w/\rho}$ is either indirectly measured or obtained from appropriate skin friction or friction factor correlations $f(\tau_w) = fct(Re, e)$ where *e* is the surface roughness. Terms representing laminar sublayer, buffer zone, and near-wake effects have all been added to the basic log-law profile (cf. Spalding, 1961; among others).

An alternative turbulent velocity profile, typically applied to pipe flow, is the power law of the form:

$$\frac{\bar{u}}{u_{max}} = \left(1 - \frac{r}{r_0}\right)^{\frac{1}{n}} \tag{1.3.61}$$

which can be constructed where *n* is a weak function of the Reynolds number, $Re = u_{av} D/\nu$ with $D = 2r_0$, and $u_{av} = Q/\pi r_0^2$ (cf. Table 1.3.1).

It has to be noted that the power law profile has a nonzero gradient at the centerline, i.e., $\left.\left(\partial u/\partial r\right)\right|_{r=0} \neq 0$. Equation (1.3.61) can be rewritten in terms of the friction velocity $u_\tau = \sqrt{\tau_w/\rho}$ and the wall coordinate $y = r_o - r$, with n = 7, as

$$\frac{u}{u_\tau} = 8.74\left(\frac{y u_\tau}{\nu}\right)^{\frac{1}{7}} \tag{1.3.62}$$

which is known as the 1/7-law. In summary, neither the log-law nor the power law fulfill all boundary conditions at $y = 0$ and $y = \begin{cases} \delta \\ r_o \end{cases}$.

Table 1.3.1 Exponent n and Velocity Ratio u_{av}/u_{max} as a Function of Reynolds Number.

Re #	4×10^3	2.3×10^4	1.1×10^5	1.1×10^6	2×10^6	3.2×10^6
n	6.0	6.6	7.0	8.8	10.0	10.0
$u_{av}/u_{\max}$	0.791	0.806	0.817	0.849	0.865	0.865

Note: After Schlichting & Gersten (2000).

<u>Example 1.9</u>: Turbulent Pipe Flow

Rather than solving numerically the Reynolds-averaged Navier-Stokes equations with appropriate turbulence model [cf. Eqs. (1.3.41 & 1.3.42) and Sect. 1.3.4.1], tailored turbulent velocity profiles are often employed to calculate flow rates, wall shear stresses, pressure drops, etc. (cf. Kleinstreuer, 1997; Schlichting & Gersten, 2000; among others).

<u>Problem</u>: Consider water at 20°C flowing through a horizontal pipe (D=0.1 m, Q=4×10^{-2} m^3/s) when $\Delta p/l$=2.59 kPa/m.

(a) Find the thickness δ_s of the viscous sublayer (cf. Fig. 1.3.4).

(b) Determine the centerline velocity u_{max} and the ratio $u_{average}/u_{max}$.

(c) Calculate the ratio τ_{turb}/τ_{lam} at the midpoint r_m=0.025 m.

<u>Sketch</u>:

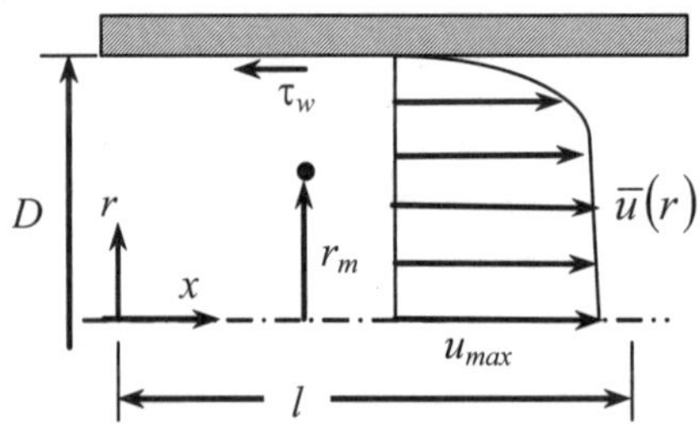

<u>Assumptions</u>:

- steady fully-developed turbulent flow
- constant fluid properties

<u>Approach</u>:

- Use empirical velocity profiles

<u>Solution</u>:

(a) Thickness δ_s: The viscous sublayer ends when the inner variable $y^+ \approx 5$, i.e.,

$$y^+\Big|_{y=\delta_s} = \frac{\delta_s u_\tau}{\nu} = 5 \tag{E1.9-1}$$

where $u_\tau \equiv \sqrt{\tau_w / \rho}$ is the friction velocity. Now, from a basic force balance

$$\Delta p = 4l\tau_w / D \tag{E1.9-2}$$

so that

$$\tau_w = 64.8 \text{ N/m}^2 \quad \text{and} \quad u_\tau = 0.255 \text{ m/s}$$

Hence,

- $\delta_s = 0.02 \text{ mm}$.

(b) Velocities u_{max} and $u_{average}$: Employing the $\frac{1}{n}$-law

$$\frac{\bar{u}}{u_{max}} = \left(1 - \frac{r}{R}\right)^{1/n} \tag{E1.9-3}$$

where (cf. Table 1.3.1)

$n = n(Re_D) := 8.4$ because $Re_D = 5.07 \times 10^5$,

$$Q = \int \bar{u}\, dA := \pi R^2 u_{av} = 2\pi R^2 u_{max} \frac{n^2}{(n+1)(2n+1)} \tag{E1.9-4}$$

Thus,

$$\frac{u_{av}}{u_{max}} = \frac{2n^2}{(2n+1)(n+1)} := 0.8432$$

or

- $u_{max} = 1.186 u_{av}$ *Note*: $u_{max}\big|_{laminar} = 2u_{av}$

Here, knowing $u_{av} = Q / \pi R^2$

- $u_{\max} = 6.04 \text{ m/s}$

(c) Shear stress ratio $\tau_{turb} / \tau_{lam}\big|_{r=r_m}$: Recall that

$$\tau_{total} = \tau_{lam} + \tau_{turb} \tag{E1.9-5}$$

From a force balance on a control volume, $\pi r^2 l$, we obtain

$\frac{\Delta p}{l} = 2\frac{\tau_{rx}}{r} := ¢$ which implies $\tau_{rx} \equiv \tau_{total} = Cr$.

Note that $\tau_t(r=0) = 0$ and that $\tau_t\left(r = D/2\right) = \tau_{wall}$. Hence,

$$\tau_{total} = 2\frac{\tau_w}{D} r \quad \text{and} \quad \frac{\Delta p}{l} = \frac{4\tau_w}{D} \tag{E1.9-6a,b}$$

Indeed, $\tau(r)$ is linear in fully developed pipe flow regardless of flow regime and type of fluid.
Now,

- $\tau_{total}(r = r_m) = \frac{2\tau_w}{D} r_m := 32.4 \text{ N/m}^2$
- $\tau_{lam} = \mu \frac{d\bar{u}}{dr} := \frac{\mu u_{max}}{nR}\left(1 - \frac{r}{R}\right)^{(1-n)/n}$

$$\overset{r=r_m}{\rightarrow} \tau_{lam}(r = r_m) := 0.0266 \text{ N/m}^2$$

- $\tau_{turb} = \tau_{total} - \tau_{lam}$, so that
- $\left.\frac{\tau_{turb}}{\tau_{lam}}\right|_{r=r_m} = 1220.$

1.3.5 Solution Techniques

The type and completeness of information given for a biofluid dynamics problem in conjunction with the justifiable physical assumptions and their mathematical consequences greatly determine the system-describing equations, boundary conditions, and submodels for closure. In an *academic setting*, what's given and required is usually well spelled out and hence the initial focus should be on a detailed system sketch, concept/approach, and assumptions/consequences rather than starting right away a solution procedure. Figure 1.3.5 outlines four basic steps with comments for academic problem analyses. The sequence of solution steps for industrial or research problems is quite similar as depicted in Fig. 1.3.5, although much more difficult for all four steps involved, as highlighted below and demonstrated in Chapter V.

In an academic environment, *clear problem statements* and *sufficient data* are typically given (Step 1). Step 2 probes an *understanding of the physics* of a given flow system and the problem requirements, which leads to the *most suitable modeling approach.* Experimental analysis not being an option and selecting the Eulerian framework, the decision for deterministic problems is between a differential or integral approach, e.g., a reduced equation of motion or a special form of the Reynolds Transport Theorem. Alternatively, the type of models or mathematical descriptions are labeled either "distributed" or "lumped." When selecting the differential approach, Step 3 in conjunction with dimensional analysis and scaling (cf. Sect. 1.3.5.3) may help to identify *key dimensionless groups*, which govern the

Problem Statement

- Check system information and data sets provided
- Study requirement(s) and/or expectations

Note: Read the problem statement carefully !

⇓

System Sketch & Concepts

- Incorporate in the schematics, a suitable coordinate system, and what's given & required
- Sketch anticipated profiles, conditions, etc.
- Determine concept(s), e.g., differential vs. integral approach, etc.

Note: Speculate about the physics of fluid flow:

(a) determine which variables and parameters play an important role in the flow problem; and

(b) how the flow system may respond to changes in operating or boundary conditions.

⇓

Assumptions & Implementations

- List assumptions based on physical insight

  Note: Any decision at this point determines the correctness, completeness, and accuracy of the anticipated solution
- List the mathematical consequences of the assumptions, or *postulates*, in case of a differential solution approach

⇓

Solution

- Formulate a complete set of equations, conditions, and submodels (e.g., turbulence or non-Newtonian fluids) employing *symbolic math*, i.e., NO NUMBERS at this stage
- Check fluid properties, equation coupling, nonlinearities, time- & space-dependencies, and type of coefficients
- Select appropriate solution method (cf. Table 1.3.2)
- Solve the final equation and verify its correctness by:
  (a) inserting numbers and verifying the result; and/or
  (b) recouping initial or boundary conditions
- Vary system parameters, i.e., operating and boundary conditions, in order to gain more physical insight and understanding
- Plot the results and comment!

Fig. 1.3.5 Outline of four basic steps with comments for academic problem analyses.

Table 1.3.2 Types of Modeling Equations and Solution Methods.

- **Type of Equations of Interest:**
  - Linear or nonlinear algebraic equations as well as integral equations;
  - IVPs or BVPs; nonlinear systems of ODEs; PDEs, i.e., parabolic,
  - parabolic/hyperbolic, parabolized, or elliptic

- **Auxiliary Conditions**
  - I.C. in time at t=0 or in space at x=0
  - Homogeneous or inhomogeneous BC
    - Neumann : $\frac{\partial u}{\partial n}$
    - Dirichlet : u
    - Cauchy : mixed

- **Solution Approaches**

Step 1: Classification of governing equations: type of equation
[algebraic, ODE, PDE, integral, mixed (integro-differential)]
Characteristics of, say, differential equations: nonlinear, variable coefficients, inhomogeneous, second-order, time-dependent; multidimensional (for PDEs).

Step 2: Employ transformations and/or approximations, e.g., parabolic PDEs can often be transformed to ODEs via separation of variables, integral method, and/or similarity theory, e.g., an asymptotic analysis using perturbation may be sufficient.

Step 3: Look up textbook solution or solution procedure for a given type and given characteristics of your equation (e.g., for ODEs: Polyanin & Zaitsev, 1995a; and for linear PDEs: Polyanin, 2001).

Step 4: Use appropriate algorithms or computer software for solving
Linear algebraic systems (Gauss-Seidel, SOR)
Nonlinear algebraic equation (Newton-Raphson)
ODE's (IVP) (Runge-Kutta) (e.g., http://www.netlib.org/odepack)
ODE's (BVP) (shooting methods) (e.g., http://www.netlib.org/odepack)
Stiff ODEs (Gear)
Nonlinear ODEs (special transformations, direct integration, Runge-Kutta)
PDEs (FDM, FEM, FVM, CVM)
Commercial PDE-solvers: CFX, FLUENT, CFD-TWOPHASE, as well as Femlab and FlexPDE.

Notes: IVP, initial value problem; BVP, boundary value problem; ODE, ordinary differential equation; PDE, partial differential equation; IC, initial condition; BC, boundary condition; SOR, successive overrelaxation; FDM, finite difference method; FEM, finite element method; FVM, finite volume method; CVM, control volume method; CFD, computational fluid dynamics.

Algebraic and ordinary differential equations, as well as linear PDEs, can also be solved with MATLAB, among other generally available software packages.

flow field, and combined (or similarity) variables, which greatly simplify the solution method. Now, the type of mathematical modeling equations and typical solution steps for differential equations are summarized in Table 1.3.2. Specific solution methods for the reduced Navier-Stokes equations are outlined thereafter. The section concludes with the methodology of how to set up the Reynolds Transport Theorem for fluid mass and linear momentum transfer.

1.3.5.1 Solution Methods for Differential Equations

A number of sophisticated but time consuming solution methods employing complex transformations have lost their appeal with the advent of affordable and powerful calculators, desktop computers, engineering workstations, and associated numerical software. Still, the *separation-of-variables methods* with special function representation, the *similarity theory*, and the *integral methods* are very useful to gain physical insight and simple problem solutions. Also, the *methods of weighted residuals* (MWR) form the foundation of several numerical methods. In reviewing the similarity theory and integral method, which is a special case of the MWR, the balanced approach of "physical insight" and "mathematical skill" can be pursued.

First, the modeling equations and related auxiliary conditions have to be understood and fully classified before any solution approach can be considered (see Table 1.3.2). Accessible texts for review include Spiegel (1971), Finlayson (1978), Greenberg (1998), and Hoffman (2001), among others.

Frequently, academic problems or simplified industrial systems can be described with reduced forms of the Navier-Stokes equations. The next subsection summarizes how the N-S equations are properly reduced based on physical insight. With Table 1.3.2, the resulting differential equation(s) can then be classified and a solution approach identified.

1.3.5.2 Solution Procedures for the Navier-Stokes Equations

The Navier-Stokes equations, describing conservation of mass and momentum of fluid flow with *constant fluid properties*, are a set of transient, *nonlinear, elliptic* (in steady state), second-order, inhomogeneous PDEs. Depending on the characteristics of the flow system (i.e., system geometry, flow patterns, boundary conditions, etc.) it might be justifiable to drastically simplify the Navier-Stokes equations. Specifically, one could justify the following on a case-by-case basis (cf. Sect. 1.2):

- Delete the nonlinear terms: $(\vec{v} \cdot \nabla)\vec{v} \equiv \vec{a}_{conv} \sim \vec{F}_{inertia} \approx 0$; e.g., for creeping flows, fully developed unidirectional flows, etc.;
- Neglect second-order terms in streamwise direction: $\frac{\partial^2 u}{\partial x^2} \approx 0$; e.g., in boundary-layer flows, etc.;

- Assume a constant pressure gradient: $-\nabla p \Rightarrow \Delta p/l$; e.g., in fully developed pipe or slit flows;
- Assume steady state: $\frac{\partial}{\partial t} = 0$ if there are no time variations;
- Check the problem's dimensionality as 1-D or 2-D flows, and/or evaluate the number of directions in which the velocity can vary;
- Neglect gravity effects, developing flow effects, end effects, etc.

A. Summary of Some Basic Exact Solutions

"Frictionless" Flow: $\left(\nu \equiv 0 \quad \text{or} \quad \nabla^2 \vec{\mathbf{v}} \approx 0\right)$ as discussed in Section 1.3.

Euler's Equation:
$$\frac{D\vec{\mathbf{v}}}{Dt} = -\frac{1}{\rho}\nabla p + \vec{\mathbf{g}} \tag{1.3.63}$$

or (see Fig. 1.2.2)

Laplace's Equation:
$$\nabla^2 \phi = 0 \tag{1.3.64}$$

where $\vec{\mathbf{v}} = \nabla \phi$ and ϕ is the velocity potential.

Solution Notes:

- Numerical solution of Eq. (1.3.63) to obtain velocity and pressure fields.
- For Eq. (1.3.64), obtain pressure field from Bernoulli's equation.
- Solve Laplace's equation analytically or numerically.

Parallel Flows: Couette, Poiseuille, suddenly accelerated plate (Stokes I or Rayleigh) flows as well as pipe flow start-up and oscillating plate (Stokes II) flow imply that

$$(\vec{\mathbf{v}} \cdot \nabla)\vec{\mathbf{v}} = 0; \quad \nabla \cdot \vec{\vec{\tau}} \sim \nabla p \quad \text{and} \quad \nabla \cdot \vec{\mathbf{v}} = 0$$

For such parallel flows v = w = 0 and the general postulate is then $u = u(y,z,t)$, i.e., Stokes' equation has to be solved:

$$\frac{\partial u}{\partial t} = -\frac{1}{\rho}\frac{dp}{dx} + \nu\left(\frac{\partial^2 u}{\partial y^2} + \frac{\partial^2 u}{\partial z^2}\right) + g_y \tag{1.3.65}$$

Typically $u = u(y,t)$ when transients are important, but if $u = u(y)$ only as in Couette and Poiseuille flows:

$$\frac{d^2 u}{dy^2} = \frac{1}{\mu}\frac{dp}{dx} + g_y \text{ :=constant} \tag{1.3.66a}$$

or

$$u'' = ¢ \tag{1.3.66b}$$

Application Schematics:

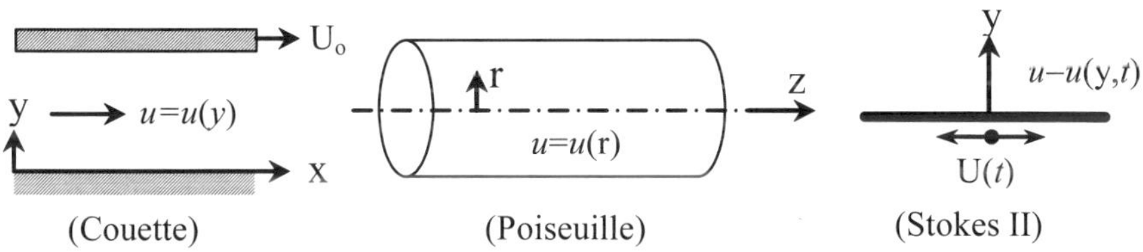

Solution Notes:

- Equation (1.3.66) can be integrated directly (see App. A). When $u = u(x,y)$ or $u = u(x,t)$, employ transformations, i.e., separation of variables or similarity method and then solve the resulting ODEs analytically or numerically (see Example 1.6 in Sect.1.3.2).
- For laminar flow in *slightly* varying conduits, e.g., tapered pipes, nonparallel slits, etc., quasiparallel flow may be assumed; however, via the "no-slip" boundary condition, the result is that $u = u(x,y)$ and $dp/dx = fct(x)$ as is demonstrated with Example 1.5 in Sect.1.3.2.

Steady Flow with Convective Acceleration: Flows around 2-D bodies flow near rotating disk, flow in conduits allow the assumption that

$$(\vec{\mathbf{v}} \cdot \nabla)\vec{\mathbf{v}} \sim \nabla p \,,\ \nabla \cdot \vec{\vec{\tau}} \tag{1.3.67}$$

As a result,

$$uu,_x + \mathrm{v}u,_y = -\frac{1}{\rho} p,_x + \nu\left(u,_{xx} + u,_{yy}\right) \tag{1.3.68a}$$

and

$$u\mathrm{v},_x + \mathrm{v}\mathrm{v},_y = -\frac{1}{\rho} p,_y + \nu\left(\mathrm{v},_{xx} + \mathrm{v},_{yy}\right) \tag{1.3.68b}$$

Solution Notes: Seek further reduction of the *y*-momentum equation; use clever postulates as in the Von Karman solution for the rotating disk (see Sect. 1.7); solve the equations numerically.

B. Summary of Some Approximate Solutions

Very Low Re#-Flows: Very slow motion around bodies and in conduits of "nonparallel" geometry (Creeping Flows), i.e., $Re \to 0$. Examples include

- Flow past a sphere (Stokes, Oseen)
- Lubrication theory (Reynolds equation)
- 1-D stretching flow (e.g., fibers, sheets, films, coats)
- Hele-Shaw flow

Neglecting or simplifying the inertia term are options, e.g.,

$$(\vec{\mathbf{v}}\cdot\nabla)\vec{\mathbf{v}} \approx \begin{cases} 0 & \text{(Stokes)} \\ U\dfrac{\partial\vec{\mathrm{v}}}{\partial x} & \text{(Oseen)} \end{cases}$$

which leads to (Stokes):

$$\nabla p = \mu\nabla^2\vec{\mathbf{v}} \rightarrow \nabla^2 p = 0 \quad \text{or} \quad \nu L^4\psi = 0 \qquad (1.3.69a, b)$$

where L is the biharmonic differential operator and y is the stream function.

For general lubrication problems (Reynolds):

$$(\vec{\mathbf{v}}\cdot\nabla)\vec{\mathbf{v}} = -\frac{1}{\rho}\nabla p + \nu\nabla^2\vec{\mathbf{v}} \quad \text{and} \quad \nabla\cdot\vec{\mathbf{v}} = 0 \qquad (1.3.70)$$

For steady 1-D stretching flows:

$$\rho w\frac{\partial w}{\partial z} = -\frac{1}{\rho}\frac{dp}{dz} + \frac{d\tau_{zz}}{dz} + \rho g_z; \quad \text{where} \quad p \approx \tau_{rr} - \frac{\sigma}{r(z)} \qquad (1.3.71a,b)$$

<u>Solution Notes:</u>

- Low Reynolds number flows around submerged bodies are discussed as part of the problem assignments in Sect. 1.7.
- For simple lubrication problems with mildly varying gap size, e.g., $h(x) << L$, the inertia term $(\vec{\mathbf{v}}\cdot\nabla)\vec{\mathbf{v}}$ in Eq. (1.3.70) may be negligible.
- Stretching flows are near-parallel flows with 1-D inertia, surface tension, viscous, and gravitational effects (see Sect. 1.7).

<u>High Re# Flows:</u> Boundary-layer flows near solid walls obey the criterion (Prandtl, 1904):

$$\frac{\delta}{l} \sim \frac{\mathrm{v}}{u} \sim \frac{1}{\sqrt{Re_l}} << 1$$

where $\delta(x)$ is the boundary-layer thickness and l is the horizontal plate length. The viscosity μ might be small but $\frac{\partial u}{\partial y} >> 1$ near the wall as shown in Fig. 1.3.6. Thus, within a thin 2-D shear layer along a flat plate Prandtl's boundary-layer equations hold:

$$\frac{\partial u}{\partial x} + \frac{\partial \mathrm{v}}{\partial y} = 0 \qquad (1.3.72)$$

$$\frac{\partial u}{\partial t} + u\frac{\partial u}{\partial x} + \mathrm{v}\frac{\partial u}{\partial y} = -\frac{1}{\rho}\frac{dp}{dx} + \nu\frac{\partial^2 u}{\partial y^2} \qquad (1.3.73a)$$

$$\frac{dp}{dy} = 0 \tag{1.3.73b}$$

where

$$-\frac{1}{\rho}\frac{dp}{dx} = \frac{\partial U_o}{\partial t} + U_o \frac{\partial U_o}{\partial x} \quad \text{and } p(y) = \text{const.} \tag{1.3.73c,d}$$

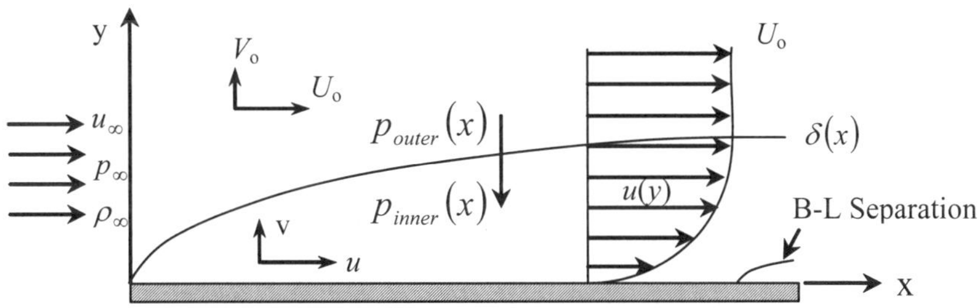

Fig. 1.3.6 Laminar flat-plate boundary-layer flow.

<u>Solution Notes:</u>

- Use the Falkner-Skan transformation for steady *self-similar* boundary-layer flows; solve resulting ODEs numerically (e.g., Runge-Kutta method with two "initial" conditions).
- Use the *integral method* for steady nonsimilar boundary-layer flows using suitable velocity profiles.
- Employ numerical parabolic equation solvers (FDM, FEM, CVM, etc.) for general cases.

1.3.5.3 Similarity Theory

Similarity solutions are of interest since they indicate that it might be possible to transform two-dimensional *partial* differential equations into *one ordinary* differential equation. Similarity theory deals with several transformation aspects.

(a) The conditions under which similar solutions exist; these include that the given system is *free of any nonuniformities* (shock waves, curvature effects, flow separation, sinks or sources, and other singularities or nonuniformities), and that the typical problem can be represented by *parabolic* PDEs (cf. Fig. 1.3.7).

(b) The selection of appropriate transformations within a particular coordinate system; specifically, a similarity variable, η, has to be found such that the explicit dependence on the former (two) independent variables disappears. Thus, η, the new independent coordinate combines, for example, x and y as $\eta = y/g(x)$ into one variable where $g(x)$ could be d(x) the boundary-layer thickness.

Associated with finding this transformation is the determination of an appropriate similarity function *f*() that, inserted into the two-dimensional PDE, converts the system equation into an ODE for *f*(*h*).

(c) The methodologies for implementing the tasks just outlined differ in what aspects they emphasize. For example, with the intuitive/physical approach, the functional forms of the similarity variable $\eta = \eta(x, y)$ or $\eta(x, t)$ or $\eta(r, z)$ and the similarity function $f(\eta)$ are postulated on physical grounds and then nondimensionalized. In more mathematically formal methods, e.g., the free parameter technique, Buckingham's pi theorem, or the group theoretic method, η and *f* are postulated with unknown exponents or the governing PDE is linearly transformed, again, with unknown exponents. Now, requirements of fulfilling the boundary conditions, nondimensionality, and/or invariance of the original and the transformed PDEs generate equations with which the exponents and hence (*x*,*y*) and *f*() can be determined (cf. Hansen, 1964; Na, 1979, etc.).

<u>Note</u>: Numerous solved example problems may be found in Kleinstreuer (1997) and other texts.

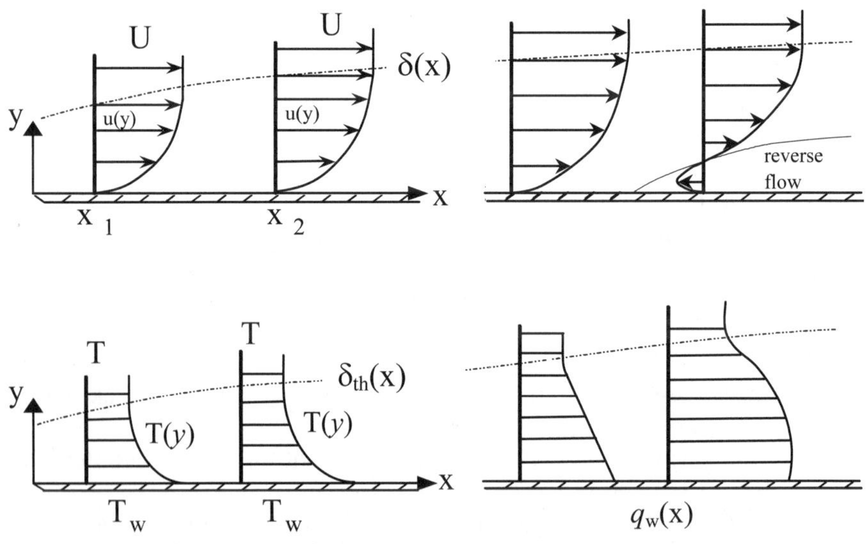

Fig. 1.3.7 Similar and nonsimilar velocity and temperature profiles.

1.3.5.4 Integral Methods

Two solution techniques dealing with integral equations are briefly discussed. The first method starts with the integration of a given set of partial differential equations that describe a given flow system, known as the integral method. The second approach starts with balance equations in integral form, i.e., the Reynolds Transport Theorem (cf. Eq. (1.3.17)), which assures the conservation laws for a control volume.

Von Karman integral method. In contrast to separation of variables and similarity theory, the integral method is an *approximation* method. The Von Karman integral method is the most famous member of the family of integral relations, which in turn is a special case of the method of weighted residuals (MWR). Specifically, a transport equation in normal form can be written as [cf. Eq. (1.3.2)]

$$L(\phi) \equiv \frac{\partial \phi}{\partial t} + \nabla \cdot (\vec{\mathbf{v}}\phi) - \nu\nabla^2\phi - S = 0 \quad (1.3.74)$$

where L (•) is a (nonlinear) operator, ϕ is a dependent variable, and S represents sink/source terms. Now, the unknown ϕ-function is replaced by an *approximate* expression, i.e., a "profile" or functional $\tilde{\phi}$ that satisfies the boundary conditions, but contains a number of unknown coefficients or parameters. As can be expected,

$$L(\tilde{\phi}) \neq 0, \text{ i.e., } L(\tilde{\phi}) \equiv R \quad (1.3.75a,b)$$

where R is the residual. In requiring that

$$\int_{\Omega} WRd\Omega = 0 \quad (1.3.75c)$$

we force the weighted residual over the computational domain to be zero and thereby determine the unknown coefficients or parameters in the assumed $\tilde{\phi}$ -function. The type of weighing function W determines the special case of the MWR, e.g., integral method, collocation method, Galerkin finite element method, control volume method, etc. (cf. Finlayson, 1978).

The Von Karman method is best applicable to laminar/turbulent similar or nonsimilar *boundary-layer type* flows for which appropriate velocity, concentration, and temperature profiles are known, i.e., thin and thick wall shear layers as well as plumes, jets, and wakes. Solutions of such problems yield global or integral system parameters, such as flow rates, fluxes, forces, boundary-layer thicknesses, shape factors, drag coefficients, etc.

In general, a two-dimensional partial differential equation is integrated in one direction, typically normal to the main flow, and thereby transformed into an ordinary differential equation, which is then solved analytically or numerically. Implementation of the integral method rests on two general characteristics of boundary-layer type problems: (i) the boundary conditions for a particular system simplify the integration process significantly so that a simpler differential equation is obtained, and (ii) all extra unknown functions, or parameters, remaining in the governing differential equation are approximated on physical grounds or by empirical relationships. Thus, closure is gained using, for example, the entrainment concept for plumes, jets, and wakes or by expressing velocity and temperature profiles with power expansions for high Reynolds number flows past submerged bodies.

Example 1.10: Integral Method applied to the Blasius Problem ($U_0 = u_\infty = ¢$, i.e., $\partial p / \partial x = 0$)

Sketch:

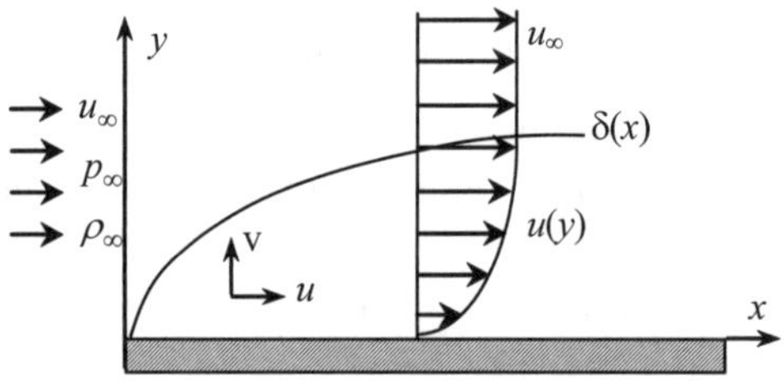

Assumptions:

- Steady laminar 2-D flow with constant fluid properties

Recall For the Blasius problem, the boundary-layer equations reduce to

$$\frac{\partial u}{\partial x} + \frac{\partial \mathrm{v}}{\partial y} = 0 \qquad \text{and} \qquad \frac{\partial u}{\partial t} + u\frac{\partial u}{\partial x} + \mathrm{v}\frac{\partial u}{\partial y} = \frac{1}{\rho}\frac{d\tau_{yx}}{dx}$$

- Solving for v in the continuity equation yields

$$\mathrm{v} = -\int \frac{\partial u}{\partial y}dy + f(x) := -\int_0^y \frac{\partial u}{\partial x}dy$$

- Integration across the x-momentum equation yields

$$\int_{y=0}^{\delta(t)} \left(u\frac{\partial u}{\partial x} + \mathrm{v}\frac{\partial u}{\partial y} \right) dy = \frac{1}{\rho}\int_0^{\delta} \frac{d\tau_{yx}}{dx}dy := -\frac{\tau_w}{\rho}$$

- Inserting v(x,y) and integrating the term $\int_0^{\delta} \mathrm{v}\frac{\partial u}{\partial y}dy$ by parts,

i.e., $\int u d\mathrm{v} = u\mathrm{v} - \int \mathrm{v} du$, yields

- $\int_0^{\delta} \frac{\partial}{\partial x}[u(U-u)]\,dy = \frac{\tau_w}{\rho}$

where $U = u_\infty$. Now, a suitable $u(x,y)$ profile has to be postulated, which matches the boundary conditions at y=0 and $y = \delta(x)$.

<u>*Solution*</u>: For laminar flow over a flat plate,

$$\frac{u}{u_\infty} = 2\frac{y}{\delta} - \left(\frac{y}{\delta}\right)^2$$

is a suitable profile where $\delta(x)$ is the key unknown.
With the previously derived *momentum integral relation*,

$$\tau_w(x) = \rho \frac{d}{dx}\int_0^\delta u(U-u)\,dy := 0.133\,\rho U^2 \frac{d\delta}{dx}$$

We also know that

$$\tau_w(x) = \mu \left.\frac{\partial u}{\partial y}\right|_{y=0} := \mu U \frac{2}{\delta}$$

Combining both results leads to an ODE for $\delta(x)$:

$$\frac{\mu dx}{0.0665\,\rho U} = \delta d\delta \text{ subject to } \delta(x=0) = 0.$$

Integration yields

- $$\delta(x) = 5.48\sqrt{\frac{\mu x}{\rho U}}$$

and hence

- $$\tau_w(x) = 0.356\,\rho U^2 \sqrt{\frac{\mu}{\rho U x}}$$

<u>*Graph*</u>:

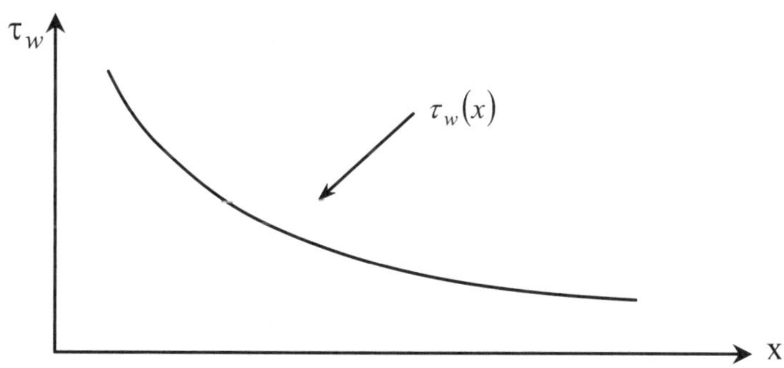

1.3.5.5 Dimensional Analysis and Scaling

Dimensional analysis (DA), in conjunction with appropriate scaling and order-of-magnitude estimation, is a powerful tool in the engineering sciences (Astarita, 1997; Middleman, 1999; among others). DA covers: (i) nondimensionalization of field equations resulting in dimensionless groups governing a given flow system; (ii) the Buckingham Pi Theorem, or the "by-inspection method," i.e., techniques that generate system-specific dimensionless groups similar to (i); and (iii) similitude, which is important in physical modeling, scale-up, etc., thereby relying on results from (ii). Clearly, DA, experimentation, and CFD (computational fluid dynamics) are the key tools for theoretical engineering analysis and design. Astarita (1997) even asserts that dimensional analysis, scaling, and estimation, if used jointly, provide "very often 90% of what one can ever hope to know about a problem, *without actually solving the problem.*" He then gives a few examples in fluid mechanics, heat, and mass transfer to underscore the point. Undergraduate fluid mechanics texts devote entire chapters to dimensional analysis and similitude (e.g., White, 2003).

Example 1.11: Mean Droplet Size in an Agitated Oil-Water Dispersion (Dimensional Analysis)

Consider a cylindrical water tank (*D*, *H*) with a rotating baffled impeller (d_I, ω_I). A finite volume fraction α of an immiscible fluid, e.g., air or oil, is initially injected and the impeller speed, i.e., angular velocity ω_I, is steadily increased. Obviously, fluid phase 2 (μ_2, ρ_2) mixes and forms air bubbles or oil droplets in the carrier fluid phase 1 (μ_1, ρ_1), larger particles, i.e., bubbles or droplets, break up and disperse throughout the tank (interfacial surface tension σ_{12}).

It would be very difficult to calculate the particle size distributions and particle concentrations as a function of space, time, and impeller speed. Dimensional analysis provides significant insight; at least how the mean particle diameter, d_p, relates to all the important system parameters listed, i.e., fluid properties, operating parameters, and design parameters.

Thus, based on logical reasoning we postulate in steps that the mean particle size decreases as

- the impeller speed increases nonlinearly [$\mathrm{v}_I = (d_I \cdot \omega_I)^n$];
- the surface tension between the fluids decreases, σ_{12}^{-1}; and

- the fluid property ratios increase, ρ_1/ρ_2, μ_1/μ_2 .

If large enough, the size of the tank (*D*, *H*) is immaterial. Thus, we have, *by inspection*, with n=2

$$d_p \sim \left(\frac{\rho_1}{\rho_2}\frac{\mu_2}{\mu_1}\right)\frac{\omega_I^2 d_I^2}{\sigma_{12}}$$

Trying to obtain *dimensionless expressions*, we note that $[\sigma] = F/L$ or M/T^2 and hence a density $[\rho] = M/L^3$ and one more length scale $[d_I] = L$ are required to nondimensionalize the RHS. The result is

$$\frac{d_p}{d_I} \sim \left(\frac{\rho_1}{\rho_2}\frac{\mu_2}{\mu_1}\right)\frac{\omega_I^2 d_I^3 \rho_1}{\sigma_{12}}$$

If the fluid properties do not vary,

- $$\frac{d_p}{d_I} \sim We = \frac{v_I^2 d_I \rho}{\sigma}$$

which is the *Weber number*. For example, experimental results for light oil droplet dispersion as depicted in the graph were given by Chen & Middleman (1967). Now, curve-fitting of the data generates a nonlinear equation for d_p/d_I .

Graph:

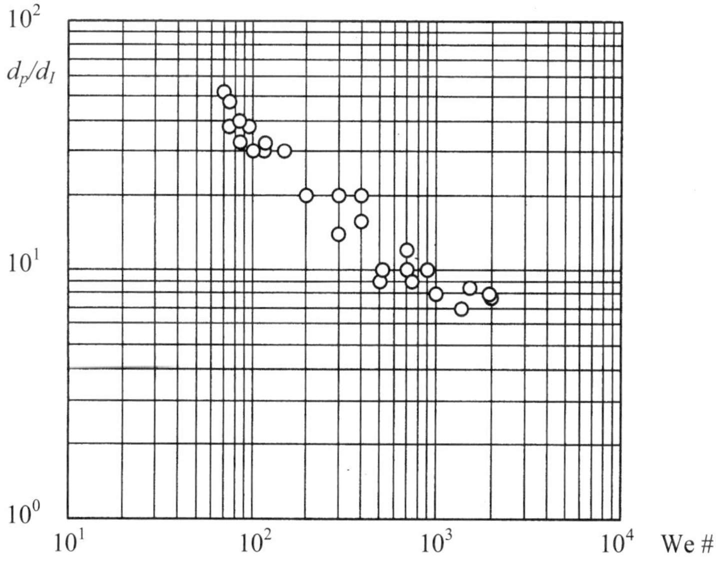

Example 1.12: Derivation of the Reynolds Number (Scale Analysis)

Scale analysis, which evolved from dimensional analysis (DA) and relative-order-of-magnitude analysis (ROMA), is a useful modeling tool to develop dimensionless groups characterizing the process dynamics at hand (cf. Sect. 1.3.4.2), to determine similarity variables for PDE-to-ODE transformations (cf. Sect. 1.6.3), and to form suitable profiles of dependent variables (cf. Kleinstreuer, 1997). The example given here is the derivation of the Reynolds number, starting with the definition

$$Re \equiv \frac{inertial\ \ forces}{viscous\ \ forces} = \frac{F_I}{F_v}$$

From the Navier-Stokes equation in vector form [cf. Eq. (1.3.19)], we know

$$\vec{F}_I \sim (\vec{\mathbf{v}} \cdot \nabla)\vec{\mathbf{v}} \quad \text{and} \quad \vec{F}_v \sim \nu \nabla^2 \vec{\mathbf{v}}$$

Choosing the scale parameters $\bar{u}$ (average or reference velocity scale) and l (system length scale), we scale the force ratio

$$\frac{(\vec{\mathbf{v}} \cdot \nabla)\vec{\mathbf{v}}}{\nu\nabla^2\vec{\mathbf{v}}} := \frac{\bar{u}l^{-1}\bar{u}}{\nu l^{-2}\bar{u}} = \frac{\bar{u}l}{\nu}$$

i.e.

- $$Re_l = \frac{\bar{u}l}{\nu}$$

1.4 TWO-PHASE FLOWS

Most biofluids are actually fluid-particle mixtures, e.g., blood as well as inhaled air with particulate matter, and contaminated body fluids. Such circulating particle suspensions fall into the research category of multiphase flow (Kleinstreuer, 2003; Crowe et al., 1998; Fan & Zhu, 1998; or Soo, 1990). Of this diverse research field, we focus on two-phase liquid-solid and gas-solid flows with quasi-spherical particles of diameters $d_p>1$ μm (fine) and $d_p<1$ μm (ultrafine). In general, two-phase flow is best described as the interactive motion of two different kinds of matter or media. The difference between the matter (or media) can be its thermodynamic state, called the phase (e.g., gas, liquid, solid) and/or its multiple chemical components, as illustrated in Table 1.4.1.

Table 1.4.1 Examples of Flow Media*: Phase of Matter vs. Component of Matter.

Component	Phase: Single	Phase: Multiple
Single	Flow of water, oil, oxygen, etc. (single-phase, single-component flows)	Flow of water & steam (two-phase, single-component flow)
Multiple	Flow of air, liquid polymer mixture, etc. (single-phase, multiple-component flows)	Flow of air-water-particles, e.g., bubbly slurries, etc. (three-phase, multicomponent flows)

After Crowe et al. (1998).
Note: Some authors (e.g., Drew & Passman, 1999) label "phases" as "components," i.e., multiphase flow becomes "flow of multicomponent fluids."

1.4.1 Modeling Approaches

There are two basic types of models, i.e., *corpuscular/molecular models* on the subcontinuum scale (kinetic theory) and *continuum models* on the micro/macroscale (continuum mechanics) (cf. Table 1.4.1 and Fig. 1.4.1). For both approaches, the conservation laws are applied, i.e., deterministic transfer equations (cf. molecular dynamics), enhanced by probability density functions (cf. Boltzmann equation with suitable collision integral), or with random perturbation functions added (e.g., turbulent particle trajectories). However, historically the development of two-phase flow analysis has not followed any rigorous approach. It has evolved from separate contributions in several application areas, such as nuclear power, chemical processing, combustion engineering, and aerosol science, with their particular definitions and modeling approaches. As a result, two-phase flow modeling, simulation, and design still relies on *system-specific* data correlations, semi-empirical equations, assumed probability density functions, or the use of random number generators, all somehow embedded in continuum mechanics laws and/or statistical mechanics concepts. Furthermore, because of the mathematical (and computational) complexities involved, many systems have been simulated with single-phase flow models or as (transient) one-dimensional two-phase flows. Examples include the study of single or monodisperse solids, droplets, or bubbles in a continuum, i.e., a gas or liquid stream. Free surface

(a) Hierarchy of multiphase flow models

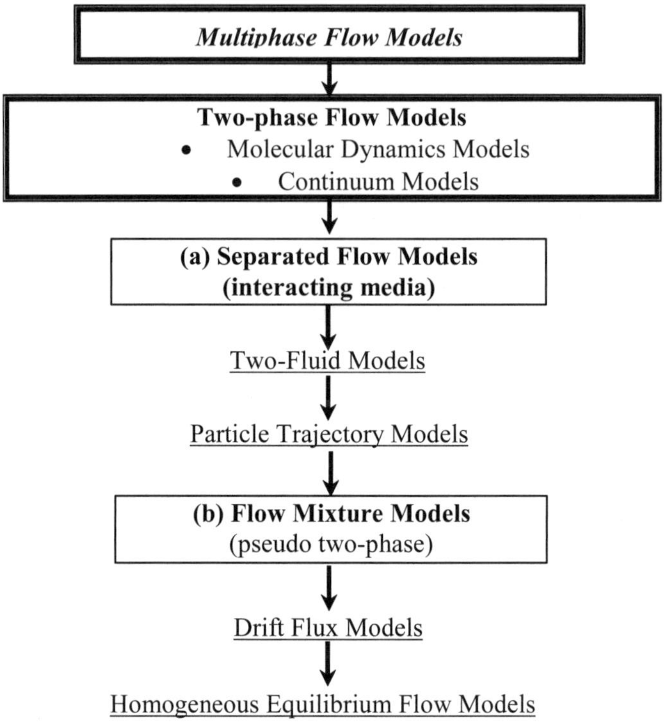

(b) Two-phase flow model applications

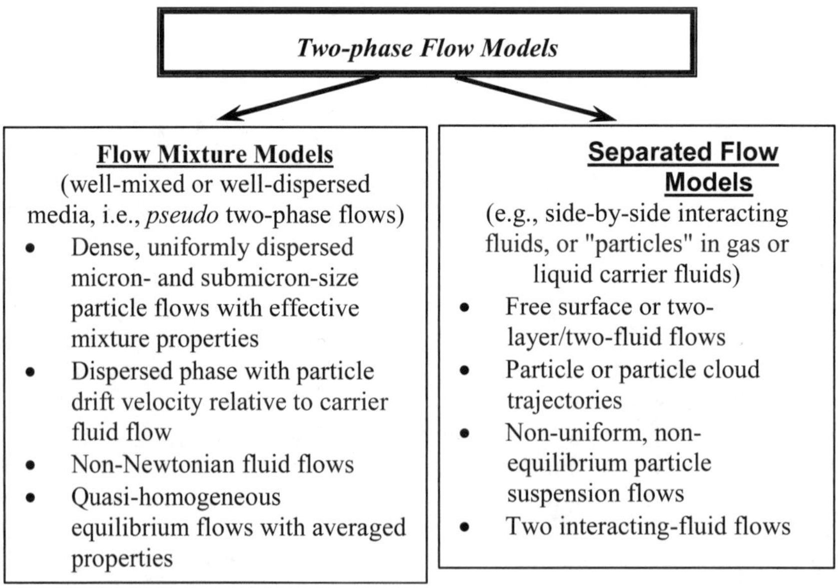

Fig. 1.4.1 Two-phase flow models: (a) hierarchy and (b) sample applications.

flow and separated multifluid flow problems have been treated as single-phase with coupled boundary conditions at the phase interface(s).

Clearly, in order to analyze multiphase (or just two-phase) flow systems, mathematical models, i.e., approximations to the actual flow phenomena, have to be employed. Their complexity increases sharply with the hierarchical level as depicted in Fig. 1.4.1a. They are further highlighted in Fig. 1.4.1b.

1.4.1.1 Definitions

A number of dispersed flow terms are briefly introduced in this section and further elaborated on in subsequent chapters. The definitions of characteristic system parameters are based on the continuum assumption discussed in Sect.1.2.1, i.e., the limiting volume, $\delta V'$, for the dispersed phase contains a sufficient number of particles so that a stationary average is warranted. Some of these parameters appear in the two-phase flow modeling equations while critical values of others are used to determine an appropriate modeling approach in the first place.

Densities, fractions, and concentrations. In order to characterize and quantify two-phase flows, especially dispersed particle flows falling into one of the two modeling categories, i.e., either mixture flows or separated flows, the dispersed phase is commonly quantified in terms of particle densities, fractions, loadings, and/or concentrations. While the nomenclature may differ from source to source, the parameter definitions are basically the same (cf. Appendix C). Specifically,

- The number of particles per unit volume

  $$n = \lim_{\delta V \to \delta V'} \frac{\delta N}{\delta V} \tag{1.4.1}$$

  is called the *number density,* where δN is the number of particles in the mixture volume δV and $\delta V'$ is the limiting mixture volume in line with the continuum assumption.

- The space occupied by the solid particles, droplets, or bubbles in the mixture

  $$\alpha_d = \lim_{\delta V \to \delta V'} \frac{\delta V_d}{\delta V} \tag{1.4.2a}$$

  is the *volume fraction* of the dispersed phase. The *volume or void fraction* of the continuous (i.e., liquid) phase is

  $$\alpha_c = \lim_{\delta V \to \delta V'} \frac{\delta V_c}{\delta V} \tag{1.4.2b}$$

  where

  $$\alpha_d + \alpha_c = 1 \tag{1.4.2c}$$

- The bulk (or effective or apparent) density of the *mixture* is

$$\rho_m = \alpha_c \rho_c + \alpha_d \rho_d := \overline{\rho}_d + \overline{\rho}_c \text{ and } \overline{\rho}_d = n m_p \tag{1.4.3a,b}$$

where m_p is the mass of a single representative particle, ρ_i is the actual density of material *i,* and $\overline{\rho}_i$ is the bulk density of phase *i=d,c.*

- The *mass concentration of the dispersed phase* is

$$c = \overline{\rho}_d / \overline{\rho}_c \tag{1.4.4a}$$

- The mass flux ratio $(\overline{\rho}_d \overline{\text{v}})/(\overline{\rho}_c \overline{u})$ is the *local loading*, while the ratio of overall mass flow rates

$$\kappa = \frac{\dot{m}_d}{\dot{m}_c} \tag{1.4.4b}$$

is the *total loading*, often approximated as $\kappa \approx c = \overline{\rho}_d / \overline{\rho}_c$.

- The mixture mass flow rate can be rewritten and applied to two-phase pipe flow as

$$\dot{m}_m = \dot{m}_c + \dot{m}_d := (\rho Q)_c + (\rho Q)_d \tag{1.4.5a}$$

Now, the mass flux of the mixture is

$$G_m = \dot{m}_m / A = G_c + G_d \tag{1.4.5b}$$

and the volume flux is

$$j_m = j_c + j_d = G_m / \rho_m := \frac{Q_c + Q_d}{A} = \text{v}_{c,s} + \text{v}_{d,s} = \text{v}_m \tag{1.4.6 a-c}$$

where v_m is the *superficial velocity* of the mixture, while the *phase* or *actual mean velocities* are

$$\text{v}_c = \frac{j_c}{\alpha_c} \text{ and } \text{v}_d = \frac{j_d}{\alpha_d} \tag{1.4.7a,b}$$

This implies that the phase "*i*" superficial velocities are

$$\text{v}_{i,s} = \alpha_i \text{v}_i;\ i = c,\ d \tag{1.4.7c}$$

Finally, the relative (or slip) velocity can be defined as

$$\text{v}_r = \text{v}_c - \text{v}_d \tag{1.4.7d}$$

so that the slip ratio is

$$\text{s} = \frac{\text{v}_c}{\text{v}_d} = 1 + \frac{\text{v}_r}{\text{v}_d} \tag{1.4.7e}$$

whereas the drift velocity, i.e., dispersed phase velocity relative to the mixture velocity, is

$$\text{v}^{\text{drift}} = \text{v}_d - \text{v}_m \tag{1.4.7f}$$

- The time it takes for a particle to respond to a (local) change in fluid velocity or fluid temperature can be estimated from the solutions of simplified forms of the equations of particle motion and particle heat transfer, respectively. Here,

$$m\frac{dv}{dt}=\frac{1}{2}C_D\frac{\pi d_p^2}{4}\rho_c(u-v)|u-v| \tag{1.4.8a}$$

and

$$mc_d\frac{dT_d}{dt}=\pi k_c d_p Nu(T_c-T_d) \tag{1.4.8b}$$

where u is the fluid velocity and v is the particle velocity. Assuming low relative particle Reynolds numbers, i.e.,

$Re_p=\frac{(u-v)d_p}{\nu}<1,\ C_D\approx 24/Re_p$ and $Nu\approx 2.0$, we have

$$\frac{dv}{dt}=\frac{18\mu}{\rho_d d_p^2}(u-v) \quad \text{and} \quad \frac{dT_d}{dt}=\frac{12k_c}{\rho c_d d_p^2}(T_c-T_d) \tag{1.4.9a,b}$$

which allows the *velocity (or momentum) response time* to be defined as

$$\tau_v=\frac{\rho_d d_p^2}{18\mu} \tag{1.4.10a}$$

and the *thermal response time*

$$\tau_{th}=\frac{\rho_d c_d d_p^2}{12k_c} \tag{1.4.10b}$$

- Another characteristic parameter of interest in dense particle suspension flows is the time between collisions, i.e.,

$$\tau_{col}=f_{col}^{-1} \tag{1.4.10c}$$

where f_{col} is the collision frequency, which is a function of the particle loading [Eq. (1.4.4b)].

1.4.1.2 Phase Coupling

There always exists some degree of interactions between the dispersed and continuous phases due to momentum exchange, heat, or mass transfer, which greatly complicate two-phase flow modeling and analysis. Fortunately, two-way coupling is occasionally negligible. While in continuum fluid mechanics the key thermodynamic property is the pressure that powers the fluid flow, in a *dilute* particle phase there is no particle pressure, i.e., no kinetic pressure due to particle velocity fluctuations or collisions. Thus, the carrier fluid determines particle motion, and hence information transfers along particle trajectories. Obviously, criteria that determine the degree of *phase coupling* are very useful. For example, particle suspensions are considered to be *dilute* (i.e., one-way coupled or "uncoupled" phases) when there are no particle-particle collisions, which also implies that the particles stay so far apart that the flow field remains unaffected and hence the fluid flow forces on the individual particles stay the same. Still, even in very dilute particle suspensions, a measurable pressure gradient around the exterior of an individual (larger) particle induces a stress field inside a particle (Patankar et al., 2000), which may alter particle shape and motion (e.g., buoyancy) and hence influences the local flow field.

In contrast, turbulent *dense* particle suspensions exhibit *strong* two-way coupling effects, known as "turbulence modulation" due to the presence of particles. These examples reflect *momentum coupling*, while hot solid particles in a cool gas stream changing the local gas properties is an example of *two-way energy coupling* via heat exchange, and droplet vaporization in a hot gas stream is an example of *two-way mass coupling*. In summary, two-way mass, momentum, and energy coupling between the two phases could affect the carrier fluid's velocity, pressure, temperature, as well as properties and the particles' velocities, temperatures, sizes, and possibly number densities. Clearly, for any given system, the number density $n = m_p / \bar{\rho}_p \approx N / V_{system}$, or loading $\kappa = \dot{m}_{disp.} / \dot{m}_{syst.}$, and the particle size, d_p, greatly determine the probability of interparticle collision in conjunction with the local flow pattern and changes in system geometry. It should be noted that the loading parameter values of κ, c, or a_d may vary regionally in a flow field due to transient or spatial changes in particle distributions. In the average:

(i) When $\tau_{vel} / \tau_{col} < 1$ and $a_d \ll 1$, the *flow is dilute* because collisions occur infrequently. Experimentally it was found that $a_d^{1/3} \ll 1$ works as a criterion for dilute particle suspensions, where $a_d \approx 0.004$ for solids and drops and $a_d \approx 0.001$ for bubbles. For example, considering gas-particle (solid or liquid) flow, where $\bar{\rho}_p / \bar{\rho}_g \geq 10^3$, $\alpha_p = \kappa \bar{\rho}_g / \bar{\rho}_p \ll 1$, and $l_{particle} / L_{system} \geq O(10^{-2})$, the mixture is dilute as long as $\kappa \leq O(1)$. Another criterion for *dilute* suspensions when $\rho_p / \rho_f \approx 10^{-3}$ can be given in terms of particle density number, n:

- $nd_p^3 \ll 10^{-3}$ for dilute case where the average particle distance is $l = n^{-1/3}$, i.e., $l \gg d_p$

  and
- $nd_p^3 \sim 10^{-3}$ for intermediate case.

Nevertheless, even in steady flows, particle pathlines differ from gas streamlines and hence particle clouds, i.e., locally denser concentrations, and particle-free regions may appear (cf. Zhang & Kleinstreuer, 2002). Particle trajectory models with appropriate point forces, especially near walls, nicely describe such phenomena.

(ii) When $\tau_{vel}/\tau_{col} > 1$, *the flow is dense* because $\tau_{col} << 1$ indicates that particles are close together, ready to collide, or because $\tau_{vel} > 1$, i.e., the particles have no time to respond to the fluid dynamic forces before the next collision; in both situations, large particles contribute to collision events because $\tau_{vel} \sim d_p^2$ and $\tau_{col} \sim d_p^{-2}$.

Associated criteria of particle behavior are the Stokes number $St = \tau_p / \tau_f$ and the density ratio. For example, *St* << 1 implies that particles follow readily the fluid flow including eddy fluctuations, whereas *St* >> 1 means that particle inertia is dominant ignoring any fluid fluctuations; i.e., trajectories are determined by the mean convective flow plus gravity. Obviously, dense particle suspension flows contribute two-way *momentum* coupling between the two phases, while dilute flows can be often treated as uncoupled flows.

For example, combining the total particle loading κ with the Stokes number yields a *momentum coupling* parameter (Crowe et al., 1998) as mentioned in Example 1.9, i.e.,

$$K_M \equiv \left.\frac{\kappa}{1+St}\right\} \begin{array}{ll} <<1 & \cdots \text{ one-way coupling} \\ >1 & \cdots \text{ two-way coupling} \end{array} \tag{1.4.11}$$

Aspects of phase coupling and proper model selection with suitable solution method are summarized in Table 1.4.2.

<u>*Example 1.13:*</u> Estimate of Phase Coupling

Because dynamically uncoupled flows are much easier to analyze, a reliable estimate of possible phase coupling is very important. One key parameter is the "minimum" particle spacing *l*, with which the ratio l/d_p can be evaluated. Considering spherical particles a distance *l* apart, each in a cube of a carrier fluid of extent *l*, the volume fraction of the dispersed phase is

$$\alpha_d = \frac{V_{sphere}}{V_{cube}} = \frac{\pi d_p^3}{6 l^3}$$

Table 1.4.2 Parameter-Range Criteria for Two-Phase Flow Model Selection and Possible Solution Method.

Two-phase Flow Models	Parameter Ranges				Solution Methods[#]
	$St = \tau_p / \tau_f$ [†]	$Re_p \sim \lvert\Delta\vec{v}\rvert$ [‡]	$\alpha = \frac{V_p}{V}$ [§]	$\kappa \cong \frac{\bar{\rho}_p}{\bar{\rho}_f}$ [§]	
Flow Mixture Models					**General:** *Single-phase* flow solution with modified fluid properties; for example, $\mu = \mu(\alpha)$; $\rho \hat{=} \rho_{mixture}$; $\mu \to \eta = \eta(\dot{\gamma})$ where $\dot{\gamma} \sim \lvert\nabla\vec{v}\rvert$; j_{drift} or ϕ_{drift}. **Specifics:** Mixture solution method depends on system's time dependence, dimensionality, flow regime, and mixture properties $\mu, \rho,$ etc.
Quasihomogeneous flows	≈ 0	≈ 0	Variable	≈ 1	
Non-Newtonian fluid flows	≈ 0	≈ 0	$0<\alpha\leq 0.5$	≈ 1	
Flows with component drift flux	$<< 1$	$<< 1$	Variable	$O(1)$	
Separated Flow Models **Layered two-fluid flows** (i) Smooth interface (ii) Wavy interface	• Stratified two-fluid flows in smooth constant layers. Main criterion: $\tau_{interface} \cdot A \leq \sigma_{critical} \cdot L$, or $Re_\delta{}^* < Re_{crit}$				Specific solutions of two moving fluids interacting at a straight or wavy interface; matching of velocity, temperature, heat flux, and stress at interface (see above).
Particle suspension flows (Trajectory model) (i) Dilute/uncoupled (ii) Dense/coupled	$0<St<O(10)$ $0<St<1$	$0<Re_p<O(10)$ $0<Re_p<1$	Finite particle number only	Limited restrictions	**One-way coupling:** Eulerian (fluid) Lagrangian (limited number of particles/points). Note: "resolved-volume" representation for a very few, nonspherical particles only, where $d_p >> \Delta h$.
Two-interacting fluid flows (i) Gas-solid (ii) Gas-liquid (iii) Liquid-solid (iv) Liquid-gas (v) Liquid-liquid	$St < 1$	$Re_p < O(10)$	$\alpha < 1$	Limited restrictions	**Two-way coupling:** • Point-volume formulation with additional collision and (random) interaction forces (i.e., Eulerian-Lagrangian approach); • Interpenetrating phases (i) –(v) (i.e., Eulerian-Eulerian approach); • Direct simulation Monte Carlo (DSMC) method for relatively high Knudsen number flows

All footnotes are given on the next page.

Footnotes to Table 1.4.2:

† $St \equiv \rho_p d_p^2 U/(18\mu D)$; where

- $St \to 0$ implies that particles readily follow fluid motion including eddy fluctuations;
- $St < 1$ implies that particles may deviate somewhat from fluid streamlines, i.e., "crossing effect" (Notes: Brownian motion has to be considered when $d_p < 1\mu m$); In lung airways with $d_p > 1\,\mu m$, the typical St number range is $0.01<St<0.2$, while
- $St >> 1$ implies that particle trajectories are due to drag force and gravity.

‡ $Re_p \equiv |\vec{v} - \vec{v}_p| d_p / \nu_f$; where for $Re_p \to 0$ see $St \to 0$; $Re_p << 1$ may imply *dilute* particle suspensions when particle-particle collisions are negligible and presence of small particles do not affect fluid flow; for $Re_p >> 1$ see $St >> 1$.

§ α, the particle volume fraction, similar to $\kappa \equiv \dot{m}_p / \dot{m}_f$ or $c = \bar{\rho}_p / \bar{\rho}_f$ may vary significantly in the flow field. Experimentally it was determined that $\alpha^{1/3} << 1$ for *dilute* particle suspensions; for example, $\alpha = 0.004$ for solids and drops and $\alpha = 0.001$ for bubbles.
In summary,

- $\alpha \le 4 \times 10^{-4}$ and $\kappa \le 1/3$ for dilute gas particle flows
- $4 \times 10^{-4} \le \alpha \le 3 \times 10^{-2}$ and $\kappa \le 5 \sim 50$ for dense suspensions
- $0.03 \le \alpha \le 0.3$ and $\kappa \le 50 \sim 500$ for highly concentrated dispersed flows
- $0.3 \le \alpha \le 0.65$ and $\kappa \le 1 \sim 50$ for moving packed bed

Resources include

- User-friendly, commercial software (e.g., Fluent or CFX for solving Flow Mixture Models, decoupled fluid-particle flows, etc.; or CFD-TWOPHASE for two-phase flows)
- CFD books with Fortran codes and review articles (e.g., Tannehill et al., 1997; Anderson et al., 1998; Roache, 1998; Ferziger & Peric, 1999; Loth, 2000; Hoffman, 2001; Cebeci, 2004)

* Critical Reynolds numbers, $Re = 4\delta\bar{v}/\nu$, for moving film: For $Re < 20$ (negligible rippling); $20 < Re < 1500$ (laminar flow with pronounced rippling); and $Re > 1500$ (turbulent flow).

or the length scale ratio is

- $$\frac{l}{d_p} = \left(\frac{\pi}{6\alpha_d}\right)^{1/3} := \begin{cases} \geq 10 & \text{for dilute particle suspensions} \\ \leq 2 & \text{for dense particle suspensions} \end{cases}$$

An alternative expression for α_d in terms of dispersed phase mass fraction relies on the definition of the bulk densities, i.e.,

$$\bar{\rho}_i = \lim_{\delta V \to \delta V'} \left(\rho_i \frac{\delta V_i}{\delta V} \right) = \rho_i \alpha_i; \; i = d, c$$

Now we estimate α_d from

$$\alpha_d = \bar{\rho}_d / \rho_d = \frac{\kappa}{1+\kappa}$$

where

$$\kappa = \frac{\bar{\rho}_d}{\bar{\rho}_c}$$

and hence

- $$l = \left(\frac{\pi}{6} \frac{1+\kappa}{\kappa} \right)^{1/3} d_p$$

In any case, if, for example,

$$\alpha_d = 10\% \text{ then } l/d_p \approx 1.7$$

which indicates that the particles are so close that the flow is considered to be dense and particle interaction as well as phase coupling effects may be important. However, when $l/d_p \geq 10$, particles can be treated individually and the suspension flow is dilute.

1.4.2 Mixture Models

The mixtures of interest are those which are either quasi-homogeneous, i.e., multi-component fluids in thermodynamic equilibrium, or those for which one phase migrates relative to, say, a bulk motion, i.e., a *drift flux* exists.

1.4.2.1 Homogeneous and Non-Newtonian Fluid Flow Models

There are two types of materials or media which may fall into the category of homogeneous equilibrium models (HEM):

(a) truly homogeneous, although possibly multicomponent, fluids such as air; and

(b) actually nonhomogeneous fluids, i.e., ultrafine particle suspensions, that are treated as "homogeneous" mixtures, such as bubbly flows (e.g., carbonated beverages) and other well-dispersed particle flows (e.g., slurries, mists, sprays, paints, blood, and certain food stuff).

Thus, quasi-homogeneous behavior can be assumed if the different substances in solution are near uniform and *well mixed.* However, as with most fluids and especially mixtures, the material behavior depends on the geometric flow scale (cf. Gad-el-Hak, 1999). For example, blood flow in large conduits with shear rates over 200 s^{-1}, can be treated as a homogeneous Newtonian fluid, in smaller bifurcating conduits as a *non-Newtonian* fluid mixture, and in capillaries as *separated* multicomponent fluid flow (see Fig.1.4.1).

In "homogeneous" mixtures, stimuli response is identical everywhere, i.e., the time scale for the transport between phases or components is very much shorter than the overall characteristic or system time scale. This assumption leads to the requirement that the two phases are in thermodynamic equilibrium, i.e., $v_1 = v_2$, $p_1 = p_2$, and $T_1 = T_2$. Thus, the HEM treats the mixture as a pseudofluid, which obeys all the laws of single-phase flow. In order to use this model, suitable thermodynamic and transport properties are needed for the mixture as well as a mixture equation of state (cf. Batchelor, 1970; Patankar & Hu, 2001, among others).

Mixture properties and solution techniques. The mean density can be expressed as a function of the static concentration, $\alpha \equiv V_d/V$, commonly called the particle volume or void fraction [see Eq. (1.4.2) in Sect. 1.4.1.1]; alternatively, the kinematic mass concentration, $\chi \equiv \dot{m}_d/\dot{m}$, is employed, commonly called the quality. Marking the two phases with indices $k = 1$ (carrier fluid) and $k = 2$ (dispersed phase), the *mixture or effective density* is (see Fig. 1.4.2)

$$\rho_m = \alpha\rho_2 + (1-\alpha)\rho_1 \tag{1.4.12}$$

or

$$\frac{1}{\rho_m} = \frac{\chi}{\rho_2} + \frac{1-\chi}{\rho_1} \tag{1.4.13}$$

where for homogeneous equilibrium flow the static void fraction $\alpha = V_2/V$ is equal to the kinetic volume fraction $\beta = Q_2/Q$ and mass fraction $\chi = c$, i.e., $\dot{m}_2/\dot{m} = m_2/m$ (cf. Appendix C).

Several theoretical expressions for the *mixture viscosity* have been derived. For example, suspensions of fluid spheres at low concentrations, i.e., $\alpha \leq 0.05$ (Taylor, 1932; Wallis, 1969; Soo, 1990), may have a mixture viscosity

$$\mu_m = \mu_1\left(1 + 2.5\alpha\frac{\mu_2 + \frac{2}{5}\mu_1}{\mu_2 + \mu_1}\right) \tag{1.4.14}$$

Equation (1.4.14) implies two special cases:

- Suspensions of solid particles where $\mu_2 >> \mu_1$ (Einstein, 1906; Brady, 1984)

$$\mu_m = \mu_1(1 + \beta\alpha); \quad \beta = 5/2 \quad \text{or} \quad \beta \lesseqgtr 1 \tag{1.4.15a-c}$$

and

- Suspensions of low-viscosity gas bubbles

$$\mu_m = \mu_1(1 + \alpha) \tag{1.4.16}$$

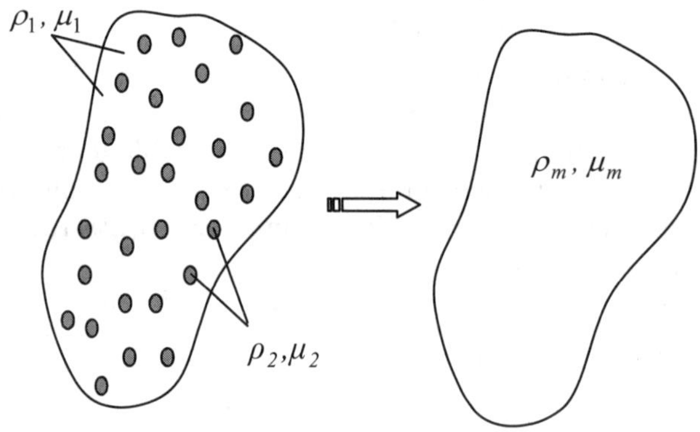

Fig. 1.4.2 Conversion of two-phase flow to a single mixture flow.

As stated, the above three equations (1.4.14, 1.4.15, 1.4.16) are valid only for relatively small void fractions, i.e., $\alpha < 0.05$. To account for mixtures with larger void fractions, several authors have proposed averaged expressions that converge to the individual phase viscosities for the limiting cases of $\alpha \to 0$ and $\alpha \to 1$. The following mixture viscosity expressions have been proposed for evenly distributed gas-liquid flows; for example, water droplets in air, sprays, etc. (see Fig. 1.4.2):

$$\frac{1}{\mu_m} = \frac{\chi}{\mu_g} + \frac{(1-\chi)}{\mu_f} \quad \text{(McAdams et al., 1942, Isbin et al., 1957)} \tag{1.4.17}$$

or

$$\mu_m = \chi\mu_g + (1-\chi)\mu_f \quad \text{(Cicchitti et al., 1960)} \tag{1.4.18}$$

or

$$\mu_m = \frac{j_f}{j}\mu_f + \frac{j_g}{j}\mu_g \quad \text{(Dukler et al., 1964)} \tag{1.4.19}$$

Additional correlations for effective viscosities of suspensions and emulsions may be found in Zapryanov & Tabakova (1998).

Example 1.14: Quasihomogeneous Mixture Flow

Consider a mixture of water ($\dot{m}_l = 0.42$ kg/s) and air ($\dot{m}_g = 0.01$ kg/s) flowing upwards in a vertical pipe (D=25 mm, L=45 cm). Given a friction factor of f = 0.079 $Re^{-1/4}$, find the total pressure drop, the volumetric flow rates, void fraction, and mean water and air velocities.

Sketch:

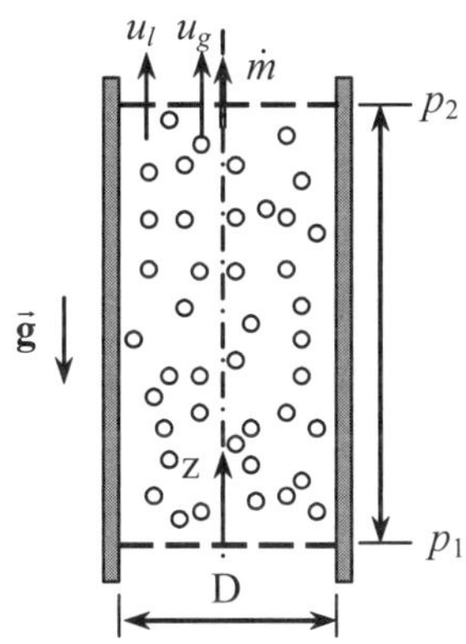

Assume:

- steady turbulent fully developed two-phase flow
- area-averaged (i.e., uniform) velocities
- thermodynamic equilibrium, i.e., slip ratio $u_g / u_l = 1$

Concepts:

- homogeneous equilibrium model (HEM)
- $Dp_{total} = Dp_{friction}$ $+Dp_{static}$ from extended Bernoulli equation
- Newtonian fluid

Solution:
Based on mass conservation,

$$\dot{m} = \dot{m}_g + \dot{m}_l \tag{E1.14-1}$$

so that with $u_g = u_l = u = \dot{m} / (\rho A)$ and with quality $x \equiv \dot{m}_g / \dot{m}$, the mixture density

$$\rho = \left(\frac{x}{\rho_g} + \frac{1-x}{\rho_l} \right)^{-1} \tag{E1.14-2}$$

and the effective viscosity

$$\mu = \left(\frac{x}{\mu_g} + \frac{1-x}{\mu_l} \right)^{-1} \tag{E1.14-3}$$

Thus, with $u = 17.65$ m/s, $x = 0.023$ and at $T = 20\ ^oC$, $\rho = 49.64$ kg/m^3, and $\mu = 4.435\times10^{-4}$ kg/m·s, the Reynolds number is $Re = \rho uD / \mu := 49{,}388$ so that

$$f = \frac{0.079}{Re_D^{0.25}} := 0.0053$$

<u>Note:</u> For smooth-walled pipes, Blasius (1910) suggested $f = 0.316\, Re_D^{-1/4}$ for 4000< Re_D <10^5.

Now, from the extended Bernoulli equation

$$\frac{p_1}{\rho g} + \frac{v_1^2}{2g} + z_1 = \frac{p_2}{\rho g} + \frac{v_2^2}{2g} + z_2 + h_f \qquad \text{(E1.14-4)}$$

where

$$h_f = f\left(\frac{L}{D}\right)\frac{v^2}{2g} \qquad \text{(E1.14-5)}$$

$$\Delta p_{total} = 3169\ N/m^2$$

The volumetric flow rates are

$$Q_i = \left(\frac{\dot{m}}{\rho}\right)_i = \begin{cases} 0.00833 & \text{m}^3/\text{s} \quad (\text{air}) \\ 0.0042 & \text{m}^3/\text{s} \quad (\text{water}) \end{cases}$$

so that the void fraction

$$\alpha = \frac{Q_g}{Q} = 0.952$$

and as stated

$$u_g = u_l = u = 17.65 \quad \text{m/s}$$

Basic non-Newtonian fluid models. Some mixtures, although treatable as single-phase flow, exhibit unusual behavior due to their component make-up or molecular structure. Stokes' hypothesis of a linear relationship between shear stress and shear rate, e.g., $\vec{\vec{\tau}} = \mu\dot{\vec{\vec{\gamma}}} = \mu\left(\nabla\vec{v} + \nabla\vec{v}^{tr}\right)$, is invalid for such non-Newtonian type fluids because they may have memory, the effective viscosity is shear-rate dependent, etc.

There are two "sources" of nonlinear viscous flow behavior:

- genuine non-Newtonian fluids such as exotic lubricants, blood at low shear rates, paints, syrups, and other food stuff, as well as polymeric liquids with high molecular weights, i.e., MW>10^4; and
- uniformly dispersed particle suspensions such as droplet sprays, slurries, pastes, etc., where the presence or absence of "non-Newtonian effects" in homogeneous mixtures depends primarily on the rate of particle diffusion vs. relative convection by an imposed flow.

Clearly, the Peclet number $Pe = \bar{u} r_0 / D_{eff}$ is indicative, where $Pe = O(1)$ in order for the flow to be able to disrupt significantly the suspension microstructure and to produce a nonlinear, time-dependent rheology (cf. Zapryanov & Tabakova, 1998). Books on single-phase non-Newtonian fluids include Macosko (1994), Churchill (1988), Bird et al. (1987), and Tanner (1985).

As in multiphase flow modeling (cf. Sect. 1.4.1), two approaches for the development of "rheological equations of state" are commonly used: the continuum mechanics theory and the molecular dynamics theory. We deal only with the continuum approach and focus on simple polymeric liquids with a viscosity that is shear-rate dependent. Flow phenomena such as rod climbing and jet swelling or viscoelastic effects such as fluid recoil, stress relaxation, and stress overshoot are discussed elsewhere (Bird et al., 1987). The steady-state shear flows of interest here can be basically described with an analytical representation of the shear-rate dependent viscosity, $\eta = \eta(\dot{\gamma})$, which fits specific experimental data sets (cf. Tanner, 1985; Bird et al., 1987; Macosko, 1994).

In Sect. 1.3.2, the shear stress tensor $\vec{\vec{\tau}}$ for Newtonian fluids was *linearly* related to the shear-rate tensor $\vec{\vec{\dot{\gamma}}} = \nabla\vec{\mathbf{v}} + (\nabla\vec{\mathbf{v}})^{tr}$. Now, $\vec{\vec{\tau}}$ has to be nonlinearly related to $\vec{\vec{\dot{\gamma}}}$. Thus, for incompressible fluids (cf. Fig. 1.4.3):

$$\tau_{ij} = \begin{cases} \mu\dot{\gamma}_{ij} & \text{for Newtonian fluids} \\ \eta(\dot{\gamma} \text{ or } \tau)\dot{\gamma}_{ij} & \text{for non - Newtonian or "generalized Newtonian" fluids} \end{cases}$$

Some practical fluid models and their applications can be summarized as follows:

- The "power-law" model of Ostwald and deWaele is a two-parameter model for a wide variety of shear-thinning ($n < 1$) or shear-thickening ($n > 1$) aqueous solutions, simple polymeric liquids, and so on.

$$\eta = m\dot{\gamma}^{n-1}, \quad \dot{\gamma} \equiv \left|\vec{\vec{\dot{\gamma}}}\right| \geq 1\ s^{-1}; \quad \left|\vec{\vec{\dot{\gamma}}}\right| = \sqrt{\frac{1}{2}\sum_i \sum_j \dot{\gamma}_{ij}\dot{\gamma}_{ji}} \qquad (1.4.20a,b)$$

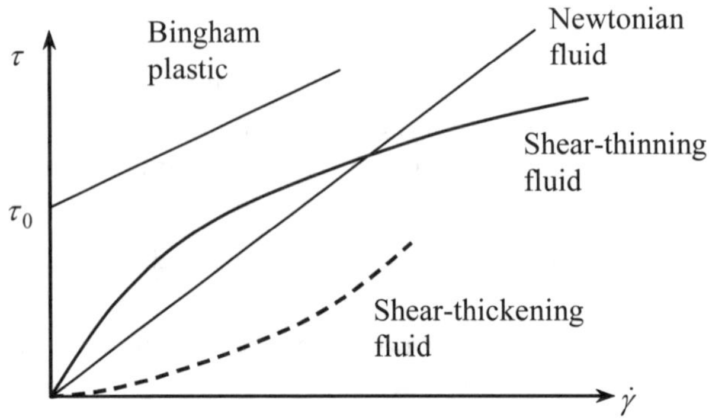

Fig. 1.4.3 One Newtonian and three non-Newtonian fluids.

For Newtonian fluids, the dimensionless exponent n is equal to 1, which implies that m becomes μ.

- The Carreau-Yasuda model is a five-parameter expression that extends the application of the power law model to concentrated polymer solutions and melts.

$$\frac{\eta-\eta_\infty}{\eta_o-\eta_\infty}=\left[1+(\lambda\dot{\gamma})^a\right]^{(n-1)/a} \tag{1.4.21}$$

where η_o is the zero-shear-rate viscosity, η_∞ is the infinite-shear-rate viscosity (typically $\eta_\infty \to 0$), λ is a time constant that represents the "fading memory" of certain polymers, i.e., $\lambda = O(10s)$ to $O(100s)$, and a (typically $a = 2$) is a dimensionless parameter describing the transition between the zero-shear-stress and the power-law regions.

- The Bingham model is a two-parameter formula for pastes, including ketchup, and slurries exhibiting a threshold or yield stress τ_o resisting motion.

$$\eta=\begin{cases}\infty & \text{for} \quad \tau\le\tau_o \\ \mu_p+\dfrac{\tau_o}{\dot{\gamma}} & \text{for} \quad \tau>\tau_o\end{cases} \tag{1.4.22a,b}$$

- The Casson model is a nonlinear extension of the Bingham model suitable for simulating suspensions of (spherical) particles in polymeric solutions.

$$\sqrt{\tau_{ij}} = \begin{cases} 0 & \text{for} \quad \tau \le \tau_o \\ \sqrt{\tau_o} + \sqrt{\mu_o}\sqrt{\dot{\gamma}_{ij}} & \text{for} \quad \tau > \tau_o \end{cases} \tag{1.4.23a,b}$$

- A modified Casson model can be employed to simulate blood rheology, which is represented by the stress tensor $\vec{\vec{\tau}}$ with the following relations:

$$\vec{\vec{\tau}} = 2\eta(II_D)\vec{\vec{D}} \tag{1.4.24a}$$

Here, $\vec{\vec{\varepsilon}} \equiv \vec{\vec{D}} = \frac{1}{2}\left[\nabla\vec{\mathbf{v}} + (\nabla\vec{\mathbf{v}})^{tr}\right] = \frac{1}{2}\vec{\vec{\dot{\gamma}}}$ is the rate-of-strain tensor, and the apparent viscosity η is a function of the shear rate

$$\eta(II_D) = \frac{1}{2\sqrt{II_D}}\left[C_1(Ht) + C_2(Ht)\sqrt{2\sqrt{II_D}}\right]^2 \tag{1.4.24b}$$

where Ht is the hematocrit and II_D is the second scalar invariant of $\vec{\vec{D}}$, that is,

$$II_D = \frac{1}{2}\left[\left(trace\vec{\vec{D}}\right)^2 + \left(trace\vec{\vec{D}}^2\right)\right] \tag{1.4.24c}$$

Writing out the right-hand side in component form yields

$$\begin{aligned} II_D &= D_{11}D_{22} + D_{11}D_{33} + D_{22}D_{33} \\ &\quad - D_{12}D_{21} - D_{13}D_{31} - D_{23}D_{32} \end{aligned} \tag{1.4.24d}$$

The coefficients C_1 and C_2 were determined for $Ht = 40\%$ as $C_1 = 0.2$ (dyne/cm^2)$^{1/2}$ and $C_2 = 0.18$ (dyne·s/cm^2)$^{1/2}$, based on Merrill's experimental data (cf. Merrill, 1968). The following graph shows the variation of blood viscosity η [Eq. (1.4.24b)] with shear rate $\dot{\gamma} = 2\sqrt{II_D}$. For human hemodynamic studies, Kleinstreuer et al. (2001) discussed μ=0.0348 (dyne·s/cm^2), that is, ν=0.033 (cm^2/s), which guarantees a smooth transition from the Casson fluid to a Newtonian fluid. At the other end of the curve, since Casson's model is suitable only for $\dot{\gamma}$ > 1 (s^{-1}), they took $\eta = \eta\big|_{\dot{\gamma}=1} = 0.1444\left(\text{dyne}\cdot\text{s/cm}^2\right)$ when $\dot{\gamma}$ < 1 (s^{-1}), which is the "zero-shear rate" condition (cf. Fig. 1.4.4). An even better representation of the blood rheology is achieved with the Quemada model (see Buchanan & Kleinstreuer, 1998).

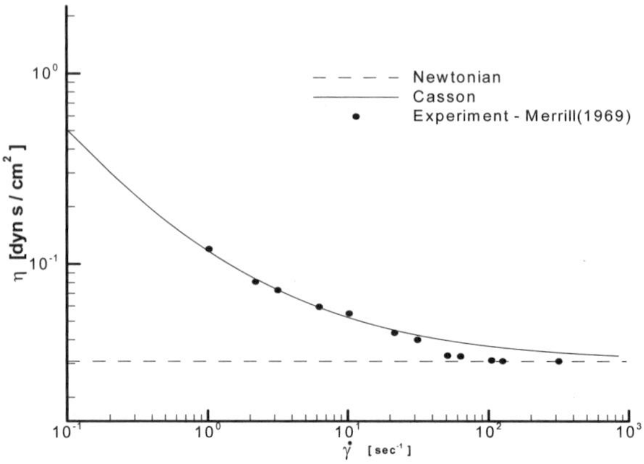

Fig. 1.4.4 Example of low-shear-rate blood rheology.

Example 1.15: Casson-Fluid Flow through an Inclined Tube

Calculate the volumetric flow rate of a Casson fluid through a slanted circular tube of radius R and length l, where $-dp/dx+\rho g\sin\theta\approx\Delta p/l=$ constant.

System:

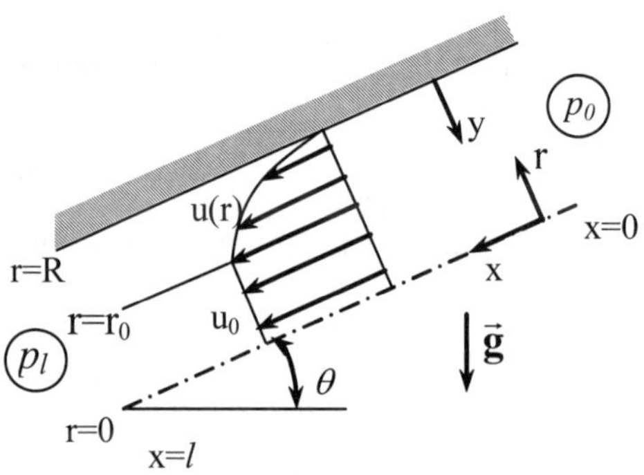

Assumptions:

- steady laminar fully developed flow;
- shear-rate-dependent viscosity;
- uniform fluid motion when $\tau_{rx}<\tau_o$ (where τ_0 is a yield stress), and
- gravitational effect is absorbed in $\Delta p/l$.

Postulates:

v = w = 0 and $u=u(r)$ only; $\dot{\gamma}_{ij}\Rightarrow\dot{\gamma}_{rx}=\dfrac{du}{dr}$; $\dot{\gamma}_{rx}=0$ for $0\le r\le r_o$

Governing Equations:

The Navier-Stokes equation is *not* applicable. Thus, the reduced equation of motion in the x-direction reads (cf. Appendix C)

$$0 = -\frac{dp}{dx} + \frac{1}{r}\frac{d}{dr}\left(r\tau_{rx}\right) \qquad \text{(E1.15-1)}$$

Integration yields

• $$\tau_{rx} = \frac{\Delta p}{2l}r \quad \text{where } \tau_{rx}(r=0)=0 \qquad \text{(E1.15-2)}$$

Invoking the Casson model [Eq. (1.4.23)], we obtain

at $r = r_o$: $$\tau_{rx}\Big|_{r=r_o} = \tau_o = \frac{\Delta p}{2l}r_o \qquad \text{(E1.15-3a)}$$

and

at $r = R$: $$\tau_{rx}\Big|_{r=R} = \tau_w = \frac{\Delta p}{2l}R \qquad \text{(E1.15-3b)}$$

so that $\tau_{rx} = \tau_w \dfrac{r}{R}$ as for Newtonian fluids. Continuity is preserved in terms of

$$Q = 2\pi\int_o^R urdr := 2\pi\left[\int_o^{r_o} u_o rdr + \int_{r_o}^R u(r)rdr\right] = const \qquad \text{(E1.15-4)}$$

Solution:

From the Casson relation [Eq. (1.4.23b)] we know that

$$-\dot{\gamma}_{rx} = -\frac{du}{dr} = \frac{1}{\mu_o}\left(\sqrt{\tau_{rx}} - \sqrt{\tau_o}\right)^2 \quad \text{for } \tau > \tau_o \qquad \text{(E1.15-5)}$$

Now we have two possibilities:

(i) Inserting $\tau_{rx} = Ar$, $A = \tau_w / R$, and integrating to find $u(r)$, or

(ii) expressing the equation for Q = ¢ in terms of r and integrating to find Q directly.

For the second approach we recall that

$$r = \frac{\tau_{rx}}{\tau_w}R \text{ or } dr = \frac{R}{\tau_w}d\tau_{rx}, \quad r_o = \frac{2l}{\Delta p}\tau_o, \quad R = \frac{2l}{\Delta p}\tau_w$$

and, applying integration by parts,

$$Q = 2\pi \int_{o}^{R} urdr = \pi \left[ur^2 \Big|_{o}^{R} - \pi \int_{o}^{R} r^2 du \right] = -\pi \int_{r_o}^{R} r^2 \left(\frac{du}{dr} \right) dr$$

After substitution,

$$Q = \frac{\pi}{\mu_o} \left(\frac{R}{\tau_o} \right)^3 \int_{\tau_o}^{\tau_w} \tau_{rx}^2 \left[\tau_{rx} - 2\sqrt{\tau_{rx}}\sqrt{\tau_o} + \tau_o \right] d\tau_{rx}$$

Introducing $\kappa = \frac{2l\tau_o}{\Delta p R}$ and using $\tau_w = \frac{\Delta p}{2l} R$, we obtain after integration

$$Q = \frac{\pi R^4}{8\mu_o} \left(\frac{\Delta p}{l} \right) \left(1 - \frac{16}{7}\sqrt{\kappa} + \frac{4}{3}\kappa - \frac{1}{21}\kappa^4 \right)$$

Note: For $\tau_o = 0$, $\kappa = 0$ and with $\mu_o \equiv \mu$ the well-known Hagen-Poiseuille solution is obtained.

Additional applications of non-Newtonian fluid flow may be found in the books cited, as well as in the articles by Kleinstreuer & Agarwal (1987) and Buchanan & Kleinstreuer (1998), among many others.

1.4.2.2 Drift-Flux Model

The drift flux model, like the pseudo-homogeneous equilibrium model, is based on the concept of analyzing the mixture as a whole, rather than as separate phases. However, the advantage of the drift flux model is that it accounts for the relative motion between the phases, making it possible to model flows in which a measurable *drift* exists between, say, a particle ensemble and the carrier fluid. Hence, particle pathlines differ from fluid element streamlines, known as the "crossing trajectory effect." Examples include *turbulent* particle suspension flows where a drift parameter can be defined as the ratio of the particle ensemble drift velocity to the fluid phase rms fluctuation velocity, cell migration toward the vessel wall in blood flow, particle settling in sedimentation processes, countercurrent flows, etc. The primary assumption associated with this model is that the mixture momentum equation, along with the constitutive equations describing the drift or relative motion between the phases, can sufficiently model the dynamics of the two-phase flow field. Thus, the drift flux model is most appropriate when the motions of the phases are strongly coupled.

1.4.3 Separated Flow Models

It should be noted that the adjective "separated" refers to (a) the physical separation of two immiscible fluids flowing in layers, as well as (b) the mathematical need for two separate sets of equations, one describing the particles dispersed in a carrier fluid, and the other describing the continuous phase (the fluid). Thus, separated flow examples include two stratified fluids such as oil on water and steam in a partially filled water pipe, or distinct particles in a carrier fluid such as solids in a gas or liquid, droplets in a gas, and bubbles in a liquid stream, or distinct but interacting fluid/solid phases such as dense, nonuniform particle suspensions. First the modeling equation for particle trajectories are based on the Eulerian-Lagrangian approach, while the generalized two-phase flow equations describing interacting gas-solid, gas-liquid, and liquid-solid media (see Kleinstreuer, 2003) require the Eulerian-Eulerian approach.

1.4.3.1 Particle Trajectory Models

There are various forms of (solid) particle interactions causing up to four-way coupling phenomena. Examples include:

- particle-particle and particle-fluid interactions influencing individual particle trajectories and fluid flow between neighboring particles, respectively;
- particle-wall interactions such as direct impacting, interception, bouncing, rolling, and resuspension;
- particle-wall interactions such as direct impacting, interception, bouncing, rolling, and resuspension;
- random interparticle collisions resulting in Brownian-type motion for submicron particles, or particle aggregation/droplet coalescence, or a drift flux, or an apparent viscosity of the particle phase; and
- fluid-particle interactions as expressed with Newton's second law of motion, or with dispersion terms in the enhanced mass transfer equation, as well as turbulence modulation in dense particle suspensions.

Furthermore, inhaled droplets may deform and in thermal flows or high humidity environments, they may shrink due to evaporation or grow due to condensation. However, quite frequently (cf. Table 1.4.2), the dispersed phase, i.e., solid particles, droplets, or bubbles, is *uncoupled* from the continuous phase and, given the particle release conditions, the trajectories of many individual particles can be directly calculated. In fact, particle tracking, i.e., the Lagrangian description of particles, avoids numerical diffusion and allows readily incorporation of particle characteristics, such as size and composition. However, difficulties arise when the particles are nonspherical, generate large wake effects, are interacting or randomly diffusing. In any case, for accurate results the velocity field of the carrier fluid, the particle characteristics, and all relevant "point" forces acting on the particles have to be known. The starting point in this section is the complete dynamics equation for a single solid sphere. A historical review of the various particle trajectory equations and their applications has been provided by Michaelides (1997) and Kim et al. (1998), who also proposed a new particle trajectory equation for wider ranges of particle Reynolds numbers $2 \leq \mathrm{Re}_{d_p} \leq 150$ and particle-to-fluid- density ratios $5 \leq \rho_p / \rho_f \leq$ 200.

Single spherical particles. The kinematics of a particle in a fluid depends on the external forces imposed on the body by the suspending medium or carrier fluid. Ideally, all surface and exchange forces should be accurately integrated from the fluid-particle and particle-particle interactions, for example, employing direct numerical simulation (DNS) of the velocity, pressure, and stress fields surrounding each particle; however, DNS is still cost prohibitive (cf. Joseph, 2001). Hence, single particle transport will involve the solution of a general equation of particle motion, i.e., Newton's second law, for its position in the fluid flow field. The initial assumptions will be that the calculated flow field is undisturbed by the presence of particles and that particles are spherical with negligible Magnus effect, i.e., particle rotation is insignificant.

Basic particle equation in laminar flow. The equation of motion for a representative particle in the Eulerian-Lagrangian framework may be written as:

$$m_p \frac{\hat{\mathrm{d}}\,\vec{\mathbf{v}}_p}{\hat{\mathrm{d}}\,t} = \sum \vec{\mathbf{F}}_{volume} + \sum \vec{\mathbf{F}}_{surface} \tag{1.4.25a}$$

where

$$\sum \vec{\mathbf{F}}_{volume} = \vec{\mathbf{F}}_{buoyancy} - \vec{\mathbf{F}}_{virtual\ mass} = (m_p - m_f)\,\vec{\mathbf{g}} - \frac{1}{2} m_f \frac{\hat{\mathrm{d}}(\vec{\mathbf{v}}_p - \vec{\mathbf{v}})}{\hat{\mathrm{d}}t} \tag{1.4.25b}$$

and

$$\sum \vec{\mathbf{F}}_{surface} = \vec{\mathbf{F}}_{drag} + \vec{\mathbf{F}}_{pressure} + \vec{\mathbf{F}}_{Basset} + \vec{\mathbf{F}}_{lift} + \vec{\mathbf{F}}_{interaction} \tag{1.4.25c}$$

Equation (1.4.25a) is simply a form of Newton's second law where the volumetric and surface forces are actually "point forces" as specified in Eqs. (1.4.25b) and (1.4.26) to (1.4.31). Note that in these equations $\hat{d}/\hat{d}\,t = \partial/\partial t + \vec{\mathbf{v}}_p \cdot \nabla$, i.e., a time derivative following the moving particle; in contrast to the substantial derivative, $D/Dt = \partial/\partial t + \vec{\mathbf{v}} \cdot \nabla$, which follows a moving fluid element. However, when $Re_p = |u - \mathrm{v}| d_p / \nu << 1$, $D/Dt \approx \hat{d}/\hat{d}t = d/dt$. In Eq. (1.4.25b), the first term accounts for the inclusion of the buoyancy effects indicated by the difference in densities and the second is the virtual mass due to the inertia of the particle. The surface forces accounted for in Eq. (1.4.25c) are the viscous drag,

$$\vec{\mathbf{F}}_{drag} = m_p \Im\left(\vec{\mathbf{v}} - \vec{\mathbf{v}}_p\right),\ \Im = \frac{3}{4} \mathrm{C_D} \frac{\rho_\mathrm{f}}{\rho_\mathrm{p}} \frac{1}{d_p} |\vec{\mathbf{v}} - \vec{\mathbf{v}}_p| \tag{1.4.26a,b}$$

where $\Im$ is the inverse time constant of momentum transfer, and for solid spheres,

$$C_D = \begin{cases} 24/\mathrm{Re}_p & \text{for } 0 < \mathrm{Re}_p \le 1 \\ 24/\mathrm{Re}_p^{0.646} & \text{for } 1 < \mathrm{Re}_p \le 400 \end{cases} \tag{1.4.26c}$$

while for liquid spheres,

$$C_D = \frac{3.05\left(783\kappa^2 + 2142\kappa + 1080\right)}{(60 + 29\kappa)(4 + 3\kappa)} \mathrm{Re}_p^{-0.74} \tag{1.4.26d}$$

for $4 < \mathrm{Re}_p \le 100$ with $\kappa \equiv \mu_p/\mu$. The friction coefficient C_D is a function of the particle Reynolds number, $Re_p = \left|\mathrm{v} - \mathrm{v}_p\right| d_p / \nu$, $d_p = 2a$, assuming *uniform* Newtonian fluid flow. Thus, nonuniform shear field, especially near the wall, requires the addition of a correction, i.e., the Faxen force (Happel & Brenner, 1973). Hence,

$$\vec{\mathbf{F}}_{drag} = \underbrace{3\pi\mu d_p\left(\vec{\mathbf{v}} - \vec{\mathbf{v}}_p\right)}_{\text{Stokes}} + \underbrace{\mu\pi \frac{d_p^3}{8} \nabla^2 \vec{\mathbf{v}}}_{\text{Faxen}} \tag{1.4.26e}$$

where $\nabla^2 \vec{\mathbf{v}}$ is evaluated at the deposition of the particle. Clearly,

$$\frac{F_{Faxen}}{F_{Stokes}} \sim \left(\frac{d_p}{L}\right)^2 \tag{1.4.26f}$$

The pressure force,

$$\vec{\mathbf{F}}_{press} = -m_f \frac{D\vec{\mathbf{v}}}{Dt} \approx -V_p\left(\nabla p + \nabla \cdot \vec{\vec{\tau}}\right) \tag{1.4.27}$$

captures pressure gradient effects. The Basset force, or time history term,

$$\vec{\mathbf{F}}_{Basset} = 6\,\pi\, a^2\, \mu \int_{t_0}^{t} \frac{dt'}{\sqrt{\pi\, \mathrm{v}(t - t')}} \frac{\hat{d}(\vec{\mathbf{v}} - \vec{\mathbf{v}}_p)}{\hat{d}t'} \tag{1.4.28}$$

arises from the acceleration of the fluid around the particle, i.e., a temporal velocity changes with time (see $\left(\dot{\mathrm{v}} - \dot{\mathrm{v}}_p\right)$-factor). Considering the history term as an additional form of resistance, a ratio can be formed that determines the relative importance of the Basset force

$$\frac{F_{Basset}}{F_{Stokes}} = \sqrt{\frac{18}{\pi}\frac{\rho}{\rho_p}a_r} \tag{1.4.29}$$

where $a_r = \tau_p / \Delta t$ is the acceleration rate with $\tau_p = \rho_p d_p^2 / (18\mu)$ being the particle relaxation time. Clearly, when a_r <<1 and/or ρ/ρ_p <<1, the history term is negligible.

Lift forces are caused by particle rotation either induced by strong shear flow where velocity gradients generate particle rotation leading to top/bottom pressure differentials (cf. Saffman lift), or due to prerotation, wall contact, etc. (cf. Magnus lift in conjunction with a particle torque equation). Clearly, when particles are large, rotate, and/or are subjected to strong shear flows, as in near-wall regions, lift forces should be considered. For example, Li & Ahmadi (1995) suggested a Saffman lift force for supramicron particles in flows with high shear rates as

$$F_i^{lift} = 5.2 m_p \sqrt{\nu}\,\frac{\rho}{\rho_d}\frac{\left(\mathrm{v}_j - \mathrm{v}_j^p\right)\varepsilon_{ij}}{d_p\left(\varepsilon_{lk}\varepsilon_{kl}\right)^{1/4}} \tag{1.4.30a}$$

where the deformation-rate tensor is

$$\varepsilon_{ij} = \frac{1}{2}\left(\frac{\partial \mathrm{v}_i}{\partial x_j} + \frac{\partial \mathrm{v}_j}{\partial x_i}\right) \tag{1.4.30b}$$

Alternative forms for F_{lift} may be found in Crowe et al. (1998) and Fan & Zhu (1998).

In wall-bounded suspensions, a modified Stokes drag, called the "lubrication force," has to account for the influence of a boundary on wall-normal and wall-tangential particle motion (see Sect.5.2).

Interaction forces, F_{inter}, are due to random processes, i.e., particle collision, flow turbulence, and/or Brownian motion effects. For example, the bombardment of molecules of the carrier fluid on submicron particles is encapsulated in the Brownian force. It can be modeled as a Gaussian white noise random process where the amplitudes of the components at every time step, Δt, are (Li & Ahmadi, 1992)

$$F_i^{Brown} = G_i^{m_p}\sqrt{\frac{\pi S_0}{\Delta t}} \tag{1.4.31a}$$

where G_i are the zero-mean, unit variance-independent Gaussian random numbers; and the spectral intensity is

$$S_0 = 216\frac{\mu k T}{\pi^2 d_p^5 S^2 C_c} \tag{1.4.31b}$$

Here, T is the fluid temperature [K], $k = 1.38 \times 10^{-23}$ J/K is the Boltzmann constant, μ is the fluid viscosity, S is the ratio ρ_p / ρ, and C_c is the Cunningham slip correction factor

$$C_c = 1 + \frac{2\lambda}{d_p}\left[1.257 + 0.4\exp\left(-1.1 d_p / 2\lambda\right)\right] \tag{1.4.31c}$$

i.e., $C_c \approx 1.0$ when $d_p >> \lambda$, the molecular mean free path.

In *turbulent flows*, an individual particle trajectory or dispersion of fine particle clouds is measurably influenced by the random velocity fluctuations of the continuous phase. Rather than creating a separate turbulent fluid-particle interaction force, turbulence effects are incorporated in the fluid velocity vector. Specifically, when employing Reynolds' decomposition of the instantaneous fluid velocity, i.e., $\vec{\mathbf{v}} = \bar{\vec{\mathbf{v}}} + \vec{\mathbf{v}}'$, only the time-smoothed or bulk flow velocity vector is calculated (cf. Sect. 1.3.4). This implies that either $\vec{\mathbf{v}}'$ or $\vec{\mathbf{v}}$ has to be recorded and integrated into a particle trajectory/dispersion equation in order to simulate the stochastic process of particle motion in a turbulent flow field. For example, the fluctuating components of $\vec{\mathbf{v}}'$ can be expressed as (Gosman & Ioanuides, 1981; Katz & Mortonen, 1999)

$$\vec{\mathbf{v}}' \equiv \mathbf{v}_i' = \lambda \left(\frac{2}{3} k_i\right)^{1/2} \tag{1.4.32}$$

where λ is a random number, i.e., $-1 \le \lambda \le 1$; and $k \equiv 1/2\overline{u_i'^2}$ is the turbulence kinetic energy. Now, the time-smoothed fluid velocity $\vec{\mathbf{v}} := \bar{\mathbf{v}}_i$ is augmented with Eq. (1.4.32) to form the instantaneous velocity $\bar{\mathbf{v}}_i + \mathbf{v}_i' = \mathbf{v}_i$. Alternatively, the instantaneous fluid velocity is (directly) obtained from a stochastic equation (cf. He & Ahmadi, 1999):

$$\frac{d\mathbf{v}_i}{dt} = -\frac{\mathbf{v}_i - \bar{\mathbf{v}}_i}{\tau_I} + \left(\frac{2\overline{\mathbf{v}'^2_{(i)}}}{\tau_I}\right)^{1/2} \varsigma_i(t) \tag{1.4.33}$$

where $\tau_I \approx 0.3k/\varepsilon$ is the particle integral time, here related to the Lagrangian fluid element integral time, which depends on the turbulence kinetic energy k and dissipation ε; $\overline{\mathbf{v}'^2_{(i)}}$ is the mean-square of the ith fluctuation velocity, i.e., the turbulence intensity, which has to be accurately simulated (or measured), especially the normal component; and $\varsigma_i(t)$ is a Gaussian vector white noise random process given by $\varsigma_i(t) = G_i / \sqrt{\Delta t}$ at every time step Δt with G_i being zero-mean, unit variance-independent Gaussian random numbers.

Another (random) interaction force is due to particle-particle collisions. Numerous models have been postulated (cf. Fan & Zhu, 1998; Crowe et al., 1998, among others); one rather simple approach given here is to construct an expression for $\vec{\mathbf{F}}_{collision}$ similar to $\vec{\mathbf{F}}_{drag}$, i.e.,

$$\vec{\mathbf{F}}_{collision} = C_I A_s \frac{\rho}{2}\left(\vec{\mathbf{v}}_p - \vec{\mathbf{v}}\right)\left|\vec{\mathbf{v}}_p - \vec{\mathbf{v}}\right| \tag{1.4.34}$$

where C_I is a semiempirical interaction coefficient that depends mainly on the probability of particle collision and intraparticle collision impact. Specifically, $C_I = C_I\left(\lambda, I_{n,t}\right)$, where $0 \leq \lambda \leq 1$ is a random number and the normal/tangential impact $I_{n,t}$ can be obtained, for example, from the "soft sphere" collision model (Crowe et al., 1996).

Alternatively, assuming appropriate (collisional) probability density functions, kinetic theory formalism can be applied to account for interparticle collisions in the model approach of the dispersed phase. However, particle and molecule systems have basic differences as discussed elsewhere (cf. Bird, 1994).

Simplified particle equation. Obtaining a numerical solution to Eq. (1.4.25) is a very challenging problem especially if the Basset (i.e., history) term must be retained. Fortunately for many problems Eq. (1.4.25) can be substantially reduced. As pointed out by Clift et al. (1978), if the unsteady motion is of type I, characterized by rapid changes in Reynolds number with the displacement modulus, then the transient effect terms (i.e., Basset and virtual mass) should be retained in Eq. (1.4.25) since the instantaneous drag may differ radically from the corresponding steady drag. This condition is commonly found in liquid-particle systems where the density ratio is low to moderate, e.g., bubbly flows. For unsteady motion of type II (characterized by slow changes in Reynolds number with the displacement modulus), it is generally acceptable to drop the transient effect terms such as the Basset and virtual mass terms in Eq. (1.4.25), thus greatly simplifying the solution process. In practice the type II condition usually corresponds to gas-particle systems with their corresponding high density ratios. As a result, the virtual mass, pressure, and Basset forces are often negligible due to the relative density ratios, while the interaction force is neglected if the suspension is dilute. Also Brownian motion may be ignored for the particles with effective diameters $d_p > 1$ μm, especially in "dirty gas" flow. Thus, Eq. (1.4.25) can be reduced to the following form:

$$m_p \frac{d^2\vec{\mathbf{x}}_p}{dt^2} = \sum\vec{\mathbf{F}}_p = \sum\vec{\mathbf{F}}_{body} + \sum\vec{\mathbf{F}}_{surface} \tag{1.4.35a}$$

where

$$\sum \vec{\mathbf{F}}_{surface} = \vec{\mathbf{F}}_{drag} = \frac{1}{8}\pi\rho d_p^2 C_{Dp}\left(\vec{\mathbf{v}} - \vec{\mathbf{v}}_p\right)\left|\vec{\mathbf{v}} - \vec{\mathbf{v}}_p\right| \qquad (1.4.35b)$$

and

$$\sum \vec{\mathbf{F}}_{body} = \vec{\mathbf{F}}_{buoyancy} = \frac{1}{6}\pi d_p^3\left(\rho_p - \rho\right)\vec{\mathbf{g}} \qquad (1.4.35c)$$

For laminar flows the carrier fluid velocity vector is $\vec{\mathbf{v}}$. In case of turbulent flows, particle dispersion has to be considered via an instantaneous system velocity $\vec{\mathbf{v}}$ defined as the sum of the mean velocity and a randomly fluctuating velocity component, as discussed (see also Sects. 3.1.4.1 and 5.3).

Example 1.16: Accelerating Particle in Uniform Steady Flow

Sketch:

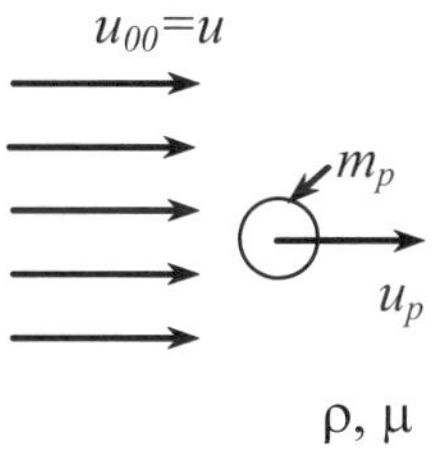

Assumptions:

- steady unidirectional flow w/constant properties
- spherical particle
- drag dominated particle motion

Concepts:

- carrier fluid $\vec{\mathbf{v}} = (u,0,0)$ where u=¢
- particle dynamics $m_p \frac{du_p}{dt} = F_D$

Solution:

Equation (1.4.35) can be reduced to

- $$m_p \frac{du_p}{dt} = F_D = \frac{\rho}{2} A C_D \left(u - u_p\right)\left|u - u_p\right| \qquad (E1.16\text{-}1)$$

Here, $C_D = 24/\mathrm{Re}_d\left(1 + 0.15 \cdot \mathrm{Re}_d^{0.687}\right)$, where $\mathrm{Re}_d = \rho|u_{rel}|d/\mu$ and $u_{rel} \equiv u_r = u - u_p$. Now, in terms of u_r, noting that $du_r/dt = -du_p/dt$, Eq. (E1.16-1) reads with $A = d_p^2\pi/4$ and $m_p = \rho_p \pi d_p^3/6$,

$$-\frac{du_r}{dt} = \frac{18\mu}{\rho_p d_p^2}\left[1 + 0.15\left(\frac{\rho d}{\mu}\right)^{0.687} u_r^{0.687}\right] u_r^2 \qquad (E1.16\text{-}2)$$

with the particle initially at rest, $u_r(t=0)=u$. Integration of this nonlinear ODE after separation of variables yields (see graph)

$$u_r \equiv u - u_p = \left[\left(a + u_r^{-0.687}\right)\cdot \exp(0.687 b\, t) - a\right]^{-1/0.687} \quad \text{(E1.16-3)}$$

where

$$a = 0.15\left(\frac{\rho d}{\mu}\right)^{0.687} \quad \text{and} \quad b = \frac{18\mu}{\rho_p d_p^2}$$

Graph:

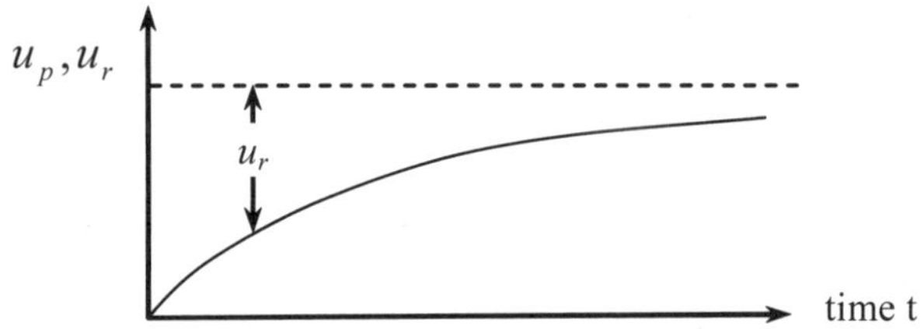

Example 1.17: Particle Trajectory in a Recirculation Zone (Karino & Goldsmith, 1977)

Sketch:

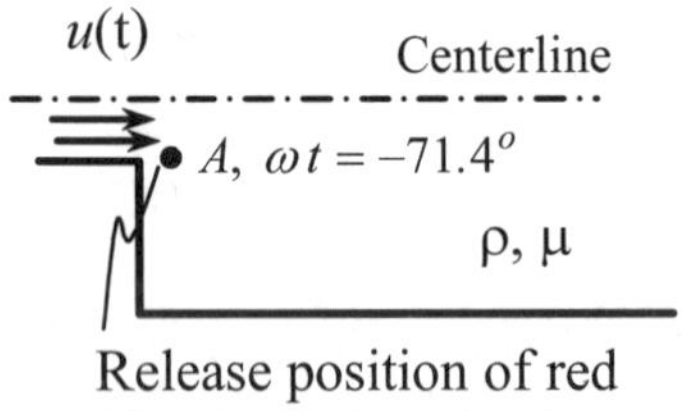

Release position of red blood cell A with phase angle –71.4° of a sinusoidal input pulse

Assumptions:

- transient axisymmetric internal flow with constant properties
- spherical particle, i.e., hardened human red blood cell (RBC)
- drag dominated particle motion

Concepts:

- carrier fluid $\vec{v}(t) = (u(t), v(t), 0)$
- 2-D particle motion due to fluid flow induced "point forces," i.e., pressure, drag, and "virtual mass."

Solution:

A one-way coupled point-force model can be applied to simulate the trajectory of the particle in the annular expansion due to the size of the particle in relation to the tubular diameter, i.e.,

$$d_p / D << 1$$

Larger irregular particles, such as an aggregate of blood cells, may not be sufficiently small to have a negligible effect on the surrounding flow field. For example, Karino and Goldsmith (1977) also studied the motion of a 55-mm aggregate in the 504-mm-diameter expansion. Resolved-volume techniques may be required to simulate such a motion.

Properties of the RBC include a particle response time of

$$\tau_p \equiv \frac{2\rho_d a^2}{9\mu_c} = 3.531 \times 10^{-6} \qquad \text{(E1.17-1)}$$

and a continuous to dispersed phase density ratio

$$\kappa \hat{=} \frac{\bar{\rho}_c}{\bar{\rho}_d} = 0.885 \qquad \text{(E1.17-2)}$$

The governing dimensional equation, assuming only pressure and drag effects, can be written as [cf. Sect. 1.4.3.1, i.e., Eq.(1.4.25a) with Eqs. (1.4.26) and (1.4.27)]:

$$\frac{d\mathrm{v}_i}{dt} = \kappa \frac{Du_i}{Dt} + \frac{1}{\tau_p}\left(u_i - \mathrm{v}_i\right) \qquad \text{(E1.17-3a)}$$

where v_i and u_i are the components of particle and local fluid element velocity, respectively. If virtual mass effects are to be included, the above equation can be written as

$$\frac{d\mathrm{v}_i}{dt} = \frac{3}{2}\beta\kappa \frac{Du_i}{Dt} + \frac{\beta}{\tau_p}\left(u_i - \mathrm{v}_i\right) \qquad \text{(E1.17-3b)}$$

where

$$\beta = \frac{1}{(1+\kappa)/2} := 0.693 \qquad \text{(E1.17-4)}$$

<u>Solution:</u> Particle A (red blood cell) trajectory plots considering different point forces.

Using CFX-4.3 on an engineering workstation, the results of sequentially adding drag, pressure, and virtual mass terms over separate runs are illustrated in the *Graph*. The first trajectory, assuming fluid element *pathline* track, is nearly identical to the RBC trajectory of Karino & Goldsmith (1977). Then, the inclusion of only the drag term under identical release conditions reduces the number of particle loops by one and slightly modifies the location at which the particle ends its recirculating trajectory. Furthermore, the computational effort required has increased 40 times, as indicated in the trajectory plots by the number of time steps required by the quality control integration routine. Including both pressure and drag effects modifies the particle trajectory such that it again very nearly resembles that of the RBC. Interestingly, both pressure and drag terms have an influence on particle trajectory; however, their net effect is essentially represented by the motion of a pathline for this system, i.e., these terms apparently cancel one another.

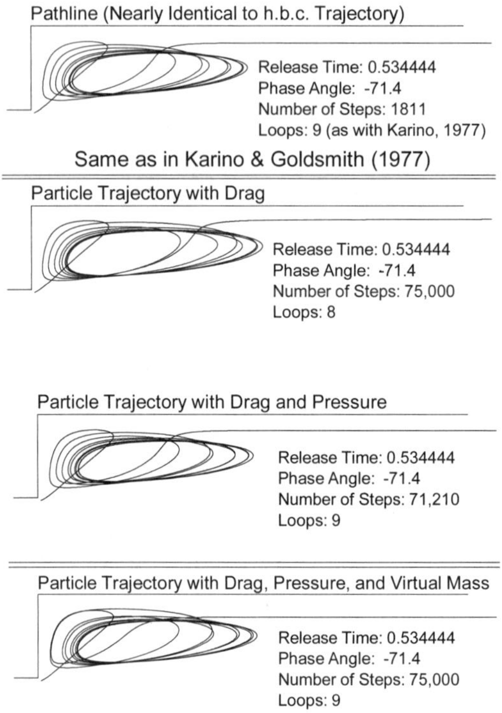

1.4.3.2 Species Mass Transfer

When submicron particles are convected by the carrier fluid, e.g., inhaled air, an Eulerian-Eulerian approach should be considered. For example, the general dynamics equation, requiring knowledge of the continuous phase velocity field u_j, particle diffusivity D^M, and particle sink/source term Q^M, can be written as [cf. Eq.(1.3.36)]:

$$\frac{\partial}{\partial t}(\rho M_k)+\frac{\partial}{\partial x_j}(\rho M_k u_j)=\frac{\partial}{\partial x_j}\left[D_k^M\frac{\partial(\rho M_k)}{\partial x_j}\right]+Q_k^M \tag{1.4.36}$$

where M_k represent the number (or fraction) of particles in section k, $k = 1, 2, \ldots, n$, i.e., the suspension is categorized into n domains according to the given particle size distribution. Clearly, for monodisperse nano-particles k=1.

Extending Eq (1.4.36) to turbulent flow, it can also be written for ultrafine particles (or vapor) in more familiar form as the convective-dispersion equation, i.e.,

$$\frac{\partial C}{\partial t}+u_j\frac{\partial C}{\partial x_j}=\frac{\partial}{\partial x_j}\left((D_p+\frac{D_T}{Pr_T})\frac{\partial C}{\partial x_j}\right) \tag{1.4.37a}$$

where C is the particle concentration, D_T is the turbulent eddy viscosity, Pr_T=0.9 is the turbulent Prandtl number, and the particle diffusivity is

$$D_p=\frac{k_B T C_{slip}}{3\pi\mu d_p} \tag{1.4.37b}$$

Here, k_B=1.38×10^{-23} J/K, C_{slip} is the Cunningham slip correction factor (Clift et al., 1978), and d_p is the particle diameter.

Applications of Sect. 1.4.3 material may be found in Sects. 3.3 and 5.6, while "two-fluid" modeling, i.e., the interaction of two separate phases, is discussed by Kleinstreuer (2003).

1.4.4 Porous Media Flow

Flow equations. A porous medium can be a solid structure with continuous interconnected conduits, particles packed or suspended in a container, geologic material such as sand, shale, limestone, clay, soil, and pebbles, as well as *tissue* modeled as a fiber matrix, sponge-like structure, or granular structure. Thus, a proper analysis of "flow through porous media" is important in chemical/petroleum engineering, environmental/civil engineering, and mechanical/biomedical engineering. Nevertheless, the reader should keep in mind that the use of Darcy's law in biomedical engineering is controversial. Although it provides mathematical convenience in describing, for example, tissue perfusion, it ignores the intricacies of different living tissues.

As with all internal flows, the pressure drop and the resulting velocity field (or flow rate) are the key unknowns. Clearly, fixed or suspended particles in the flow path may greatly increase the resistance encountered by the pressure- or gravity-driven fluid. Specifically, a functional relationship for the dimensionless "pressure gradient" can be stated as

$$\left(-\frac{\Delta p}{l}\right)\frac{d_p}{\rho u_0^2} = \text{fct}\left(Re, \alpha, \xi, \varphi, \frac{e}{d_p}, etc.\right) \tag{1.4.38}$$

where d_p is the effective particle (or cell) diameter, $Re = \frac{u_0 d_p}{\nu}$, $\xi = \frac{d_p}{e_{system}}$, α is the medium porosity (or void fraction), φ is the sphericity, and e/d_p is the relative surface roughness. With $d_p \ll 1$ or the equivalent conduit or open passage hydraulic diameter $D_h \ll 1, Re \to 0$ and Stokes flow can be assumed [see Eq.(1.3.22)], i.e.,

$$\partial(\alpha\rho)/\partial t + \nabla \cdot (\alpha\rho\vec{\mathbf{v}}) = 0 \quad \text{and} \quad \rho\frac{\partial\vec{\mathbf{v}}}{\partial t} = -\nabla p + \mu\nabla^2\vec{\mathbf{v}} + \rho\vec{\mathbf{g}} \qquad (1.4.39a,b)$$

A simpler version of the momentum equation for steady incompressible flow through an isotropic porous medium was suggested by Darcy in 1856, i.e., in 3-D it reads

$$\vec{\mathbf{v}}_0 = -\frac{k}{\mu}(\nabla p - \rho\vec{\mathbf{g}}) \qquad (1.4.40)$$

where $\vec{\mathbf{v}}_0$ is the superficial velocity vector which is averaged over a local, relatively small volume containing both the solid matrix and the fluid, and $k/\mu \equiv K$ is the hydraulic conductivity, where k is the medium permeability.

Example 1.18: Evaluation of the Hydraulic Conductivity $K = k/\mu$

Assuming a porous medium to be a solid structure with n parallel tubes, employ the Hagen-Poiseuille result for steady laminar pipe flow to estimate K.

Sketch:

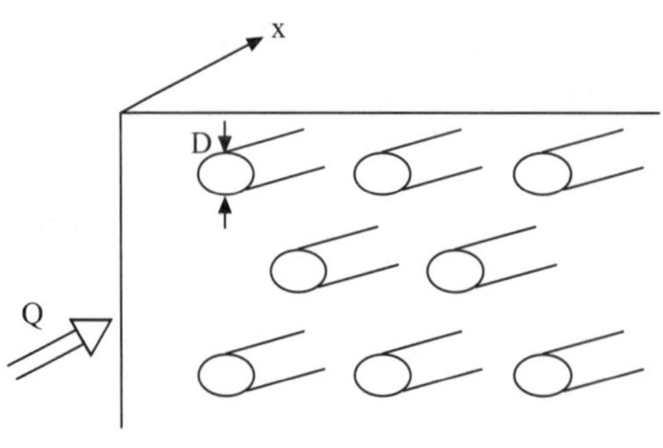

Concept:

Compare steady laminar fully-developed pipe flow result with Darcy's law in 1-D, i.e.,

$$\frac{Q}{A} = v_0 = -K\frac{dp}{dx}$$

<u>Solution:</u>

Considering Poiseuille flow (cf. Example 1.6), i.e.,

- $$v(r) = v_{max}\left[1 - \left(\frac{r}{r_0}\right)^2\right], \quad v_{max} = -\frac{r_0^2}{4\mu}\left(\frac{dp}{dx}\right),$$

we obtain for the volumetric flow rate per pore/tube:

$$\frac{Q}{n} = 2\pi\int_0^{D/2} v r dr = -\frac{\pi D^4}{128\mu}\left(\frac{dp}{dx}\right)$$

Thus, with $v_0 = -K\frac{dp}{dx} = \frac{Q}{A}$, $A = n\frac{D^2\pi}{4}$, we set $Q_{H-P} = Q_{Darcy}$ and the hydraulic conductivity turns out to be

- $$K = \frac{D^2}{32\mu} \tag{E1.18-1}$$

In order to solve Eq.(1.4.39), all fluid passway geometries have to be known as well as inlet/outlet conditions and the conduit wall roughness. Further complications may arise when the solid phase moves, the void spaces containing two fluids (e.g., air and water as in unsaturated geologic media) or a non-Newtonian fluid. When the Reynolds number is greater than one, nonlinear inertia effects appear and when $Re \geq 100$, the flow is turbulent.

Extended Darcy equation. Historically, such insurmountable challenges were answered with (one-dimensional) empirical correlations, starting with Eq. (1.4.40) which was extended by Forchheimer in 1901 and Brinkman in 1947. Specifically, for steady fully-developed laminar flow in a 2-D isotropic porous medium

$$\frac{d\hat{p}}{dx} = \mu\frac{d^2u_0}{dy^2} - \frac{\mu}{k}u_0 + \frac{\rho C}{\sqrt{k}}u_0^2 \tag{1.4.41}$$

[Stokes/Brinkman] [Darcy] [Forchheimer]

where $\hat{p} = p + \rho g x$, u_0 is the volume-averaged (i.e., superficial) axial velocity, k is the medium-dependent permeability and the empirical factor C varies from 0.1 to 0.7, where $C = 0.55$ is typical (see Nield & Bejan, 1999).

<u>Example 1.19:</u> Stokes' Flow in Interstitial Spaces

Consider a porous-medium channel, e.g., the uniform space between two cells where lymph flows through an extracellular matrix.

Sketch:

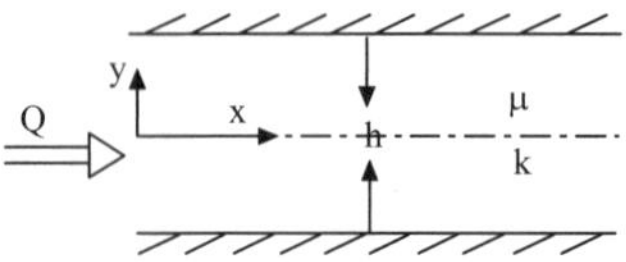

Assume:
- Creeping flow, i.e., $u_0=u_0(y)$ only
- Constant pressure gradient
- Constant properties

Concept:

Use of Brinkman equation to invoke no-slip wall conditions

Solution:

Equation (1.4.41) can be written without the "high-speed" Forchheimer term as :

- $$\frac{dp}{dx} = \mu\frac{d^2u_0}{dy^2} - \frac{\mu}{k}u_0 = \text{constant} \tag{E1.19-1a}$$

subject to

$$u_0(y = h/2) = 0 \text{ and } \left.\frac{du_0}{dy}\right|_{y=0} = 0 \tag{E1.19-1b,c}$$

This linear, inhomogeneous second-order ODE with constant coefficients has the solution

- $$u_0(y) = -\frac{k}{\mu}\left(\frac{dp}{dx}\right)\left[1 - \frac{\cosh\left(\frac{y}{\sqrt{k}}\right)}{\cosh\left(\frac{h}{2\sqrt{k}}\right)}\right] \tag{E1.19-2}$$

The flow rate per unit depth $\hat{Q} = 2\int_0^{h/2} u_0 dy$ is

- $$u_0(y) = -\frac{kh}{\mu}\left(\frac{dp}{dx}\right)\left[1 - \frac{2\sqrt{k}}{h}\tanh\left(\frac{h}{2\sqrt{k}}\right)\right] \tag{E1.19-3}$$

and the dimensionless *effective* hydraulic conductivity is

- $$\frac{K}{k/\mu} \equiv \frac{\overline{u_0}}{dp/dx} = 1 - \frac{2\sqrt{k}}{h}\tanh\left(\frac{h}{2\sqrt{k}}\right) \tag{E1.19-4}$$

<u>Graphs</u>:

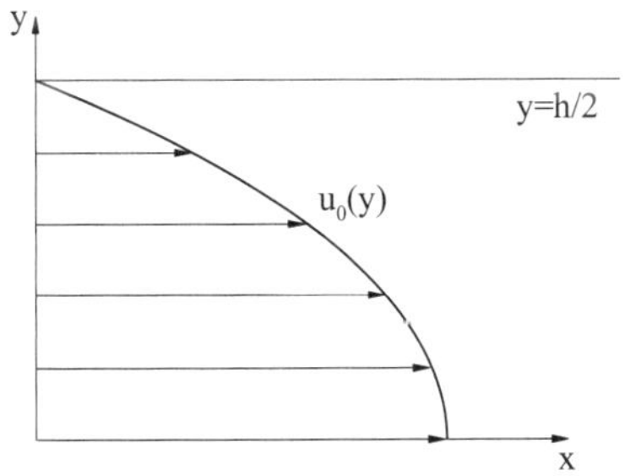

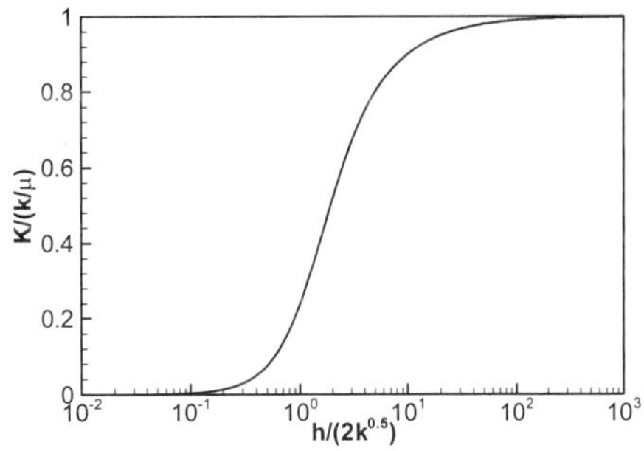

A celebrated special case of Eq. (1.4.40) is the Carman-Kozeny equation describing laminar ($Re \leq 1$) fully-developed (Poiseuille-type) flow in a tube filled with spheres of diameter d_p, creating a constant void fraction α:

$$\frac{\Delta\hat{p}}{L} = \frac{180\mu(1-\alpha)^2}{(\alpha d_p)^2}u_0 \qquad (1.4.42)$$

where $\hat{p} = p + \rho g L - p_0$, L being the (vertical) tube length, and $u_0 = Q/A_{tube} = \alpha u_n$ is the superficial velocity (or volumetric flux density) while u_n is the corresponding intrinsic average or mean velocity through the pores, i.e., void space, passages, or channels.

For a packed bed of spheres, permeability, k, and porosity, α, have been related as

$$k = \frac{d_p^2\alpha^3}{180(1-\alpha)^2} \qquad (1.4.43)$$

where it has to be recalled that the porosity at a wall may double.

In turbulent flow through packed beds, the pressure drop is largely due to inertial effects which leads to the Burke-Plummer equation

$$\frac{-\Delta\hat{p}}{L}\left(\frac{\alpha^3}{1-\alpha}\right)\frac{d_p}{\rho u_0^2} = C_f \qquad (1.4.44)$$

where the friction coefficient is in the range $1.2 \le C_f < 4.0$. Ergun (1952) combined Eqs.(1.4.42) and (1.4.44), replacing the factor 180 by 150 and setting $C_f = 1.75$, after successful comparisons with experimental data sets, i.e.,

$$\left(\frac{-\Delta\hat{p}}{L}\right)\left(\frac{\alpha^3}{1-\alpha}\right)\frac{d_p}{\rho u_0^2} = 150\frac{\mu(1-\alpha)}{\rho u_0 d_p} + 1.75 \tag{1.4.45}$$

Example 1.20: Radial Flow through a Porous-Walled Tube

Consider pressure-driven flow through a porous-walled tube; for example, seepage into a lymph vessel. Find the *p(r)* and $v_0(r)$ profiles as well as the added mass flow rate over the tube length *L*.

Sketch:

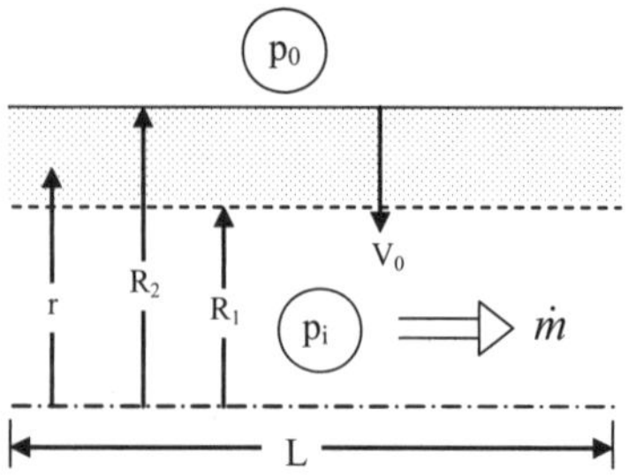

Assumptions:

- steady radial seepage through $R_1 \le r \le R_2$
- constant pressures and system properties

Concept:

- Darcy's equation
- mass balance

Solution: Taking the divergence of Eq. (1.4.40), i.e.,

$\nabla \cdot \vec{\mathbf{v}}_0 = -\frac{k}{\mu}\nabla^2 p$. and recalling that for an incompressible fluid $\nabla \cdot \vec{\mathbf{v}}_0 = 0$, we have to solve in cylindrical coordinates:

$$\frac{1}{r}\frac{d}{dr}\left(r\frac{dp}{dr}\right) = 0 \tag{E1.20.1}$$

subject to $p(r = R_1) = p_i$ and $p(r = R_2) = p_o$. Double integration yields

$$p(r) = C_1 \ln r + C_2$$

and ultimately

• $$\frac{p - p_i}{p_o - p_i} = \frac{\ln(r/R_1)}{\ln(R_2/R_1)} \tag{E1.20.2}$$

For purely radial flow $\vec{\mathbf{v}}_0 = (0, \mathrm{v}_0, 0)$ and Eq. (1.4.40) reduces to

$$\mathrm{v}_0 = -\frac{k}{\mu}\left(\frac{dp}{dr}\right) \tag{E1.20.3}$$

so that with Eq. (E1.20.2)

• $$\mathrm{v}_0 = -\frac{k}{\mu}\frac{\Delta p}{\ln(R_2/R_1)}\frac{1}{r} \tag{E1.20.4}$$

A 1-D (radial) mass balance yields

$$\Delta \dot{m} = -\rho \mathrm{v}_0 A_{surface}$$

where $A_{surf} = 2\pi R_1 L$, so that

• $$\Delta \dot{m} = \frac{2\pi k \rho \Delta p}{\mu \ln(R_2/R_1)} L \tag{E1.20.5}$$

Heat and mass transfer in porous media. For a volume-averaged homogeneous medium where the solid-phase temperature, T_s, differs locally from the fluid temperature, T_f, two heat transfer equations can be written as [see Eq. (1.3.35) and HWA in Sect. 1.7]:

$$(1-\alpha)\left(\rho c \frac{\partial T}{\partial t}\right)_s = (1-\alpha)\nabla \cdot (k \nabla T)_s + h \frac{A_{s-f}}{\forall_M}\left(T_f - T_s\right) + (1-\alpha)\dot{q}_s \tag{1.4.46a}$$

$$\alpha\left(\rho c_p \frac{\partial T}{\partial t}\right)_f + (\rho c_p \vec{\mathbf{v}}_0 \cdot \nabla T)_f = \alpha \nabla \cdot (k \nabla T)_f + h \frac{A_{s-f}}{\forall_M}\left(T_s - T_f\right) + \alpha \dot{q}_f \tag{1.4.46b}$$

where α is the porosity, h is the convective heat transfer coefficient, A_{s-f} is the solid-fluid contact area, $\forall_M$ is the porous medium volume, and $\dot{q}$ is the local heat generation rate per unit volume $[J/(m^3 \cdot s)]$. Clearly, determining system parameters such as α, h, and A_{s-f} for, say, soft tissue constitutes a major challenge. Recent porous medium applications may be found in Alazmi & Vafai (2002), Nield et al. (2002), and Nakayama et al. (2001).

Similar to the scalar equations (1.4.46b and 1.4.37), the species mass transfer equation describing transient convection, diffusion, and conversion can be written as

$$\alpha \frac{\partial c}{\partial t} + (\vec{\mathbf{v}}_0 \cdot \nabla) c = -\nabla \cdot \vec{\mathbf{j}} + \sum S_c \tag{1.4.47}$$

where c is the solute concentration of the homogeneous mixture flowing [Eq. (1.4.40)] in a porous medium of porosity α, $\vec{\mathbf{j}}$ is the flux vector, accommodating potentially (Bird et al., 2002):

- Fickian diffusion: $\vec{\mathbf{j}} = -D\nabla c$ (1.4.48a)
- Turbulent dispersion: $\vec{\mathbf{j}} = -D_T \nabla c$ (1.4.48b)
- Soret mass flux due to temperature gradients:

$$\vec{\mathbf{j}} = -D_S \nabla T \tag{1.4.48c}$$

- Nernst-Planck ion mass flux due to an electrical potential gradient across a barrier B:

$$\vec{\mathbf{j}} = -D_B \frac{zF}{RT} c \nabla \psi \tag{1.4.48d}$$

For solute flux across a neutral (i.e., noncharged) biomembrane (or filter), Starling's hypothesis (Michel, 1997) could be formulated as

$$\vec{\mathbf{j}}_s = (1-\sigma)\vec{\mathbf{v}}_0 \bar{c}_s - D_M \nabla c_s \tag{1.4.49}$$

where σ is the reflection coefficient [$0 \le \sigma \le 1$; $\sigma = 0$ no restriction for solute passage through pores and $\sigma = 1.0$ rejection of solute passage], $\vec{\mathbf{v}} = -\frac{k}{\mu}(\nabla p - \sigma \Pi)$ is the average mixture velocity, p is the fluid static pressure and Π is the osmotic pressure, $\bar{c}_s$ is the average solute concentration within the membrane, and D_M is the membrane diffusion coefficient, where D_M=æD_p with æ<1 and D_p given as Eq. (1.4.37b).

<u>*Example 1.21*</u>: 1-D Diffusion across a Neutral Biomembrane

The selective exchange of material between fluids and compartments occurs through two types of biomembranes, i.e., plasma (or cell) membranes and capillary walls. Both are very thin (w = 7 -500 nm) with tiny pores ($d \leq 6nm$) where solute and water transport are caused by diffusion, osmosis, static pressure difference, electro-chemical potential, etc.

Consider transient 1-D diffusion across a thin neutral membrane and develop an equation to obtain the membrane diffusion coefficient for different solute-membrane conditions.

Sketch:

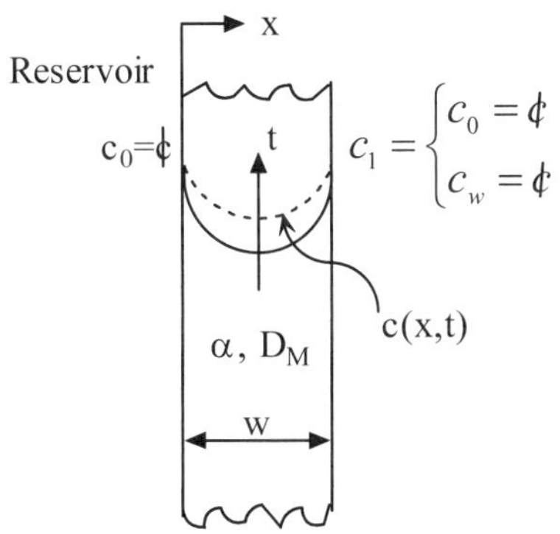

Assumptions:

- Simple 1-D diffusion only
- Constant properties
- No solute sinks or sources inside membranes

Approach:

- Analytical solution of the diffusion equation
- Plotting of c/c_0 vs. time

Solutions: Based on the assumptions, Eq. (1.4.47) with Eq. (1.4.48a) can be reduced.

(a) Analytic solution when $c(x = 0, w) = c_0 = const$:

$$\alpha\frac{\partial c}{\partial t} = D_M\frac{\partial^2 c}{\partial x^2} \tag{E1.21.1}$$

subject to $c(t = 0) = 0$ for $0 < x < w$, $c(x = 0, w) = c_0$, and $\partial c/\partial x = 0$ at $x = w/2$.

Separation-of-variables method yields (Özisik, 1993 or Gates & Newman, 2000):

$$\frac{c(x, t)}{c_0} = 1 + \sum_{n=1}^{\infty}\left[\frac{\cos n\pi - 1}{n\pi}\right]\sin\left(\frac{n\pi x}{w}\right)\exp\left(-\frac{n^2\pi^2 D_M}{\alpha w^2}t\right) \tag{E1.21.2}$$

<u>*Graph:*</u>

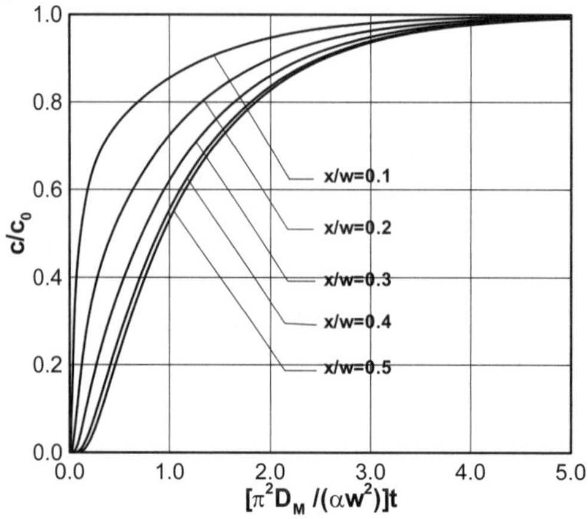

Once $c(x, t)/c_0$ is known, we can read $\frac{\pi^2 D_M}{\alpha w^2}t$ from the chart and then calculate the diffusion coefficient D_M.

(b) Approximate solution when $c(x = 0) = c_0(t)$ and $c(x = w) = c_w(t)$:

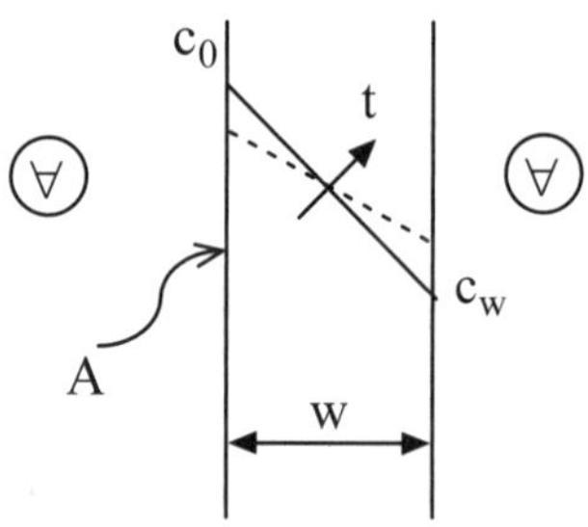

For thin membranes the solute flux can be linearized as $j_x = D_M(c_0 - c_w)/w$ so that for a 1-D mass balance, i.e.,

$$-\forall \frac{\partial c_0}{\partial t}\bigg|_{x=0} = \underbrace{AD_M \frac{c_0 - c_w}{w}}_{0<x<w} = \forall \frac{\partial c_w}{\partial t}\bigg|_{x=w}$$

so that with $A/\forall := w$ <system length scale>

$$\frac{d}{dt}(c_0 - c_w) = -2\frac{A}{\forall} j_x = -2\frac{D_M}{w}\frac{c_0 - c_w}{w} \qquad \text{(E1.21-3a,b)}$$

Integration, subject to $c_0(t = 0) = c_{initial} = ¢$ and $c_w(t = 0) = 0$, yields

$$c_0(t) = \frac{c_i}{2}\left[1 + \exp\left(-\frac{2D_M}{w}t\right)\right] \qquad \text{(E1.21-4)}$$

Equation (E1.21-4) indicates a characteristic diffusion time for the membrane

$$t_M = w^2 / D_M \tag{E1.21-5a}$$

whereas the characteristic times for solute concentration changes at both sides of the membrane are

$$t_0 = t_w = \frac{\forall}{2\forall_M} t_M \tag{E1.21-5b}$$

where $\forall >> \forall_M$ and hence $t_M \ll 1$.

<u>Graph:</u>

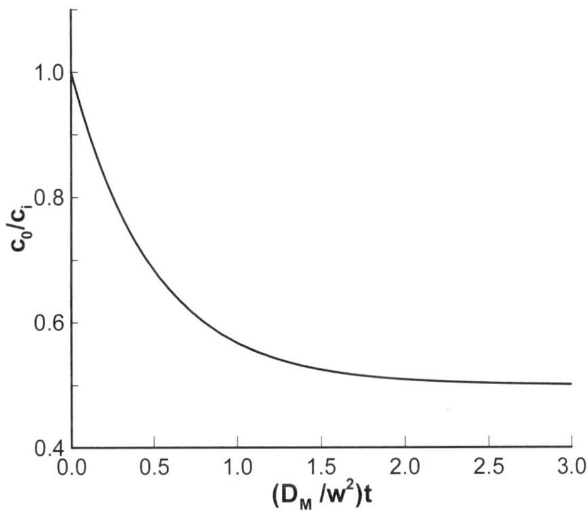

Again, once $c_0(t)/c_i$ is known, we can read $\frac{D_M}{w^2}t$ from the chart and then calculate the diffusion coefficient D_M.

1.5 BIOMECHANICS REVIEW

Fluid dynamics theory applied to biological systems or biomedical devices cannot ignore the frequent coupling of momentum transfer with structural mechanics. Examples include pulsatile blood flow interacting with moving vessels, contracting/expanding heart chambers as well as opening/closing heart valves, joint lubrication and deformable cartilage, airflow in flexible bronchioles, lymph flow and tissue motion, blood flow interacting with implants (e.g., stents, vessel inserts, artificial valves, synthetic grafts, etc.), and fluid flow interacting with moving parts of medical devices (e.g, pumps, filters, valves, drug delivery systems, etc.).

Most of the biological structures, e.g., blood vessels and airways, are made out of layered viscoelastic materials, while soft tissue may experience finite strain and hence relatively large deformations (see Fig.1.5.4). However, as a first approximation it is reasonable to model some materials as anisotropic, linearly elastic solids with Hooke's law as the appropriate constitutive equation. More realistic models are introduced and applied in Sect. 5.4 where nonlinear interactions between blood flow, stent-graft, stagnant cavity blood, and arterial wall are discussed. In this section, elements of fluid kinematics and dynamics of Sects. 1.2 and 1.3 are extended with a review of solid deformation (strain) caused by forces (stress), acting on body continua. Continuum mechanics books catering to biomedical engineers include Fung (1994) and Humphrey & Delange (2004).

1.5.1 Introduction

As mentioned, fluid-structure interactions (FSI) are an integral part of most transport phenomena in biological systems, ranging from solute transfer across cell membranes to heart-valve actions, heart-lung bypass machines, and low-drag body motion, to name a few examples.

In this section, the basic solid mechanics equations are reviewed which, in conjunction with the conservation laws (Sect. 1.3), prepare to solve FSI problems. Typically, a pressure due to fluid flow generates a load on a structure which results in wall stress, strain, and deformation (see Sect. 5.4). Other FSI examples are dilute suspension flow across a deflecting cell membrane, plasma-cell migration into a moving arterial wall, and ultrafine particle deposition in the expanding/contracting alveoli. Clearly, under the umbrella of continuum mechanics, material of Sects. 1.3 and 1.4 may be interactively coupled with the material of Sect. 1.5.

1.5.2 Principal Stresses

As reviewed in Sect. 1.3.2, the state of material interaction at any point in a body is specified by the stress tensor

$$\sigma_{ij} = \begin{bmatrix} \sigma_{11} & \sigma_{12} & \sigma_{13} \\ \sigma_{21} & \sigma_{22} & \sigma_{23} \\ \sigma_{31} & \sigma_{32} & \sigma_{33} \end{bmatrix} := \begin{bmatrix} \sigma_x & \tau_{xy} & \tau_{xz} \\ \tau_{yx} & \sigma_y & \tau_{yz} \\ \tau_{zx} & \tau_{zy} & \sigma_z \end{bmatrix} \qquad (1.5.1a,b)$$

where σ_{ii} are the normal stresses, i.e., generated by forces F_i perpendicular to surfaces A_i, and the shearing stresses $\sigma_{ij} \hat{=} \tau_{ij}$ are caused by tangential forces F_j acting on surfaces A_i with normal vector $\hat{n}_i$ (see Fig. 1.5.1).

As alluded to in Sect.1.3.2, the stress tensor is symmetric, i.e., $\tau_{ij} = \tau_{ji}$ (cf. Example 1.22). This equality of shearing stresses can be readily shown in applying the statics conditions $\sum \vec{\mathbf{F}} = m\vec{\mathbf{a}} \equiv 0$ and $\sum \vec{\mathbf{M}} = I\vec{\boldsymbol{\alpha}} \equiv 0$ to the solid cube of Fig.1.5.1a, where m is the material mass, $\vec{\mathbf{a}}$ its acceleration, $\vec{\mathbf{M}} = \vec{\mathbf{r}} \times \vec{\mathbf{F}}$ is the applied moment or torque, i.e., force times distance, $\boldsymbol{I}$ is the moment of inertia , and $\vec{\boldsymbol{\alpha}}$ is the angular acceleration vector. The normal and shearing stresses relate to the following forces and moments (Fig.1.5.1b):

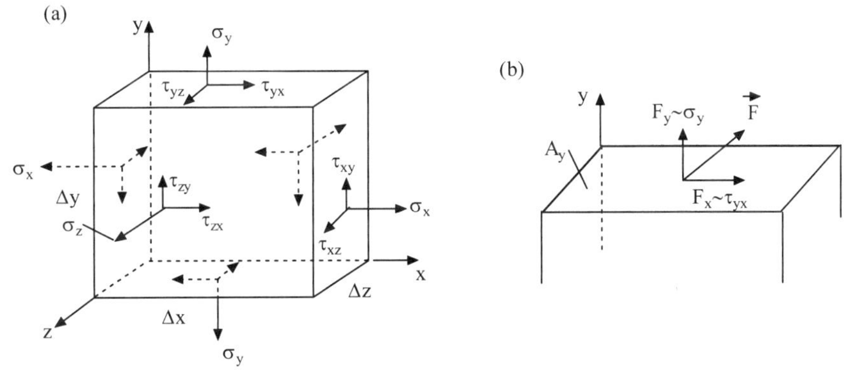

Fig. 1.5.1 Stresses and forces: (a) 3-D solid element with positive stresses; (b) 2-D force components and stresses in A_y - plane.

- *axial force* $\sim \int \sigma_{ii} dA$, where σ_{ii} are either tensile or compressive stresses;
- *shear force* $\sim \int \tau_{ij} dA$, which try to move (shear) adjacent parts of a solid; and

- *moments* (or torques), which may bend a beam, e.g., $M_y = \int \sigma_x z dA$, $M_z = -\int \sigma_x y dA$, etc., or may twist a body, e.g., $T = \int (\tau_{xz} y - \tau_{xy} z) z dA$.

Example 1.22: Simple Bending and Torsion Formulas for Rods of Linearly Elastic Material

Consider a 2-D beam subject to pure bending and then a cylinder subject to simple torsion. Show that $\sigma_x = -M_z y / I$, where I is the moment of inertia, and $\tau_{r\theta} = Tr/J$, where J is the polar moment of inertia.

Sketches:
(a) Bending

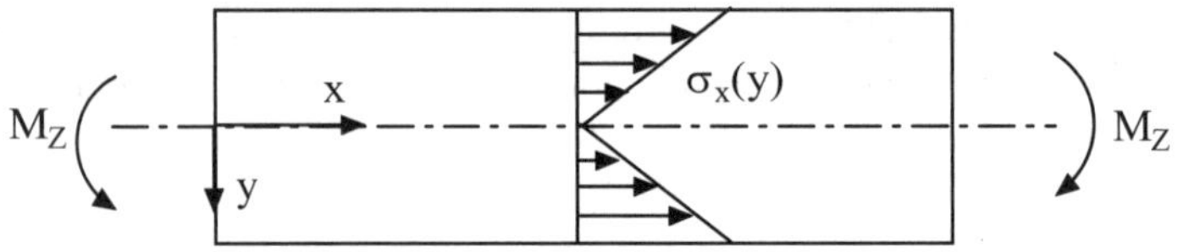

(b) Torsion

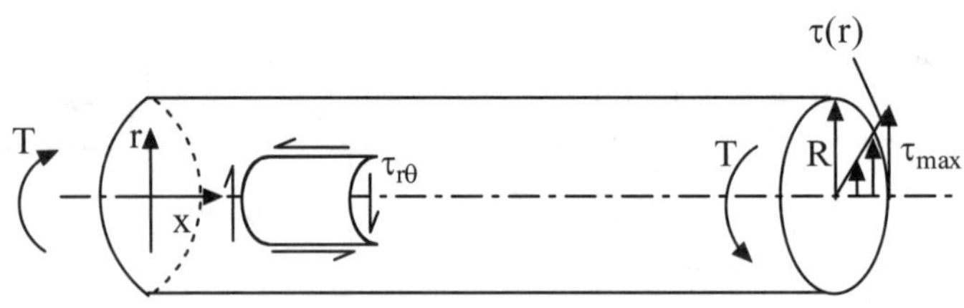

Assumptions:

- The normal stress $\sigma_x(y)$ is linear, and all other stresses in the beam are zero (pure bending).
- The resultant of the internal force of the beam is zero, i.e., $\int_A \sigma_x dA = 0$.

- The torsional loading does not reshape or reorientate the cross-sectional planes of the cylinder.

<u>Solution</u>:

(a) Bending

The moments of the internal forces about the beam's centerline (i.e., normal axis) are equal to the applied moment around the z-axis (see Sketch (a)):

- $$M_z = -\int_A y\sigma_x dA \qquad \text{(E1.22-1)}$$

With $\sigma_x = ky$, where k is a constant, we have

$$M_z = -k\int_A y dA := -kI_z$$

where I_z is the moment of inertia of the cross section A about the z-axis, e.g., $I_z = bh^3/12$ for $A_{\square} = bh$ and $I_z = \pi R^4/4$ for $A_{\circ} = \pi R^2$.

Thus, with $k = -M_z/I_z$,

- $$\sigma_x = -M_z y/I_z \qquad \text{(E1.22-2)}$$

(b) Torsion

The resultant of the shearing stress distribution being equal to the applied torque yields (see Sketch (b)):

- $$T = r\tau dA; \quad \tau(r) = \frac{r}{R}\tau_{max} \qquad \text{(E1.22-3)}$$

Thus,

$$T = \frac{\tau_{max}}{R}\int_A r^2 dA := \frac{\tau_{max}}{R}J$$

where $J = \pi R^4/2$ is the polar moment of inertia of the cylinder's cross section.

Thus, with $\tau_{max}/R = \tau/r$, we have

- $$\tau \equiv \tau_{r\theta} = Tr/J \qquad \text{(E1.22-4)}$$

Returning to Eq.(1.5.1), due to symmetry the stress tensor has six unknown components. These can be further reduced to three *principal stresses* which act in mutually perpendicular, *principal directions.* They are normal to three principal planes in which all shearing stresses are zero (see Example 1.23). Thus, Eq. (1.5.1) now reads after the coordinate transformation from σ_{ij} to σ_{ii}:

$$\sigma_{ij} = \begin{bmatrix} \sigma_1 & & \varnothing \\ & \sigma_2 & \\ \varnothing & & \sigma_3 \end{bmatrix} \tag{1.5.2}$$

The principal stresses can be combined according to Von Mises' deformational theory in terms of an effective stress σ_e, known as the *Von Mises stress* (see HWA in Sect.1.7).

$$\sigma_e = \frac{1}{\sqrt{2}}[(\sigma_1 - \sigma_2)^2 + (\sigma_2 - \sigma_3)^2 + (\sigma_3 - \sigma_1)^2]^{1/2} \tag{1.5.3}$$

Equation (1.5.3) is used in Sect. 5.4 to determine the locations of maximum stresses in both aneurysms and stent-grafts.

Example 1.23: Stresses on an Inclined Plane of a Bar under Uniaxial Tension

Consider a bar ($A = 10^3 mm^2$) subject to a tensile load of N=100kN. Determine the stress in a plane slanted by $\alpha = 35°$ and verify stress symmetry and the existence of a principal plane.

Sketch:

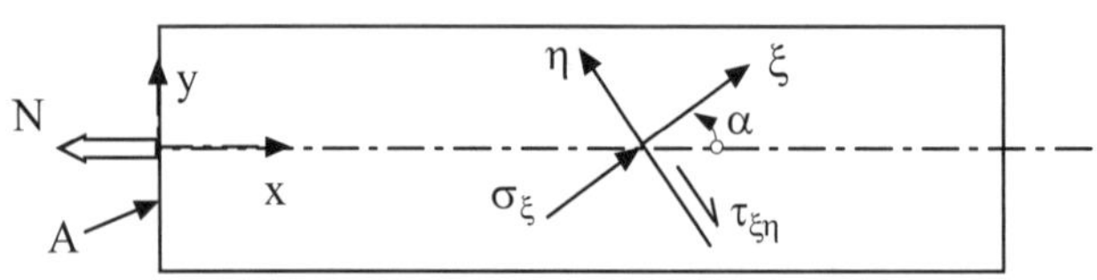

Assumption:

- uniform stress distributions in/on every plane

Solution: With respect to the incline, the loading force N generates a tensile stress σ_ξ because of the normal $N\cos\alpha$ and a shearing stress $\tau_{\xi\eta}$ because of the tangential $N\sin\alpha$. Thus, with $\sigma_x = N/A$,

- $$\sigma_\xi = \frac{N\cos\alpha}{A/\cos\alpha} = \sigma_x \cos^2\alpha \qquad \text{(E1.23-1)}$$

and

- $$\tau_{\xi\eta} = -\frac{N\sin\alpha}{A/\cos\alpha} = -\sigma_x \sin\alpha\cos\alpha \qquad \text{(E1.23-2)}$$

In letting the "cutting angle" vary, i.e., $0° \le \alpha \le 180°$, σ_ξ and $\tau_{\xi\eta}$ undergo maxima and minima. For example,

$$\sigma_\xi(\alpha = 0°, 180°) = \sigma_x = \sigma_{max} \text{ when } \tau_{\xi\eta} = 0$$

and

$$\tau_{\xi\eta}(\alpha = 45°, 135°) = \tau_{max} \quad \text{when } \sigma_\xi = \pm\frac{1}{2}\sigma_x$$

Graph:

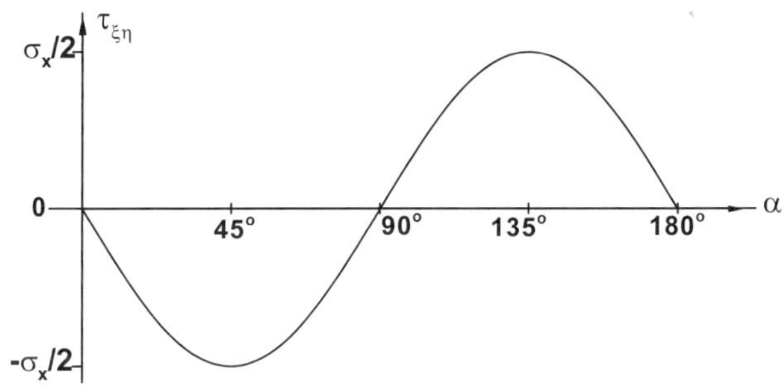

Comments:

When plotting Eq. (E1.23-2), it is revealed that both symmetry of the stress tensor (see Eq. (1.5.1)) and principal planes or axes [see Eq.(1.5.2)] actually exist. Specifically,

$$|\tau_{\xi\eta}(\theta)| = |\tau_{\xi\eta}(\theta + 90°)|$$

and at $\alpha = 0°$ & $180°$

$$\sigma_\xi = \sigma_{max} \equiv \sigma_x \text{ when } \tau_{\xi\eta} = 0$$

With the given data,

$$\sigma_x = \frac{N}{A} = 100 \text{ MPa}, \ \sigma_\xi = \sigma_x \cos^2\alpha = 67.11 \text{ MPa}, \text{ and}$$

$$\tau_{\xi\eta} = -\sigma_x \sin\theta\cos\theta = -47 \text{ MPa}$$

1.5.3 Equilibrium Conditions

So far we assumed uniformly distributed stresses across each surface caused by an external load. More likely, stress components vary from point to point in a body, but, nevertheless, all external forces and internal stresses have to be in static equilibrium, i.e.,

- $$\sum \vec{\mathbf{M}} = 0 \text{ and } \sum \vec{\mathbf{F}} = 0 \qquad (1.5.4, 1.5.5)$$

For example, considering a thin 2-D solid element of differential area $dA_z = dxdy$ (see Fig. 1.5.2), we take the moments about the origin [see Eq. (1.5.4)] and can show again that $\tau_{xy} = \tau_{yx}$. Now, employing $\sum F_x = 0$ [see Eq.(1.5.5)], we obtain

$$\left(\sigma_x + \frac{\partial \sigma_x}{\partial x} dx\right) dy - \sigma_x dy + \left(\tau_{xy} + \frac{\partial \tau_{xy}}{\partial y} dy\right) dx - \tau_{xy} dx + f_x dxdy = 0$$

or

$$\frac{\partial \sigma_x}{\partial x} + \frac{\partial \tau_{xy}}{\partial y} + f_x = 0$$

Similarly,

$$\frac{\partial \sigma_y}{\partial y} + \frac{\partial \tau_{xy}}{\partial x} + f_y = 0$$

In 3-D, using tensor notation with the summation convection of repeated indices, we have

$$\frac{\partial \sigma_{ij}}{\partial x_j} + f_i = 0 \qquad (1.5.6a)$$

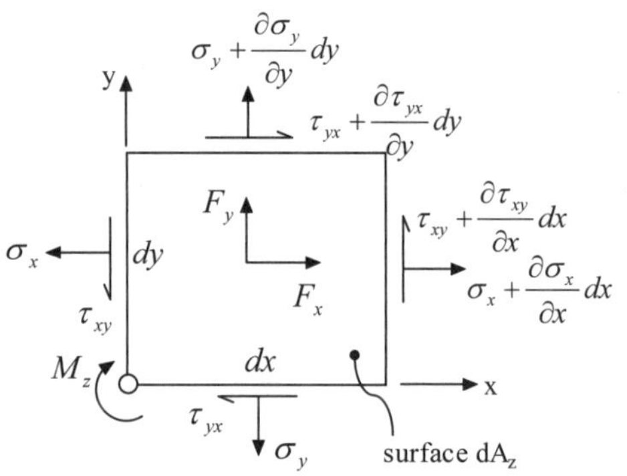

Fig.1.5.2 Differential 2-D solid element with varying forces and stresses

or

$$\sigma_{ij,j} + f_i = 0; \quad i, j = x, y, z \qquad (1.5.6b)$$

Clearly, given the external forces per unit volume f_i, Eq. (1.5.6) contains six unknown stress components, i.e., three additional equations have to be found. This brings us to the three basic principles for solving solid mechanics problems:

(i) *Conditions of Equilibrium* [see Eq. (1.5.6)]

(ii) *Stress-Strain Relations* (see Sect.1.5.4), where material properties in terms of constitutive equations correlate forces causing stresses in a given solid with body displacements.

(iii) *Conditions of Compatibility*, where body continuity is everywhere preserved, consistent with local strain distributions and deformations.

Equation (1.5.6) can be readily extended to dynamic structures where in general [cf. Eq.(1.3.9) or (1.3.18)]:

$$\rho a_i = \sigma_{ij,j} + f_i \qquad \text{in } \Omega(t) \tag{1.5.7}$$

with $a_i = d\hat{u}_i/dt$ being the acceleration of a material point in the domain Ω at time t. Equation (1.5.7) is subject to appropriate boundary conditions and necessary stress-strain relations as discussed in the next section [cf. Eq.(1.5.16)].

1.5.4 Deformation Analysis and Stress-Strain Relationships

When a force and/or moment is applied to an object, material deformation may occur in form of body elongation/contraction, beam bending, rod twisting, and/or simple shearing. In all these cases, body elements undergo shape changes. In contrast, during rigid-body motion, e.g., translatory displacement due to an axial force, no stress is induced (see Fluid Statics). Thus, deformation relates directly to stress when body element distortion, i.e., shape changes occur, say, due to stretching and rotation.

Deformation analysis. As illustrated in Fung (1994) and other texts, a line element in a 3-D body, $\overline{AA'} = ds_A$, translates, stretches, and rotates, because of body deformation, to $\overline{BB'} = ds_B$ (see Fig. 1.5.3).

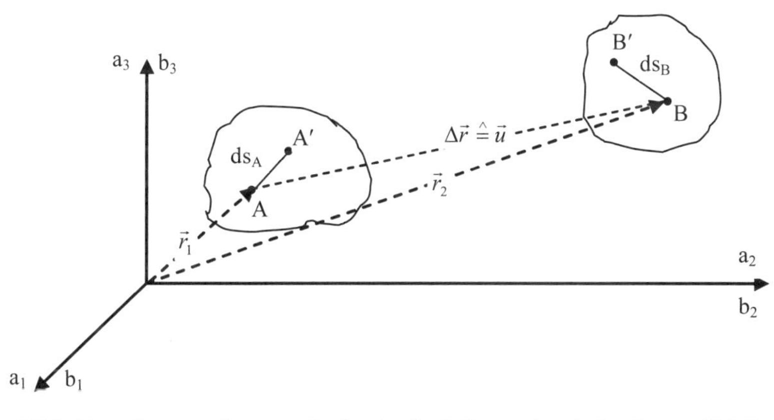

Fig. 1.5.3 Line element changes during body deformation (after Fung, 1994).

Clearly, with point A at (a_1, a_2, a_3) and point B at (b_1, b_2, b_3) the distance to the neighboring point $A'(a_1 + da_1, a_2 + da_2, a_3 + da_3)$ is after Pythagoras

$$(ds_A)^2 = (da_1)^2 + (da_2)^2 + (da_3)^2 \tag{1.5.8a}$$

Similarly,

$$(ds_B)^2 = (db_1)^2 + (db_2)^2 + (db_3)^2 \tag{1.5.8b}$$

Now, determination of any continuous body deformation requires known mapping functions

$$b_i = b_i(a_1, a_2, a_3) \text{ or } a_i = a_i(b_1, b_2, b_3) \tag{1.5.9a, b}$$

for every point in the body. The components of the displacement vector $\vec{\mathbf{u}}$ (see Fig. 1.5.3) are then

$$u_i = b_i - a_i \tag{1.5.10}$$

Given the mapping functions (1.5.9 a & b), we can write

$$db_i = \frac{\partial b_i}{\partial a_j} da_j \text{ and } da_i = \frac{\partial a_i}{\partial b_j} db_j \tag{1.5.11a, b}$$

so that Eqs. (1.5.8a & b) can be expressed, using the Kronecker delta, as

$$ds_A^2 = \delta_{ij} da_i da_j = \delta_{ij} \frac{\partial a_i}{\partial b_l} \frac{\partial a_j}{\partial b_m} db_l db_m \tag{1.5.12a}$$

and

$$ds_B^2 = \delta_{ij} db_i db_j = \delta_{ij} \frac{\partial b_i}{\partial a_l} \frac{\partial b_j}{\partial a_m} da_l da_m \tag{1.5.12b}$$

Forming the difference, *a measure of body deformation*, we obtain

$$ds_B^2 - ds_A^2 = \begin{cases} 2E_{ij} da_i da_j \\ \text{or} \\ 2\varepsilon_{ij} db_i db_j \end{cases} \tag{1.5.13a, b}$$

where the symmetric Lagrangian E_{ij} is Green's strain tensor and the Eulerian ε_{ij} is Cauchy's strain tensor. Clearly, if the length of each line element stays the same, $ds_B^2 - ds_A^2 = 0$, implying that $E_{ij} = \varepsilon_{ij} = 0$, and the body is at rest or undergoing rigid-body motion.

Using the displacement vector (see Eq. (1.5.10)), we can form

$$\frac{\partial b_i}{\partial L_j} = \frac{\partial u_i}{\partial a_j} + \delta_{ij} \text{ and } \frac{\partial a_i}{\partial b_j} = \delta_{ij} - \frac{\partial u_i}{\partial b_j} \tag{1.5.14a, b}$$

where $u_i = (u_1, u_2, u_3)$ and $b_j = (x_1, x_2, x_3)$. Thus, inserting (1.5.14a & b) into (1.5.12a & b) and neglecting squares amd products of the derivatives of the displacement components u_i, the Cauchy infinitesimal strain terms in Eq. (1.5.13b) reduce to (cf. Sect. 1.3.2):

$$\varepsilon_{ij} = \frac{1}{2}\left(\frac{\partial u_j}{\partial x_i} + \frac{\partial u_i}{\partial x_j}\right) \quad (1.5.15a)$$

or in terms of the shearing strain (i.e., the shear-rate tensor of Sect.1.3.2)

$$\gamma_{ij} = \frac{\partial u_i}{\partial x_j} + \frac{\partial u_j}{\partial x_i} \quad (1.5.15b)$$

Equation (1.5.15a) can be written as

$$[\varepsilon_{ij}] = \begin{bmatrix} \varepsilon_x & \frac{1}{2}\gamma_{xy} & \frac{1}{2}\gamma_{xz} \\ \frac{1}{2}\gamma_{yx} & \varepsilon_y & \frac{1}{2}\gamma_{yz} \\ \frac{1}{2}\gamma_{zx} & \frac{1}{2}\gamma_{zy} & \varepsilon_z \end{bmatrix} \quad (1.5.15c)$$

Clearly, for infinitesimal displacements $E_{ij} \approx \varepsilon_{ij}$ because it is immaterial if the derivatives $u_{i,j}$ are calculated at the position of a point before or after deformation (see Fig. 1.5.3).

Stress-strain relationships. A stress-strain relationship describes the mechanical property of a material and is therefore a *constitutive equation.* For linear elastic materials, for example bones for which *Hooke's law* holds,

$$\sigma_{ij} = C_{ijkl}\varepsilon_{kl} \quad (1.5.16)$$

where the tensor of rank four, C_{ijkl}, is symmetric, i.e., the $3^4 = 81$ elastic constants (or moduli) can be reduced to 36 (Fung, 1994). For an *isotropic* elastic solid, i.e., when the properties are identical in all directions, Eq. (1.5.16) reduces to (cf. Sect.1.3.2):

$$\sigma_{ij} = \lambda\varepsilon_{kk}\delta_{ij} + 2\mu\varepsilon_{ij} \quad (1.5.17)$$

Clearly, instead of 36 material values, only two, i.e., the *Lame constants*, are needed. Writing Eq. (1.5.17) out and solving for the six ε-components, we obtain

$$\varepsilon_{ij} = \frac{1+\nu}{E}\sigma_{ij} - \frac{\nu}{E}\sigma_{kk}\delta_{ij} \quad (1.5.18)$$

where E is *Young's modulus*, $\nu = -\frac{\varepsilon_y}{\varepsilon_x} = -\frac{\varepsilon_z}{\varepsilon_x}$ is the *Poisson ratio* $\triangleq \frac{\text{lateral strain}}{\text{axial strain}}$, and $G \equiv \mu$ is the *shear modulus*. Since only two independent constants are needed for homogeneous isotropic elastic materials [see Eq.(1.5.17)], E, ν, and G are related, i.e.,

$$G = \frac{E}{2(1+\nu)} \tag{1.5.19}$$

Actual materials, such as biological soft and hard tissues, rubber, shape memory alloys (SMAs), and non-Newtonian fluids, do not follow Hooke's law. They are *nonlinear viscoelastic*, i.e., they may exhibit creep, hysteresis, memory, and other complex behaviors. For example, Fig. 1.5.4 depicts qualitatively stress-strain graphs for metal, soft tissue, and rubber. Clearly, compared to metals, the other materials exhibit nonlinear relationships and hystereses, i.e., the loading and unloading curves differ. Mechanical models, a combination of linear springs and viscous dashpots, mimic some viscoelastic behaviors. For example, the *Kelvin model*, an improvement over the Maxwell model and Voigt model, relates the load F to the deflection (or displacement) *u(t)* as (Fig.1.5.5):

$$F + a\dot{F} = b(u + c\dot{u}) \tag{1.5.20a}$$

subject to

$$aF(t=0) = bcu(t=0) \tag{1.5.20b}$$

where *a* to *c* are constant system parameters.

Clearly, using nonlinear springs k_1 and k_2 as well as an exotic fluid η in the dashpot and assembling several of these models in series, some biological tissues can be approximated (see Fung, 1994; Humphrey & Delange, 2004).

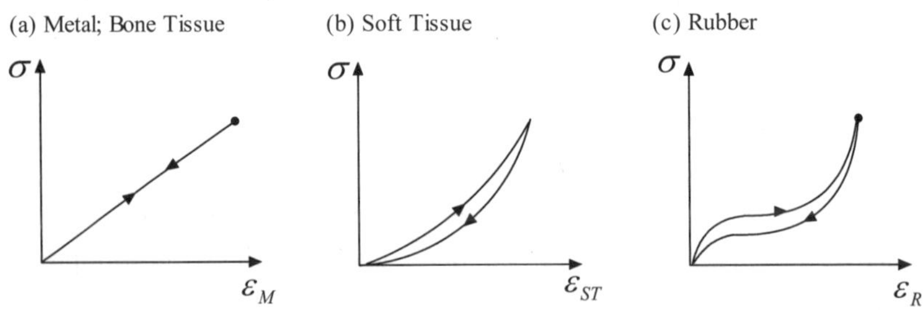

Fig. 1.5.4 Stress-strain behavior of different materials. Note: $\varepsilon_M \ll \varepsilon_{ST} \ll \varepsilon_R$ (after Humphrey & Delange, 2004).

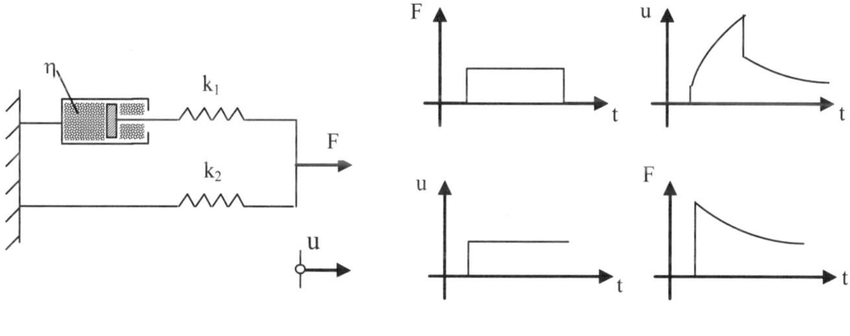

Fig. 1.5.5 Kelvin viscoelasticity model with creep function due to force pulse $F \to u(t)$ and relaxation function due to sudden deformation $u \to F(t)$.

1.5.5 Simplifications

Hooke's law, i.e., Eq. (1.5.18), can be rewritten as:

$$\varepsilon_x = \frac{1}{E}[\sigma_x - \nu(\sigma_y + \sigma_z)] \tag{1.5.21a}$$

$$\varepsilon_y = \frac{1}{E}[\sigma_y - \nu(\sigma_x + \sigma_z)] \tag{1.5.21b}$$

$$\varepsilon_z = \frac{1}{E}[\sigma_z - \nu(\sigma_x + \sigma_y)] \tag{1.5.21c}$$

$$\varepsilon_{xy} = \frac{1+\nu}{E}\tau_{xy} = \frac{1}{2G}\tau_{xy} := \frac{1}{2}\gamma_{xy} \tag{1.5.21d}$$

$$\varepsilon_{yz} = \frac{1+\nu}{E}\tau_{yz} = \frac{1}{2G}\tau_{yz} := \frac{1}{2}\gamma_{yz} \tag{1.5.21e}$$

$$\varepsilon_{zx} = \frac{1+\nu}{E}\tau_{zx} = \frac{1}{2G}\tau_{zx} := \frac{1}{2}\gamma_{zx} \tag{1.5.21f}$$

Clearly, for plane stress analysis, $\sigma_z = 0$, and the remaining normal stresses are:

Rectangular Coordinates:

$$\sigma_x = \frac{E}{1-\nu^2}(\varepsilon_x + \nu\varepsilon_y)$$

$$\sigma_y = \frac{E}{1-\nu^2}(\varepsilon_y + \nu\varepsilon_x)$$

Polar Coordinates:

$$\sigma_r = \frac{E}{1-\nu^2}(\varepsilon_r + \nu\varepsilon_\theta)$$

$$\sigma_\theta = \frac{E}{1-\nu^2}(\varepsilon_\theta + \nu\varepsilon_r)$$

(1.5.22a-d)

Provided that Young's modulus E and Poisson's ratio ν are known, each of these equations contain extra unknowns; thus, equilibrium equations have to be established first which relate forces to stresses, as indicated in Sect. 1.5.4 and illustrated in Examples 1.24 and 1.25.

Example 1.24: Displacement of a Non-uniform Rod under Axial Stress

Consider a vertical, axially loaded rod with suddenly changing cross section and material property as shown (cf. bone with implanted metal segment). Find the total axial displacement *u*.

Sketch:

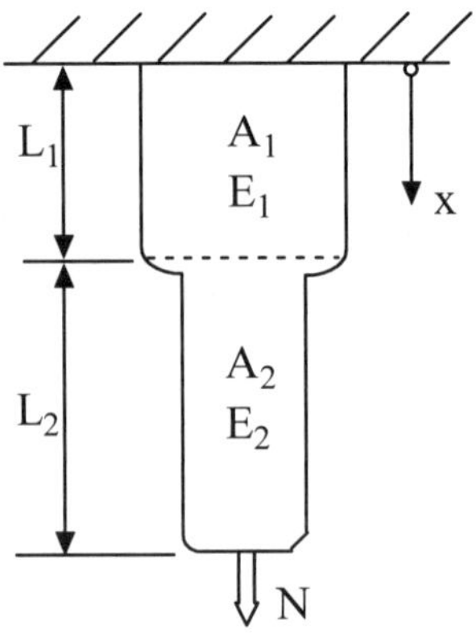

Assumptions:

- Segmentally constant A_i and E_i
- Only normal, i.e., axial force N
- Hooke's law holds

Concept:

Use of Eq. (1.5.21a) where $\sigma_y = \sigma_z = 0$

Solution:

The reduced governing equations read

- $\varepsilon_x = \frac{\sigma_x}{E}$ and $\Delta u = \int_0^L \varepsilon_x dx$, where $\sigma_x = \frac{N}{A}$

Specifically, with $u(x=0) = 0$,

- $u_{total} = \int_0^{L_1} \frac{N}{A_1 E_1} dx + \int_{L_1}^{L} \frac{N}{A_2 E_2} dx$; $L = L_1 + L_2$

- $u_{total} = N\left(\frac{L_1}{A_1E_1} + \frac{L_2}{A_2E_2}\right) = \frac{NL_1}{A_1E_1}\left(1 + \frac{L_2}{L_1}\frac{A_1}{A_2}\frac{E_1}{E_2}\right)$

Example 1.25: Blood Vessel Applications

Some large blood vessels can be approximated as circular thin-walled or thick-walled tubes where σ_z is the axial or *longitudinal stress* and σ_θ is called the circumferential or *hoop stress*. The transmural pressure, the difference between blood pressure and outside pressure, generates the load.

Sketches:

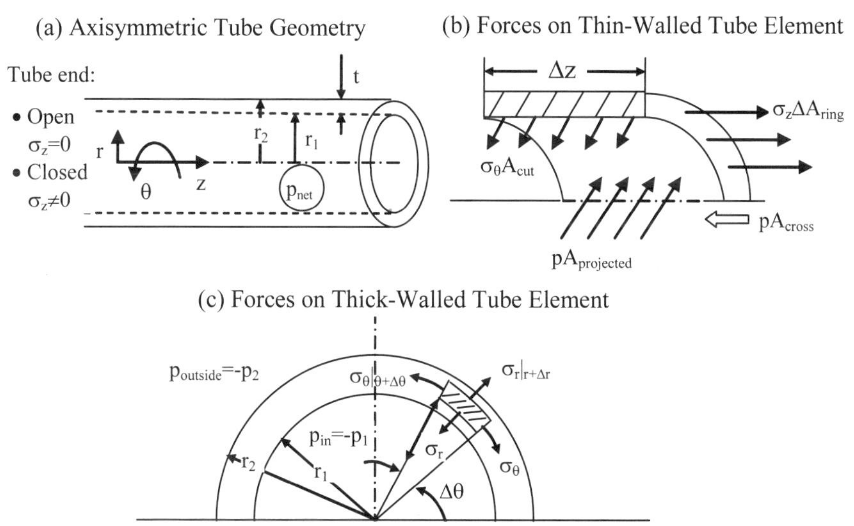

Solution:

Thin-walled implies that $t/r_1 < 0.1$ and radial stresses can be neglected, while perfect axisymmetry precludes any shear stresses. Thus, a 1-D force balance yields (see Sketch (b)):

$$2(\sigma_\theta t\Delta z) - p(2r_1\Delta z) = 0$$

and the hoop stress is

• $$\sigma_\theta = \frac{pr_1}{t} \tag{E1.25-1}$$

Also, in case one vessel end is closed and the pressure force exerts a longitudinal stress, we have

$$\sigma_z(2\pi r_1 t) - p(\pi r_1^2) = 0$$

or

• $$\sigma_z = \frac{pr_1}{2t} \tag{E1.25-2}$$

Both stress equations were originally derived by Laplace nearly 200 years ago.

Most blood vessels are *thick-walled*, i.e., $r_1 \leq tr_2$, and hence for an open-ended tube ($\sigma_z = 0$), a radial force balance on the material element $(\Delta r\theta)\Delta r$ yields (Sketch (c)):

$$\sigma_r\big|_{r+\Delta r}(r+\Delta r)\Delta\theta\Delta z - \sigma_r r(\Delta\theta\Delta z) - 2\sigma_\theta \Delta r \Delta z \sin\frac{\Delta\theta}{2} = 0$$

Taking $\sin\frac{\Delta\theta}{2} \approx \frac{\Delta\theta}{2}$ and $\sigma_r\big|_{r+\Delta r} \approx \sigma_r + \frac{\partial\sigma_r}{\partial r}\Delta r$ while neglecting higher-order terms, we obtain in the limit

• $$\frac{d\sigma_r}{dr} + \frac{\sigma_r - \sigma_\theta}{r} = 0 \tag{E1.25-3}$$

In order to solve this ODE, we need an expression for the hoop stress $\sigma_\theta(r)$. Alternatively, we employ appropriate stress-strain relations (see Eq.(1.5.21)), expressing the strains in terms of the radial displacement $u_r = u$. Specifically,

$$\varepsilon_r \equiv \frac{du}{dr}$$

and

$$\varepsilon_\theta = \frac{u}{r}$$

Thus, Eqs. (1.5.22c,d) can be written as

$$\sigma_r = \frac{E}{1-\nu^2}\left(\frac{du}{dr} + \nu\frac{u}{r}\right) \tag{E1.25-4a}$$

and

$$\sigma_\theta = \frac{E}{1-\nu^2}\left(\frac{u}{r} + \nu\frac{du}{dr}\right) \tag{E1.25-4b}$$

so that Eq. (E1.19.3) reads

• $$\frac{d^2u}{dr^2} + \frac{1}{r}\frac{du}{dr} - \frac{u}{r^2} = 0 \tag{E1.25-5a}$$

or

$$\frac{d}{dr}\left[\frac{1}{r}\frac{d}{dr}(ur)\right] = 0 \tag{E1.25-5b}$$

subject to two boundary conditions, which are best expressed in terms of the radial stresses, i.e.,

$$\sigma_r = \begin{cases} p_{in} = -p_1 \text{ at r=r}_1 \\ p_{out} = -p_2 \text{ at r=r}_2 \end{cases} \tag{E1.25-6a,b}$$

Thus, with the solution of Eq. (E 1.17.5b), i.e.,

$$u(r) = C_1 r + \frac{C_2}{r}$$

Eqs. (E1.25-4a,b) read

$$\sigma_r = \frac{E}{1-\nu^2}\left[(1+\nu)C_1 - \frac{1-\nu}{r^2}C_2\right]$$

and

$$\sigma_\theta = \frac{E}{1-\nu^2}\left[(1+\nu)C_1 + \frac{1-\nu}{r^2}C_2\right]$$

So, after invoking the boundary conditions (E1.25-6a,b), we obtain Lamé's relationships for $r_1 \le r \le r_2$:

• $$\sigma_r = \frac{p_1 r_1^2}{r_2^2 - r_1^2}\left(1 - \frac{r_2^2}{r^2}\right) - \frac{p_2 r_2^2}{r_2^2 - r_1^2}\left(1 - \frac{r_1^2}{r^2}\right) \tag{E1.25-7a}$$

• $$\sigma_\theta = \frac{p_1 r_1^2}{r_2^2 - r_1^2}\left(1 + \frac{r_2^2}{r^2}\right) - \frac{p_2 r_2^2}{r_2^2 - r_1^2}\left(1 + \frac{r_1^2}{r^2}\right) \tag{E1.25-7b}$$

which are plotted below (see Graph).

<u>*Graph:*</u>

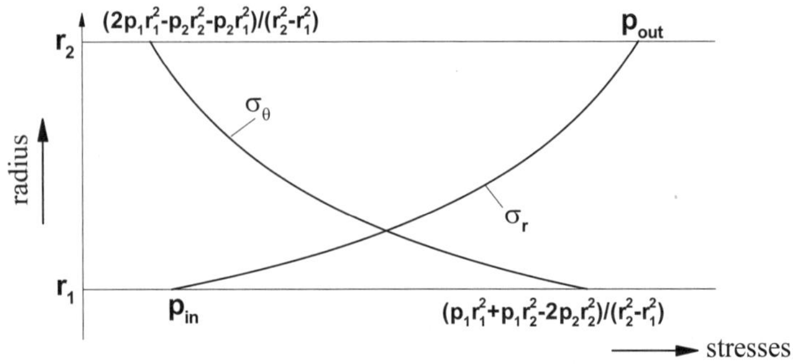

Plane stress analysis. In case the body force F_z=0, then $\sigma_z = \tau_{xz} = \tau_{yz} = 0$ (cf. "thin plate problem"). Thus, three equations are necessary to calculate σ_x, σ_y, and τ_{xy}, subject to *surface tractions*, i.e.,

$$p_x = \sigma_x l + \tau_{xy} m \tag{1.5.23a}$$

and

$$p_y = \tau_{xy} l + \sigma_y m \tag{1.5.23b}$$

where $l = \cos(\hat{n}, x)$ and $m = \cos(\hat{n}, y)$ are the direction cosines for the normal vector $\hat{n}$. The first two equations for σ_x, σ_y, and τ_{xy} are obtained from Eq. (1.5.6):

$$\frac{\partial \sigma_x}{\partial x} + \frac{\partial \tau_{xy}}{\partial y} + F_x = 0 \tag{1.5.24a}$$

$$\frac{\partial \sigma_y}{\partial y} + \frac{\partial \tau_{xy}}{\partial x} + F_y = 0 \tag{1.5.24b}$$

The third one is the equation of compatibility in terms of stress, i.e.,

$$\left(\frac{\partial^2}{\partial x^2} + \frac{\partial^2}{\partial y^2}\right)(\sigma_x + \sigma_y) = -\frac{1}{1-\nu}\left(\frac{\partial F_x}{\partial x} + \frac{\partial F_y}{\partial y}\right) \tag{1.5.25a}$$

Equations (1.5.25a,b) can be derived from the compatibility condition for strain components which directly relate to the displacements u and v (see Sect. 1.7).

$$\frac{\partial^2 \varepsilon_x}{\partial y^2} + \frac{\partial^2 \varepsilon_y}{\partial x^2} = \frac{\partial^2 \gamma_{xy}}{\partial x \partial y} \tag{1.5.25b}$$

When the body forces F_x and F_y are zero, the task of evaluating planar stress, strain, and displacement subject to (1.5.23a,b) is less challenging. Similar to the stream function $\Psi(x, y)$ satisfying the continuity condition (1.3.21a) automatically, a *stress function* $\Phi(x, y)$ is introduced where

$$\sigma_x \equiv \frac{\partial^2 \Phi}{\partial y^2}, \ \sigma_y \equiv \frac{\partial^2 \Phi}{\partial x^2}, \text{ and } \tau_{xy} \equiv -\frac{\partial^2 \Phi}{\partial x \partial y}$$

so that Eqs. (1.5.24a,b) balance, and Eq. (1.5.25a) yields

$$\frac{\partial^4 \Phi}{\partial x^4} + \frac{2\partial^4 \Phi}{\partial x^2 \partial y^2} + \frac{\partial^4 \Phi}{\partial y^4} = 0 \tag{1.5.26}$$

Traditionally this biharmonic equation (1.5.26) has been approximately solved, subject to (1.5.23a,b), with polynomial functions of various degrees (see Ugural & Fenster, 1995).

1.6 SUMMARY AND OUTLOOK

It is transparent that "biofluid dynamics" deals with rather challenging transport phenomena, including fluid-particle dynamics and fluid-structure interactions, as well as disease etiologies/effects and tissue restructuring. This is because of the inherent complexities of biological systems in man (Sect. 1.1). Taking arterial blood flow as an example, in general one has to consider pulsatile 3-D laminar/turbulent flow of a non-Newtonian fluid interacting with deforming cells in irregular branching tubes with viscoelastic composite walls and wall mass transfer. As the blood vessels decrease in mean diameter, capillary effects on the blood rheology, tubular elasticity, and cell deformation plus biological response become important as well. While the general mathematical models, typically in terms of partial differential equations, are in place, specific submodels for closure, fluid, and material properties, as well as feasible solution techniques (i.e., complementary to the continuum mechanics approach), have to be designed, measured, and developed for an accurate simulation of biological systems in man.

Now, what should be done in light of such complexities? First of all, a basic understanding of the human physiology is a prerequisite (see Vander, 2001; Fox, 2002; Silverthorn, 2004, among others). Then, in order to understand the principles of biofluid dynamics and to be able to utilize them, one has to first comprehend the fundamental case studies as they apply to biomechanics in man and medical devices. Such fluid mechanics case studies, with an engineering math refresher (see App. A), include:

- Fully-developed flows (see Examples 1.4, 1.5 <nearly>, 1.7 and 1.9 plus HPAs in Sect. 1.7), where Poiseuille flow representing simple blood flow in straight vessels and planar (joint) lubrication are the prime examples.
- Thin-film flows (see Example 1.5 plus HWAs in Sect. 1.7) are important in joint lubrication and solid surface coating of implants.
- Porous-layer flows (see Examples 1.18-1.20 plus HWAs in Sect. 1.7) are applicable to convection across membranes, cell or tissue layers, etc.
- Curved-duct flows (see HWAs in Sect. 1.7) relate to respiratory airways and blood vessels.
- Pulsatile flows (see Example 1.6 plus HPAs in Sect. 1.7) due to sudden inlet pressure changes or an oscillating input pressure (e.g., cardiac cycle) generate transient velocity fields.
- Developing flows (see HPAs in Sect. 1.7) occur in entrance regions, behind partial occlusions, after bifurcations, etc.
- Moving boundary flows (see Sect. 5.4) are important in coronary arteries, alveoli, deforming cells, etc.
- Particle-suspension flows (see Examples 1.13-1.17 plus HWAs in Sect. 1.7) come into play when considering arterial and respiratory diseases.
- Boundary-layer flow of biofluids containing platelets, hormones, growth factors, etc., interacting with vessel walls (see Sects. 1.1.2 and 1.3.5).

Basic solid mechanics case studies include:

- Forces on vessel walls and implants (see Example 1.3 and Sect. 5.4).
- Stress & strain due to rod bending, twisting, and/or stretching (see Examples 1.22 amd 1.23 as well as Sect. 5.4).
- Elastic material displacements (see Examples 1.24 and 1.25 as well as Sect. 5.4).

Physical and mathematical insight gained from thoroughly studying the dynamics of such basic fluid flow, fluid-particle, and fluid-structure phenomena in the context of biological systems forms the launching pad to tackle more realistic biomedical engineering problems. It is of great importance to solve the HWAs of all chapters independently; however when help beyond this textbook is needed, prime sources include:

- Cengel & Cimbala (2006) or White (2003) for basic fluid flow and Schlichting & Gersten (2000) for boundary-layer and pipe flows.
- Truskey et al. (2004) and Bird et al. (2002) for transport phenomena, especially mass transfer.

- Kleinstreuer (1997, 2003) for momentum transfer and convection heat transfer as well as two-phase flow (see also Crowe et al., 1997 and Clift et al., 1977).
- Humphrey & Delange (2004) for introductory biomechanics.
- Fung (1990, 1994, 1997) for continuum mechanics and mechanical properties of biological systems.

In order to advance beyond these resources, it is necessary to come up with creative research work accompanied by state-of-the-art literature contributions, typically published in top biomedical engineering and related journals, starting selectively with overviews given in Annual Review of Fluid Mechanics and Annual Review of Biomedical Engineering <http://www.AnnualReviews.org>, as well as other book series.

1.7 HOMEWORK ASSIGNMENTS

This section consists of two groups of homework problems (HWPs), i.e., assignments that focus on open questions, unproven statements, or omitted derivations in the text, and HWPs that further help to illustrate the physics of: basic fluid flows, suspension flows, and fluid-structure interactions, as well as associated solution techniques. A few HWPs were inspired by assignments given in Truskey et al. (2002), Humphrey & Delange (2004), and Fung (1990).

<u>Note</u>: Whenever possible, follow the four basic solution steps outlined in Fig. 1.3.5! <u>Recall:</u> $\cancel{c} \hat{=}$ constant

<u>*Visualization problems:*</u>

1.1 Figure enhancements and new graphs:
  - (a) Fig. 1.1.1 shows a single cell. Sketch and describe distinct cell layers lining the arteries and lung airways; specifically, what are the interactions with fluid flow under normal and pathological conditions?
  - (b) In Fig. 1.1.2 vital organs such as the brain, lungs, liver, and/or kidneys are indicated as "boxes." Select your favorite organ and show detailed fluid pathways and organ functions.
  - (c) Concerning Fig. 1.1.3, sketch the alveolar region in more detail and describe the local air/blood flow with gas exchange.
  - (d) Figure 1.2.1 depicts flow regimes and modeling approaches as a function of the global Knudsen number. Incorporate the Reynolds number

$Re_L = \bar{u}L/\nu$ (e.g., $Re_L\ Kn = \bar{u}\lambda/\nu$) and rework the figure. Discuss differences between gases and liquids.

(e) Figure 1.2.2 lists real fluid flow characteristics. Focusing on internal vs. external flows, depict and state mathematically all possible initial and boundary conditions for both cases.

(f) Contrast the Lagrangian vs. the Eulerian approach by sketching closed vs. open systems with associated variables and parameters helpful in deriving/explaining Eq. (1.3.2) and Eq. (1.3.1).

(g) Starting with a fixed and non-inertial coordinate system for an accelerating object, illustrate all terms of Eq. (1.3.3). Then let the object rotate and add angular momentum terms in your description.

(h) Sketch boundary-layer developments and associated axial velocity profiles for steady laminar incompressible flow over a horizontal flat plate and in the entrance region of a circular tube and compare.

(i) Provide sketches for Eqs. (1.3.24) and (1.3.27) and explain all terms.

(j) Considering turbulent boundary-layer flow, sketch coherent structures and indicate random phenomena in a Δx-slice for $0 \le y \le \delta$.

(k) Concerning Sect. 1.3.4.2, depict in two bar-diagrams the relative time scales and length scales of turbulent air as well as water flow past a flat plate.

(l) Develop illustrative graphs explaining the (physical) meaning of Eqs. (1.4.2c), (1.4.4b), (1.4.6), (1.4.7d), and (1.4.7f).

Sect.1.1 problems:

1.2 Find the effective diffusion coefficient for diffusion through multiple layers of tissue.

(a) Consider a two-layer model of an artery. The radius of the cylindrical lumen, thickness of inner layer, and outer layer are R_i, R_0-R_1, and R_1-R_i, respectively. The solute concentration in the lumen and the outside wall ($r=R_0$) are C_i and C_0, respectively. The diffusivity in the inner and outer layers are D_i and D_0, respectively.

(b) Consider a rectangular laminate consisting of two layers with thickness of L_1 and L_2, and diffusivity of D_1 and

D_2. Also determine conditions when the two-layer model behaves as an effective one-layer model.

1.3 Species boundary-layer flow: Consider flow of a fluid over a flat membrane with permeability P_i. Given a constant species inlet, i.e., bulk, concentration C_0, determine an expression for the flux in terms of bulk fluid concentrations on either side of the membrane.

1.4 Considering one-dimensional diffusion of a nonelectrolyte molecule A in a membrane with thickness L, derive the permeability coefficient (P) of this molecule as a function of L, its effective diffusivity (D_A), and the available volume fraction of A in the membrane (K_A). P is defined as the ratio of solute flux to the concentration difference across the vessel wall.

1.5 Revisiting Example 1.8, now assume that the concentration at x=h is maintained at zero.

(a) Assess the time required for the membrane to reach steady state and compare this result with that obtained from quasi-steady analysis.

(b) Consider that the side x=0 with concentration c_0 is connected to a tissue (chamber) with volume V_1, find the concentration $c_1(t)$ in the chamber. Assume that the $c_1=c_0$ at t=0. Use this result to evaluate the accuracy of quasi-steady analysis.

1.6 The diffusivity of platelets in blood can be enhanced by increasing mixing due to the local fluid motion generated by individual red cell rotation. Thus, the effective diffusivity can be the sum of Brownian molecular diffusivity (D) and the "rotation" induced diffusivity (D_p). Using dimensional analysis, show that D_p may be written as

$$D_p = Ca^2\dot{\gamma}$$

where C is a constant, a is the red cell radius, and $\dot{\gamma}$ is the shear rate. Following Keller (1971), assess the effect of red cells on the diffusion of O_2, protein, and platelet.

1.7 Derive the mass balance equation for steady one-dimensional diffusion through a funnel of varying cross section (i.e., A=A(x)). Find the distribution of mass flux along the funnel whose radius varies linearly with distance following the formula

$$r(x) = r_0\left(1 + \frac{x}{L}\right)$$

where L is the length of the funnel. The material concentrations at the inlet and outlet are c_0 and c_L, respectively.

1.8 Consider a selective semi-permeable membrane of negligible thickness. Water can penerate easily but solute is somewhat hindered, i.e., a solute build-up can be observed very near the membrane. Find c(x,t) as well as the region δ of elevated solute concentration under steady-state conditions. Take c_0=2 mol/litre, $v_{water}=30\times10^{-9}$m/s, $D_{AB}=2\times10^{-10}$ m^2/s.

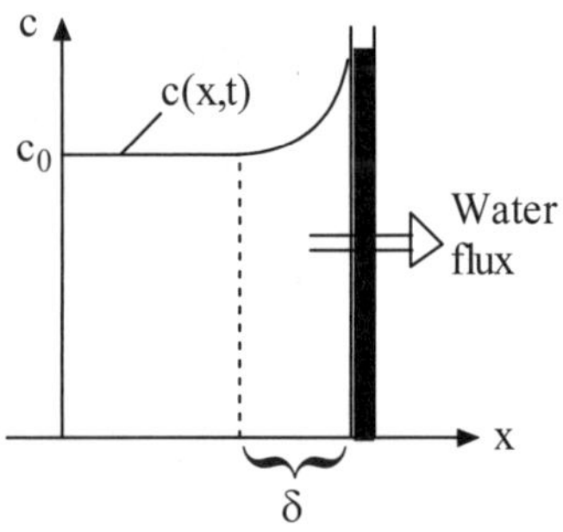

Sect.1.2 problems:

1.9 Consider a rotated and deformed fluid element at time $t+\Delta t$ (cf. Fig. 1.2.3). Derive the vorticity vector $\vec{\zeta}$ and the rate-of-deformation tensor $\vec{\vec{\varepsilon}}$.

1.10 In Section 1.2, three basic questions (i)–(iii) (page 17) are raised and some flow assumptions are listed to help to identify and categorize a fluid dynamics problem. Using Fig.1.2.2 and Table 1.2.1 as a guide, *set up* the following problems (cf. Fig. 1.3.5) for future solutions:

(a) Species diffusion and/or convection in and around cells, including transport across cell membranes.
(b) Microcirculation of blood in tissue, e.g., arterial wall or lung.
(c) Lymph transport in tissue and vessels.
(d) Perfusion, i.e., fluid flow and species transport, in organs such as liver and kidney.

(e) Laminar flow through a porous tube such as a hollow filter, perforated shunt, leaky pipe, etc.
(f) Droplet formation after jet break-up.
(g) Wire-coating in a die or sheet-coating with a blade.
(h) Pulsatile flow in an arterial segment.
(i) Airflow in a lung segment.
(j) A flow system of your choice.

Note: Set up initial problem solution steps by providing a detailed system sketch, answering the three questions (i)–(iii) and listing key assumptions with their mathematical consequences.

1.11 Explain with sketches and equations/proportionalities the following counterintuitive flow phenomena:

(a) Keeping the tailgate of a pick-up truck *up* reduces aerodynamic drag and hence saves gasoline.
(b) Under otherwise identical conditions, it is easy to *blow out* a candle but nearly impossible to suck it out.
(c) Certain non-Newtonian fluids when stirred in a cylinder *climb up* the rotating rod.
(d) Chunks of metal are *torn out* from ship propellers at high speeds.
(e) When pouring a "heavy" beer, e.g., a Guinness stout, into a glass, small bubbles *float down* along the glass wall.
(f) When bringing a spoon near a jet, e.g., faucet stream, it gets *sucked into* the jet.
(g) A snow storm leaves a cavity *in front* of a pole or tree and deposits snow *behind* the vertical cylinder.
(h) Flapping *butterfly wings* in China may cause a *monsoon* in India.
(i) Very high winds, e.g., tornados, rip certain roofs *off* houses.
(j) Particles of a suspension settle faster in a *tilted* container than in a vertical one.
(k) 3-D effects in river bends create unusual (axial) velocity profiles right after the bend and lateral material transport results in shifting riverbeds.

1.12 Consider steady 1-D (plug) flow in a slightly converging nozzle where

$$A(x) = A_0 \Big/ \left(1 + \frac{x}{l}\right) \text{ and } u(x = 0) = u_0 .$$

Find the total acceleration in the x-direction:

(a) $a_x(x)$ (Euler)

(b) $a_x(t)$ (Lagrange)

Sect.1.3 problems:

1.13 Derive: (a) $\frac{D\vec{\mathbf{v}}}{Dt} = \frac{\partial \vec{\mathbf{v}}}{\partial t} + (\vec{\mathbf{v}} \cdot \nabla)\vec{\mathbf{v}} = \vec{\mathbf{a}}_{total} = \vec{\mathbf{a}}_{local} + \vec{\mathbf{a}}_{convection}$ starting with the total differential $d\vec{\mathbf{v}} \mathrel{\hat{=}} D\vec{\mathbf{v}}$; and (b) Eq. (1.3.1) using an open system.

1.14 Using "scale analysis" (see Sect. 1.3.5.5), derive the Reynolds, Peclet, Stokes, Strouhal, and Womersley numbers.

1.15 Three-step derivations:
(a) Starting with Eqs. (1.3.20), derive Eqs. (1.3.21).
(b) Starting with Eq. (1.3.23), derive Eqs. (1.3.24) and (1.3.25).

1.16 Consider a cylindrical control volume $\Delta r \cdot r\Delta\theta \cdot \Delta z$; derive the continuity equation. If $\rho = ¢$, $\nabla \cdot \vec{\mathbf{v}} = 0$; does this equation hold for transient flows as well?

1.17 Derive the angular momentum equation

$$\frac{\partial}{\partial t}(\vec{r} \times \rho\vec{\mathbf{v}}) + \nabla \cdot (\vec{r} \times \rho\vec{\mathbf{v}}\vec{\mathbf{v}}) = \nabla \cdot \left(\vec{r} \times \vec{\vec{T}}\right) + \vec{r} \times \rho\vec{f}_B$$

1.18 Starting with Eq. (1.3.33), derive Eq. (1.3.35a).

1.19 Use a sketch similar to Fig.1.3.1 and a differential mass-balance approach to derive Eq. (1.3.36).

1.20 Considering direct numerical simulation (DNS), do the Navier-Stokes equations hold for incompressible turbulent flow as well? Explain!

1.21 Show that $\nabla \cdot \vec{\mathbf{v}} \neq \vec{\mathbf{v}} \cdot \nabla$ and $\nabla\vec{\mathbf{v}} \neq \vec{\mathbf{v}}\nabla$.

1.22 (a) Given:

$$u = x^2 - y^2, \quad \mathrm{v} = -2xy;$$

and

$$\mathrm{v}_r = -A/r, \quad \mathrm{v}_\theta = -B/r.$$

Characterize the flows, find the streamline equations [Recall: Streamline equation to be obtained from $\frac{dy}{dx} = \frac{\mathrm{v}}{u}$, etc.; or from the stream function $\psi(x,y)$=¢, where $u \equiv \frac{\partial \psi}{\partial y}$ and $\mathrm{v} \equiv -\frac{\partial \psi}{\partial x}$, etc.], and plot the results. Comment!

(b) Derive Eq. (1.3.20) in cylindrical coordinates and discuss biofluids applications.

1.23 Continue Example 1.5 by finding $p(x)$, plotting $p(x)$ for different design parameters, and computing load L for an "optimal" case.

1.24 Consider a smooth steady liquid film of quasi-constant thickness δ on an inclined wall. The viscosity changes normal to the surface as $\mu = \mu_0 \, exp[-\alpha(y/\delta)]$ where α is a constant. Find the velocity distribution and the flow rate per unit depth.

1.25 Solve the Blasius problem [Recall: Steady laminar flat plate flow with $u_\infty = u_{outer}$=¢, i.e., $\partial p/\partial x = 0$]:

(a) using $u = u_\infty \, sin(ay/\delta)$ for the integral method; and

(b) employing $\eta = y/\delta$ using similarity theory (see App. A and /or page 154 in Kleinstreuer, 1997).

Obtain $u(x,y)$ and $\mathrm{v}(x,y)$ and plot $u(y)$ at two x-locations and $\mathrm{v}(x)$ along the B-L edge, $\delta(x)$.

1.26 Consider draining of a liquid film on a vertical plate. Avoiding initial and end effects, show that the film thickness

$$\delta(y,t) = \left(\frac{\nu y}{gt}\right)^{1/2}$$

Use a combined variable, η=η(y,t), to find the velocity profile and flow rate per unit depth.

1.27 Consider a long very slender cylinder of radius r_0 rotating at $\omega_0 = ¢$ in a large container filled with a liquid of viscosity ν.

(a) Under steady-state conditions, show that $v_\theta = r_0^2 \omega_0 / r$.

(b) Suddenly, the cylinder is stopped. Find $v_\theta(r,t)$.

1.28 Repeat Problem 1.27 (a) for turbulent fluid flow, assuming a log-law with "radial" coordinate $y = r - r_0$. Estimate the "radius of influence."

1.29 Consider the history of "drop-spreading" on a horizontal surface, where approximately,

$$-\sigma\left(R_1^{-1} + R_2^{-1}\right) + \tau_{nn}\Big|_{inside} = \tau_{nn}\Big|_{outside}$$

Perform a dimensional analysis of this problem and find the key (dimensionless) system parameters.

1.30 A balloon is being filled through inlet area A_{in} with a fluid velocity v_{in} of density ρ.

(a) Find an expression for the rate-of-change of system mass within the balloon at that instant.

(b) Can the RTT be used to solve a shrinking, i.e., vaporizing droplet? Set up (the) problem and find a macroscale solution.

1.31 For Poiseuille-type flow (ρ, μ, dp/dx=¢) in an airway (radius a) with a concentrically-placed cylindrical catheter (radius b), where b=æa, æ<1, consider three cases w.r.t. the catheter: (i) stationary; (ii) translating with u_0=¢; and (iii) translating and rotating with u_0 and ω_0. <u>Hint</u>: Solve for Case (iii) where the other ones are subsets, including the basic Poiseuille case (b=0).

1.32 Consider steady laminar flow (ρ, μ) in a parallel-plate device (graph) where one wall is lined with endothelial cells to demonstrate elongated cell alignment with substantial shear flow. Given ρ_{water} = 10^3 kg/m^3, μ_{water} = 10^{-3} Ns/m^2 and h=250μm as well as measued τ_{water}=10, 50, 100 dyn/cm^2, calculate the operational conditions, i.e., associated Reynolds numbers (check if $Re_{max} < Re_{critical} \approx 2000$) and entrance length (for cell placement in the fully-developed flow region).

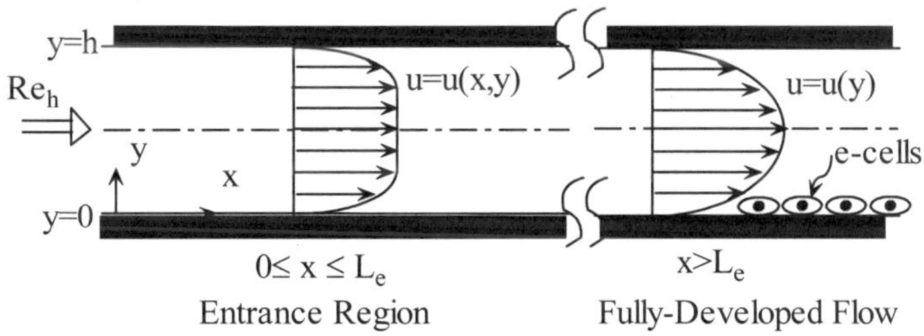

1.33 Red blood cells (RBCs) in tubular flow appear to congregate in the tube's central core, leaving a (nearly) cell-free plasma layer along the vessel wall. Assuming a constant layer thickness δ characterized by a plasma viscosity μ_p, and a homogeneous mixture region, $0 \leq r \leq (R-\delta)$ with μ_m, find suitable fluid property expressions/values and derive the velocity profiles, volumetric flow rates, and shear stress distributions.

1.34 Due to tissue/muscle actions or near vessel junctions, blood vessels are noncircular. As a first approximation, assume an elliptic cross section and find the volumetric flow rate Q(a, b; μ; dp/dz=¢) and the wall shear stress at any point (x_i, y_i) on the elliptic boundary.

1.35 Analyze the 3-D parallel plate solution of Usami et al. (1993) in Annals of Biomedical Engineering, Vol. 22, pp. 77-83, and then:

(a) calculate the wall shear stress vector $\vec{\tau}_w$; and

(b) recover the basic Poiseuille flow solution, i.e., $v_z=v_z(y)$.

1.36 Consider a syringe-balloon system applicable for temporary vessel occlusion, balloon angioplasty, metered medical dose delivery, etc. Employ the Reynolds Transport Theorem to find $R_b(t; d_s; v_s)$, where $R_b(t)$ is the balloon radius, d_s is the syringe diameter, and v_s is the average fluid velocity in the syringe, i.e., piston velocity.

1.37 In the deep lung region, O_2 from the inhaled air diffuses from the alveolus across the alveolar wall into the blood,

i.e., RBCs, moving in parallel-plate like capillaries (see Graph). In turn, CO_2 diffuses into the alveolus and is exhaled (see Fig. 1.1.3).
Focusing on the pressure drop for blood flow in one of the capillaries, develop dimensionless groups on which $\Delta p/\rho$ depends.

Graph:

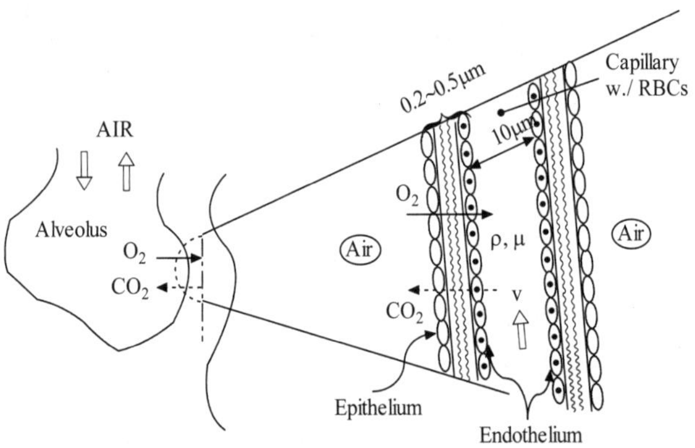

1.38 Some blood vessels are tapered, i.e., u=u(x, r) rather than just u(r) as in Poiseuille flow. Using Example 1.5 as a guide, find u(x, r), Q and τ_{wall} for a slightly tapered tube $R_1 \leq r \leq R_0$, where $r(x=0)=R_1>R_0= r(x=L)$.
Compare the theoretical results with experimental observations where $D_1=260\times10^{-6}$m and $D_0=78\times10^{-6}$m

Q [cm^3/s]$\times10^4$	15.58	9.64	6.57	3.175
τ_{wall} [dyn/cm^2]$\times10^{-3}$	541	330	225	120

Note: For microcirculation, the movement of a peripheral plasma layer should be considered.

1.39 Design a laboratory set-up and develop a mathematical model to test "flow through a stenotic tube." Of interest are the pressure drop across the stenosis and the centerline velocity as a function of inlet Reynolds number.

1.40 Oscillatory blood flow in straight vessels: Consider a circular pipe of radius R, filled with a viscous liquid of density ρ and dynamic viscosity μ, find the oscillatory, laminar, fully developed velocity profile as a function of radial poistion (r) and time (t), i.e., u(r,t) (Womersley, 1955), and calculate the volumetric flow rate Q(r,t). Assume that axial pressure gradient term can be decomposed into temporal harmonics of the form $\frac{\partial p}{\partial z} = \sum_{n=0}^{\infty} A_n e^{in\omega t}$ and the velocity can be decomposed as

$u(r,t) = \sum_{n=0}^{\infty} u_n(r) e^{in\omega t}$, where $i = \sqrt{-1}$ is the imaginary number.

1.41 The following graph is a representation of a carotid artery inlet flow waveform in terms of the centerline velocity. The waveform repeats with a frequency of 1 Hz (pulse rate 60 beats/min). The waveform can be represented by a Fourier decomposition:

$$u_{in}(t) = 40 + 20\sin(\omega t) + 18\sin(2\omega t - 0.5236) + 4\sin(3\omega t + 2.618) \quad \text{[cm/s]}$$

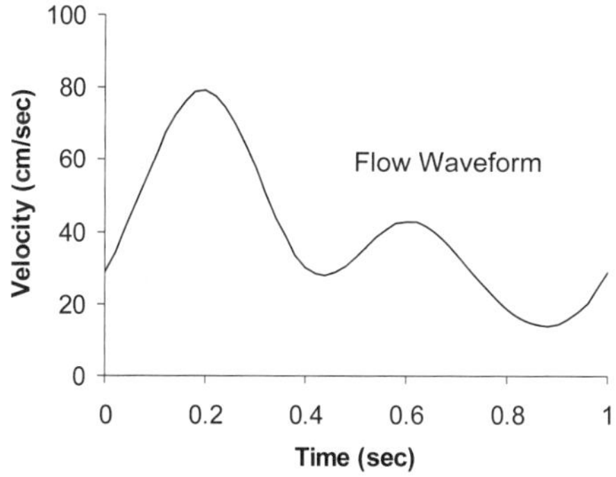

Calculate and plot the fully developed velocity profile as a function of radial position for the following time levels of the waveform: (i) early acceleration (t=0); (ii) peak systole (t = 0.2 sec); (iii) mid deceleration (t=0.3 sec); and (iv) end deceleration (t=0.4 sec). Assume that the tube radius is 0.4cm, the dynamic viscosity of blood is 0.035 g/(cm-sec), and that the density of blood is 1.05 g/cm^3.

1.42 Cone-and-plate viscometer: Derive a more accurate solution for $v_\theta(r,z)$, i.e., following Sdougos et al. (1984), it is postulated that

$$v_\theta = \omega r \left\{ \left(\frac{z}{r\alpha} \right) + \varepsilon^2 \left[A\left(\frac{z}{r\alpha} \right) + B\left(\frac{z}{r\alpha} \right)^4 + \cdots \right] \right\}$$

where α is the plate-cone angle and $\varepsilon = \rho r^2 \alpha^2 \omega / \mu << 1$.
(a) Derive the circumferential velocity and interpret ε.
(b) Find an expression for the wall shear stress and calculate ω and $\tau_{z\theta}$ (z=0) for R=30cm, α=3°, $\mu = 0.01 g/cm \cdot s$, and ρ=1g/cm^3, such that the shear stress does not differ by more than 10% between location R_1=10cm and R_2=20cm.
(c) Describe BME applications of this device.

1.43 A bed of capillaries in a tissue can be approximately represented as a repetitive arrangement of capillaries surrounded by a cylindrical layer of tissue (see graph). Assume that the blood flows through the capillary with a uniform velocity U and an inlet solute concentration C_0. Calculate the solute concentration distribution in the tissue space. The consumption (or production) of the solute by the cells within the tissue space is the driving force for diffusion of the solute, which can be described by an effective diffusivity D_T. Assume that the metabolic consumption of the solute is a constant (R_0) and the combined resistance of the fluid flowing through the capillary and the permeability of the solute in the capillary wall can be represented by an overall mass transfer coefficient (K_0).

Sect.1.4 problems:

1.44 Draw two time-sequence profiles of fluid flow start-up in simple Couette flow for:
(a) Newtonian fluids; and
(b) power-law fluids.
Justify your graphs with physical/ mathematical arguments.

1.45 Derive an expression (i.e., equation) to describe the relationship between the velocity and pressure gradient for oscillatory flow of a Casson fluid (blood) in a rigid cylindrical tube.

1.46 Non-Newtonian flow in a stenosed artery: Assume that a stenosis develops in the arterial wall in an axially symmetric manner. Following Young (1968) and considering steady, laminar, fully developed blood flow in an artery with mild stenosis ($\delta_h \ll R$ with δ_h being the maximum height of the stenosis and R being the radius of the normal artery), find the effects of stenosis on resistance to flow and wall shear stress. That is, develop the expressions for ratio of λ/λ_n and τ/τ_n as a function of δ_h/R (where λ and τ are the resistance and wall shear stress in the presence of the stenosis, respectively; and λ_n and τ_n are the resistance and wall shear stress in the case of no stenosis, respectively). Consider the blood to be a power-law fluid, where the effect of flow behavior index n should be discussed as well.

1.47 Micro-particle deposition in a 2-D bend: Consider a spherical aerosol flowing in a two-dimensional bend with θ in radians, as shown in the sketch. Derive an expression to calculate the deposition probability of the particle. Assume that the airflow is uniform with a velocity of U and $\delta \ll R$. The deposition probability is the ratio of δ and $2r_0$.

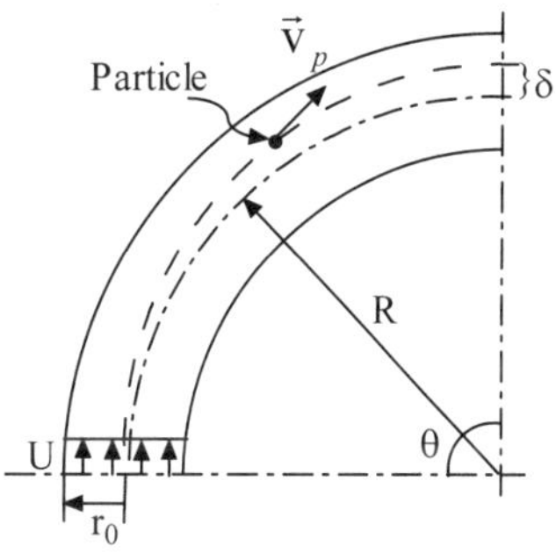

1.48 The suspended leukocytes in a tissue culture medium may flow through a parallel-plate channel and adhere to the endothelial cells which make up the lower surface of the channel. Assume that the channel height is 250μm, width is 2cm, length is 7.5cm, the fluid viscosity is 0.009 $gcm^{-1}s^{-1}$, density is 1gcm^{-3}, and the average velocity is 2000 μms^{-1}.

(a) Calculate the drag force acting on a leukocyte of diameter (d) 12μm that adheres to the endothelium. The density of leukocyte is 1.07gcm^{-3}. When the spherical particles move near a solid surface, the additional drag force due to the presence of surface should be considered besides the Stokes drag force (Eq. (1.4.26)). The additional drag force is tabulated by Goldman et al.

(1967) as a function of H/d with H being the height of the particle above the surface.

(b) Find the distance the above leukocyte travels through the channel before it adheres to the endothelial cells.

1.49 Calculation of the respiratory mass transfer coefficient, h_m, is helpful in quantitatively predicting the regional uptake of inhaled vapor or nanoparticles. Considering an airway segment with length *L* and diameter *D*, discuss how to calculate h_m. Assume that the concentration distributions of vapor or nanoparticles are known. Also derive an equation to calculate the uptake coefficient (i.e., deposition efficiency) of vapor or nanoparticles from h_m with the assumption of zero wall concentration.

1.50 The capillary wall can be modeled as a porous membrane. Assume that such a porous medium consists of n parallel tubes per unit surface area (see Sketch in Example 1.18) where each tube has length of *h* and diameter of d. Find the vascular permeability coefficient *P* as a function of *n*, *h*, *d,* and *D* (the diffusion coefficient in the pore). Assume that diffusion through each tube is identical and the solute size is much less than *d.*

1.51 Tissue as a porous medium surrounding capillaries: The medium between endothelial cells in the capillary wall could be considered as channels with random cylindrical fibers. Assume that the area fraction of such porous clefts on the endothelial surface is A_p, the channel length is *h*, the average diameter of fiber is d_f, and the fiber length per unit volume of the porous region is *l*.

(a) Find the porosity of the clefts and the hydraulic conductivity of vessel wall using the Carmen-Kozeny theory (cf. Eq. (1.4.42)).

(b) Find the fluid flux across the capillary wall assuming that the transmural pressure difference is Δp and the wall thickness is *h*.

1.52 Similar to Example 1.20, now consider steady laminar flow in a channel of length *L* and height 2*h* with porous walls. Find the mass flow into the wall. Sketch the velocity profiles in the channel and compare to those in a channel with a solid wall.

1.53 As part of atherogenesis, the cholesterol and plasma may pass across the damaged region in the endothelial layer to the sub-endothelial space. Assume that the damaged region is a circular hole with radius a and the sub-endothelial space is a semi-infinite porous medium with hydraulic conductivity K. The fluid and cholesterol transport radially and symmetrically about the center of the hole. Find the steady distribution of fluid velocity, pressure, and cholesterol concentration in the sub-endothelial space. The cholesterol concentration and the fluid pressure at the wall of the hole (i.e., $r=a$) is c_0 and p_0, respectively, while they are zero at the outside wall of the sub-endothelial space (i.e., $r \rightarrow \infty$). The diffusion coefficient of cholesterol in the tissue is D.

Sect.1.5 problems:

1.54 Many soft tissues often conserve the volume when deforming. Considering such a soft tissue with original length L=10mm, compare the variations of true and nominal stresses for the current lengths (l) of tissue from 10.001 to 10.7mm. Recall: The true stress is a measure of forces acting over a deformed oriented area A, while the nominal stress is forces over the undeformed oriented area A_0. The stress and strain are assumed to be uniform.

1.55 Consider a 2-D loaded body with planar element $\Delta x, \Delta y$ (see Graph). Show that, indeed, there are axes where all shear stresses vanish.

Graph:

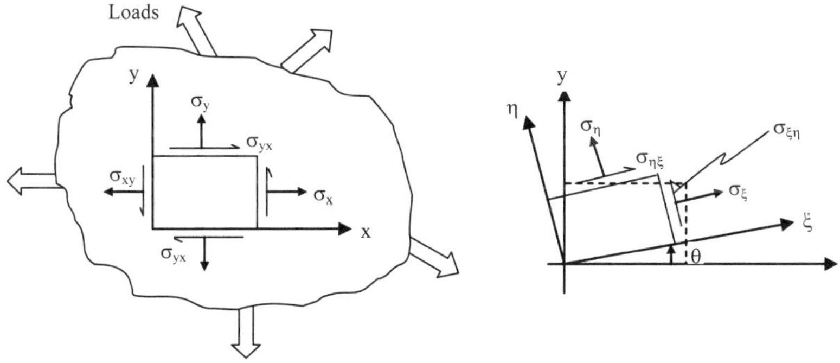

(a) Basic Material Element (b) Arbitrary Element Rotation

1.56 Given the state of stress in Problem 1.55, σ_x=120 kPa, σ_y=150 kPa, and σ_{xy}=0 kPa, calculate the values of σ_η for all values of θ from 0° to 90° and plot as a function of θ.

1.57 Assuming that σ_x=0, σ_y=0, and $\sigma_{xy}=\sigma_{yx}$=5 MPa, find the values of the principal stresses and denote them on an infinitesimal element with orientation given by σ_p.

1.58 Material deformation analyis (see Sect.1.5.4): the component of a 2-D displacement vector are given by $u_x=(\Lambda-1)a_1$ and $u_y=0$. Calculate and compare the exact (E_x) and the approximate/linearized (ε_x) strains for Λ=1.001, 1.01, 1.1, 1.5, and 2.0. Compute the error introduced by the linearization in each case and evulate the values of Λ for obtaining the reasonable approximation. The calculation of E_x can be found in Humphrey & Delange (2004).

1.59 Given σ_x=20 MPa, σ_y=-10 MPa, and σ_{xy}=-20 MPa, find the principal stresses and principal strains with linear, elastic, homogeneous, and isotropic behaviors where E=16 GPa and ν=0.325.

1.60 The strain energy function of a material can be

$$\phi = \frac{\mu}{k}\left(\lambda_1^k + \lambda_2^k + \lambda_3^k - 3\right)$$

where $\lambda_1, \lambda_2, \lambda_3$ are the principal extensions and μ and k are material constants.

(a) Show that the inflation pressure p for a spherical balloon can be given by

$$p = \frac{2h_0\mu}{R_0}\left(\lambda^{k-3} - \lambda^{-2k-3}\right)$$

where λ is the circumferential extension, and h_0 and R_0 are the initial (zero-pressure) balloon thickness and radius, respectively.

(b) Demonstrate the possibility of blow out for a rubber balloon with k =2 and show that a ventricle will not blow out ($k \approx 18$ for myocardium).

(c) Plot the pressure as a function of k and comment.

1.61 Example 1.25 revisited:

(a) Show that Eq. (E1.25-1) derived for a thin-walled cylinder also provides a reasonable estimate for the mean circumferential stress in a thick-walled tube.

(b) In hypertension, the aorta distends (i.e., r_1 increases) and the wall thins (i.e., τ decrease) so that σ_θ increases dramatically. Consider a luminal pressure of $p_L=np$, where the constant n can be varied, and the luminal radius returns to r_1 due to smooth muscle contraction and a shear-stress-mediated vasoconstriction. How much does the aorta need to thicken to restore σ_θ back to its original value?

1.62 Estimate the axial, partitioned stress in a cylindrical bone with a concentrically implanted prosthesis subjected to different loads $N=N^{bone}+N^{prosthesis}$ assuming equal axial strain components, i.e., $\varepsilon_x{}^b=\varepsilon_x{}^p$, and different but uniform properties, i.e., A_b, E_b and A_p, E_p, for bone and prosthesis, respectively.

1.63 Consider a spherical thin-walled aneurysm, e.g., at an intracranial location, assuming that its radius R=2.5mm, wall thickness h=15μm, and a varying mean net blood pressure p≥120 mmHg. Setting the aneurysm rupture stress at $\sigma_{critical}$ ≈2MPa, calculate the maximum mean pressure when rupture may occur.

References

Alazmi B., and Vafai K. (2002) Int. J.Heat & Mass Transfer, 45: 3071-87.

Alberts, B., Johnson, A., Lewis, J., Raff, M., Roberts, K., and Walter, P. (2002) "Molecular Biology of the Cell," Garland Publishing, New York, NY.

Anderson, D. A., Tannehill, J. C., and Pletcher, R. H. (1998) "Computational Fluid Mechanics and Heat Transfer," McGraw-Hill, New York, NY.

Arpaci, V. S., Kao, S. H., and Selamet, A. (1999) "Introduction to Heat Transfer," Prentice Hall, Upper Saddle River, NJ.

Astarita, G. (1997) Chem. Eng. Sci., 52 (24): 4681-98.

Batchelor, G. K. (1970) J. Fluid Mech., 41: 545.

Bejan, A. (1995) "Convection Heat Transfer," Second Edition, John Wiley & Sons, New York.

Bird, G. A. (1994) “Molecular Gas Dynamics and the Direct Simulation of Gas Flows,” Oxford University Press, New York, NY.

Bird, R. B., Armstrong, R. C., and Hassager, O. (1987) “Dynamics of Polymetric Liquids,” Wiley-Interscience, New York, NY.

Bird, R. B., Stewart, W. E., and Lightfoot, E. N. (2002) “Transport Phenomena,” John Wiley, New York, NY.

Blasius, H. (1910) Z. Math. Phys., 58: 90.

Boussinesq, J. (1877) Mem. Pres. Acad. Sci., 23: 46.

Brenner, H., and Edwards, D. A. (1993) “Macrotransport Processes,” Butterworth-Heinemann, Boston, MA.

Buchanan, J. R., and Kleinstreuer, C. (1998) ASME J. Biomech. Eng., 120: 446-54.

Cengel, Y.A. and Cimbala, J.M. (2006) “Fluid Mechanics: Fundamentals and Applications,” McGraw-Hill, New York, NY.

Chandler, D. (1987) “Introduction to Modern Statistical Mechanics,” Oxford University Press, New York, NY.

Chen, H. T., and Middleman, S. (1967) AIChE J., 13 (5): 989.

Churchill, S. W. (1988) "Viscous Flows: The Practical Use of Theory," Butterworth-Heinemann, Boston, MA.

Cicchitti, A., Lombardi, C., Silvestri, M., Soldaini, G., and Zavattarelli, R. (1960) Energia Nucl., 7: 407-525.

Clift, R., Grace, J. R., and Weber, M. E. (1978) “Bubbles, Drops, and Particles,” Academic Press, New York, NY.

Crowe, C., Sommerfeld, M., and Tsuji, Y. (1998) “Multiphase Flows with Droplets and Particles,” CRC Press, Boca Raton, FL.

Drew, D. A., and Passman, S. L. (1999) “Theory of Multicomponent Fluids,” Springer-Verlag, New York, NY.

Dukler, A. E., Wicks, M., and Cleveland, R. G. (1964) AIChE J., 10: 44- 51.

Fan, L. S., and Zhu, C. (1998) “Principles of Gas-Solid Flows,” Cambridge University Press, New York, NY.

Ferziger, J. H., and Peric, M. (1999) “Computational Methods for Fluid Dynamics,” Springer-Verlag, New York, NY.

Finlayson, B. A. (1978) “Nonlinear Analysis in Chemical Engineering,” McGraw-Hill, New York, NY.

Fox, S. I. (2002) “Human Physiology,” McGraw-Hill, Boston, MA.

Fox, R. W., McDonald, A.T., and Pritchard, P. J. (2004) “Introduction to Fluid Mechanics,” Wiley, New York, NY.

Fung, Y. C. (1990) “Biomechanics: Motion, Flow, Stress, and Growth,” Springer-Verlag, New York, NY

Fung, Y. C. (1993) “Biomechanics: Mechanical Properties of Living Tissues,” Springer-Verlag, New York, NY.

Fung, Y. C. (1994) “A First Course in Continuum Mechanics: For Physical and Biological Engineers and Scientists,” Prentice Hall, Englewood Cliffs, NJ.

Fung, Y. C. (1997) “Biomechanics: Circulation,” Springer, New York, NY.

Gad-el-Hak, M. (1999) ASME J. Fluids Eng., 121: 5-12.

Gates, C.M., and Newman J. (2000) AICHE J., 46 (10): 2076-2085.

Gebhart, B., Jalurice, Y., Mahajae, R. C., and Sammakice, B. (1988) "Buoyancy-Induced Flows and Trasport," Hemisphere Publishing Corp., New York, NY.

Greenberg, M. D. (1998) "Advanced Engineering Mathematics," Prentice Hall, Upper Saddle River, NJ.

Goldman, A.J., Cox, R.G., and Brenner, H. (1967) Chemical Engineering Science, 22: 637-651.

Gosman, A. D., and Ioanuides, E. (1981) AIAA Paper, 81-03223.

Hansen, A. G. (1964) "Similarity Analyses of Boundary Value Problems in Engineering," Prentice Hall, Englewood Cliffs, NJ.

Happel, J., and Brenner, H. (1983) "Low Reynolds Number Hydrodynamics with Special Applications to Particulate Media," Martinus Nijhoff Publ., Boston, MA.

He, C. H., and Ahmadi, G. (1999) J. Aerosol Sci., 30 (6): 739-58.

Hoffman, J. D. (2001) "Numerical Methods for Engineers & Scientist," Marcel Dekker, Inc., New York.

Humphery, J.D., and Delange, S.L. (2004) "An Introduction to Biomechanics," Springer, New York, NY.

Isbin, H. S., Moy, J. E., and Da Cruz, A. J. R. (1957) AIChE J., 3: 361-5.

Joseph, D. D. (2001) "Interrogations of Direct Numerical Simulation of Solid - Liquid Flow," Res. Report UMSI 2001/26, University of Minnesota, Minneapolis, MN.

Karino, T., and Goldsmith, H. L. (1977) Phil. Trans. R. Soc. Lond. Ser. B, 279 (967): 413.

Kashiwa, B. A., and Van der Heyden, W. B. (2000) "Toward a General Theory for Multiphase Turbulence: Part I," Los Alamos National Labs, Los Alamos, NM.

Katz, I. M., and Martonen, T. B. (1999) J. Aerosol Sci., 30: 173-83.

Kaviany, M. (2002) "Principles of Heat Transfer," Wiley-Interscience, New York, NY.

Kays, W. M., and Crawford, M. E. (1993) "Convective Heat and Mass Transfer," McGraw-Hill, New York, NY.

Keller, K.H. (1971) Fed. Proc., 30: 1591.

Kim, I. C., Elghobashi, S., and Sirignano, W. A. (1998) "On the Equation for Spherical-Particle Motion: Effect of Reynolds and Acceleration Numbers," MAE Dept., University of California, Irvine, CA.

Kleinstreuer, C. (1997) "Engineering Fluid Dynamics: An Interdisciplinary Systems Approach," Cambridge University Press, New York, NY.

Kleinstreuer, C. (2003) "Two-Phase Flow: Theory and Applications," Taylor & Francis, New York, NY.

Kleinstreuer, C., and Agarwal, S. S. (1987) Int. J. Eng. Sci., 25 (5): 597-607.

Kleinstreuer, C., Hyun, S., Buchanan, J. R., Longest, P. W., Archie, J. P., and Truskey, G. A. (2001) Crit. Rev. Biomed. Eng., 29 (1): 1-64.
Kolmogorov, A. N. (1942) Izv. Akad, Nauk S.S.S.R., Ser. Fiz., 6: 56.
Li, A., and Ahmadi, G. (1992) J. Aerosol Sci. Tech., 16: 209-26.
Li, A., and Ahmadi, G. (1995) J. Aerosol Sci. Tech., 23 (2): 201-23.
McAdams, W. H. (1942) Trans. ASME, 64: 193.
Macosko, C. W. (1994) "Rheology: Principles, Measurements and Applications," VCH Publ., New York, NY.
Merrill, E. W. (1968) J. Physiol. Rev., 26 (4): 863-88.
Michaelides, E. E. (1997) J. Fluids Eng. Trans. ASME, 119 (2): 233-47.
Michel C. C. (1997) Exp. Physiol., 82: 1-30.
Middleman, S. (1998) "An Introduction to Fluid Dynamics," John Wiley, New York, NY.
Loth, E. (2000) Prog. Energy Combust. Sci., 26 (3):161-223
Murray, C.D. (1926) Proc. National Academy of Science, 12: 207-14.
Na, T. Y. (1979) "Computational Methods in Engineering Boundary Value Problems," Academic Press, New York, NY.
Nakayama A., Kuwahara F., Sugiyama M., and Xu G.L. (2001) Int. J. Heat & Mass Transfer, 44 (22): 4375-4379.
Naterer, G. F. (2003) "Heat Transfer in Single and Multiphase Systems," CRC Press, Boca Raton, FL.
Nichols, W. W., and O'Rourke, P. J. (1998) "McDonald's Blood Flow in Arteries: Theoretical, Experimental and Clinical Principles," Lea and Febiger, Philadelphia, PA.
Nield, D.A., and Bejan, A. (1999) "Convection in Porous Media," Springer, New York, NY.
Nield, D.A., Kuznetsov A.V., and Xiong M. (2002) Int. J. Heat & Mass Transfer, 45 (25): 4949-4955.
Oran, E. S., Oh, C. K., and Cybyk, B. Z. (1998) Ann. Rev. Fluid Mech., 30: 403-41.
Özisik, M. N. (1993) "Heat Conduction," Wiley-Interscience, New York, NY.
Panton, R. L. (2005) "Incompressible Flow," (3rd edition) Wiley, New York, NY.
Patankar, N. A., and Hu, H. H. (2001) J. Non-Newtonian Fluid Mechanics, 96: 427-443.
Patankar, N. A., Singh, P., Joseph, D. D., Glowinski, R., and Pan, T. W. (2000) Int. J. Multiphase Flow, 26 (9): 1509-24.
Pnueli, D., and Gutfinger, C. (1992) "Fluid Mechanics," Cambridge University Press, Cambridge, UK.
Polyanin, A. D., and Zaitsev, V. F. (1995a) "Exact Solutions for Ordinary Differential Equations," CRC Press, Boca Raton, FL.
Polyanin, A. D., and Zaitsev, V. F. (1995b) "Handbook of Solutions for Partial Differential Equations," CRC Press, Boca Raton, FL.

Pope, S. B. (2000) “Turbulent Flows,” Cambridge University Press, Cambridge, UK.

Potter, M. C., and Wiggert, D. C. (1997) “Mechanics of Fluids,” Brooks/ Cole, Pacific Grove, CA.

Prandtl, L. (1904, and 1928) “Über Flüssigkeitsbewegung bei sehr kleiner Reibung,” Teubner Verlag, Leipzig and NACA Memo N. 452.

Roache, P. J. (1998a) J. Fluids Eng. Trans. ASME, 120 (3): 635.

Roache, P. J. (1998b) AIAA J., 36 (5): 696-702.

Rodi, W. (1980) “Turbulence Models and Flow Applications in Hydraulics,” IAHR, Delft, The Netherlands.

Sabersky, R. H., Acosta, A. J., Hamptmann, E. G., and Gates, E. M. (1998) “Fluid Flow,” Prentice Hall, Upper Saddle River, NJ.

Schlichting, H., and Gersten, V. (2000) “Boundary-Layer Theory,” Springer-Verlag, Berlin, Germany.

Sdougos, H.P., Bussolari, S.R., Dewey, C.F. (1984) J. Fluid Mechanics, 138: 379-404.

Silverthorn, D.U. (2004) “Human Physiology: An Integrated Approach,” Benjamin Cummings, San Francisco, CA.

Smits, A. J. (2000) “A Physical Introduction to Fluid Mechanics,” John Wiley, New York, NY.

Soo, S. L. (1990) "Multi-Phase Fluid Dynamics," Science Press, Beijing, P.R.C. and Gower Technical, Brookfield, MA.

Spalding, D. G. (1961) J. Appl. Mech., 28: 455-57.

Spiegel, M. R. (1971) “Advanced Mathematic for Engineering & Scientists,” Schrum’s Outline Press, McGraw-Hill, New York, NY.

Tanner, R. I. (1985) "Engineering Rheology," Oxford University Press, Oxford, UK.

Taylor, G. I. (1932) Proc. R. Soc., Ser. A, 138: 41.

Tannehill, J. C., Anderson, D. A., and Pletcher, R. H. (1997) "Computational Fluid Mechanics and Heat Transfer," Taylor & Francis, Washington, DC.

Truskey, G. A., Yuan, F., and Katz, D.F. (2004) “Transport Phenomena in Biological Systems,” Pearson Prentice Hall, Upper Saddle River, NJ.

Ugural, A.C., and Fenster, S.K. (1995) “Advanced Strength and Applied Elasticity,” Prentice Hall, Upper Saddle River, NJ.

Vander, A., Sherman, J., and Luciano, D. (2001) “Human Physiology: The Mechanisms of Body Function,” McGraw-Hill, Boston, MA.

Wallis, G. B. (1969) “One Dimensional Two-Phase Flow,” McGraw-Hill, New York, NY.

Warsi, Z. U. A. (1999) “Fluid Dynamics,” CRC Press, Boca Raton, FL.

White, F. M. (1991) “Viscous Fluid Flow,” McGraw-Hill, New York, NY.

White, F. M. (2003) “Fluid Mechanics,” McGraw-Hill, Boston, MA.

Wilcox, D. C. (1998) “Turbulence Modeling for CFD,” DCW Industries, La Canada, CA.

Wilkes, J. O. (1999) “Fluid Mechanics for Chemical Engineers,” Prentice Hall, Upper Saddle River, NJ.

Womersley, J. R. (1955) J. Physiology, 127:553-563.

Yakhot, V., and Orszag, S. A. (1986) J. Sci. Computing, 1: 3-51.

Young, D.F. (1968) ASME J. Eng. Industry, 90: 248-254.

Zamir, M. (2000) “The Physics of Pulsatile Flow,” AIP Press, Springer-Verlag, New York, NY.

Zapryanov, Z., and Tobakova, S. (1998) "Dynamics of Bubbles, Drops, and Rigid Particles," Kluwer Academic Publ., New York, NY.

Zhang, Z., and Kleinstreuer, C. (2002) Phys. Fluids, 14 (2): 862-80.

Chapter II

BIOFLUID DYNAMICS CONCEPTS

Biofluid dynamics encapsulates fluid flow, including fluid-structure interaction (FSI), heat transfer and mass transfer in mammalian systems as well as in medical devices. Focusing on the human body, such transport and FSI processes occur at the cellular, tissue, organ, and whole-body levels (Truskey et al., 2004; Datta, 2002; Berger et al., 1996; Charney, 1992, among others). As discussed in Sect. 1.1.2, water, solutes, proteins, ions, etc. are driven across the cell (or plasma) membrane (Fig. 1.1.1) by passive diffusion, hydraulic, and osmotic pressure differentials, as well as other more complex transport mechanisms, such as carrier-mediated, passive-ion, and active transport (see Pollack, 2001; or Weiss, 1996). For the metabolic needs of the cells at the tissue and organ level, oxygen and nutrients are provided by the respiratory and circulatory systems, relying on effective air and blood flow on the macro-level as well as diffusive oxygen and species transfer on the micro-level. At the whole-body level, thermoregulation via body movement and sweating is necessary to achieve a constant (body) temperature. For thermal therapy, hyperthermia or cryogenics can be employed to destroy diseased tissue or deactivate cells and body parts.

Chapter I, in conjunction with Appendices A-C, reviewed the essentials of transport phenomena, summarized basic solution techniques, and provided standard exercises (see Sect. 1.7), in order to establish a sufficient background for studying biofluid dynamics topics of the remaining chapters. For decoupled fluid-structure and/or fluid-particle systems, solving the conceptual and mathematical complexities of fluid mechanics is most challenging. Once that has been accomplished, the scalar transport phenomena of living structures and biomedical devices typically encountered can be readily simulated and analyzed, provided that initial/

boundary conditions, as well as reliable data for material properties, transfer parameters, and computer model validation are available. Now, Chapter II deals with a representative example of biotransport phenomena as well as biofluids aspects of the cardiovascular system. Chapter II material functions as a stepping stone to understand the mechanisms and implications of arterial diseases (Ch. III) as well as that of respiratory, filtering, and regulatory organs (see Ch. IV).

2.1 Transport Phenomena

Through various mechanisms, a healthy body regulates fluid flow, mass transfer, and heat exchange in such a way that "extremes" are avoided. For example, blood flow and air flow are laminar except in severely stenosed arteries and in the largest airways during rapid breathing. For blood flow, this is achieved with the controlled pumping action of the heart and the viscoelasticity of the blood vessels reacting locally to variations in blood pressure. In the lung, the local Reynolds number drops rapidly because after the trachea the airways bifurcate swiftly into smaller and smaller branches. The body temperature varies only between, say, 34°C at extremities and 37°C in the core due to the thermo-regulatory activities of the blood and skin.

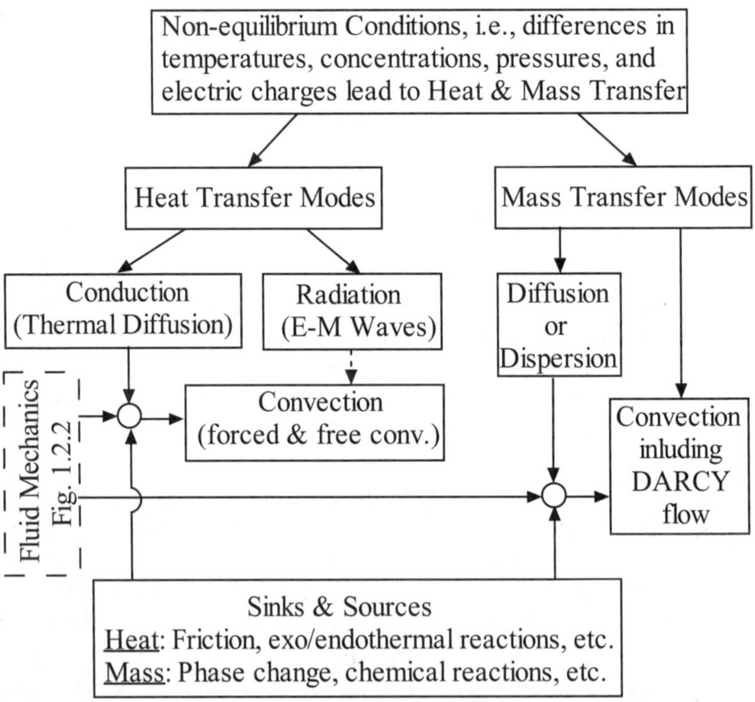

Fig. 2.1.1 Concepts of transfer processes.

Figure 2.1.1 lists key driving forces and concepts associated with heat and mass transfer. It should be noted that heat convection is not a separate energy transfer mode but actually conduction and possibly thermal radiation in a moving fluid. As mentioned, fundamental to heat and mass transfer, beyond diffusional processes in and around cells, is momentum transfer (Fig. 1.2.2 and Sect. 1.3), including fluid-particle dynamics (Sect. 1.4) and fluid-structure interactions (Sect. 1.5). These topics are further elucidated in subsequent sections when applied to basic living structures.

2.1.1 Biofluid-compartment Models

Because of the complexities involved in solving biofluid flow problems and modeling micro-scale to macro-scale interactions (say, membrane → cell → tissue → vessel → organ → body-segments → whole-body), transport phenomena were traditionally analyzed with a lumped-parameter (or body-compartment) approach, relying on coupled first-order rate equations or greatly simplified transient diffusion/dispersion equations (cf. Middleman, 1972; among others). In general, a compartment is a "well-mixed box," e.g., a homogeneous, isotropic tissue region or capillary bed, organ, or body part, which has uniform inlet and outlet streams carrying species, i.e., O_2, CO_2, ions, nutrients, toxins, or drugs of different concentrations. Of interest is the time-rate-of change of species concentrations inside the box as well as species deposition and conversion. Examples include:

- Single or multiple body compartments in various series, loop, and parallel arrangements. They are employed when tracking a chemical through the body from intake to discharge, with an emphasis on its conversion and deposition in a specific organ (see Finlay, 2001, for example, on physiologically based pharmaco-kinetics modeling). Alternatively, inert tracers could be injected and their transient concentrations could be measured in order to evaluate mass transfer coefficients or volumetric flow rates with an appropriate compartmental transport model (see Fig. 2.1.2). Needless to say, these basic units can be combined into different formations.
- Heat and/or mass transfer of the human body in a polluted environment, e.g., a workplace; a scenario which could be modeled as "a cylinder in cross flow with uniform point inhalation" (see Hyun & Kleinstreuer, 2001).
- The human respiratory system could be lumped into a "collapsible-tube model" consisting of a tube with axially varying cross-section and material properties as well as a balloon-like end, representing the alveolar region (see Fig. 2.1.3). An inlet pressure wave $p_{in}(t)$ changes the lung pressure, p_e, plus lung volume which depends on the surrounding tissue, i.e., the pleural pressure p_s (see Sects. 4.1 and 4.2).

- The weight of a human body, depending on age and sex, could be subdivided into three compartments, consisting of: (i) fat; (ii) extra-cellular mass with extra-cellular water (ECW); and (iii) body cell mass with intra-cellular water (ICW). Figure 2.1.4 depicts the distribution of the total body fluids, i.e., basically water, which makes up 60% of the body weight in adult males.

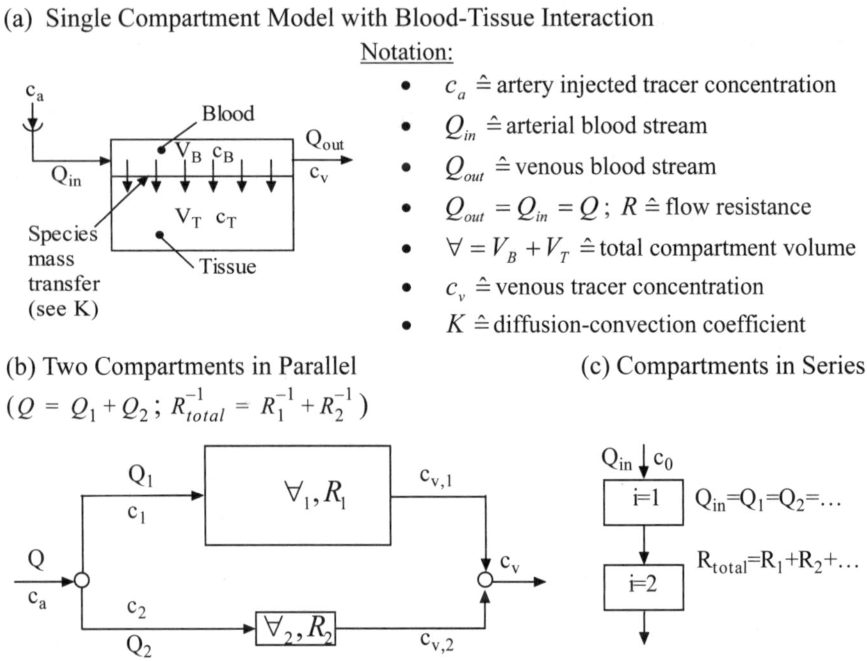

Fig. 2.1.2 Single- and multiple-compartment models.

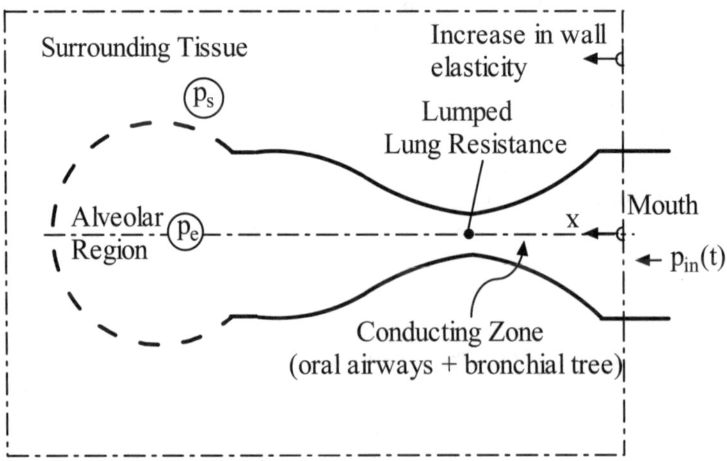

Fig. 2.1.3 The lung as an elastic-tube model.

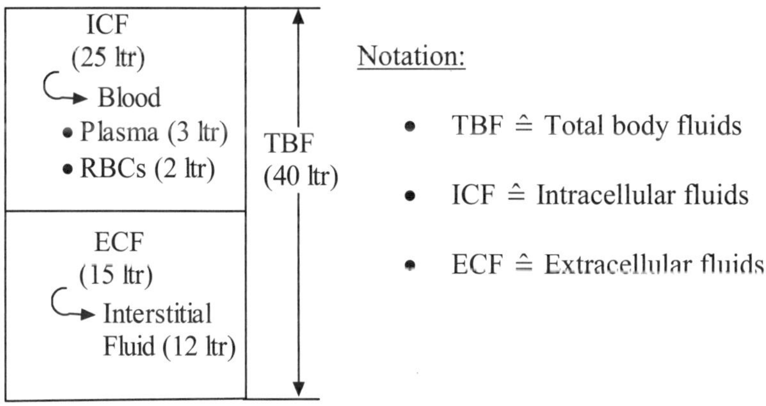

Fig. 2.1.4 Example of total body-fluid distribution.

In general, the material conservation law for compartments in series (see Fig.2.1.2c) can be expressed as:

$$\frac{d\,c_i}{dt} = \sum_{j=1}^{N} (q_{ji} - q_{ij}) + S_i(t) \tag{2.1.1}$$

where c_i is the amount of material per unit volume accumulating in compartment i, q_{ij} is the rate of material being transferred from compartment i to compartment j, and $S_i(t)$ is the rate of material injected from the system's exterior (or generated due to chemical reaction in side compartment i).

A constitutive relationship is necessary to express the transfer rates q_{ij} with the principal unknown c_i, i.e.,

$$q_{ij} = f(c_i) \tag{2.1.2a}$$

where a Taylor series expansion around point *c* yields

$$f(c) = a_o + a_1\, c + a_2 c^2 + \cdots \tag{2.1.2b}$$

For *c* being positive and much less than unity, *f(c)* can be linearized to

$$f(c) \approx a_1 c \tag{2.1.2c}$$

and hence

$$q_{ij} = k_{ij}\; c_i \tag{2.1.2d}$$

where k_{ij} are the rate constants.

Thus, for linear compartmental models with constant coefficients, Eq. (2.1.1) can be written for each compartment *i* as:

$$\frac{dc_i}{dt} = \sum_{j=1}^{N} (k_{ij}c_j - k_{ji}c_i) + S_i(t) \qquad (2.1.3)$$

Equation (2.1.3) has a homogeneous solution of the form

$$c_i(t) = \sum_{k=1}^{n} A_k e^{-\alpha_k t} \qquad (2.1.4)$$

where the coefficients A_k are proportional to k_{ij}, $c_i(t=0)$, and $\mathrm{S_i}(t)$, while the eigenfrequencies $\alpha_k \sim k_{ij}$.

The particular solution for $c_i(t)$ depends on the information for k_{ij}, $c_i(t=0)$, and $S_i(t)$. Linearity of Eq. (2.1.3) allows one to compute $c_i(t)$ in steps and then sum up all the (independent) solutions (see Hoffman, 2001, among others).

<u>Example 2.1</u>: Steady-state Solution for Three Compartments in a Loop Configuration

Take $k_{12} = 1, k_{23} = 1, k_{31} = 4, k_{32} = 2, k_{21} = 1, \text{and } \mathrm{k}_{13} = 2,$ while initially $c_1 = (0) = 1, c_2(0) = 0, \text{and } \mathrm{c}_3(0) = 0.$

<u>Sketch:</u>

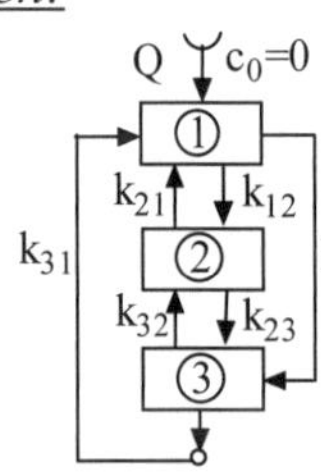

<u>Assumptions:</u>

- Well-mixed compartments
- Steady-state
- No external input nor sinks/sources

<u>Concepts:</u>

- Lumped parameter approach
- First-order rate equation

<u>Solution:</u> With $dc_i/dt = 0$ and $S_i(t) = 0$ and using the k_{ij} values, Eq. (2.1.3) reads for *i=1*:

$$\frac{dc_1}{dt} = 0 = -(k_{21} + k_{31})c_1 + k_{12}\,c_2 + k_{13}\,c_3 \qquad \text{(E2.1-1)}$$

or with the *k*-values for *i=1,2,3*:

$$-5c_1 + c_2 + 2c_3 = 0$$

$$c_1 - 3c_2 + c_3 = 0 \qquad \text{(E2.1-2a-c)}$$

$$4c_1 + 2c_2 - 3c_3 \ = 0$$

This matrix has the solution

$$c_1 = c_2 = 0.5\,c_3$$

Thus, a second equation is needed. The total species mass per unit volume in the system was at time t = 0:

$$c_{\text{initial}}^{\text{total}} \equiv c_1(0) + c_2(0) + c_3\,(0) = 1.0 \qquad \text{(E2.1-3a)}$$

Now, mass conservation requires that at all times

$$\sum_{i=1}^{3} c_i(t) = c_{\text{initial}}^{\text{total}} \ = 1.0 \qquad \text{(E2.1-3b)}$$

Hence, the steady-state values of c_i in the three compartments are:

$$c_1 = 0.25,\, c_2 = 0.25, \text{and}\, c_3 = 0.5$$

Comments: At steady-state an equilibrium has been reached after initially c_1 was equal to 1.0 while c_2 and c_3 were at zero concentration. The strong transfer out of compartment ① (see Eq. (E2.1-1)) favors influx to compartment ③ for which k_{13}=2.0.

Example 2.2: Single-Compartment Model

Returning to the single-compartment model as depicted in Fig. 2.1.2a, it is of interest to find an equation describing the outlet species concentration $C_v(t)$ for a given inlet tracer concentration C_a.

Sketch:

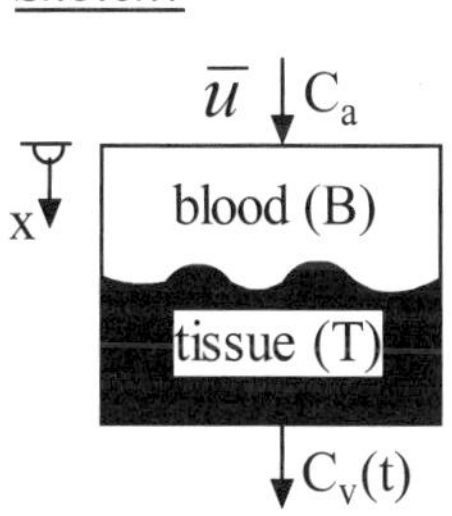

Assumptions:

- Dual well-mixed reservoirs within single compartment
- Constant flow rate and system parameters

Concepts:

- Lumped parameter approach
- First-order rate equation

To record the changes in concentration of a tracer C_B in blood and tissue, we assume the convective stream to be the average velocity

$\bar{u} = Q/A$, $D_{B,T}$ are the axial blood/tissue diffusion coefficients, $\dot{m}$ is the species mass flow rate across the blood-tissue interface, i.e., $\dot{m} = K(C_B - C_T)$, $\forall_{B,T}$ are the blood/tissue volumes, and R is the rate of species consumption by metabolism. Thus, specifying Eq. (1.3.36) for species transport in blood and tissue, we have:

$$\frac{\partial C_B}{\partial t} + \bar{u}\frac{\partial C_B}{\partial x} = D_B \frac{\partial^2 C_B}{\partial x^2} - \frac{\dot{m}}{\forall_B} \tag{E2.2-1a}$$

and

$$\frac{\partial C_T}{\partial t} = D_T \frac{\partial^2 C_T}{\partial x^2} + \frac{\dot{m}}{\forall_T} - R \tag{E2.2-1b}$$

Neglecting diffusion and biochemical reaction and assuming that K, C_a,and Q are constant, the simplified equations for $C_B(t)$ and $C_T(t)$ are then

$$\forall_B \frac{dC_B}{dt} = -Q\,(C_v - C_a) - \mathrm{K}(C_B - C_T) \tag{E2.2-2a}$$

and

$$\forall_T \frac{dC_T}{dt} = \mathrm{K}(C_B - C_T) \tag{E2.2-2b}$$

Assuming a well-mixed blood flow zone, i.e., $C_v \approx C_B$, the solution of Eq. (E2.2-a,b) is of the form

$$\frac{C_v(t)}{C_o} = A_1 e^{-\alpha_1 t} + (1 - A_1)e^{-\alpha_2 t} \tag{E2.2-3}$$

where $C_o = C_v(t = 0)$ and A_1, α_1, and α_2 are functions of Q, V_B, and κ (the partition or solubility coefficient), and K (the diffusion-convection mass transfer coefficient), is discussed in Bird et al. (2002). A detailed solution and graphing of Eqs. (E2.2-2a,b) are left as an HWA (see Sect. 2.3).

For a somewhat more detailed compartmental transport analysis of the microcirculation region, we consider a representative capillary with a thin membrane wall, surrounded by a homogeneous soft tissue of constant concentration C_T (see Fig. 2.1.5). We assume that C = C(x,t) only, Q/A = $\bar{u}$ is the area-averaged constant blood velocity, axial diffusion is negligible, and the transfer coefficient $B \equiv PS/\forall_B$ is constant, where P is the membrane permeability, $\mathrm{S} = d\pi\ell$ is the surface area of species mass transfer between the capillary and tissue, and $\forall_B$ is the blood volume. Then, the governing equation can be written as

$$\frac{\partial C}{\partial t} + \bar{u}\frac{\partial C}{\partial x} = B(C - C_T) \tag{2.1.5}$$

subject to appropriate initial and boundary conditions (cf. Fig. 2.1.2). With the "constants" $\bar{u}$, B, and C_T known, Eq. (2.1.5) can be numerically solved using readily available math software such as Matlab, Mathcad, Femlab, FlexPDE, etc.

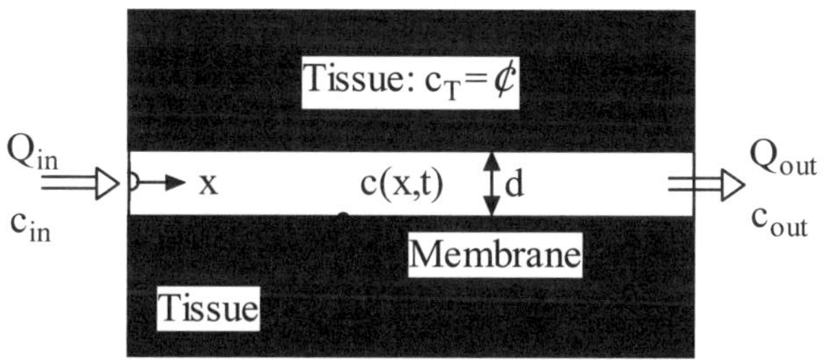

Fig. 2.1.5 Mass transfer between tissue and embedded capillary of diameter *d* and length $0 \le x \le l$.

In case a more complex tissue region or an organ (e.g., a kidney) has to be subdivided, two compartments in parallel may form an appropriate model (see Fig. 2.1.2). A fluid mass balance yields:

$$Q = Q_1 + Q_2 \tag{2.1.6}$$

and a species mass balance dictates:

$$QC_v = Q_1\, C_1 + Q_2\, C_2 \tag{2.1.7}$$

Using this solution to Eq. (E2.2-1a) with $\dot{m} = 0$, i.e., $K \equiv 0$ describing tracer depletion in a single compartment, Eq. (2.1.7) can be written as

$$C_v(t) \;=\; \frac{Q_1}{Q_2}C_1 e^{-k_1 t} + \frac{Q_2}{Q}C_2 e^{-k_2 t} \tag{2.1.8}$$

where $k_i = (\kappa Q/\forall)_i$, i =1,2. Measuring $C_v(t)$, Eq. (2.1.8) can be employed to find k_i or Q_i/Q, where κ is the partition coefficient.

Example 2.3: Single Mass Transfer Compartment

Consider a homogeneous tissue region perfused by a steady blood stream. Given the volumetric flow rate Q, tissue volume ∀, and partition (or solubility) coefficient κ for tracer C in the blood stream, find the outflow concentration $C_v(t)$ for Case (a): constant tracer inflow concentration C_a, when $C_v(t=0)=0;$ and Case (b): tracer depletion, where $C_v(t=0)=C_0$.

Sketch:

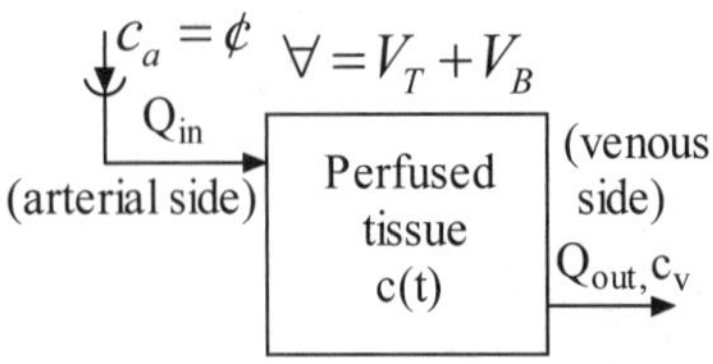

Assumptions:

- constant C_a, Q
- $\forall_B << \forall_T \approx \forall$
- $C_v = \kappa C$, C being the $\forall$-averaged tracer concentration

Concepts:

- compartmental transport
- first-order rate equation

Solution:

Fluid flow: $-Q_{in} + Q_{out} = \dfrac{d\forall}{dt} := 0$ from which follows

$$Q_{in} = Q_{out} = Q \tag{E2.3-1}$$

Species mass balance: $$Q(C_v - C_a) = -\forall \frac{dC}{dt} \tag{E2.3-2a}$$

or

$$C_v(t) = -\frac{\forall}{\kappa Q}\frac{dC_v}{dt} + C_a \tag{E2.3-2b}$$

Case (a): With $C_v(t=0)=0$ and $k = \kappa Q/\forall$,

$$C_v(t) = C_a(1-e^{-kt}) \tag{E2.3-3}$$

Case (b): With $C_v(t=0)=C_0$ and $C_a = 0$, we can simulate tracer depletion with time, i.e.,

$$QC_v = -\forall \frac{dC}{dt} \text{ or } C_v = -\frac{1}{k}\frac{dC_v}{dt} \tag{E2.3-4a,b}$$

which has the solution

$$C_v(t) = C_0 e^{-kt} \tag{E2.3-5}$$

Graphs:

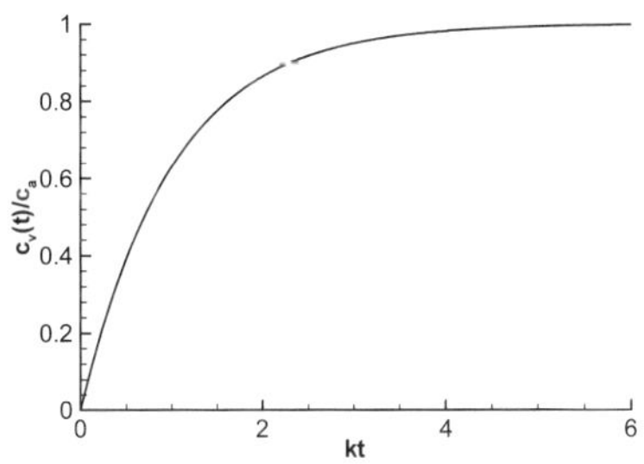

(a) Eq. (E2.3-3)

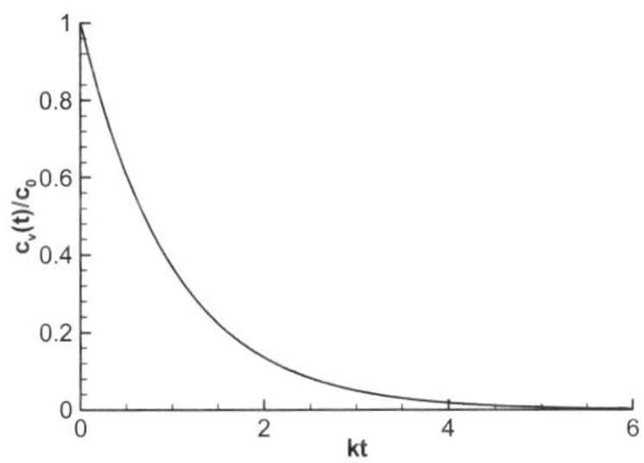

(b) Eq. (E2.3-5)

Comments:
Clearly, when comparing Case (a) with Case (b) it is apparent that a system's initial/inlet conditions greatly determine the tracer output $C_V(t)$. How rapidly steady-state is reached, i.e., $C_V \rightarrow C_o$ in Case (a) and $C_V \rightarrow 0$ in Case (b), is determined by $k=\kappa Q/\forall$.

Example 2.4: Compartmental Transport Model for Circulatory System

Develop a set of coupled first-order rate equations for species mass transfer (e.g., O_2, CO_2, etc.) between the right heart chambers, pulmonary artery, lungs, and pulmonary vein (see Fig. 1.1.2).

Sketch:

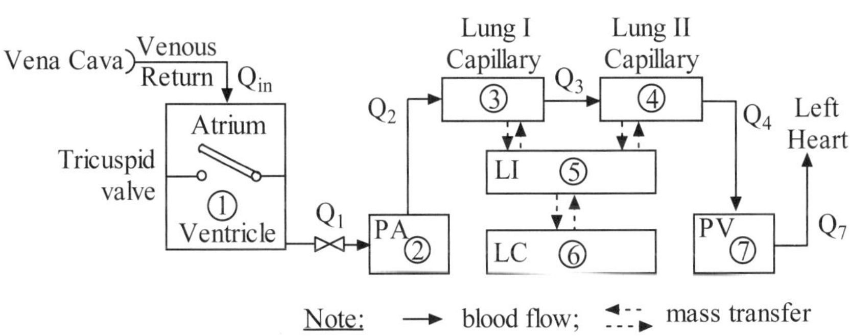

Assume: Well-mixed compartments; constant volumetric blood streams Q, volumes $\forall$, mass transfer areas A, permeabilities p_i, and reaction rate coefficients k_j.

<u>Approach:</u> Coupled rate equations for multi-compartment system in series

<u>Solution:</u>

Right heart: $$V_1 \frac{dC_1}{dt} = (CQ)_{in} - Q_1 C_1 \qquad \text{(E2.4-1a-g)}$$

Pulmonary artery: $$V_2 \frac{dC_2}{dt} = Q_1 C_1 - Q_2 C_2$$

Lung I capillary: $$V_3 \frac{dC_3}{dt} = (CQ)_2 - (CQ)_3 + (pA)_3 (C_5 - C_3)$$

Lung II capillary: $$V_4 \frac{dC_4}{dt} = (CQ)_3 - (CQ)_4 + (pA)_4 (C_5 - C_4)$$

Lung interstitial:

$$V_5 \frac{dC_5}{dt} = (pA)_3 (C_3 - C_5) + (pA)_4 (C_4 - C_5) +$$
$$(pA)_6 (C_6 - C_5)$$

Lung cellular: $$V_6 \frac{dC_6}{dt} = (pA)_6 (C_5 - C_6) - k_6 C_6$$

Pulmonary vein: $$V_7 \frac{dC_7}{dt} = (CQ)_4 - (CQ)_7$$

With a given set of system parameters and coefficients, a Runge-Kutta routine can be used to solve the coupled set of rate equations, subject to some type of initial conditions. For example, a sudden tracer pulse injection $C_{in}(t)$ in form of a Dirac delta function,

$$\delta(t) = \lim_{\lambda \to \infty} \left[\frac{2\lambda}{\sqrt{\pi}} \exp(-\lambda^2 t^2) \right], \text{ where } \int_0^\infty \delta(t) dt = 1$$

(E2.4-2a,b)

allows us to observe the tracer distribution $C_i(t)$ in various body compartments (see Graphs).

Graphs:

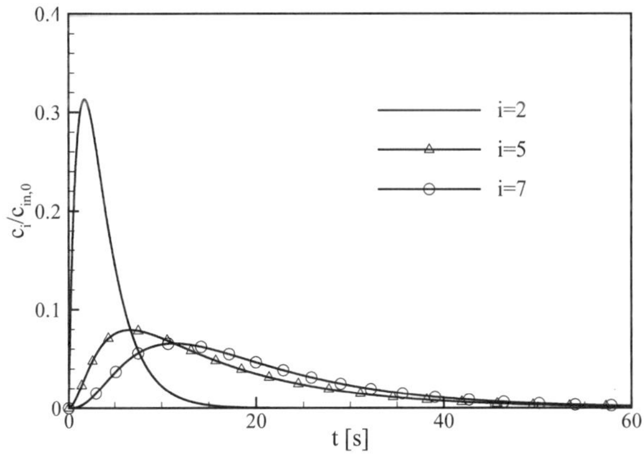

Note: The parameters used are: $C_{in}(t) = C_{in,0}\delta(t)$, $C_i(0)=0$, V_1, V_2, V_3, V_4, V_5, V_6, V_7 = 125, 250, 50, 50, 200, 1000, 400 ml, respectively; $Q_1 = Q_2 = Q_3 = Q_4 = Q_7$ = 100 ml/s, $(pA)_3=(pA)_4=(pA)_6=10^4$ ml/s, k_6 = 1000ml/s.

Comment: Given the inlet condition, $C_{in}(t) = C_{in,0}\delta(t)$, for the heart chambers (see Compartment ①) and the noted system parameters, the three graphs depict the transient species concentrations in Compartments ②, ⑤, and ⑦. As a result of species dispersion and net efflux, the maximum concentrations in each chamber continuously reduce and shift, i.e., experience a time lag. Diminished by tissue uptake, the species reaches the left heart after about 60 seconds.

Most of the remaining examples given in Chapter II rely on the lumped-parameter approach, i.e., biofluid compartment models are employed. Additional case studies using the compartmental transport approach may be found in UG heat/mass transfer textbooks as well as in Middleman (1972), Cooney (1976), Bird et al. (2002), and Truskey et al. (2004).

2.1.2 Tissue Heat and Mass Transfer

Transport phenomena associated with tissue include blood flow with heat transfer in the circulatory system (see Fig. 1.1.2) and lymph flow in the lymphatic network, as well as diffusion-dominated mass transfer on the cellular/capillary level when oxygen and nutrients carried by the blood are exchanged with carbon dioxide and other waste products (see Fig. 1.1.3, Sect.1.1.2, and Chapter IV). The processes can be categorized into convection-diffusion mass transfer and convection-conduction heat transfer (see Sects. 1.3.3.1 and 1.3.3.2). Clearly, one has to distinguish between:

- species (or tracer) mass transfer expressed as concentration C; and
- fluid-mass flow, indicated by the local velocity vector $\vec{v}$, or expressed as bulk flow in terms of average velocity $\bar{v}$, or flow rate $Q = \bar{v}A$, or mass flow rate $\dot{m} = \rho Q$.

<u>*Convection mass transfer examples:*</u>

- Blood flow takes place in two sets of vessels, i.e., arteries and veins, communicating with each other via the microcirculatory bed's arterioles, capillaries, and venules (see Fig. 2.1.6a).
- Pressure-driven blood filtration occurs from the micro-circulation into the interstitial tissue space where the fluid (lymph) is collected by the lymphatic network, called lymphatics.
- Muscle tissue motion and lymphatic pumping cause lymph flow ultimately to the vena cava, while excess fluid and proteins migrate to the blood via the thoratic duct.

Most species transport processes on the cellular and tissue levels are diffusion driven (see Sect. 1.1.2 and Ch. IV). Clearly, Sects. 1.3 to 1.5 provide the mathematical framework to model and simulate these momentum and mass transfer processes. Supplementary references on mass transfer include Truskey et al. (2004) and Bird et al. (2002).

<u>*Heat transfer examples:*</u>

- The blood temperature in the heart's ventricles and the major arteries remains essentially constant, i.e., when body parts are suddenly being overheated or subcooled, tissue temperature equilibration occurs as the blood passes through the smaller arteries (see Fig. 2.1.6b). Both local blood and tissue temperatures are the same until blood mixes at various confluences as well as in the vena cava and the heart's right atrium.

(a) Schematic of blood flow in the circulatory system

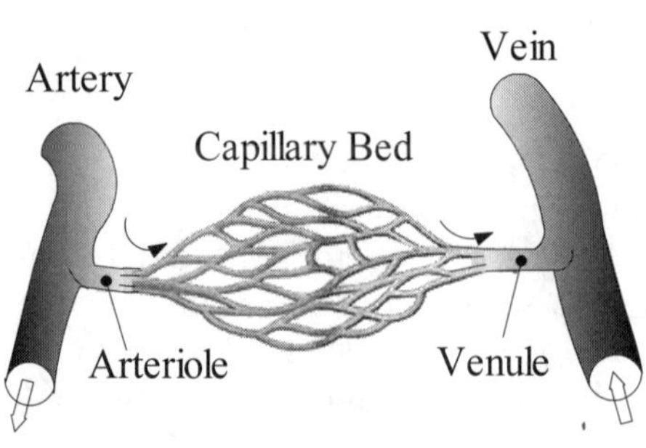

(b) Schematic of arterial temperature (T_A) adjustments to sudden tissue temperature changes ($T_i - T_A$); i = H, C (after Datta 2002)

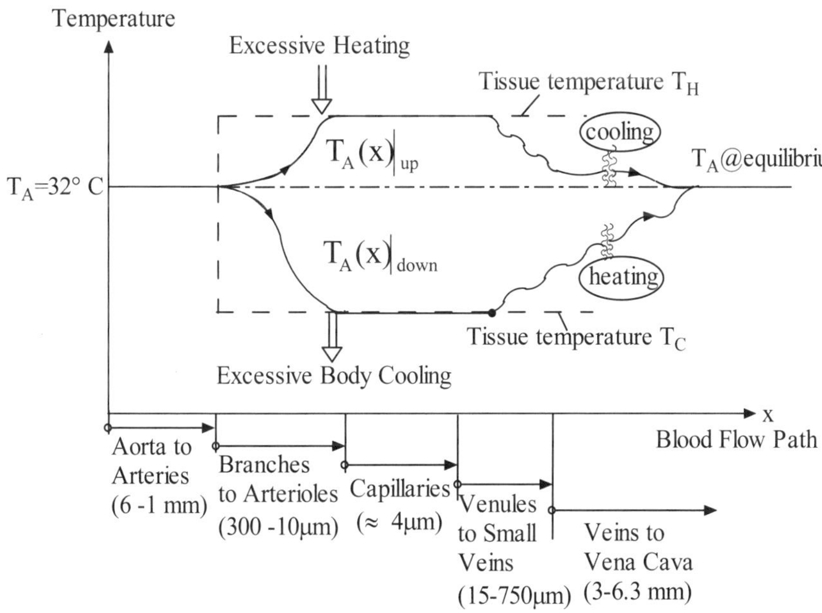

Fig. 2.1.6 Schematics of blood flow: (a) in the circulatory system; and (b) with body heat transfer included.

- A mathematical description of the thermal exchange in tissue is complicated by two sets of blood vessels in the millimeter to micrometer range, sharply varying material property values, geometric irregularities, metabolic activities, etc. Nevertheless, Eq. (1.3.35) can be rewritten as (Pennes, 1948):

$$\underbrace{\rho c \frac{\partial T}{\partial t}}_{\text{accumul.}} = \underbrace{k\nabla^2 T}_{\text{conduction}} + \underbrace{\rho_b c_b \dot{\forall}_b (T_A - T)}_{\text{convection}} + \underbrace{S_T}_{\text{heat source}} \tag{2.1.9}$$

where T is the tissue temperature, T_A is the arterial blood temperature, $\dot{\forall}_b$ is the volumetric flow rate of blood per unit volume of tissue, and S_T represents a heat generation rate due to metabolic processes.

Tissue heat transfer equation. Equation (2.1.9) is known as the "bioheat equation for mammalian tissue." Its underlying assumptions include constant material properties, uniform distribution of blood capillaries in the tissue volume, constant metabolic heat generation, and constant arterial blood temperature. Assuming such an idealized tissue volume, Eq. (2.1.9) has been used to predict the tissue temperature in space and time due to excessive body surface cooling (e.g., cryosurgery or frost bites), surface heating (e.g., skin burning or hyperthermia), and whole body freezing (see Datta, 2002, among others).

Improvements of Pennes' model and special applications have been provided by several researchers, as reviewed by Charney (1992). Clearly, because of the lack of our knowledge in terms of local blood (or lymph) velocity fields, material properties and boundary conditions, different forms of bioheat equations have to be applied to specific soft-tissue regions as well as small and large blood vessels. For example, in order to capture the effects of blood perfusion with heat transfer through tissue, which implies that thermally significant blood vessels are distributed throughout the tissue, Chen & Holmes (1980) postulated an advancement over Eq. (2.1.9), i.e.,

$$\underbrace{\rho_t c_t \frac{\partial T_t}{\partial t}}_{\substack{\text{Heat} \\ \text{accumulation}}} + \underbrace{\rho_b c_b \bar{u} \nabla T_t}_{\substack{\text{Heat convection} \\ \text{in pervious tissue}}} = \underbrace{\nabla(k_t \nabla T_t) + \nabla(k_c \nabla T_t)}_{\substack{\text{Thermal diffusion in} \\ \text{tissue and capillaries}}} + \underbrace{\rho_b c_b \dot{\forall}(T_j - T_t)}_{\substack{\text{Heat transfer from} \\ \text{blood vessel j}}} + \underbrace{\dot{q}_s}_{\substack{\text{Internal} \\ \text{heat source}}} \tag{2.1.10}$$

As always, with the increase in the number of equation terms and necessary parameters, a more realistic description of tissue heat transfer may be gained; however, with it comes the need for more relevant physiological information and powerful computational tools.

Example 2.5: Application of the Bioheat Equation

Consider blood perfusion of a tissue layer of thickness h where at the fat-tissue interface $T = T(x = 0) = T_1$ and at the tissue-core interface $T = T(x = h) = T_2$. The blood (ρ, c_p) enters the tissue with a constant flow rate $\dot{\forall}$ and temperature $T_A < T_1 < T_2$.

Sketch:

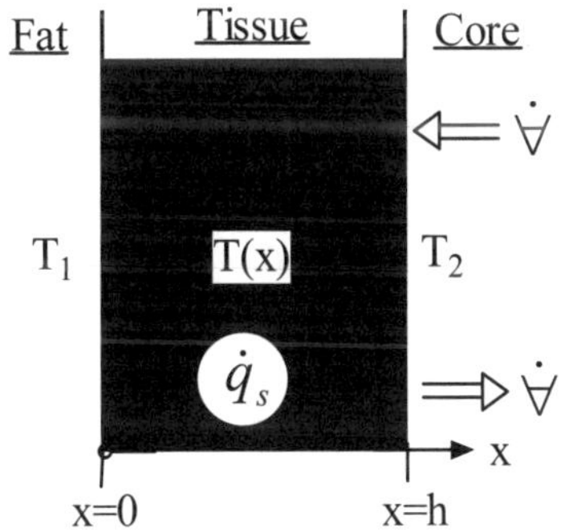

Assumptions:

- Steady 1-D uniform flow
- Negligible metabolic rate
- Constant properties

Approach:

- Reduced Bioheat Equation
- Direct integration

Solution:

Based on the stated assumptions, Eq. (2.1.9) can be reduced to:

$$k\frac{d^2T}{dx^2} - (c_p\dot{\forall})_b(T - T_A) = 0 \tag{E2.5-1}$$

where T(x) is the local tissue temperature. Introducing $c_p\dot{\forall}/k \equiv m^2$, we have to solve an ODE of the form:

$$T'' - m^2(T - T_A) = 0 \tag{E2.5-2}$$

subject to $T(0) = T_1$ and $T(h) = T_2$. The analytic solution in terms of the dimensionless temperature is (Özisik, 1993):

$$\Theta \equiv \frac{T - T_A}{T_1 - T_A} = \frac{T_2 - T_A}{T_1 - T_A}\frac{\sinh(mx)}{\sinh(mh)} + \cosh(mx) - \coth(mh)\sinh(mx) \tag{E2.5-3}$$

Graph:

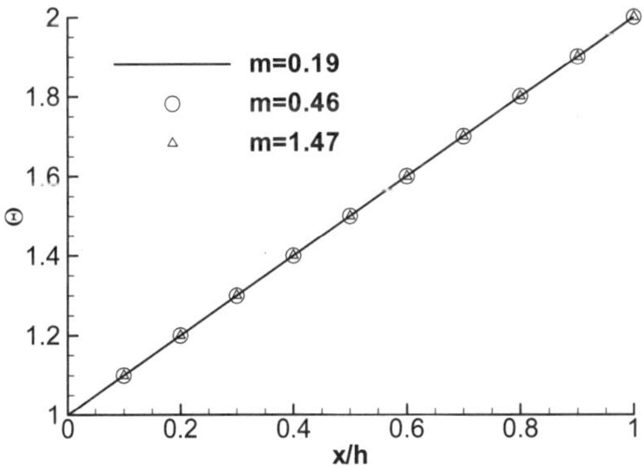

Note: The parameters used for calculation are: h=1cm, T_A=32°C, T_1=34.5°C, T_2=37°C , c_p=3.6×10^{3}J/(Kg·K), k=0.67 W/(m·K), $\dot{\forall}$ =400, 2400, 24000 ml/min (i.e., m=0.19, 046, 1.47).

Comments:
Surprisingly, the spatial temperature exhibits in this case approximately linear variations. The reasons are that the blood flow rate has only a minor effect on the variations in temperature T(x) under normal physiological conditions when assuming steady 1-D heat transfer across a rather thin tissue slice.

Example 2.6: Cell Hyperthermia

Temperature variations have a profound effect on cell functions. For example, the rate of thermo-chemical reactions (i.e, metabolism) in a cell increases with temperature up to a point. Then, outside the biokinetic zone, say, $10°C \leq T_{cell} \leq 45°C$, the rate of activities declines and in case of excessive hyperthermia the cell dies (cf. Yang, 1989). Exposing cells to a medium of elevated temperature, e.g., $T_\infty = 50^\circ C$, causes cell-death by hyperthermia.

Assuming the cell to have a very thin spherical membrane, develop a heat transfer equation and solve it for $T(r,t)$ subject to appropriate initial and boundary conditions.

Sketch:

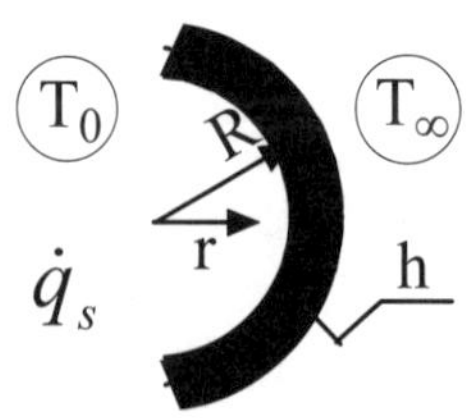

Assumptions:

- Transient radial conduction only
- Negligible membrane thickness and heat production rate $\dot{q}_s$
- Constant properties

Approach:

- Reduced heat equation in spherical coordinates
- Infinite series solution

Solution:
With the assumption that T = T(r,t) only and ignoring any fluid convection or heat source, Eq. (2.1.9) can be reduced to (see Appendix B for spherical coordinates):

$$\frac{\partial T}{\partial t} = \frac{\alpha}{r^2}\frac{\partial}{\partial r}\left(r^2 \frac{\partial T}{\partial r}\right) \qquad \text{(E2.6-1a)}$$

where $\alpha \equiv k/(\rho c_p)$ is the thermal diffusivity. Equation (E2.6-1a) has to be solved subject to:

$$T(r,t=0)=T_0;\ \left.\frac{\partial T}{\partial r}\right|_{r=0}=0;\ and$$
$$\left.k\frac{\partial T}{\partial r}\right|_{r=R}=h[(T_\infty - T(R)] \qquad \text{(E2.6-1b-d)}$$

With $h/k \equiv H$, where h is the convection heat transfer coefficient and k is the thermal conductivity, and defining $\Theta = \dfrac{T_\infty - T}{T_\infty - T_0}$ the infinite series solution to (E2.6-1), based on the separation-of-variables approach, is given by Carslaw & Jaeger (1959) or Özisik (1993).

$$\Theta = 2\frac{H}{r}\sum_{n=1}^{\infty}\frac{(\beta_n R)^2+(R\cdot H-1)^2}{\beta_n^2[(\beta_n R)^2+H\cdot R(R\cdot H-1)]}\sin(\beta_n R)\sin(r\beta_n)\cdot e^{-\alpha\beta_n^2 t} \qquad \text{(E2.6-2a)}$$

where β_n (n = 1,2…) are the roots of

$$\beta R\cot(\beta R)+R\cdot H-1=0 \qquad \text{(E2.6-2b)}$$

Graphs:

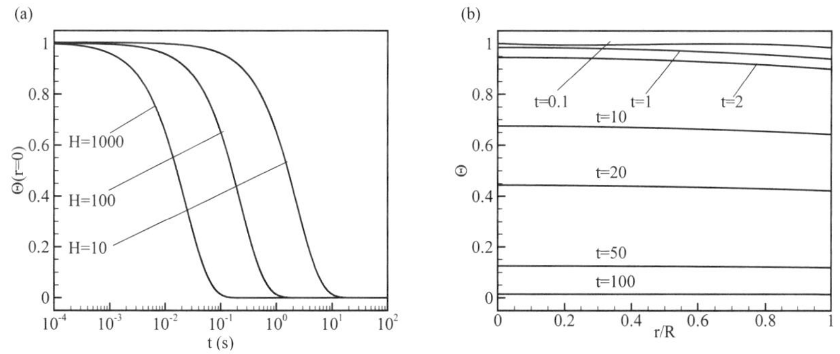

Notes: For graph (a), R=10μm, and $\alpha=1.43\times10^{-7}\text{m}^2/\text{s}$; and for graph (b), R=1mm, H=100 m^{-1}, and $\alpha=1.43\times10^{-7}\text{m}^2/\text{s}$.

Comments:

The temperature variation inside the sphere with a small radius (say, R<1mm) can be negligible. The rate of temperature rise in the cell increases with larger H- and smaller R- values.

Transvascular transport. The systemic blood circulation delivers fluids, nutrients, and drugs into the microcirculation of individual organs, first mainly by convection and then almost exclusively by diffusion across microvessel walls. The material transport across such capillaries, which feature high porous surface areas, is characterized by the hydraulic conductivity and permeability coefficient for specific solutes (see Sect. 1.4.4 and Table 2.2.1). Although the microvessel wall, consisting of the glycocalyx, endothelium, and basement membrane, is a composite, for transvascular transport calculations it is considered to be a single porous membrane.

Fluid flow across such a membrane is described by a modified form of Starling's law of filtration similar to Darcy's law (see Sects. 1.4.4 and 1.4.5):

$$j_v = L_p A_s (\Delta p - \sigma_s \Delta\pi) \tag{2.1.11a}$$

where L_p is the effective hydraulic conductivity of the membrane, A_s is the surface area, Δp is the static pressure difference, $\Delta\pi$ is the osmotic pressure difference (missing in Darcy's law), and σ_s is the osmotic reflection coefficient, $0 \leq \sigma_s \leq 1$, which depends on the types of solute and membrane structure. Clearly, actual values for these phenomenological parameters L_p and σ_s have to be determined experimentally on a case by case basis. Furthermore, Eq. (2.1.11a) is limited to membranes with macroscopically uniform structures and ideal solutes (see Hu & Weinbaum, 1999). Equation (2.1.11a) can be rewritten in more practical terms with $\sigma_s \approx 1$ as:

$$\hat{m} \equiv \frac{\dot{m}}{A_s} = k[(p_c - \pi_c) - (p_i - \pi_i)] \tag{2.1.1b}$$

where,

$\hat{m} \triangleq$ mass flux $[\frac{kg}{cm^2 \cdot s}]$

$p_c \triangleq$ hydrostatic capillary blood pressure

$\pi_c \triangleq$ osmotic pressure of the plasma proteins

$p_i \triangleq$ interstitial pressure outside capillary wall

$\pi_i \triangleq$ osmotic pressure of proteins in interstial fluid

$k \triangleq$ filtration constant [s/cm] indicating the degree of permeability of the capillary wall to water.

Filtrated water which passes into the surrounding tissue is either reabsorbed into the capillary blood or returned to the blood via the lymphatic system (see Sect. 4.3). As $p_c = p_c(x)$, so does $\hat{m}$ vary axially. As a simplification, it is suggested to set $p_c := p_{mean} = 0.5(p_{arteriolar} + p_{venular}) \approx$ constant.

Example 2.7: Plasma Flow in Filtrating Capillary

Introducing $\alpha = p_i + \pi_c - \pi_i = ¢$, $\Delta p = p_a - p_v$, $\Delta\alpha = p_a - \alpha$, $\varepsilon = \mu k / R$, $\beta = R/L$, $\zeta = r/R$, $\xi = z/L$, and assuming steady laminar "fully-developed" flow; i.e., $\mathrm{v} << u$, but, because of filtration $\mathrm{v} \neq 0$ in a straight microtube, and constant parameters $\mu, k, p_i, \pi_c, \pi_i, p_a, and\ p_v$, solve for *u*, v, and *p*. Take $\mu = 1cP$ and $k = 2.5\ x\ 10^{-8}\ s/cm,$ so that with $R = 5\mu m$, $\varepsilon = 1\ x\ 10^{-7}$.

Sketch:

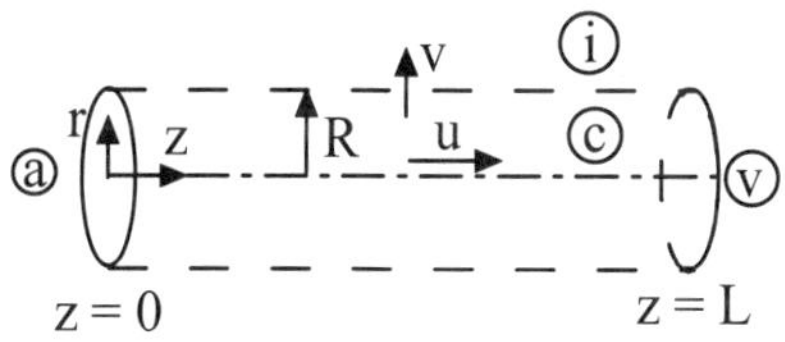

Assumptions:

- As stated

ⓐ ≙ arterial
ⓒ ≙ capillary
ⓘ ≙ interstitial
ⓥ ≙ venous

Concepts:

- Reduced N-S equation
- Starling's law

Solution: The appropriate boundary conditions are:

at the centerline $\left.\dfrac{\partial u}{\partial r}\right|_{r=0} = 0$ and v(r =0) = 0; (E2.7-1a-d))

at the membrane surface u(r = R)=0 but v (r = R) = k (p - a). Specifically, because of the curved streamlines in the capillary due to radial fluid flux, v≠0 and the Poiseuille flow solution has to be modified. Solving the reduced N-S equations and neglecting $\varepsilon^2 - terms$, we obtain (Oka, 1981):

$$u = \frac{R^2}{4\mu}(\frac{\Delta p}{L})(1-\varsigma^2)[1+\varepsilon\ f\ (\varsigma,\xi)] \tag{E2.7-2}$$

$$\mathrm{v} = \frac{\varepsilon R}{\mu}(\frac{\Delta p}{L})\ (2\varsigma - \varsigma^3)\ (\frac{\Delta\alpha}{\Delta p} - \xi) \tag{E2.7-3}$$

and

$$p = p_a - \xi \Delta p + \varepsilon g\,(\varsigma, \xi)\,\Delta p \tag{E2.7-4}$$

where

$$f(\varsigma,\xi) = \left\{ \frac{8}{\beta^2}\left[\xi - \frac{\Delta\alpha}{\Delta p}\right]^2 - (\frac{\Delta\alpha}{\Delta p})^2 + \frac{\Delta\alpha}{\Delta p} - \frac{1}{3} + \frac{1}{2}\beta^2 - \frac{1}{4}\beta^2\varsigma^2 \right\} \tag{E2.7-5a}$$

and

$$g(\varsigma,\xi) = -\frac{8}{3\beta^2}\left[\begin{array}{l} \xi^3 - 3\frac{\Delta\alpha}{\Delta p}\xi^2 - \left\{1 - 3\frac{\Delta\alpha}{\Delta p} - \frac{3}{4}\beta^2(1-2\varsigma^2)\right\}\xi \\ -\frac{3}{4}\beta^2\frac{\Delta\alpha}{\Delta p}(1-2\varsigma^2) \end{array}\right] \tag{E2.7-5b}$$

Note when $k \to 0$ and hence $\varepsilon = 0$, the Poiseuille solution is recovered.

With $\mathrm{v}(r = R) = \mathrm{v}(1, \xi) = \frac{\varepsilon R}{\eta}(\frac{\Delta\alpha}{\Delta p} - \xi)$, plasma/water outflow and inflow depends on the sign of the radial velocity, i.e.,

when $\xi < \frac{\Delta\alpha}{\Delta p} \Rightarrow$ outflow (filtration) and when $\xi > \frac{\Delta\alpha}{\Delta p} \Rightarrow$ inflow (absorption).

Streamline Graph:

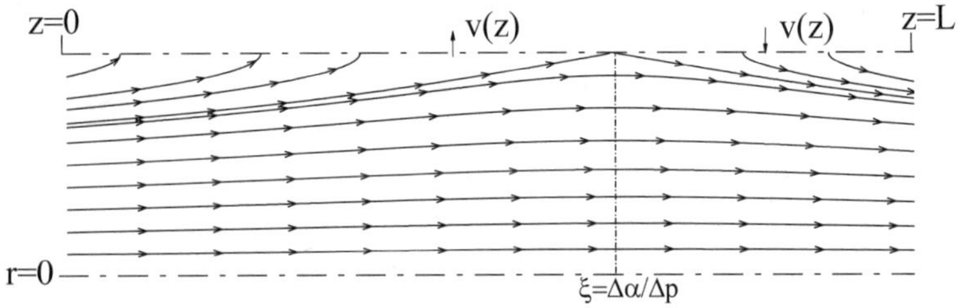

Integrating Eq. (E2.7-2) over dA = 2πrdr yields the local volumetric flow rate, i.e.,

$$Q = \frac{\pi R^4}{8\eta}\left(\frac{\Delta p}{L}\right)\left[1 + \varepsilon\, f\left(\frac{1}{\sqrt{3}}, \xi\right)\right] \tag{E2.7-6}$$

Clearly $Q = Q(f)$, i.e., $Q = Q_{min}$ when $\xi = \frac{\Delta\alpha}{\Delta p}$. Thus Q (x) can be plotted for three cases as shown in the flow-rate graphs.

Flow-rate Graphs:

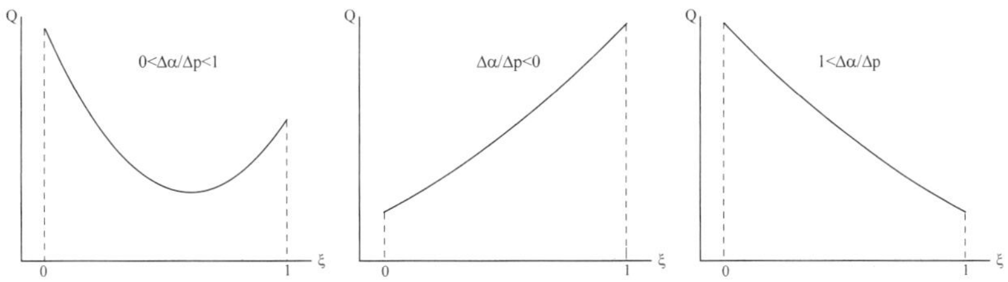

The net outflow across the membrane is:

$$\hat{m} = Q(\xi = 0) - Q(\xi = 1) = \varepsilon \frac{2\pi R^4}{\mu\beta^2}\left(\frac{\Delta p}{L}\right)\left(\frac{\Delta\alpha}{\Delta p} - \frac{1}{2}\right) = k\, A_s \left(\frac{\Delta\alpha}{\Delta p} - \frac{1}{2}\right)$$

Clearly, when $\frac{\Delta\alpha}{\Delta p} = \frac{1}{2}$, $\hat{m} = 0$, i.e., we have a balanced inflow and outflow.

Microcirculation and capillary bed mass transfer. In contrast to the single-membrane transvascular-transport approach, we now consider an entire capillary bed. It consists of tiny, membrane-walled blood vessels (5~10μm) which are embedded in heterogeneous tissue, both capillaries and tissue "connecting" arterioles, and venules for gas (O_2, CO_2) and material exchange (see Fig. 2.1.2a). A similar exchange of gases, nutrients, proteins, and wastes occurs through the walls of sinusoids (30~40μm) and venules which are contained in the liver, spleen, and bone marrow. Blood flow in these distensible microvessels is subject to arterial pressure pulsations and is controlled via contracting/expanding ring-muscle actions of arterioles and *precapillary* sphincters. Clearly, *microcirculation* is another challenging research area.

In its simplest form, the capillary bed can be modeled as an assembly of soft tissue cylinders with capillaries at their centers connecting the arterial to the venous end (cf. Yang, 1989; Middleman, 1972, among others). Figure 2.1.7 depicts a single cylinder, known as the Krogh model. Assuming transient convection-diffusion of a species, *c*, in the capillary lumen and transient diffusion in the capillary membranes plus tissue, the simplified governing equation reads (see Eq. (1.4.37a)) or Eq. (1.4.47)):

$$\frac{\partial c}{\partial t} + \mathrm{v}\frac{\partial c}{\partial z} = \frac{D_r}{r}\frac{\partial}{\partial r}(r\frac{\partial c}{\partial r}) + D_z\frac{\partial^2 c}{\partial z^2} + S_c \tag{2.1.12a}$$

where v is v(r,z) or just Q/A (the averaged axial velocity in $0 \le r \le R_1$), D_r and D_z are the radial and axial diffusion coefficients in each region, respectively, and S_c represents a species source or sink due to biochemical reactions.

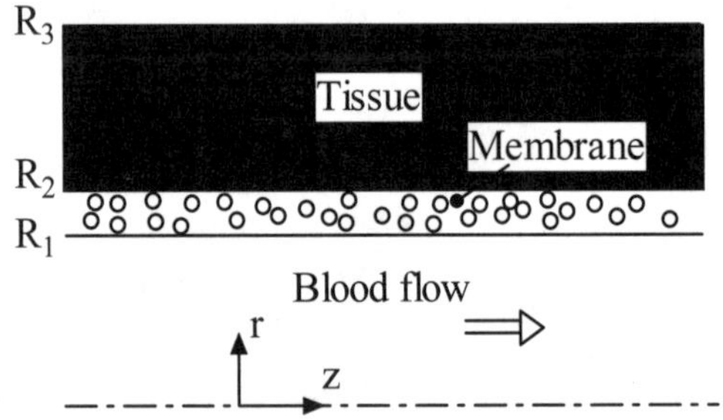

Fig. 2.1.7 Schematic of cylindrical tissue-capillary model.

Equation (2.1.12) has to be formulated for each of the three regions (see Fig. 2.1.7). The associated boundary conditions are (where $\mathrm{B} \triangleq \text{blood},\ \mathrm{M} \triangleq \text{membrane, and}\ \mathrm{T} \triangleq \text{tissue}$):

Capillary: $$\left.\frac{\partial C_B}{\partial r}\right|_{r=0} = 0 \text{ for } t > 0 \text{ and } C_B(t=0) = C_{B0} \text{ for all z}$$

Membrane/Tissue:

$$\left.D_B\frac{\partial C_B}{\partial r}\right|_{r=R_1} = \left.D_M\frac{\partial C_m}{\partial r}\right|_{r=R_1} \quad and$$

$$\left.D_M\frac{\partial C_M}{\partial r}\right|_{r=R_2} = \left.D_T\frac{\partial C_T}{\partial r}\right|_{r=R_2}$$

$$C_B(r=R_1) = C_M(r=R_1) \text{ and } \mathrm{C_M}(r=R_2) = C_T$$

Tissue: $$\left.\frac{\partial C_T}{\partial r}\right|_{r=R_3} = 0$$

(2.1.12b-h)

Solution of Eq. (2.1.12) subject to the appropriate boundary conditions, plus its interpretation, is left as an HWA (see Sect. 2.3).

Example 2.8: Convective-diffusive Mass Transfer

Assume steady fully-developed blood flow in a capillary and diffusion without any biochemical reaction in the surrounding membrane and tissue. Develop the governing equations plus boundary conditions for the blood and tissue species concentrations when C_o is the blood/tissue inlet concentration, while mass transfer at the tissue/capillary interface is $S_B \sim h\frac{P}{A}\Delta C$ and in the tissue $S_T = \alpha C_T/(\beta + C_T)$, where α is the rate of metabolic species consumption, and β is the Michaelis-Menten constant.

Sketch:

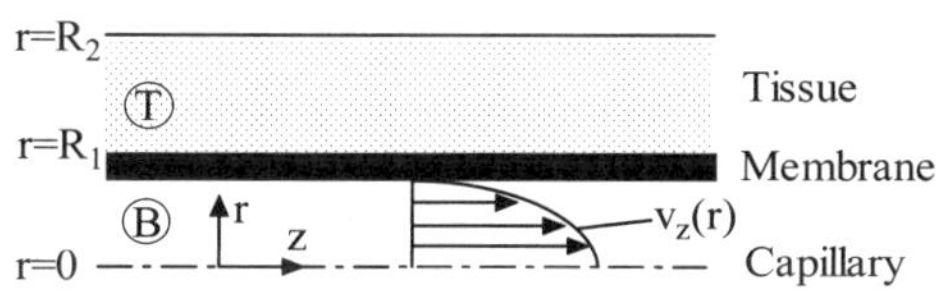

Assumptions:

- Neglect membrane
- Constant parameter

Approach:

- Analogous heat transfer solution (see Cebeci, 2004)

Solution: Reduced forms of Eq. (2.1.12) have to be written for both the capillary and the tissue. The velocity field is described by Poiseuille flow (see Sect.1.3.2).

Capillary: $$\mathrm{v}_z \frac{\partial C_B}{\partial z} = -h\frac{P}{A}[C_B(z) - C_T(r = R_1, z)] \tag{E2.8-1}$$

Tissue: $$0 = \frac{D_r}{r}\frac{\partial}{\partial r}(r\frac{\partial C_T}{\partial r}) + D_z\frac{\partial^2 C_T}{\partial z^2} - \frac{\alpha C_T}{\beta + C_T} \tag{E2.8-2}$$

subject to:

$$C_T(r,0) = C_B(r,0) = C_0; \; -D_r\frac{\partial C_T(R_1,z)}{\partial r} = h[C_R(z) - C_T(R_1,z)];$$

(E2.8-3a-e)

$$\left.\frac{\partial C_T}{\partial r}\right|_{R_2} = 0; \frac{\partial C_T(r,0)}{\partial z} = 0; \frac{\partial C_T(r,1)}{\partial z} = 0$$

2.1.3 Joint Lubrication

Joint lubrication is a subset of biotribology, the theory of friction, lubrication, wear, damage, and replacement of biological systems, such as movable joints – natural or artificial. In order to slide two solids relative to each other, a tangential resistance, i.e., a frictional force, F, has to be overcome:

$$F = fN \tag{2.1.13}$$

where N is the normal load, and f is the friction coefficient with a typical range of $0.01<f<1.0$.

Low *friction coefficients* can be achieved when the two solid body surfaces stay at all times separated via a suitable lubricant, e.g., the *synovial* (i.e., egg-whitish) fluid in human joints (Fig.2.1.8a). Such a case of *hydrodynamic lubrication* greatly reduces wear and hence surface damage. In general, as pointed out by M. J. Furey in Schneck & Bronzino (2003), friction and wear should be considered as separate phenomena, as in the case of *"boundary" lubrication* (e.g., cartilage rubbing a cartilage) when friction and wear are determined by the surface properties of the *solids and the chemical nature of the lubricant* rather than its viscosity. Specifically, wear is the removal of one material by another due to sliding contact. Wear debris may cause tissue reaction in an attempt to remove the particulate matter, potentially leading to bone tissue loss and hence joint failure. Other lubrication modes include squeeze-film, hydrodynamic and elasto-hydrodynamic lubrication, where cartilage deformation is considered as well.

The human synovial joint is deceptively complicated because of the biomechanical-biochemical interactions of articular cartilage with the synovial fluid which separates load-bearing bones. For example, certain lubricant components reduce friction while others may enhance cartilage wear (see Furley, 2003). Furthermore, during body motion, e.g., running, the mode of lubrication can change from squeeze-film to hydrodynamic to boundary lubrication, potentially causing cartilage wear and surface damage. The idealized lubrication model of Fig. 2.1.8c was discussed in Sect. 1.3.2, Example 1.5, while squeeze-film solutions are discussed next. Not surprising, theoretical work has been confined to joint lubrication, while the analysis of cartilage damage requires experimental contributions.

Joint squeeze-film lubrication. As demonstrated with Example 1.5, for load-carrying lubrication relying on the *wedge-film* effect to work, a minimum fluid (or angled plate) velocity is required to generate a load-carrying pressure field. For the common hip replacement, i.e., a ball-in-a-cup prosthesis, *squeeze-film* lubrication is the basic mechanism (Fig. 2.1.9). A major challenge is the right selection of cup-and-ball material, perfect sphericity, and surface finish. Also important are the minimum sustained radial clearance and constant properties of the pseudo-synovial fluid in light of realistic load changes during walking, etc. Problems may arise when particles are entrained due to wear, or boundary lubrication occurs when the fluid film breaks. A fluid-structure-interaction (FSI) analysis of the coupled squeeze-film and cup-and-ball deformation problem, labeled *elasto-hydrodynamic lubrication*, can provide shape design solutions which reduce wear of joint prostheses. Modern cup-and-ball materials include polyethylene, metal alloys, and ceramics, where the latter are superior. Typically, the ball diameter is 28mm. While larger sizes apparently reduce wear, a large head may necessitate a greater bone volume loss around the acetabular cup. The clearance between head and cup surfaces is in the range of $5 \leq h \leq 20 \mu m$, where the lower limit is related to the surface roughness, e.g., $\varepsilon = 0.005$ mm for ceramics (Mabuchi et al., 2004).

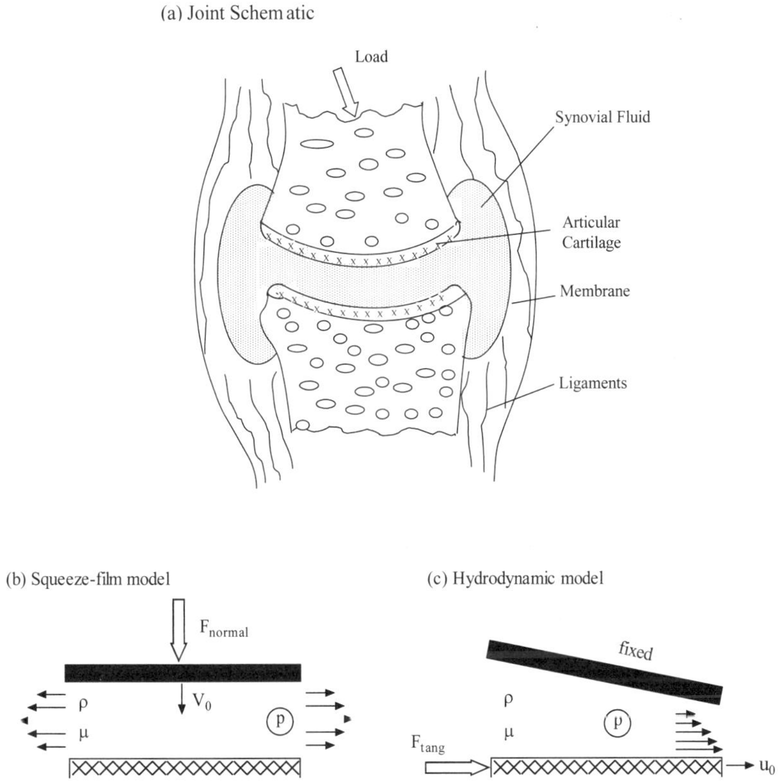

Fig 2.1.8 Joint lubrication: (a) Schematic of the knee; (b) Squeeze-film model; and (c) Hydrodynamic model.

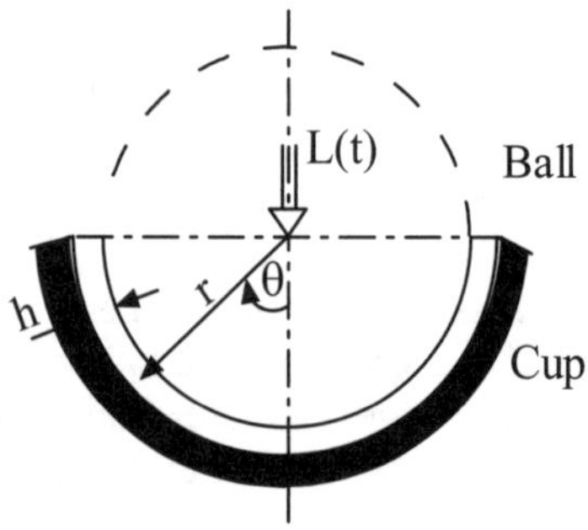

Fig. 2.1.9 Schematic of idealized joint prosthesis with squeeze-film lubrication.

In order to estimate the film pressure and critical gap height, a Poiseuille-type analysis of fluid flow in a rigid hemispherical ball-and-cup configuration is considered (see Fig. 2.1.9). A more realistic FSI description is given in Mabuchi (2004). Assuming a sinusoidal load, L(t), resulting from walking, with a period of $T=1s$ and start-up time (or half-period) $T_s = 0.5s$ we can write:

$$L(t) = \begin{cases} L_{\max} \sin(\frac{\pi t}{2T_s}) & \text{for } 0 \le t < T_s \\ \frac{L_{\max} + L_{\min}}{2} + \frac{L_{\max} - L_{\min}}{2} \cos\left[\frac{2\pi(t - T_s)}{T}\right] & \text{for } t \ge T_s \end{cases}$$

(2.1.14a,b)

Typical measured values are L_{min} = 0.1kN and L_{max} = 1.6kN. Because of axisymmetry, the film pressure is $p = p(\Theta, t)$ only, i.e.,

$$L(t) = 2\pi r^2 \int_0^{\pi/2} p(\Theta, t) \sin\Theta \cos\Theta \, d\Theta \tag{2.1.15}$$

Equation (2.1.15) can be employed to find $p(\Theta, t)$ which can be also estimated from the volumetric flow rate, assuming Poiseuille flow in the curved conduit, where $\mu = 15 mPa \cdot s$. Specifically (see Sect.1.3.2),

$$Q = -\frac{\pi h^3 \sin\Theta}{6\mu} (\frac{\partial p}{\partial \Theta}) \tag{2.1.16}$$

Now, Q is also the flow rate during squeezing action, i.e.,

$$Q = 2\pi \int_0^{\Theta} (\frac{\partial h}{\partial t}) r^2 \sin\Theta \, d\Theta \tag{2.1.17}$$

where $\partial h/\partial t$ is the "squeeze velocity." With this set of equations, the varying film thickness under the assumption of rigid boundaries can be evaluated as:

$$h(\Theta,t) = h_o\left[1 - \varepsilon(t)\cos\Theta\right] \tag{2.1.18}$$

where h_o is the initial radial clearance, $h_o < 30\mu m$, and ε is the eccentricity with $\varepsilon(t=0)=0$. If both the cup and the ball experience (elastic) deformations, expressed as $2 \cdot \Delta s(\Theta, t)$, the elastohydrodynamic film thickness is (see Dowson & Higginson,1977):

$$h_{total} = h(\Theta,t) + 2\,\Delta s(\Theta,t) \tag{2.1.19}$$

Of special interest is the minimum film thickness, $h_{\min}$, to check full film lubrication. It can be estimated from the correlation

$$\frac{h_{\min}}{\sqrt{e_b^2 + e_c^2}} \approx 2.5 \tag{2.1.20}$$

where e_b and e_c are the surface roughness of ball and cup, respectively.

Example 2.9: Joint Lubrication Mode: Squeeze-film

Consider squeeze-film lubrication under a constant load L for different fluids, i.e., Newtonian, power-law, and viscoelastic.

Model Schematic:

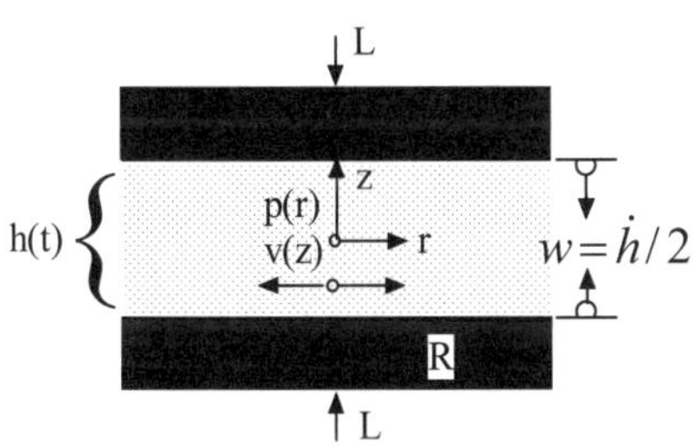

Assumptions:

- Transient 2-D axisymmetric incompressible Stokes flow
- Gap height $h << R$ and $\dot{h} << \overline{v}$
- Normal stresses are negligible

Concepts:

- Reduced continuity and equation of motion
- Insert appropriate rheology models for τ_{rz}

Solution:

Based on the assumptions, we reduce the continuity equation to

$$\frac{1}{r}\frac{\partial(rv)}{\partial r} + \frac{\partial w}{\partial z} = 0 \tag{E2.9-1}$$

and the equation of motion to

$$\frac{\partial}{\partial r}p(r,t)=\frac{\partial}{\partial z}\tau_{rz} \tag{E2.9-2}$$

Integration of Eq. (E2.9-1) across the gap $-\frac{h}{z}\leq z\leq\frac{h}{2}$ yields

$$\dot{h}\equiv\frac{dh}{dt}=-\frac{1}{r}\frac{d}{dr}(rh\overline{\mathrm{v}}) \tag{E2.9-3}$$

where

$$\int_{-h/2}^{h/2}dw=w(\frac{h}{2})-w(-\frac{h}{2})=\frac{\dot{h}}{2}-(-\frac{\dot{h}}{2})=\dot{h} \tag{E2.9-4a}$$

$$-\int_{-h/2}^{h/2}\frac{1}{r}\frac{\partial(r\mathrm{v})}{\partial r}dz=-\frac{1}{r}\frac{d(r\overline{\mathrm{v}})}{dr}\int_{-h/2}^{h/2}dz=-\frac{1}{r}\frac{d}{dr}(r\overline{\mathrm{v}})h \tag{E2.9-4b}$$

and

$$\overline{\mathrm{v}}=\frac{2}{h}\int_{0}^{h/2}\mathrm{v}(r,z;t)dz \tag{E2.9-4c}$$

Before integrating Eq. (E2.9-2) across the gap, τ_{rz} has to be defined.

(i) <u>Newtonian fluid</u>: $\tau_{rz}\approx\mu\frac{\partial\mathrm{v}}{\partial z}$ (E2.9-5a)

Thus, Eq. (E2.9-2) reads:

$$\mu\frac{\partial^2\mathrm{v}}{\partial z^2}=\frac{dp}{dr} \tag{E2.9-5b}$$

subject to $\mathrm{v}\,(z=\pm\,h/2)=0$. Integration yields:

$$\mathrm{v}=\frac{h^2}{2\mu}\left(\frac{dp}{dr}\right)\left[\left(\frac{z}{h}\right)^2-\frac{1}{4}\right] \tag{E2.9-5c}$$

Evaluating the average radial velocity as needed in Eq. (E2.9-3), we have:

$$h\overline{\mathrm{v}}=\frac{1}{\mu}\left(\frac{dp}{dr}\right)\int_{0}^{h/2}(z^2-\frac{h^2}{4})\,dz:=\frac{-h^3}{12\mu}\frac{dp}{dr} \tag{E2.9-6}$$

so that:

$$\frac{1}{r}\frac{d}{dr}\left(r\frac{dp}{dr}\right)=\frac{12\mu}{h^3}\frac{dh}{dt} \tag{E2.9-7a}$$

subject to $p(r{=}R){=}0$ and $dp/dr{=}0$. Thus, double integration yields:

$$p = \frac{3\mu R^2}{h^3}\dot{h}\left[(\frac{r}{R})^2 - 1\right] \qquad \text{(E2.9-7b)}$$

Now the load can be evaluated as:

$$L = \int_0^R p dA = 2\pi \int_0^R prdr := \frac{3\pi}{2}\mu R^4 \, \dot{h} \Big/ h^3 \qquad \text{(E2.9-8)}$$

(ii) Power-law fluid: $\tau_{rz} = K(\frac{\partial u}{\partial z})^{n-1}$, (E2.9-9)

leading to (see HWA in Sect. 1.7):

$$L = \frac{2\pi(2+1/n)^n}{n+3} K \,\mathrm{sgn}\,(\dot{h})\left|\dot{h}\right|^n h^{-(2n+1)} R^{n+3} \qquad \text{(E2.9-10)}$$

(iii) Linear viscoelastic fluid: $\tau_{rz} = \int_{-\infty}^{t} G(t-t')\frac{\partial u(t')}{\partial z}dt'$ (E2.9-11a)

leads to:

$$L = \frac{3\pi R^4}{2}\int_0^t Q(t-t')(\dot{h}/h^3)dt' \qquad \text{(E2.9-11b)}$$

Graphs:

(a) Profile of radial velocity component

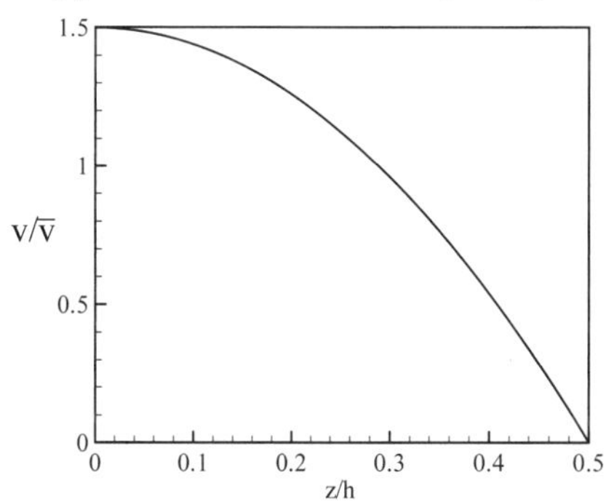

(b) Radial pressure distribution

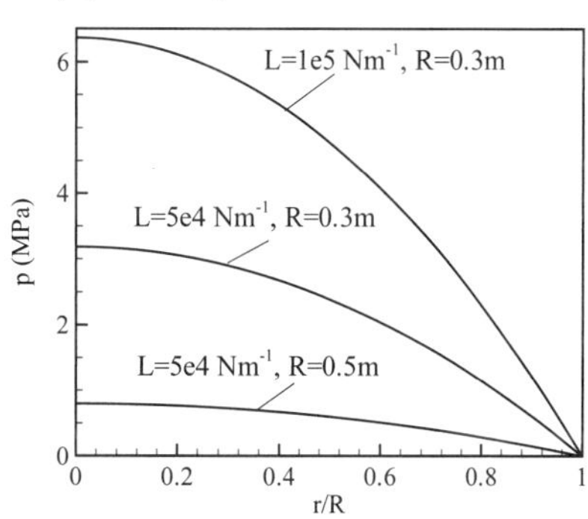

Note: Case (i) Newtonian fluid

Comments:

As expected, the radial squeeze-film velocity v(z) is parabolic due to the approaching disks creating at time *t* with $\dot{h}_{disk}$ a pressure gradient dp/dr driving the fluid outwards. The pressure distribution (see Eq.

(E2.9-7b)) clearly depends on the load, causing the disk approach speed $\dot{h}_{disk} = \dot{h}(L)$, and the disk radius.

2.1.4 Cell Transport and Microvascular Beds

Overview. As large blood vessels or airways bifurcate in a tree-like fashion into arterioles and venules or alveolar ducts with mean diameters in the micrometer range ($10 \leq D \leq 100\mu m$), the local Reynolds and Womersley numbers drop well below unity and Stokes' equation apparently holds (see Eq. (1.3.22) in Sect. 1.3.2). Other microvascular beds, e.g., the pulmonary capillary blood vessels, form a sheet-like network. While on the first glance, the conditions $Re \leq 1$ and $Wo < 1$ imply a major simplification, complications of microcirculation, i.e., flow in a network of capillaries, arise from:

- fluid-particle dynamics, e.g., deforming red blood cells;
- abnormal flow-reduced cell functions;
- fluid-structure interactions, e.g., coupled blood flow with viscoelastic, multi-layer "rough" walls; and
- passive as well as active wall mass transfer.

Less complicated are medical devices with microchannels, e.g., bio-MEMS (microelectro-mechanical systems for biofluid flow and biochemical conversion), because the conduits are rigid and geometrically well defined. However, in both application areas, i.e., biological capillaries and microchannels, the Knudsen number is relatively high (see Fig. 1.2.1) and molecular dynamics models may be appropriate whenever the continuum mechanics assumption is invalid.

Cell transport. Of the numerous blood cells, i.e., more than 10^9 particles per mm^3 of blood, erythrocytes or red blood cells (RBCs), monocytes (a member of the leukocyte or white blood cell (WBC) family), and platelets are most important in health (i.e., homeostasis) as well as in the genesis of arterial diseases (see Ch. III). Table 2.1.1 summarizes characteristics of such particles. As outlined in Sect. 1.4.1, when the particle mean diameter is below $1\mu m$, a species mass transfer equation (see Eq. (1.4.36)) may be appropriate to describe the cell concentration in a carrier fluid. For larger cells, particle tracking (see Sect. 1.4.3.1), known as the Euler-Lagrange approach, is most suitable. The modeling equations provided in Sect. 1.4.3.1 assume spherical or quasi-spherical particles. Blood particles, especially RBCs and most inhaled drugs are nonspherical; such problems have been addressed by Chhabra et al. (1999), Rosendahl (2000), and Finley (2001), among others. Alternatively, we consider blood as a whole to be a mixture of a Newtonian fluid, the water-like plasma, and highly deformable, liquid-like cells either slipping past one another or aggregating, depending on the local shear rate and particle potential. Thus, at shear rates below $200s^{-1}$, blood exhibits non-Newtonian and for $\dot{\gamma}$ > $200s^{-1}$ Newtonian behavior (see Fig. 1.4.2).

Table 2.1.1 Properties of Key Blood Constituents.

Material	Red Blood Cells	Platelets	Leukocytes (all)	Plasma	Whole Blood
Density (g/cm^3)	1.09	$1.069^{\dagger}$	1.07-1.09	1.03	1.054
Viscosity ($dyn \bullet s/cm^2$)	--	--	--	0.014*	0.0309**
Blood cell count ($\#/cm^3$)	5×10^9	3×10^8	7.0×10^6 (total)	--	--
Volume (μm^3)	88	$5.17^{\dagger,\dagger\dagger}$	460	--	--
Size (μm)	7.7×2.8	$3.0^{\dagger\dagger}$	9.5	--	--
Volume fraction	0.42 – 0.46	1.9×10^{-3}	1.2×10^{-3} (total)	--	--
τ_p (s)	1.45×10^{-6}	1.12×10^{-6}	$3.35 \times 10^{-6\blacklozenge}$	--	--
$\alpha = \rho_p / \rho_p$.	0.967	0.983	$0.964^{\blacklozenge}$	--	--

† Representative mean value for unactivated platelet
†† Platelet dimensions vary considerably
* at 37°C (may vary from 0.012 to 0.018)
** Newtonian limit at H = 40%
♦ Value for monocytes

Unfortunately, macroscopic blood rheology models, such as the power law, Casson, Quemada, or Carreau-Yasuda (see Kleinstreuer, 2003), do not provide any information on cell transport, local concentration, and wall deposition. Fortunately, monocytes and platelets appear at low concentrations and Stokes numbers so that these cells do not affect the blood flow field and hence their trajectories may be computed independently. RBCs, however, with volume concentrations of 40 to 50%, and being highly deformable, interact with the other cells and change their motion. As a result, collision-induced blood cell dispersion generates a "mixing motion" allowing cells and even large proteins to interact with the vessel surfaces. Such a mixing effect requires at the very least a shear-rate dependent dispersion coefficient, e.g., in Eq. (1.4.36), of the form:

$$D_{eff} = (a^2\dot{\gamma}) \cdot fct(Ht) \tag{2.1.21}$$

where a is the cell radius, $\dot{\gamma}$ is the shear rate, and Ht = 40-50% is the hematocrit. Specific functional forms of Eq. (2.1.21) are discussed by Zydney & Colton (1988), Phillips et al., (1992), and others. An alternative approach, focusing on shear-induced drift of platelets in non-uniform blood flow (see Sect.1.4.2.2), is given by Buchanan & Kleinstreuer (1998) and discussed by Kleinstreuer (2003).

Microvascular beds. Microcirculation of blood deals with a non-Newtonian fluid of changing characteristics flowing in micro-vessels of complex structure due to a pressure gradient, osmotic pressure difference, and/or wall pumping action (Fig. 2.1.10). Because each organ, major blood vessel, and muscle group have unique microvascular beds, there are numerous geometric variations, endothelium types, and capillary wall characteristics.

Blood, a suspension of cells in plasma, can be characterized by an apparent viscosity, η, which increases at low shear rates ($\eta > 3.5$ cP for $\dot{\gamma} < 200\text{s}^{-1}$) due to red blood cell (RBC) aggregation (see Fig. 1.4.4) and decreases in very small vessels with diameters less than 300 μm to a minimum of 1cP at D = 8μm, a phenomenon called the Fahraeus-Linquist effect. Then, in capillaries of D ≤ 8 μm, η increases rapidly while RBCs, surrounded by plasma, have to get streamlined in order to squeeze through (see Sect. 1.4.3). In addition to RBCs, circulating leukocytes (i.e., white blood cells, WBCs) are of great interest in cellular fluid mechanics (cf. Kamm, 2002). Well known for combating infection, some WBCs (e.g., monocytes) in turn play a key role in the onset of the arterial disease "atherosclerosis" (see Sect. 3.1.1). Realistic structural and biological cell responses to fluid dynamics forces and the influence of cell behavior, especially local cell aggregation, on the blood rheology may require novel, future-oriented nano-scale analyses, e.g., direct numerical simulation, molecular dynamics simulation, or statistical simulation techniques.

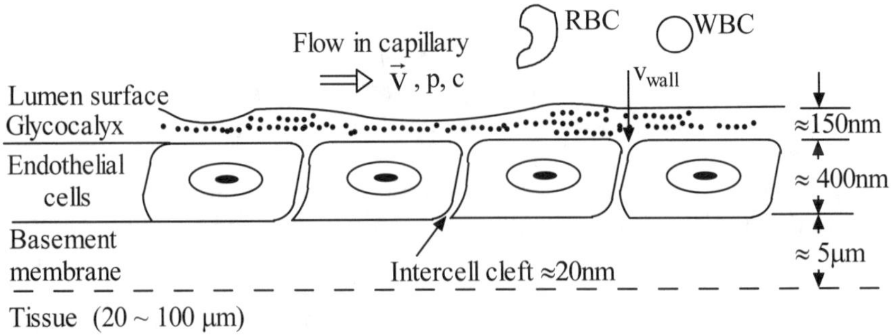

Fig. 2.1.10 Schematic of capillary wall layers.

The walls/surfaces of capillaries are complex as well (see Fig. 2.1.10), producing tubular flow results different from those formed in large blood vessels. For example, Pries et al. (1994, 1997) (see review by Kamm, 2002) demonstrated that the flow resistance measured *in vivo* was about double compared to measurements obtained with glass tubes. Differences in terms of a negatively charged macrolayer, i.e., the (glycocalyx) over the endothelium and the surface roughness created by the deforming e-cells may explain the observations. Furthermore, under natural and pathological conditions, mass transfer of fluid, nutrients, and cells takes place between the lumen and tissue. That could be simply expressed with Starling's equation (see Sects. 1.2.2 and 1.4.4):

$$v_w = K\,[p_\ell - p_t + (\pi_\ell - \pi_t)] \tag{2.1.22}$$

where v_W is the fluid volume flux (or wall velocity), K is a permeability constant, p is the fluid static pressure and π the osmotic pressure for the lumen (ℓ) and tissue (t).

<u>*Example 2.10:*</u> Blood Flow in a Capillary: The Fahraeus Effect

Considering steady laminar fully-developed blood flow in a small tube, $r_0 \le 150\mu m$, whole blood separates into a cell-free plasma-layer, $\delta \approx 5\mu m$, along the tube wall and an enriched central core, $0 \le r \le (r_0 - \delta)$. As a result, the tube hematocrit, H_t, is smaller than the outflow hematocrit, H_O, i.e., $H_t < H_o$, called the Fahraeus effect.

Sketch:

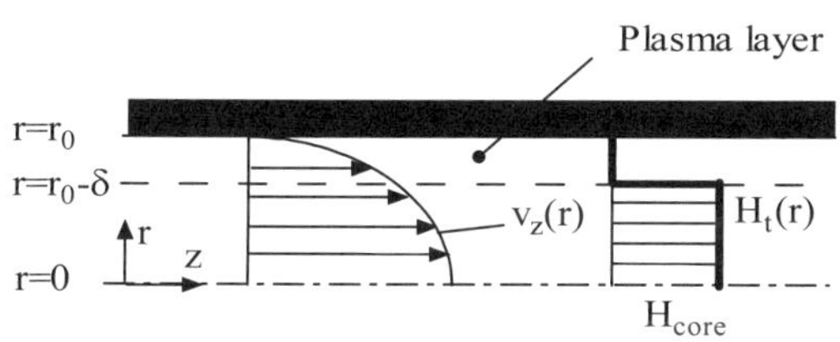

Assumptions:

- As stated
- Presence of RBCs do not perturb $V_z(r)$-profile
- Poiseuille flow

Solution: The averaged tube and outlet hematocrits are defined as follows:

Tube hematocrit $H_t = \dfrac{2}{r_0^2}\displaystyle\int_0^{r_0} H_t(r)\, r dr$ (E2.10-1)

Outlet hematocrit $H_o = \int_0^{r_0} H_t(r) v_z(r)\, r dr \Big/ \int_0^{r_0} v_z(r)\, r dr$ (E2.10-2)

With

$$H_t(r) = \begin{cases} H_{\text{core}} & \text{for } 0 \le r \le r_0 - \delta \\ 0 & \text{for } r_0 - \delta < r \le r_0 \end{cases} \quad \text{(E2.10-3a,b)}$$

and

$$v_z(r) = v_{\max}\left[1 - \left(\frac{r}{r_0}\right)^2\right] \quad \text{(E2.10-4)}$$

Eqs. (E2.10-1, 2) become:

$$H_t = H_{core}\left(1 - \frac{\delta}{r_0}\right)^2 \quad \text{(E2.10-5)}$$

and

$$H_o = H_{core}\left(1 - \frac{\delta}{r_0}\right)^2\left[2 - \left(1 - \frac{\delta}{r_0}\right)^2\right] \quad \text{(E2.10-6)}$$

Thus,

$$\frac{H_t}{H_o} = \frac{1}{2 - \left(1 - \frac{\delta}{r_0}\right)^2} \quad \text{(E2.10-7)}$$

Now assuming $\delta = 5\mu m$, $H_t < H_o$ when $r_0 < 500\ \mu m$, i.e., the critical radius when the Fahraeus effect becomes apparent (see Graph).

<u>Graph:</u>

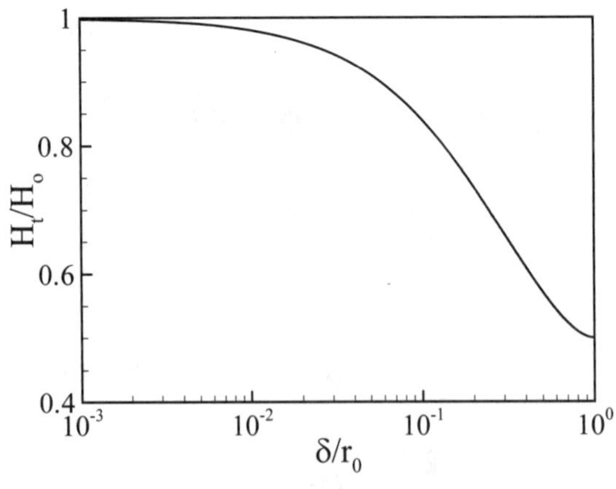

Comments:

Associated with the Fahraeus effect is a decrease in blood viscosity in capillaries with $30\mu m < d < 300\mu m$ (i.e., the Fahraeus-Lindquist effect). However, for $d < 10\mu m$ the hematocrit and apparent viscosity increase again, until even deformable RBCs cannot pass vessels with $d \leq 2.7\mu m$. It should be noted that these mechanical resistance parameters change with time because of blood flow induced remodeling, i.e., long-term changes in vessel material and geometric properties. For example, hypertension may cause thicker and stiffer arteries and a chronic artery-wall decrease may lead to localized lumen constrictions (see Sect. 3.1). Thus, in contrast to pipe networks, arteries respond biodynamically to significant changes in blood flow conditions.

2.2 The Cardiovascular System

As mentioned in Sect. 1.1, the cardiovascular system consists of the heart, blood vessels, and the circulating blood. The heart pumps blood which transports oxygen-carrying hemoglobin to the tissue as well as nutrients, hormones, and waste products to different organs. The heart itself consists of four pumping chambers, i.e., the left and right atria and ventricles (see Fig. 1.1.2), which communicate via the tricuspid valve (right) and mitral valve (left). The ventricular wall motions create the pumping function of the heart. Specifically, the ventricles are three-dimensional, thick-walled pressure vessels with greatly changing wall thickness, say, 8-15mm, and curvatures during the cardiac cycle. As a result of the pumping action, the heart undergoes translation and rotation which significantly affects the mechanics of the coronary arteries, which perfuse the muscle wall, and the four heart valves, i.e., the aortic artery (AV) and pulmonic artery (PV) valves as well as the mitral (MV) and tricuspid (TV) valves. Specifically, the mitral and aortic valves control the flow of oxygenated blood from the heart's left side via the aorta to the body, while the tricuspid and pulmonic valves regulate the blood flow to the lungs for oxygenation (see Fig. 1.1.2). The left part of the heart performs the major pumping with significantly higher pressure levels (i.e., 100 to 150 mmHg) than in the right ventricle. Hence, concerning biomedical engineering, including pathological aspects, present research efforts are focusing on the aortic and mitral valves as well as the left coronaries.

2.2.1 Cardiovascular Transport Dynamics

The cardiovascular transport system assures via convection in vessels (i.e., blood flow) and diffusion (i.e., species mass transfer) between capillaries and cells a "constancy" of the body's internal environment, known as homeostasis. On a microscale level, the regulation of species composition (e.g., oxygen, ions, pH, temperature, osmolarity, etc.) of the interstitial fluid (or lymphs) takes place via diffusion which is here a very effective and rapid process, moving large species quantities because of the microscopic distances and huge surface areas involved (see Sect. 1.2.4 and Sects. 4.3 and 4.4). On the meso- and macro-scale levels, species transport between organs occurs via convection, where the arterial blood flow through the tissue capillaries and all species concentrations have to be adequate. This implies that the nutrients in the blood stream have to be continuously reconditioned, which is done by the organs (see Chapter IV). For example, the lungs supply oxygen from the ambient and discharge carbon dioxide. The kidneys, which receive over 20 percent of the cardiac output, adjust the electrolyte composition of the blood and hence control the electrolyte balance in the tissue. In contrast, the brain, heart muscle, and skeletal muscles do not recondition blood but need the blood's supply of oxygen and nutrients, without any interruption, for their metabolism. For example, unconsciousness can occur within a few seconds after stoppage of cerebral flow, similar to cardiac arrest, and permanent brain damage can occur within four minutes without blood supply.

Pulmonary and systemic circulations. As indicated in Figs. 1.1.2 and 1.1.3, blood with increased CO_2-content from tissue metabolism returns to the heart's right atrium. From there, blood enters the right ventricle which pumps it into the pulmonary trunk/arteries. The pulmonary arteries bifurcate and transport blood to the lungs where the CO_2 -O_2 gas exchange occurs between the lung capillaries (CO_2) and the air sacs (O_2) of the alveolar region. Thus, the returning blood from the lungs via the pulmonary veins to the left atrium is oxygen enriched. This loop, i.e., blood pathway from the heart's right ventricle, through the lungs and back to the heart's left atrium, is called the *pulmonary circulation*. Now, oxygen-rich blood enters the left ventricle and is pumped into the aorta, which ascends, makes a U-turn and then descends through the thoracic (i.e., chest) cavity, across the diaphragm and into the abdominal cavities. Branches from the aorta supply blood to all the organ systems, and via arterioles/capillaries supply oxygen and nutrients to the tissue. Tissue metabolism generates CO_2 dissolved in blood which drains via the vessels into veins. These veins ultimately empty into two large veins, the superior and inferior vena cavae, that return the CO_2-rich blood to the heart's right atrium. This loop is called the *systemic circulation*.

In summary, both the systemic circulation and the pulmonary circulation are powered by the heart, receiving the same blood flow with equal frequency (cf. Nichols & O'Rourke, 1998). However, a host of geometric and hemodynamic differences distinguish the two systems. For example, the major pulmonary arteries have elliptical cross-sections rather than circular ones, the tubular elasticity is higher and the wall thickness is thinner than in systemic arteries. While the mean pulmonary artery pressure is much lower, the pulmonary vascular resistance is equally lower, i.e., about one-sixth of the systemic one in both cases (see Fig. 2.2.1). In general, the pulmonary circulation has attracted somewhat less interest because of the many cardiovascular diseases occurring in the systemic circulation, such as atherosclerosis, hypertension, thrombosis, aneurysms, etc. (see Chapter III).

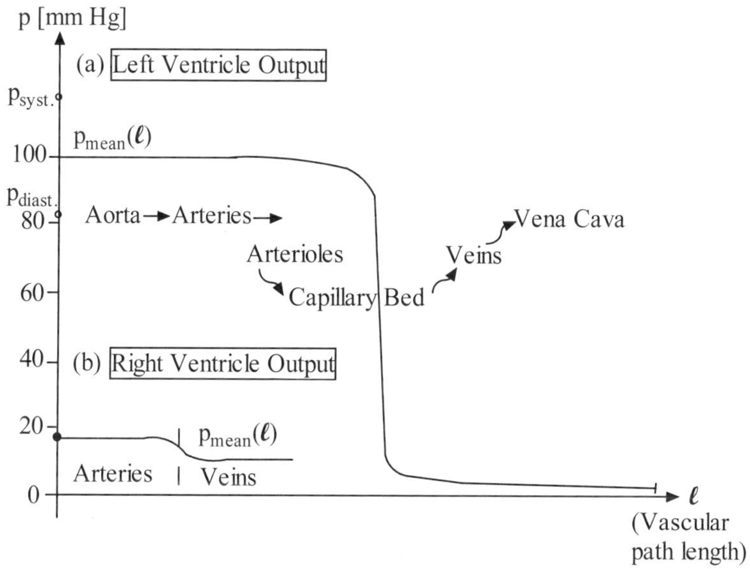

Fig. 2.2.1 Mean pressure changes along vascular segments between (a) left ventricle, ascending aorta, and vena cava; and (b) right ventricle, pulmonary arteries, and pulmonary veins (after Nichols & O'Rourke, 1998).

2.2.2 The Heart

As mentioned, the heart contains four chambers, i.e., two (upper) atria and two (lower) ventricles. Specifically, CO_2-rich blood is received in the right atrium from the inferior and superior vena cavae while O_2-rich blood arrives in the left atrium from the pulmonary veins. In turn, the right ventricle pumps the blood for reoxygenation to the lungs while the left ventricle pumps the O_2-rich blood via the aorta throughout the body. The right and left parts of the heart are separated by a muscular wall, called the septum, which prevents mixing of the blood.

The atria and ventricles (or pumps) are separated by a sheet of connective tissue, the fibrous skeleton, in which one-way atrio-ventricular (AV) valves are embedded. The AV-valve on the right side has three flaps and, hence, is called tricuspid valve. In contrast, on the left side the fibrous skeleton has two flaps, called the bicuspid (or mitral) valve. Clearly, the pressure-regulated blood flows from the atria to the ventricles, but not in the reverse direction. When the ventricles are relaxed and venous blood fills up the atria, the pressure increases and the AV valves open, allowing blood to stream into the empty chambers. While the ventricles contract, the blood pressure rises and closes the AV valves (see Fig. 1.1.2).

Aortic valve. When the aortic valve opens, which takes only 20-30 msec, blood rapidly accelerates through the valve at peak velocity after about 200 msec and then swiftly decelerating (1.35-0.35 m/s for healthy adults). The valve is closed due to an axial pressure difference which pushes the valve leaflets toward the center of the aorta. The fast closing action is assisted by vortices which develop behind the valve leaflets. Nevertheless, at the end of systole there is some reverse flow. The fluid-structure interactions are further complicated by heart muscle translation and rotation during pumping, which move the aortic valve and change its size. The pulmonic valve flow is similar to the aortic one, although with lower peak velocities (say, from 0.75 to 0.15 m/s).

Mitral valve. While the aortic valve connects the left ventricle with the aorta, the mitral valve separates the left ventricle from the left atrium as it receives oxygenated blood from the lungs via the pulmonary veins (see Fig. 1.1.2). During diastole, blood flows from the atrium to the ventricle because of the positive pressure differential. Then, the atrium contracts causing a second flow acceleration through the partially closed mitral valve, filling the left ventricle completely. The tricuspid valve flow exhibits a similar velocity profile, although at lower speeds. As with the aortic valve, vortices behind the leaflet, generated during ventricular filling, and the basic pressure drop cause the closing of the mitral valve. The pulsating heart motion moves the valve axially and reduces its effective opening which helps in the swift closing by the valve leaflets.

Another set of one-way valves is located at the origins of the pulmonary artery and the aorta. These open during ventricular pumping action, allowing blood to enter the pulmonary and the systemic circulations, and they close during ventricular relaxation to prevent the backflow of blood into the chambers.

Cardiac cycle. The cardiac cycle is the repeating pattern of heart-muscle contraction (i.e., the systolic phase) and relaxation (i.e., the diastolic phase). Contraction of the ventricles in systole ejects about two-thirds of the blood from these chambers, called the stroke volume $\forall_{st}$. The volumetric flow rate (or cardiac output) Q can be estimated as:

$$Q = \forall_{st} \cdot N \tag{2.2.1}$$

where the stroke volume, $\forall_{st}$, is measured in [ltrs/beat], the heart rate N in [beats/min], and the cardiac output, Q, in [ltr/min]. The contractility of the heart muscle and high ventricular filling pressure (i.e., preload) both increase $\forall_{st}$, while elevated arterial pressure tends to decrease $\forall_{st}$ by increasing the afterload on cardiac muscle fibers. The "preload" effect, i.e., the phenomenon that the heart contracts more forcefully when it is filled to a greater degree during diastole, is called the *Frank-Starling law* of the heart.

Because the entire heart is controlled by a single electrical impulse, both sides of the heart (i.e., the left and right pumps) act synchronously, i.e., the stroke volumes, systolic/diastolic periods, and valve actions are all the same. However, the pressures generated by the right heart pump are considerably lower because the lungs produce much less flow resistance than the systemic organs (see Fig. 1.1.2). In numbers, a typical systolic/diastolic pressure ratio for the aorta is 120/80 mmHg, while it is 24/8 mmHg for the pulmonary artery. The difference between these maximum and minimum values is the pulse pressure, e.g., $\Delta p_{LV} = 40$ mmHg. Figures 2.2.2a,b depict the left and right pressure waveforms as well as distributions of the left ventricular volume and aortic flow rate.

Specifically, with a focus on the left heart's pumping action (Fig. 2.2.2a), during the diastolic phase the mitral valve opens when the ventricular pressure falls below the atrial pressure, and the left ventricle begins to fill up from the left atrium where the pressure initially drops. Later, the atrium and ventrical pressures slowly rise again because the chambers are passively filled with blood from the veins. Then, the following events occur:

- As the ventricles start to contract, the rising chamber pressure causes the AV valves to close. With all valves closed for about 0.04s, the blood volume stays the same, i.e., it is the phase of isovolumetric contraction.
- When the pressure in the left ventricle exceeds that in the aorta, the semilunar valves open and blood ejection begins, where the peak systolic pressure is $p_{max} \approx 120$ mmHg in both ventricle and aorta.
- As the ventricular contraction diminishes, the pressure falls below the aortic pressure, the aortic valve suddenly closes, and a "notch" in the aortic pressure wave occurs due to the elasticity of the valve leaflets and because some blood flows backward to fill the aortic valve leaflets as they close. The ventricular muscle then relaxes and all valves are again shut during the short isovolumetric relaxation phase. Ultimately $p_{ventricle} = 0$ mmHg, while $p_{aorta} \approx 80$ mmHg.
- When pressure in the ventricles dips below that in the atria, a phase of rapid filling takes place, supported by atrial contraction which pushes the rest of the blood into the ventricle.
- Immediately after that, a new cardiac cycle begins.

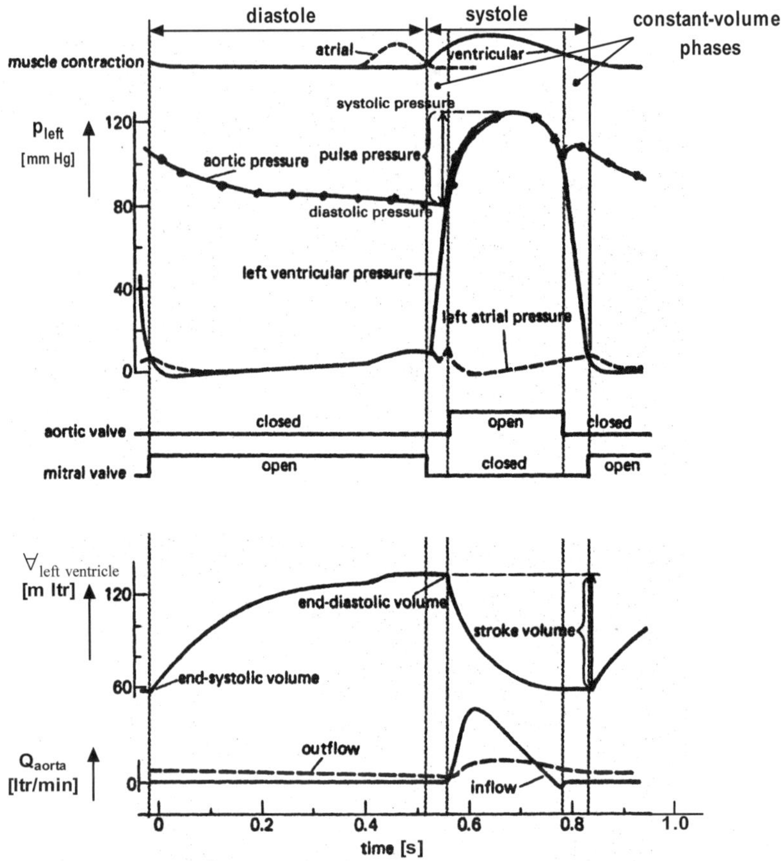

Fig. 2.2.2a Cardiac cycle of the left heart.

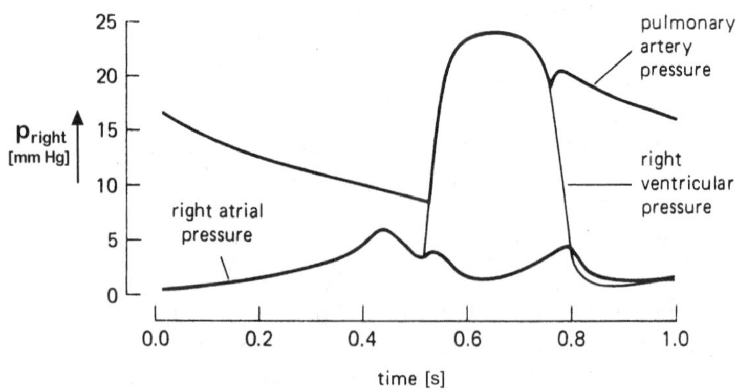

Fig 2.2.2b Cardiac cycle of the right heart.

Example 2.11: Heart Valve Dynamics (Adapted from Bellhouse & Talbot, 1969)

Assume that the aortic valve cusps bound a cone-shaped moving surface (see sketch) when pulsatile, inviscid flow passes through the aortic valve. Develop an equation to describe the pressure difference between the cusp root and tip as a function of geometry parameters as well as valve open/closure conditions. Note, a recent review of the fluid mechanics of heart valves has been provided by Yoganathan et al. (2004).

Sketch:
(a) Aortic Valve

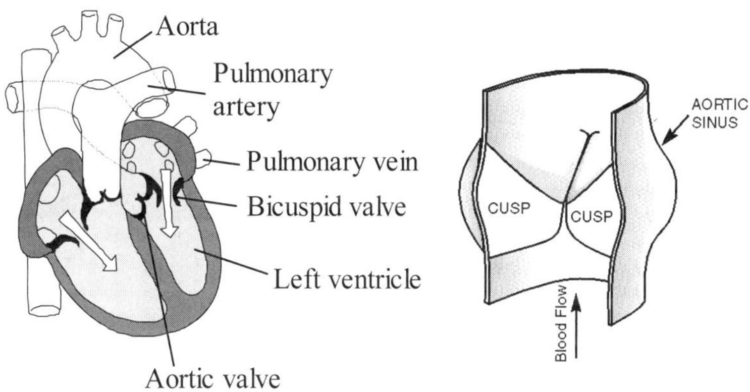

(b) Math Model

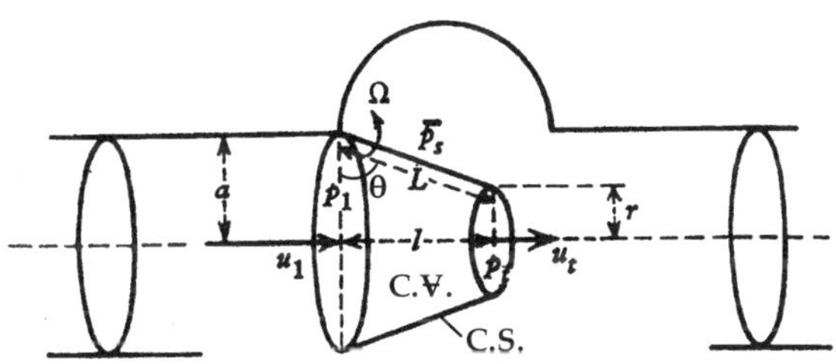

$0 \le x \le l$ and $r \le y \le a$

Assumptions:

- Inviscid flow
- The velocity variations across planes perpendicular to the axis are negligible

Concepts:

- Conservations of mass and momentum
- Unsteady Bernoulli equation

Solution:

Considering the mass conservation for the conical control volume $\forall$, we have over C.S. S:

$$\int_S \rho(\vec{v}\cdot d\vec{S}) = 0 \tag{E2.11-1}$$

i.e.,

$$\pi a^2 u_1 - \pi r^2 u_t = \frac{d\forall}{dt} \tag{E2.11-2}$$

where $\forall$ is the volume contained by S, i.e., $\forall = \frac{1}{3}\pi l(a^2 + ar + r^2)$.

Given the angular velocity of the cusps $\Omega = \frac{d\theta}{dt} = \frac{1}{l}\left(\frac{dr}{dt}\right)$ and $\beta \equiv r/a$, Eq. (E2.11-2) can be rewritten as

$$u_t = \frac{u_1}{\beta^2} - \frac{\Omega a}{3\beta^2}\left(\frac{l}{a}\right)^2 (1+2\beta) \tag{E2.11-3}$$

In fact, Eq. (E2.11-3) is valid at any cross-section x with a radius of y, so that

$$u(x,t) = \frac{a^2}{y^2}u_1 - \frac{a\Omega x^2}{3y^2}\left(1+\frac{2y}{a}\right) \tag{E2.11-4}$$

Conservation of momentum implies

$$F_x = \frac{\partial}{\partial t}\int_\forall \rho u d\forall + \int_S \rho u(\vec{v}\cdot dS) \tag{E2.11-5a}$$

Here the axial force is

$$F_x = \pi a^2 p_1 - \pi r^2 p_t - \pi(a^2 - r^2)\overline{p}_A \tag{E2.11-5b}$$

where $\overline{p}_A$ is the average pressure on the aortic-side of the cusps.

The conical tube axis is always a streamline because of symmetry; thus the Bernoulli equation reads:

$$(p_1 + \tfrac{1}{2}\rho u_1^2) - (p_t + \tfrac{1}{2}\rho u_t^2) = \rho \int_0^l \frac{\partial u}{\partial t}\, dx \qquad \text{(E2.11-6)}$$

Substituting Eq. (E2.11-4) in Eq. (E2.11-5) and Eq. (2.11-6) yields

$$\frac{p_1 - p_t}{\rho a} = C_1 \frac{du_1}{dt} + C_2 a \frac{d\Omega}{dt} + C_3 \frac{u_1^2}{a} + C_4 a \Omega^2 + C_5 \Omega u_1 \qquad \text{(E2.11-7)}$$

where

$$C_1 = \frac{1}{\beta}\left(\frac{l}{a}\right)$$

$$C_2 = -\frac{1}{3\beta}\left(\frac{l}{a}\right)^3$$

$$C_3 = \frac{1}{2}\left(\frac{1}{\beta^4} - 1\right)$$

$$C_4 = \frac{(1 + 4\beta + 10\beta^2)}{18\beta^4}\left(\frac{l}{a}\right)^4$$

$$C_5 = -\frac{(1 + 2\beta + 3\beta^2)}{3\beta^4}\left(\frac{l}{a}\right)^2$$

Comments: The velocity at the cusp tip u_t and the pressure difference during the deceleration phase can be calculated with Eqs. (E2.11-3) and (E2.11-7) if the time-dependent velocity through the valve $u_1(t)$ and closure rate of the valve are known. An example was given in Bellhouse & Talbot (1969). Their results showed that the velocity at the cusp tip was relatively constant during closure and the flow pattern established in the early stages of systole would persist until forward flow ceased when the valve began to close.

Arterial pressure. The time-averaged arterial pressure, $\bar{p}_A$, actually $\Delta\bar{p} = \bar{p}_A - \bar{p}_v$ (where $\bar{p}_v = 0$), is the driving force for blood flowing through the systemic organs. It relates to the cardiac output (Eq. (2.2.1)) as

$$\bar{p}_A = QR_{\text{total}} \qquad \text{(2.2.2a)}$$

where Rtotal is the sum of all peripheral resistances caused by the vasculature. A reasonable estimate of $\overline{p}_A$ is given by:

$$\overline{p}_A = p_D + \frac{1}{3}(p_S - p_D) \qquad (2.2.2b)$$

where p_D is the diastolic pressure and p_S-p_D is the pulse pressure, i.e., systolic minus diastolic pressure. Normal mean values for a person at rest are $\overline{p}_A = 100\,\text{mmHg}$, Q = 5L/min, and hence R_{total} = 20 mmHg/ L/min. Clearly, the heart can maintain a $\overline{p}_A$ of 100 mmHg by varying the blood flow rate in light of changing flow resistance downstream. Because proper operation of the cardiovascular system is so important, especially blood supply to the brain and the heart, arterial blood pressure is continuously monitored by various sensors in the body. Specifically, multiple reflex responses are initiated whenever $\overline{p}_A$ deviates from normal to adjust the cardiac output and/or total peripheral resistances as needed.

Example 2.12: Pressure-Stress Relation in Left Ventricle

Assuming that the left ventricle is a thin-walled, semi-ellipsoidal shell, develop a transient relationship between left ventricular pressure and ventricle volume (see Sketch), employing wall stress equations postulated by Laplace and Carlson (see "Concepts").

Sketch:

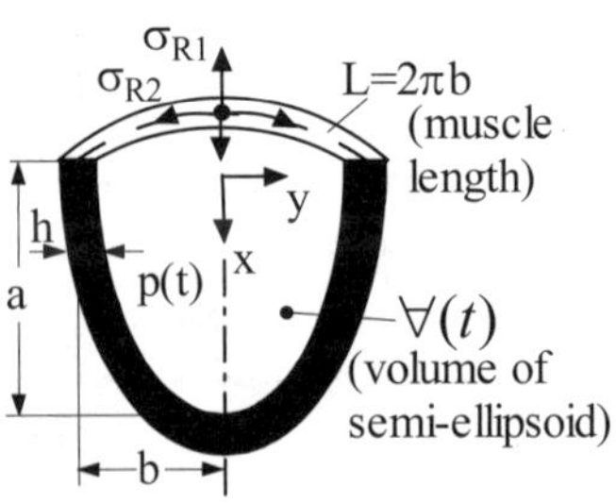

Assumptions:

- Idealized geometry w/constant wall thickness and apical lengths

Concepts:

- Laplace's eq'n for thin-walled ellipsoids:

$$p = \left(\frac{\sigma_{R_1}}{R_1} + \frac{\sigma_{R_2}}{R_2}\right)h$$

- Carlson's wall-stress eq'n:

$$\sigma_w(t) = \sigma_{iso} - c\,(dL/dt)$$

where $\sigma_w \sim \sigma_{R_2}$

Solution: The geometric parameters for an ellipsoid are:

- semi-ellipsoid volume $\forall = \dfrac{2}{3}\pi\,a\,b^2$;
- largest circumference, i.e., sequential muscle length L = 2πb;

- radii of curvatures at the stress locations (see Sketch) of σ_{R_1} and σ_{R_2} are $R_1 = a^2/b$ and $R_2 = b$, respectively.

Thus, with a = constant and b = b(t),

$$\forall(t) = \frac{a}{6\pi} L^2(t) \tag{E2.12-1}$$

With $R_1 >> R_2$, Laplace's equation can be simplified to:

$$p = \frac{\sigma_{R_2}}{R_2} h \tag{E2.12-2a}$$

or,

$$\sigma_{R_2} = \frac{b}{h} p := \frac{L}{2\pi h} p \tag{E2.12-2b}$$

which implies the assumption that $\sigma_{R_1} < \sigma_{R_2}$.
In Carlson's equation,

$$\sigma_w(t) = \sigma_{iso} - c\left(\frac{dL}{dt}\right) \tag{E2.12-3}$$

σ_w is the segmentally averaged wall stress, σ_{iso} is the isometric stress, c is a muscle-tissue constant, and L = L(t) is the muscle segment length.

Averaging the circumferential stress, σ_{R_2}, over the semi-ellipsoid, $0 \le x \le a$, yields an average wall stress, i.e.,

$$\sigma_w = \frac{1}{a}\int_0^a \sigma(x)\,dx := \frac{\pi}{4}\sigma_{R_2} \tag{E2.12-4a}$$

and using Eq. (E2.12-2b) we have

$$\sigma_w = \frac{L}{8h} p \tag{E2.12-4b}$$

Inserting Eq. (E2.12-4b) into (E2.12–3) and using Eq. (E2.12-1) yields:

$$p(t) = 8h\sqrt{\frac{a}{6\pi}}\frac{\sigma_{\text{iso}}}{\sqrt{\forall(t)}} - \frac{4ch}{\forall(t)}\left(\frac{d\forall(t)}{dt}\right) \tag{E2.12-5}$$

where $d\forall/dt \triangleq Q(t)$ is the inflow rate and ∀(t) (or L(t) in Eq. (E2.12-1)) has to be given.

Graph:

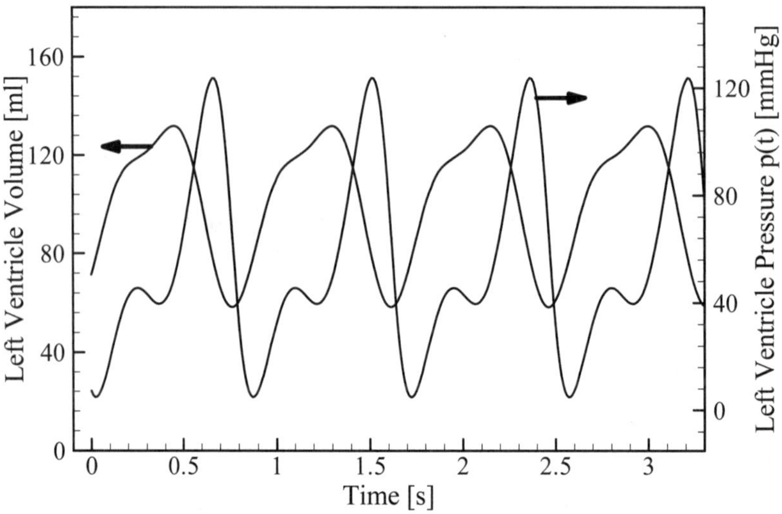

Note: The parameters used are: h=12mm, a=5cm, σ_{iso}=1.40×10^4 N/m^2, c=3.0×10^{2}N·s/m^3. The left ventricle volume $\forall(t)$ is approximately correlated from data in the literature, i.e.,

$$\forall(t) = 101.49 - 29.86\cos(7.395t) + 15.15\sin(7.395t) - 0.15\cos(14.79t) + 10.63\sin(14.79t)$$

Comments:
Clearly, the periodic heart/vessel expansion and contraction (see given $\forall(t)$) causes blood inflow Q(t) and significant pressure build-up, respectively. The Graph depicts the $\forall(t)$ and resulting p(t) interaction, while Fig.1.1.2 provides the broader framework for aortic flow generation.

Heart disease detection. The heart's sounds and electrical activity provide non-intrusive measures of its health status. The dual "lub-dub" sounds occur when the AV valves shut at the beginning of systole and then when the semilunar valves close, i.e., at isovolumetric relaxation. The heart sounds are the result of blood-induced vibrations of the cardiac structures, and are an example of fluid-structure-interaction. For example, heart murmurs are caused by defective valves, which generate abnormal blood flow patterns, and can lead to pulmonary hypertension and fluid in the lungs (i.e., edema). The distinct heart sounds are timewise correlated with the intensity of the heart's electrical signals, recorded as an electrocardiogram (ECG). Because tissue fluids contain a high concentration of ions, potential differences generated by the heart are conducted to the body surface where electrodes (or leads) pick up the conduction of electrical impulses (in millivolts) through the heart. Each cardiac cycle produces three distinct ECG waves, which represent changes in potential between two regions on the heart's surface that are produced by the combined effects of myocardial-cell activities. Clearly, deviation of a patient's ECG from the age-related standard ECG spells trouble. In summary, healthy heart function requires that:

- the contractions of all cardiac muscle cells occur at regular intervals in a synchronized fashion;
- the valves must open fully and close without leaks;
- the muscle contractions must be strong; and
- the ventricles must fill adequately during diastole.
- In addition to the electrical excitatory impulse, the cardiac function can be influenced by neural inputs from the autonomic nervous system. Such signals modify the heart's pumping action as is appropriate to meet changing homeostatic needs of the body.

2.2.3 The Blood Vessels

As indicated in Fig. 1.1.2, blood flow is propelled from the left heart chamber into the aorta and then passes consecutively through different types of vessels before it returns via the vena cava to the right heart chamber. Table 2.2.1 provides some geometric characteristics of morphological characteristics. All components of the circulatory system, including the blood vessels and heart chambers, are lined with a layer of endothelial cells, which have major regulatory functions. A dysfunctional endothelium can lead to several arterial diseases as outlined in Chapter III.

Table 2.2.1 Geometric Blood Vessel Characteristics.

Vessel Type	Number	Internal Diameter	Wall Thickness	Surface Area
Aorta	1	2.5 cm	2 mm	4.5 cm^2
Arteries	160	0.4 cm	1 mm	20 cm^2
Arterioles	5×10^7	30 μm	20 μm	400 cm^2
Capillaries	10^{10}	5 μm	1 μm	4,500 cm^2
Venules	10^8	70 μm	7 μm	4,000 cm^2
Veins	200	0.5 cm	0.5 mm	40 cm^2
Venae Cavae	2	3 cm	1.5 mm	18 cm^2

Arteries are viscoelastic multi-layered conduits, consisting of the intima, media, and adventitia from the inside-out. The very important intima and media are comprised of the endothelium lining, the lumen surface plus layers of smooth muscle cells, and membranes as well as collagen and elastin fibers. The large arteries expand when the blood pressure rises and contract during diastole, thereby supplying blood to the organs downstream.

Arterioles, known as resistance vessels, are much smaller and more muscular, implying that they can work as regulators of peripheral blood flow through organs by changing the diameter and hence flow resistance.

Capillaries are the narrowest of blood vessels and form the end points of the circulatory system, exchanging liquids, gases (O_2, CO_2), and nutrients between the blood and the tissue. A capillary wall is just a layer of endothelial cells, $1\mu m$ thick, and about 0.5mm long, where a capillary bed constitutes an enormous surface area.

After leaving the capillaries, blood is collected first in venules and then in veins (see Fig. 1.1.3). Venous vessels are thin-walled and quite distensible. Large veins feature one-way valves to prevent backflow, which is important while standing and during exercise. Peripheral venules and veins contain most of the total blood volume. Thus, electrical signal, i.e., nerve-triggered, venous contractions as well as skeletal muscle pumping may strongly affect the venous blood volume and flow to the heart. Thus, venous capacity exerts considerable control in terms of cardiac pumping and output. While skeletal muscles along veins can contract and squeeze venous blood from the lower limbs toward the heart, breathing can aid blood flow from the abdominal to the thoracic veins. Specifically, when a person inhales, the diaphragm, which is a muscular sheet separating the thoracic and abdominal cavities, contracts. During contraction, the normally dome-shaped diaphragm becomes flat or even protrudes into the abdomen. As a result, the pressure in the thoracic cavity decreases but increases in the abdomen, thereby creating a pressure drop between abdominal and thoracic vein segments which leads to blood flow back to the heart.

The Windkessel. In the *Windkessel* model, the arterial system is viewed as a reservoir which cushions the pulsatile blood flow similar to an air-filled compression chamber (i.e., a “Windkessel” in German) which smooths the intermittent pumping action of an old fire truck. Thus, the blood pressure waveform generated by the heart is assumed to be transmitted instantaneously throughout the arterial system and vanishes before the next heart beat so that each pressure waveform is the same in all major arteries. The arterioles then act as a high-resistance bed (or network), causing near-uniform creeping blood flow in the tissue and hence maintaining a high pressure throughout the cardiac cycle. The Windkessel model ignores the facts that pressure waves require small travel times and that they are being reflected in somewhat short elastic tubes, especially with downstream nonuniformities such as bifurcations, bends, diameter reductions, partial occlusions (e.g., stenoses) or changes in tube material properties, etc.

The differential change in storage capacity of such an elastic chamber with outflow resistance can be simply expressed as:

$$d\forall = D_E dp \tag{2.2.3}$$

where D_E is the chamber wall distensibility which is directly related to the bulk modulus $k = \forall / D_E := E/[3(1-2\nu)]$, where E is Young's modulus and ν is Poisson's ratio. Thus, the storage rate $d\forall/dt$ is the difference between blood influx Q_{in} (e.g., from ventricular ejection) and the outflow Q_{exit}, determined by the pressure drop or downstream resistance R_{total}. Clearly, for D_E=constant,

$$\frac{d\forall}{dt} = D_E \frac{dp}{dt} = Q_{in} - Q_{exit} := Q(t) - \frac{\Delta p}{R_{total}} \tag{2.2.4a-c}$$

<u>Example 2.13</u>: A Windkessel Application

Consider constant blood flow Q_0 during the systolic phase, $0 \le t \le t_s$, into the arterial network (i.e., the Windkessel) and zero inflow during diastole, $t_s < t \le T$. The vascular bed is described by $\Delta p = p - p_{venous}$ and the lumped resistance R_t. Find $p(t)/(Q_0 R_t)$ for the systolic and $p(t)/p_T$ for the diastolic phase, $t_s < t < T$, where T is the cycle duration and p_T is the pressure at the end of diastole. Calculate the heart stroke volume $\forall_s$.

Sketch:

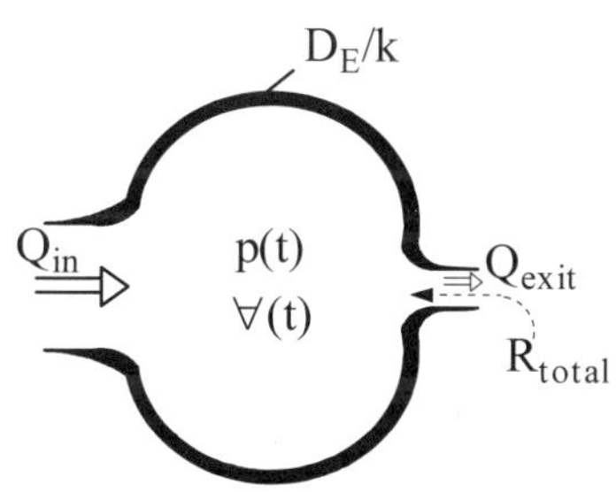

Assumptions:

- As stated
- $p_{venous} \approx 0$
- $Q_{in} = Q_0 = ¢$ for $0 \le t \le t_s$

Approach:

- Lumped-parameter Windkessel model
- Conditions: $p(t=0) = p_0$; $p(t=T) = p_T$

Solution: Based on Eq. (2.2.4) we can write:

(i) Systolic phase:
$$Q_0 - \frac{p}{R_t} = D_E \frac{dp}{dt} \tag{E2.13-1}$$

(ii) Diastolic phase:
$$0 - \frac{p}{R_t} = D_E \frac{dp}{dt} \tag{E2.13-2}$$

Separation of variables and integration yields for

Phase (i): $p_{syst} \equiv \frac{p(t)}{Q_0 R_t} = 1 - \left(1 - \frac{p_0}{Q_0 R_t}\right) \exp\left(-\frac{t}{D_E R_t}\right)$ (E2.13-3)

and for Phase (ii): $p_{diast} \equiv \frac{p(t)}{p_T} = \exp[(T-t)/(D_E R_t)]$ (E2.13-4)

Related to Eqs. (2.2.1) and (2.2.3) we have in general

$$\forall_s = \int_0^{t_s} Q_{in}(t)dt \qquad \text{(E2.13-5a)}$$

For the present case

$$\forall_s = Q_0 t_s \qquad \text{(E2.13-5b)}$$

<u>Graph:</u>

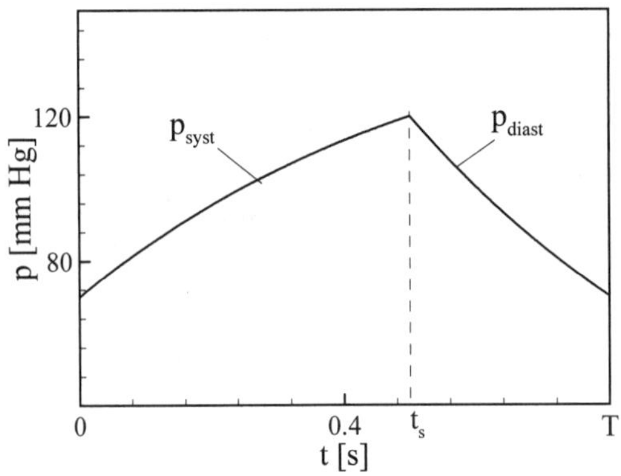

<u>Note:</u> The parameters used are: $p_0 = p_T = 70$mmHg, T=0.8s, t_s=0.5s, $D_E R_t$ = 0.557s, $Q_0 R_t$=154 mmHg.

<u>Comments:</u> As expected, the blood pressure peaks at $t = t_{systole} = 0.5$s when blood flow into the vessel system stops. Then, during outflow and vessel relaxation the pressure drops to the prescribed p_T=70mmHg. The windkessel p(t) curve is a coarse approximation of an actual waveform; for example, as shown in Fig.2.2.3.

Pressure waves. The Windkessel model is inappropriate for pulsating blood flow in (short) elastic vessels with natural constrictions or pathological obstructions (e.g., stenoses) at their ends. In such cases, we have to deal with pressure wave reflections which lead to blood pressure build-up causing peak pressure and peak blood flow to be out of sync (see Fig. 2.2.3). The effect of pressure wave reflections generated by a single inlet pressure wave created by liquid bolus injection into a straight elastic liquid-filled tube with a partial opening located at different tube ends is illustrated in Fig. 2.2.4. The reflected waves occur at longer intervals as the partially occluded tube ends are located further away from the tube inlet. Thus, the inlet (or incident) wave moves back and forth in the tube while it is reduced in strength due to viscous flow and wall material effects. Nevertheless, the mean pressure rises as a result of successive reflections while the tube slightly expands by the volume of liquid injected.

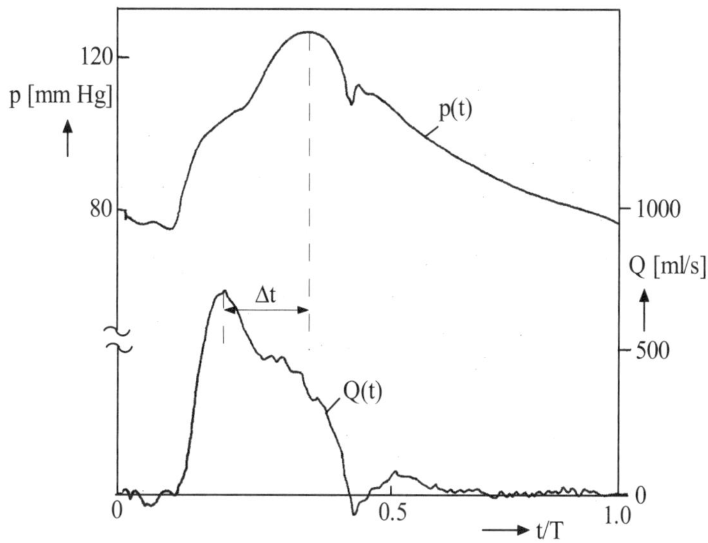

Fig. 2.2.3 Schematic pressure and flow waveforms during ventricular ejection in the ascending aorta showing delayed pressure peak after maximum flow rate due to pressure wave reflections.

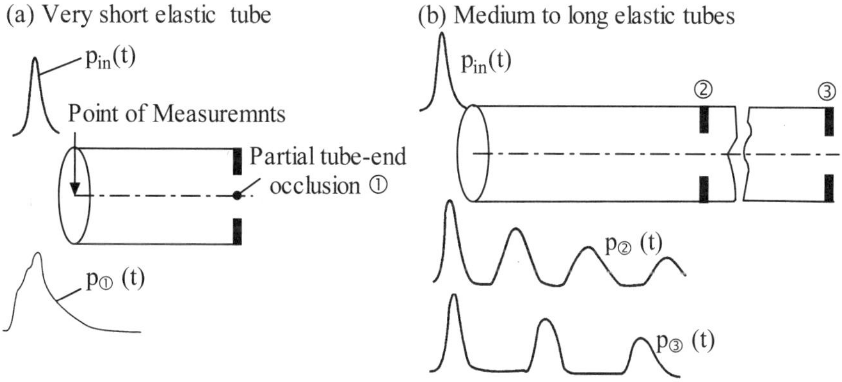

Fig. 2.2.4 Schematic of single inlet pressure wave, $p_{in}(t)$, and resulting wave trains, $p_{①}$ - $p_{②}$ due to reflections at different tube ends and attenuations due to fluid viscosity and tube-wall damping.

For all practical purposes, pressure and flow waveforms, being periodic functions, can be readily analyzed with a Fourier series expansion, i.e., a finite superposition of sine- and cosine-functions (see Spiegel, 1971 or Greenberg, 1998, for example). Specifically, y(t) = y(t+T), where $y \hat{=} p$ (pressure) or Q (flow rate), t is the time, and T is the period, e.g., T ≈ 1 sec for the cardiac cycle, after which y(t) repeats itself. Also, $T^{-1} = f$ is the frequency and $2\pi f = \omega$ is the angular frequency, in radians per second. The postulate for any periodic function is the Fourier series:

$$y(t) = \frac{A_0}{2} + A_1 \cos \omega t + A_2 \cos 2\,\omega t + \ldots + A_n \cos(n\,\omega t) + B_1 \sin \omega t + B_2 \sin 2\,\omega t + \ldots\ldots + B_n \sin(n\,\omega t) \tag{2.2.5a}$$

where:

$$A_0 = \frac{1}{2\pi}\int_0^{2\pi} y(t)dt, \quad A_n = \frac{1}{\pi}\int_0^{2\pi} y(t)\cos(n\,\omega t)\,dt \tag{2.2.5b,c}$$

and

$$B_n = \frac{1}{\pi}\int_0^{2\pi} y(t)\sin(n\,\omega t)\,dt \tag{2.2.5d}$$

Equation (2.2.5b) indicates that $A_0 = y_{average}$ and Eqs. (2.2.5c,d) are known as Euler's formulae.

Example 2.14: The Aorta as an Elastic Reservoir

Consider the sketched feedback control loop where the given flow rate during the systolic phase, $Q_{in} = A\sin\omega t$, supplies blood to the aorta, which acts like an elastic reservoir. Find the blood pressures $p_s(t)$ and $p_d(t)$ during systole and diastole, respectively.

Sketch:

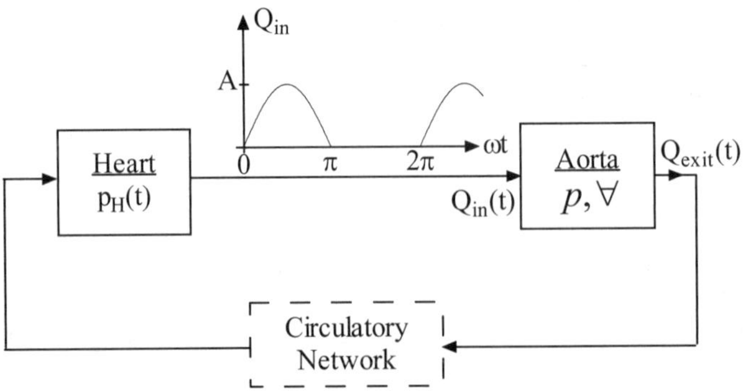

Assumptions:

- Systemic circulation consists of the heart, aorta, and circulatory network
- Heart provides $Q_{in}(t)$ to aorta during systolic phase
- Aorta expands and accommodates the blood volume

Postulates:

- $$Q_{in}(t) = \begin{cases} A\sin\omega t \text{ for } 0 \le \omega t \le \pi \text{ (systole)} \\ 0 \quad \text{for } \pi < \omega t < 2\pi \text{(diastole)} \end{cases} \tag{E2.14-1a,b}$$
- $\forall = \forall_0 + a\,p$ inside the aorta
- $Q_{exit}(t) = b\,p$

Solution: Compartmental, lumped-parameter approach, i.e.,

$$\frac{d\forall}{dt} = Q_{in} - Q_{exit} \tag{E2.14-2}$$

Using the $\forall$(p)-relative for the aorta and $Q_{exit} = b\,p$,

$$\frac{d\forall}{dt} = Q_{in} - b\,p = a\frac{dp}{dt} \tag{E2.14-3}$$

or

$$a\frac{dp}{dt} + b\,p = \begin{cases} A\sin\omega t \text{ during systole} \\ 0 \qquad \text{during diastole} \end{cases} \tag{E2.14-4}$$

subject to p (t = 0) = p_0 and p(t = π/ω) = p_1. Thus, for the systolic phase, $0 \le t \le \pi/\omega$:

$$p = p_s(t) = p_0 e^{-bt/a} + A[a\omega e^{-bt/a} + b\sin\omega t - a\omega\cos\omega t]/(b^2 + a^2\omega^2) \tag{E2.14-5a}$$

and during the diastolic phase, $\pi/\omega \le 2\pi/\omega$, where T is the cycle duration,

$$p = p_d(t) = p_1 \exp[-b(t-T)/a] \tag{E2.14-5b}$$

The parameters a, b, p_0, and p_1, are also subject to the following constraints:

$$p_s(t = t_s = \pi/\omega) = p_1 \text{ and } p_d(t = T - 2\pi/\omega) - p_0 \tag{E2.14-6}$$

Graphs:

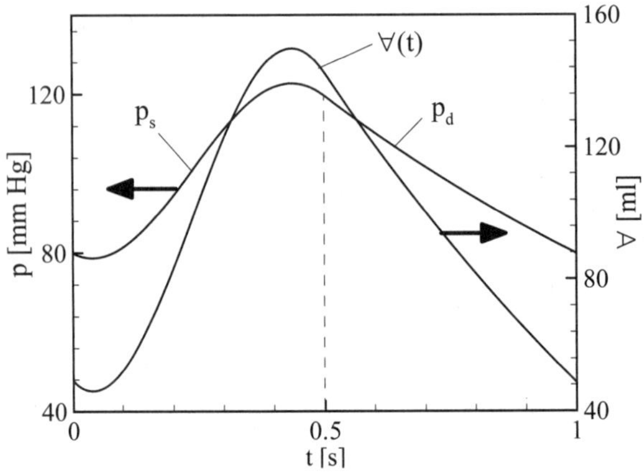

Note: The parameters used are: p_0=80 mmHg, p_1=120 mmHg; A=600 ml/s; ∀(t=0)=50 ml; plus the coefficients a=2.363 and b=1.916 according to Eqs. (E2.14-5) and (E2.14-6).

Comments:

As net blood flow, i.e., Q_{in} delivered by the heart and Q_{exit} leaving the aorta, fills the expanding vessel, p(t) drops marginally and then shoots up to p_{max} (t≈0.42s). While blood is still being supplied the aortic pressure slightly decreases before the end of the systolic phase $0 \le t \le 0.5$. Based on the outflow condition $Q_{exit}(t) \sim p$, which depends on the downstream resistance of the circulatory network, the aortic pressure decreases almost linearly during diastole $0.5 < t \le 1.0$. These curves are essentially similar to actual waveforms (e.g., as shown in Fig.2.2.3), but they are more smooth.

Example 2.15: Transient Flow in an Elastic Capillary

Consider Poiseuille-type flow in response to a sudden change in inlet pressure for an elastic micro-tube. An empirical constitutive equation (cf. Schmid-Schönbein et al., 1989), i.e.,

$$p(t) = E\varepsilon = \frac{E}{2}\left[\left(\frac{R}{R_0}\right)^2 - 1\right] \qquad \text{(E2.15-1)}$$

relates lumen pressure to wall strain, where E is the elasticity coefficient, R = R(t) is the tube radius, and R_0 is the base radius. Find the volumetric flow rates Q (x = 0) and Q (x = L) due to p(t,x), where $p = p_0$ at t = 0 for all x and then $p = p_1 > p_0$ at x = 0 for all t.

Sketch:

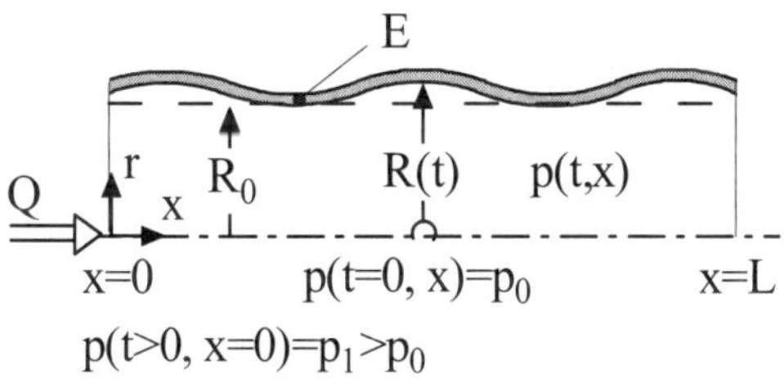

Assumptions:

- Quasi-steady flow because $R_0 << 1$ and hence $Wo \equiv \frac{\omega R_0{}^2}{\nu} << 1$
- Quasi-unidirectional flow because $\Delta R << 1$
- Linearly elastic tube wall and incompressible fluid

Approach:
- 1-D mass balance
- Derive pressure equation for p(t, x)

Solution: A 1-D mass balance yields:

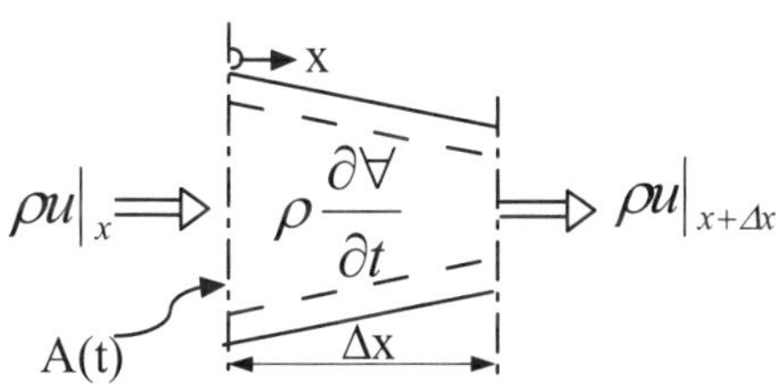

$$\rho u A|_x - \rho u A|_{x+\Delta x} = \rho \frac{\partial \forall}{\partial t}$$

$$-\frac{\partial (uA)}{\partial x}\Delta x = \frac{\partial A}{\partial t}\Delta x$$

$$\therefore \frac{\partial A}{\partial t} = -\frac{\partial Q}{\partial x} \qquad \text{(E2.15-2)}$$

In light of Wo << 1, we assume Poiseuille-type flow at any time t, i.e., as shown in Sect. 1.3.2 and in Example 2.7 (see Eq. (E2.7-6)) with ε=0:

$$Q = -\frac{\pi [R(t)]^4}{8\mu}\frac{\partial p(t)}{\partial x} \qquad \text{(E2.15-3)}$$

With $A = \pi R^2$ and $p = \frac{E}{2}\left(\frac{R}{R_0}\right)^2 - \frac{E}{2}$ from (E2.15-1), we can use Eqs. (E2.15-2, 3) to write:

$$\frac{\partial p}{\partial t} = \frac{E}{2\pi R_0^2}\frac{\partial (\pi R^2)}{\partial t} = \frac{E}{2\pi R_0^2}\frac{\partial A}{\partial t}$$

and hence

$$\frac{2\pi R_0^2}{E}\frac{\partial p}{\partial t} = -\frac{\partial Q}{\partial x} \tag{E2.15-4}$$

Using Eq. (E3.2.15-3) in the form:

$$Q = -\underbrace{\frac{1}{8\mu\pi}}_{\kappa} A^2 \frac{\partial p}{\partial x}$$

we can write

$$\frac{\partial Q}{\partial x} = \kappa\, A^2 \frac{\partial^2 p}{\partial x^2} + \kappa \frac{\partial p}{\partial x}(2A)\frac{\partial A}{\partial x} \tag{E2.15-5}$$

With $\frac{\partial A}{\partial x} = O(\pm\varepsilon)$, the second term is negligible and Eq. (E2.15-4) can be approximately written as:

$$\frac{\partial p}{\partial t} - \frac{E\,A_0}{16\pi\mu}\frac{\partial^2 p}{\partial x^2} = 0 \tag{E2.15-6a}$$

In dimensionless form with p = EP, x = ξL, and $t = \tau\frac{16\pi\mu L^2}{E\,A_0}$:

$$\frac{\partial P}{\partial \tau} - \frac{\partial^2 P}{\partial \xi^2} = 0 \tag{E2.15-6b}$$

As shown in Example 1.6 in Sect. (1.3.2), the separation of variables method leads to an infinite series representation for p(t, x). Here,

$$\frac{P_1 - P(\xi,\tau)}{P_1 - P_0} = \xi + \frac{2}{\pi}\sum_{n=1}^{\infty}\frac{1}{n}\sin\,(\pi n\xi)\cdot\exp\,(-n^2\pi^2\tau) \tag{E2.15-7}$$

with which the volumetric flow rates at ξ = 0 and ξ = 1 can be estimated from Eq. (E2.15-3), using Eq. (E2.15-1) (see Graph).

Graph:

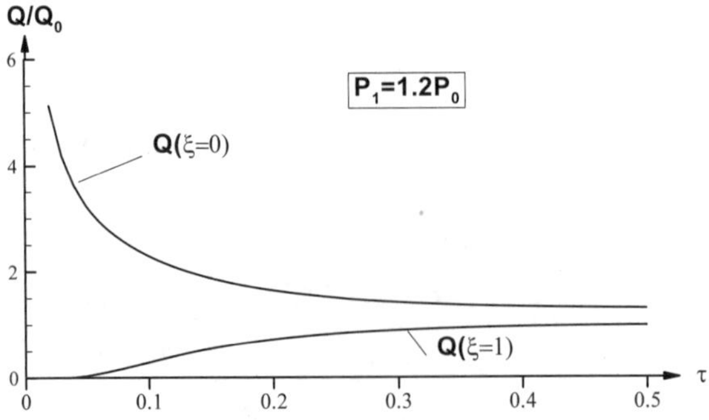

Note: $Q_0 = Q(\xi = 1, \tau \to \infty) = \frac{\pi E R_0^4}{8\mu L}(1+2P_0)^2(P_1 - P_0)$.

Comments:

With the simplification $\partial A/\partial x \approx 0$, the influence of R(t)-variations is eliminated and as a result the flow rate is determined by the pressure in space and time, i.e., Q=Q[p(x,t)]. With a sudden jump at the inlet from p_0 to $p_1=1.2p_0$, Q(ξ=0) is high and then decreases exponentially. With the assumed outlet pressure $p_{out}=p_0$, the effect of p(t>0, ξ=0)=p_1 is not felt at the tube exit, i.e., Q(ξ=1)=0 for 0≤τ≤0.04 and then after the time lag, Q_{exit} approaches Q_0, the equilibrium flow rate.

Peristaltic motion. While pressure-wave propagation in previous examples was caused by upstream forces, e.g., heart-pumping action, peristaltic fluid motion is caused by periodic contractions of the tube walls. Specifically, peristalsis is the process of wavelike muscular wall contractions that propel (axially) contained matter in tubular organs or other conduits, as in the alimentary canal, glandular ducts, capillaries, and the alveolar region. For a basic elastic conduit of radius r_0 or half-width a, the peristaltic pumping parameters include: h_0, the maximum wave amplitude; λ, its wavelength; and c, the wave speed. Based on these parameters, we now can form:

- a dimensionless wave number $N = 2\pi\, a/\lambda$ (2.2.6a)
- an amplitude ratio $\kappa = h_o / a$ (2.2.6b)
- a Reynolds number $\mathrm{Re} = Nca/v$ (2.2.7)
- the wave frequency $\omega = 2\pi c/\lambda$ (2.2.8)

In order to solve peristalsis problems, a reduced form of the Navier-Stokes equations is set up with a sinusoidal displacement wave, as a boundary condition, traveling in the conduit wall with constant speed c. The key objectives are to compute the fluid axial pressure gradient resulting from the pressure difference at both conduit ends and the superimposed traveling wall wave, the velocity field (or volumetric flow rate), and the conditions of reflux. For most applications we have small parameter values, i.e., Re < 1.0, $\kappa << 1$, and N ≤ O(1). An exception is the gastrointestinal tract where $\mathrm{Re} \approx 10, N = 2,$ and $\kappa \approx 0.3$. Example 2.16 illustrates a peristaltic problem solution.

<u>*Example 2.16:*</u> Peristaltic Pumping (after Shapiro et al., 1969)

Consider Stokes flow in a 2-D channel representing the ureter (see Sect. 4.3.1), where Re < 1 and $\lambda >> a$, with $a \approx 2mm$, $\lambda = 12cm$, length L=30cm, and $c \approx 3cm/s$. Find the velocity profile, volumetric flow rate, and overall pressure drop.

<u>*Sketch (Actual Frame):*</u>

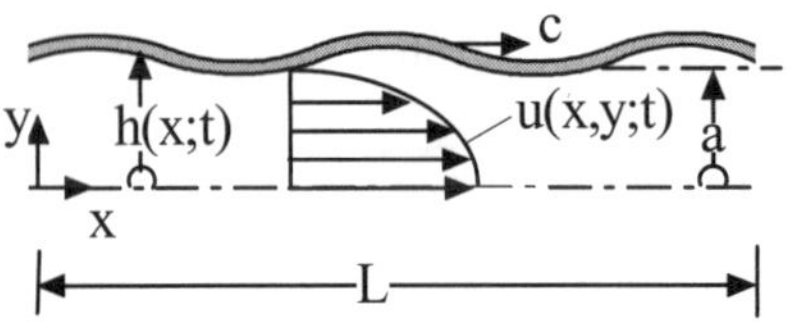

<u>*Assumptions:*</u>

- Steady-state fluid motion in wave frame
- Poiseuille type flow

<u>*Concepts:*</u>

- Observer moves with wave speed
- Reduced N-S equations

<u>*Solution:*</u> The stationary and moving frames of reference are connected as $x = \hat{x} + ct,\ y = \hat{y},\ and\ u = \hat{u} + c$, where capped values refer to the "wave frame." Thus, the x-momentum equation (1.3.20b) reduces to:

$$0 = \frac{dp}{d\hat{x}} + \mu \frac{\partial^2 \hat{u}}{\partial \hat{y}^2} \tag{E2.16-1}$$

subject to:

$$\hat{u}\,(\hat{y} = h) = -c \quad \text{and} \quad \left. \frac{\partial \hat{u}}{\partial \hat{y}} \right|_{\hat{y}=0} = 0 \tag{E2.16-2a,b}$$

Double integration yields after transformation back to the stationary frame,

$$u = \frac{2\pi a^2}{2\mu\lambda} (\frac{dp}{d\xi})(\eta^2 - H^2) \tag{E2.16-3a}$$

where,

$$\xi = 2\pi \frac{x}{\lambda} \tag{E2.16-3b}$$

$$\eta = y/a \tag{E2.16-3c}$$

$$H = \frac{h(x,t)}{a}: = 1 + \frac{h_0}{a}\cos(\xi - \tau) \tag{E2.16-3d}$$

and

$$\tau = 2\pi ct/\lambda \tag{E2.16-3e}$$

The volumetric flow rate, $Q = \int_A u \, dA,$ can be calculated as:

$$Q(\xi, \tau) = -\frac{4}{3} \frac{\pi a^2}{\mu c \lambda} (\frac{dp}{d\xi}) H^3 \qquad \text{(E2.16-4a)}$$

where $2\pi a^2 (\frac{dp}{d\xi})/(\mu c \lambda) := dP / d\xi$ is a dimensionless pressure gradient. Integration of Eq. (2.16-4a) over one wavelength yields the time-averaged flow rate:

$$\overline{Q} = \frac{1}{2\pi} \int_0^{2\pi} Q(\xi, \tau) \, d\tau = \hat{Q} + 1 \qquad \text{(E2.16-4b)}$$

In order to eliminate $\overline{Q}$, we notice that Eq. (E2.16-4a) can be split into $\hat{Q}(\tau) + H,$ so that the dimensionless pressure gradient can be expressed as:

$$\frac{dP}{d\xi} = -\frac{3[\hat{Q}(\tau) + H]}{H^3} \qquad \text{(E2.16-5)}$$

Now, integration of Eq. (E2.16-5) over the channel length L yields the pressure drop:

$$\Delta P(\tau) = -3 \int_o^L \frac{d\xi}{H^2} - 3\hat{Q}(\tau) \int_o^L \frac{d\xi}{H^3} \qquad \text{(E2.16-6)}$$

For one wavelength, i.e., $\lambda = 2\pi ct/\tau$, and using $\hat{Q} = \overline{Q} - 1$ from Eq. (E2.16-4b), we obtain:

$$\overline{Q} = \frac{3\kappa^2}{2 + \kappa^2} - \frac{(1 - \kappa^2)^{5/2}}{3\pi(2 + \kappa^2)} \Delta P\big|_\lambda \qquad \text{(E2.16-7)}$$

where $\Delta P\big|_\lambda$ is the pressure change per wavelength and $0 \le \kappa = h_o / a \le 1$. Clearly, when $\overline{Q} = 0$, $\Delta P\big|_\lambda$ is a maximum, and when $\Delta P\big|_\lambda = 0$,

$$\overline{Q}_{\max} = \frac{3\kappa^2}{2 + \kappa^2} \qquad \text{(E2.16-8)}$$

In the stationary frame, the velocity can be expressed as:

$$u(\eta; \overline{Q}, H) = \frac{3(H^2 - \eta^2)}{2 H^3} (\overline{Q} + H - 1) \qquad \text{(E2.16-9)}$$

Graphs:

(a) Time-averaged flow rate

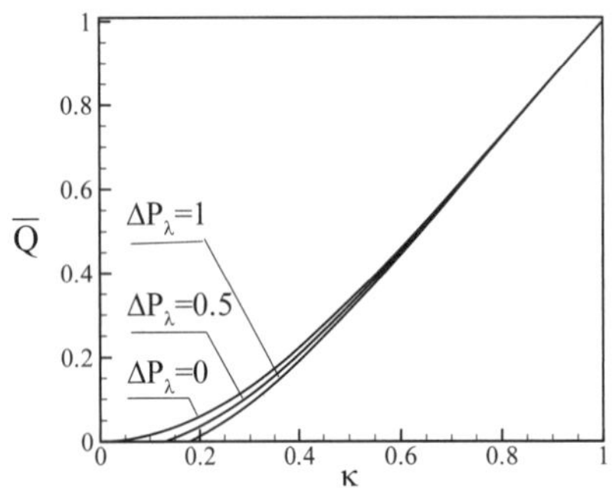

(b) Profiles of axial velocity

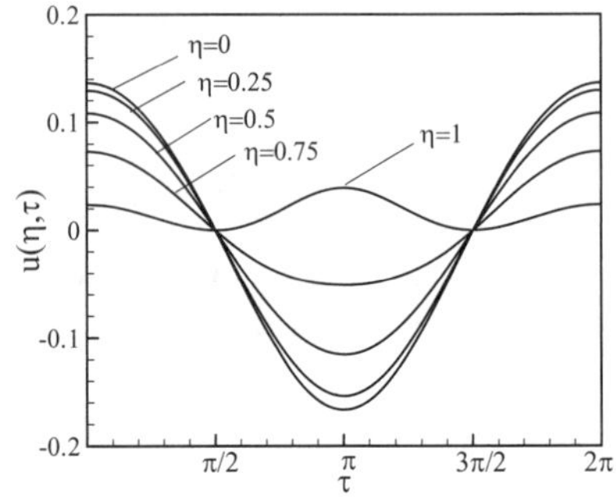

Note: For figure (b), x=12cm, $\overline{Q} = 0$, and κ=0.1.

Comments: Clearly, the time-averaged flow rate $\overline{Q}$ depends mainly on the degree of tube-wall contraction, i.e., κ~ h_0 (see Eq. (E2.16-3d)), and only weakly on the pressure drop per wavelength, $\Delta p|_\lambda$.

Assuming a peristaltic action which generates per wavelength/cycle zero net flow, the local axial velocity oscillates, i.e., fluid moves back and forth over $0 \le \tau \le 2\pi$. As expected, at the centerline, η=0, the velocity is the highest, and near/at the moving wall (η=1.0) it is the lowest.

Blood Flow in Bifurcating Arteries. As discussed (see Fig.1.4.4), blood is a shear-rate dependent fluid flowing in pulsatile fashion through non-rigid bifurcating nonplanar tubes with variable cross sections and "surface roughness." While steady or pulsatile flow of an incompressible fluid in a straight rigid or slightly elastic tube are occasionally reasonable starting points when modeling the hemodynamics, in actual arterial trees (or stenosed blood vessels) wave reflections are caused by geometric viscoelastic non-uniformities. As mentioned, such traveling pressure waves may alter the (axial) pressure distributions and hence the velocity field and any particle transport. When assuming medium-to-large *noncoronary* arteries to be quasi-rigid due to age or disease, wave reflections do not exist because Young's arterial wall modulus, *E*, is "infinite" and so is the wave speed, *c*. As a result, pressure changes at the tube inlet for thin-walled tubes (*h*<<*D*) and negligible viscous effects are instantaneously propagated, occurring simultaneously at every point along the tube. The application criterion is the magnitude of

$$c^2 = hE/(\rho D) \tag{2.2.9}$$

which is known as the *Moens-Korteweg formula* (see Milnor, 1989 and Example 2.17). Pulsatile fluid flow in elastic conduits, i.e., blood vessels as well as lung airways, was illustrated in previous Examples and is further discussed with solutions to wave reflection problems by Zamir (2000) and Nichols & O'Rourke (1998).

Example 2.17: Derivation of the Moens-Korteweg Formula

Consider transient axisymmetric flow in a very long, thin-walled elastic tube of radius $R(t)=r_0+\eta(z,t)$ and wall thickness h. The flow is powered by a pressure pulse of wavelength λ and period τ where the wave speed $c \equiv \lambda/\tau \gg v_z$, the axial flow velocity. Furthermore, we assume that $\frac{v_r}{v_z} \sim \frac{r_0}{\lambda} \ll 1$. For the (dominant) hoop stress in a thin-walled linearly elastic tube with radial displacement $\eta=\eta(z,t)$ we assume

$$\sigma_\theta = E\varepsilon_0 := E\frac{\eta}{r_0} \tag{E2.17-1}$$

Sketch:

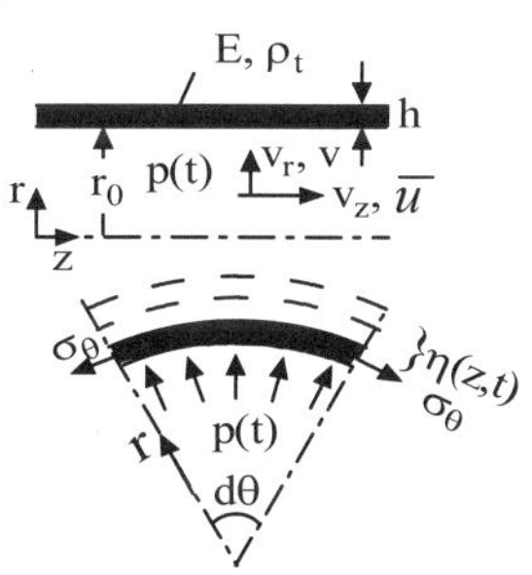

Assumptions:

- As stated
- Frictionless flow
- Displacement acceleration $\ddot{\eta}$ is negligible

Approach:

- Reduced N-S equations
- Hooke's law
- Newton's 2nd law

Solution: Based on the assumptions, the reduced Navier-Stokes equations read

(continuity)
$$\frac{\partial v_z}{\partial z} + \frac{1}{r}\frac{\partial}{\partial r}(r v_r) = 0 \tag{E2.17-2}$$

(r-momentum)
$$\rho\frac{\partial v_r}{\partial t} = -\frac{\partial p}{\partial r} \tag{E2.17-3}$$

(z-momentum) $$\rho\frac{\partial \mathrm{v}_z}{\partial t}=-\frac{\partial p}{\partial z} \quad \text{(E2.17-4)}$$

Integrating the continuity equation over $0 \le r \le r_0$ with $dA = 2\pi r dr$ and defining the volumetric flow rate as:

$$Q=\bar{u}A=2\pi\int_0^{r_0}\mathrm{v}_z r dr \quad \text{(E2.17-5)}$$

where $\bar{u}=\bar{u}(z,t)$, we obtain

$$\frac{\partial}{\partial z}\left[\pi r_0^2\bar{u}(z,t)\right]+2\pi v r\Big|_0^{r_0}=0 \quad \text{(E2.17-6)}$$

Here, v≡v_r is the radial velocity component subject to v(r=0)=0 and v(r=r0)=v_0(z,t). Clearly, v_0(z,t)=dη/dt, the radial wall velocity, and hence with Eq. (E2.17-6)

$$\frac{d\eta}{dt}=-\frac{r_0}{2}\frac{\partial\bar{u}}{\partial z} \quad \text{(E2.17-7)}$$

Integration of Eq. (E2.17-4), assuming that the pressure does not vary across the tube, i.e., $p(r)=\cent$, yields

$$\frac{\partial\bar{u}}{\partial t}=-\frac{1}{\rho}\frac{\partial p}{\partial z} \quad \text{(E2.17-8)}$$

Now, in order to relate the fluid mechanics and induced wall motion to the lumen pressure, we employ Newton's Second Law of Motion for the tube-wall segment of area h(r_0dθ) (see Sketch).

$$m_{tube}a_r=\sum F_r \quad \text{(E2.17-9a)}$$

or

$$(\rho_t h r_0 d\theta)\frac{d^2\eta}{dt^2}=\underbrace{p r_0 d\theta}_{F_{pressure}}-\underbrace{\sigma_\theta h d\theta}_{F_{stress}} \quad \text{(E2.17-9b)}$$

Inserting Eq. (E2.17-1) and solving for p(z,t), we obtain

$$p(z,t)=\frac{hE}{r_0^2}\eta+h\rho_t\ddot{\eta} \quad \text{(E2.17-10)}$$

Differentiation of Eq. (E2.17-10) and neglecting $d^3\eta/dt^3$ we have

$$\frac{\partial p}{\partial t}=\frac{hE}{r_0^2}\frac{\partial\eta}{\partial t} \quad \text{(E2.17-11)}$$

Combining Eq. (E2.17-11) with Eq. (E2.17-7) yields

$$\frac{\partial \eta}{\partial t} = \frac{r_0^2}{hE}\frac{\partial p}{\partial t} = -\frac{r_0}{2}\frac{\partial \bar{u}}{\partial z} \tag{E2.17-12a}$$

or

$$\frac{2r_0}{hE}\frac{\partial p}{\partial t} = -\frac{\partial \bar{u}}{\partial z} \tag{E2.17-12b}$$

Differentiation with respect to time and using Eq. (E2.17-8) yields

$$\frac{2r_0}{hE}\frac{\partial^2 p}{\partial t^2} = \frac{\partial}{\partial t}\left(-\frac{\partial \bar{u}}{\partial z}\right) \equiv \frac{\partial}{\partial z}\left(-\frac{\partial \bar{u}}{\partial z}\right) = \frac{1}{\rho}\frac{\partial^2 p}{\partial z^2}$$

so that

$$\frac{\partial^2 p}{\partial z^2} = \frac{2\rho r_0}{hE}\frac{\partial^2 p}{\partial t^2} \tag{E2.17-13}$$

where the coefficient is the Moens-Korteweg formula, i.e.,

$$\frac{hE}{2\rho r_0} \equiv c^2 \tag{E2.17-14}$$

Actual internal pressure and wall displacement solutions of Eq. (E2.17-10) may be obtained with postulates such as Eq. (2.2.5); for example,

$$p(z,t) = A\sin\left(\frac{z}{\lambda} - \omega t\right) \text{ and } \eta(z,t) = B\sin\left(\frac{z}{\lambda} - \omega t\right)$$

(E2.17-15a, b)

If also viscous effects for (blood) flow in thin tubular elastic arteries are important, the celebrated *Womersley model* as discussed by Nichols & O'Rourke (1998) should be considered (see Sect.2.3).

In order to provide blood, i.e., oxygen and nutrients, to different organs and tissues, main or parent vessels bifurcate into smaller daughter vessels. In turn, they bifurcate to even smaller ones, thereby forming a tree structure (cf. Karch et al., 1999). This is very efficient because it eliminates the need for parallel blood vessels starting, say, from the left ventricle (see Fig. 1.1.2), and it reduces the heart's required pumping power. While a tree structure lacks redundancy—for example, blockage of (the) parent tube cuts off blood to all subsequent daughter tubes—the total power can be optimized for an arterial tree. The same holds for the bronchial trees in the lungs.

When estimating the total power, $P_{total} = P_{dyn} + P_{bio}$ required for both fluid dynamics and biological purposes, a major simplification is to assume that fluid flow is steady, laminar, incompressible, and fully developed in rigid circular tubes, i.e., Poiseuille flow (see Eq. (1.3.66) where $g\psi \equiv 0$). Specifically,

$$\frac{1}{r}\frac{d}{dr}\left(r\frac{du}{dr}\right) = \frac{1}{\mu}\frac{dp}{dx} = ¢ \tag{2.2.10}$$

subject to $u(r = R) = 0$ and $\left.\frac{du}{dr}\right|_{r=0} = 0$ where $0 \le r \le R$ and

$$-\frac{dp}{dx} = (p_{in} - p_{out})/L = ¢.$$

The pumping power P_{dyn} is requested to maintain a given flow rate Q, i.e., to overcome the flow resistance employing a pressure difference $\Delta p = p_{in} - p_{out}$. Hence,

$$P_{dyn} = \Delta p Q \tag{2.2.11a}$$

where:

$$\Delta p = \frac{8\mu L}{\pi R^4} Q \tag{2.2.11b}$$

from the solution of Eq. (2.2.10) for a given Q. Thus,

$$P_{dyn} = \frac{8\mu L}{\pi R^4} Q^2 \tag{2.2.11c}$$

Clearly, the pumping power $P_{dyn} \sim R^{-4}$, i.e., with Q known, large blood or air conduits should be beneficial. However, filling and maintaining a vessel with fluid requires $P_{bio} \sim V_{fluid} \sim R^2 L$ which implies:

$$P_{total} = P_{dyn} + P_{bio} = AR^{-4} + BR^2 \tag{2.2.11d}$$

From Eq. (2.2.11c), $A \equiv 8\mu LQ^2/p$ and B is a positive constant independent of Q. Thus, to find the optimal radius which minimizes P_{total}, we set $dP_{total}/dR := 0$, i.e., $R^6 = 2A/B$; or,

$$Q \sim R^3 \tag{2.2.12}$$

Equation (2.2.12) is known as *Murray's law* or the "cube law," indicating that minimum energy expenditure is achieved under this correlation. Indeed, as the tube diameters get smaller in the arterial or bronchial tree, so are the flow rates reduced according to the proportionality (2.2.12). It is of interest to note that the wall shear stress (in Poiseuille flow) relates to Q/R^3 as well, i.e.,

$$\tau_{wall} = \frac{4\mu}{\pi}\left(\frac{Q}{R^3}\right) \quad (2.2.13)$$

The cube-law can be employed when describing geometric aspects of bifurcations and ultimately trees. Clearly, as demanded by local air or blood supply, bifurcations may be asymmetric, out-of-plane, and with different daughter-vessel diameters. Thus, three system-indicative parameters are a bifurcation's angle θ, index $\alpha = R_2/R_1$, and area ratio $\beta = (R_1^2 + R_2^2)/R_0$, where $R_0 \triangleq$ parent-tube radius, and $R_{1,2} \triangleq$ daughter-tube radii.

Recalling mass conservation:

$$Q_0 = Q_1 + Q_2 \quad (2.2.14a)$$

and hence with the associated cube law (2.2.6):

$$R_0^{\,3} = R_1^{\,3} + R_2^{\,3} \quad (2.2.14b)$$

Then the area ratio b can be expressed as:

$$\beta = \left(\frac{R_1}{R_0}\right)^2 + \left(\frac{R_2}{R_0}\right)^2 \quad (2.2.15a)$$

or in terms of the bifurcation index α as:

$$\beta = (1+\alpha^3)^{-2/3} + (1+\alpha^{-3})^{-2/3} \quad (2.2.15b)$$

i.e.,

$$\beta = \frac{1+\alpha^2}{(1+\alpha^3)^{2/3}} \quad (2.2.15c)$$

For example, for a symmetric bifurcation, $\alpha \equiv 1.0$ and hence $\beta = 1.26$ and $R_1/R_0 = R_2/R_0 = 0.794$. These numerical values may serve as a guide in the analyses of arterial and bronchial tree structures, provided that the underlying assumptions are met.

Example 2.18: Steady Blood Flow in a Bifurcating Tube and the Evaluation of a Potential Stent Migration Force

Consider Poiseuille-type flow in an axisymmetric bifurcation. The parent tube is characterized by p_1, v_1, and R_1 at the inlet (x=0) and at the bifurcation ($x = \ell$) by p_x and 2Θ. Both daughter tubes have the radius R_2, length $L_D = (L-\ell)/\cos\Theta$, outlet pressure p_2, and averaged velocity v_2. Apply the model to: Case (i) a bifurcating blood vessel and estimate the necessary pumping power $P = \Delta pQ$ as well as the best bifurcation point $x = \ell_{opt}$ when $P = P_{\min}$; Case (ii) a bifurcating stent-graft and evaluate the axial force potentially leading to stent-graft migration.

<u>Sketch:</u>

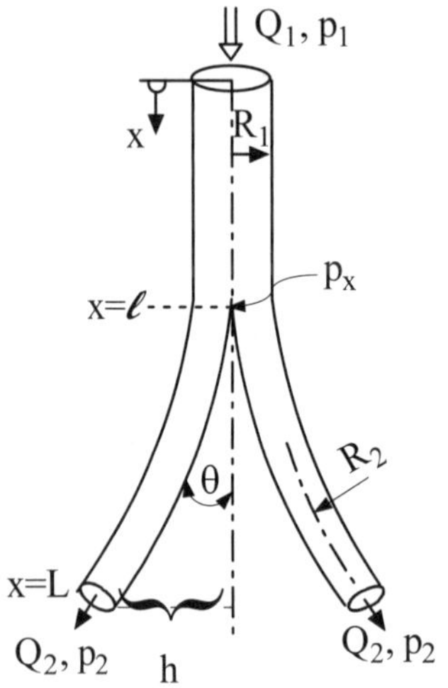

<u>Assumptions:</u>

- Steady laminar fully-developed flow
- Constant fluid properties
- Axisymmetric bifurcations with smooth rigid walls
- $P_{bio} \approx 0$

<u>Concepts:</u>

Case (i):
- $P_{total} = \sum_i (\Delta pQ)_i$
- Hagen-Poiseuille solution $Q \sim R^4$

Case (ii):
- Reynolds Transport Theorem and Bernoulli's Equation

<u>Solutions:</u> For Case (i), an equation for the total pumping power as a function of the parent-tube length has to be derived, i.e., P = fct $(\ell; R_1, R_2, h; p_1, p_2)$. Then, setting the first derivative to zero, i.e., $dP/d\ell := 0$, leads to $P(x = \ell_{\text{opt}}) = P_{\min}$. For Case (ii), of interest is the pulling (or migration) force exerted by the net momentum change on the blood vessel or equivalently on a bifurcating stent-graft (see Sect.5.4).

Case (i): The pumping power required to maintain the flow rates through the bifurcation is (see Eq. (2.2.11)):

$$P_{\text{total}} = Q_1\, \Delta p_1 + 2Q_2\,(p_x - p_2)$$

i.e.,

$$P_{\text{total}} = Q_1\,(p_1 - p_x) + 2Q_2\,(p_x - p_2) \qquad \text{(E2.18-1)}$$

where

$$Q_1 = 2Q_2 \qquad \text{(E2.18-2)}$$

so that (see Eq. (2.2.11a)),

$$P_{\text{total}} = Q_1\,(p_1 - p_2) \qquad \text{(E2.18-3)}$$

Considering Poiseuille flow with $-\partial p/\partial x = \Delta p/\ell = ¢$ and $\vec{v} = [u(r),0,0]$, Eq. (2.2.10) reduces to:

$$\frac{\mu}{r}\frac{d}{dr}\left(r\frac{du}{dr}\right) = \frac{\partial p}{\partial x} = ¢ \tag{E2.18-4}$$

subject to $u(r=R)=0$ and $u(r=0)$ is finite (or $\left.\frac{\partial u}{\partial r}\right|_{r=0} = 0$).

Double integration and invoking the boundary conditions yields:

$$u(r) = \frac{1}{4\mu}\left(\frac{\Delta p}{\ell}\right)\left(R^2 - r^2\right), \text{ where } \frac{1}{4\mu}\frac{\Delta p}{\ell} = u_{max}$$

(E2.18-5a, b)

so that in general

$$Q = \int_A \vec{v}\cdot\hat{n}\,dA = 2\pi\int_0^R u(r)rdr := \frac{\pi}{8\mu}\left(\frac{\Delta p}{\ell}\right)R^4 \tag{E2.18-6}$$

In the present case,

$$Q_1 = \frac{\pi}{8\mu}\frac{p_1 - p_x}{\ell}R_1^{\ 4} \text{ and } Q_2 = \frac{\pi}{8\eta}\frac{p_x - p_2}{\sqrt{h^2 + (L-\ell)^2}}R_2^{\ 4}$$

(E2.18-7a, b)

Note that $\sqrt{h^2 + (L-\ell)^2} = L_D = (L-\ell)/\cos\Theta$ is the length of a daughter tube.

Inserting Eqs. (E2.18-7a,b) into Eq. (E2.18-2) yields:

$$p_x(\ell) = \frac{p_1 R_1^{\ 4}\sqrt{h^2 + (L-\ell)^2} + 2p_2 R_2^{\ 4}\ell}{R_1^{\ 4}\sqrt{h^2 + (L-\ell)^2} + 2R_2^{\ 4}\ell} \tag{E2.18-8}$$

so that Eq. (E2.18-3) can be written as:

$$P_{total} = \frac{\pi}{4\mu}\frac{R_1^{\ 4}R_2^{\ 4}(p_1 - p_2)^2}{R_1^{\ 4}\sqrt{h^2 + (L-\ell)^2} + 2R_2^{\ 4}\ell} \tag{E2.18-9}$$

Now, $P_{\text{total}}(x = \ell_{\text{opt}}) = P_{min}$, i.e., with $d^2P/d\ell^2 > 0$, we obtain ℓ_{opt} by setting $dP/d\ell := 0$, from which

$$\ell_{\text{opt}} = L - \frac{2R_2^{\ 4}}{\sqrt{R_1^8 - 4R_2^8}}\underbrace{(L-\ell)\tan\Theta}_{h} \tag{E2.18-10}$$

Recalling that

$$\cos\Theta = \frac{L-\ell}{\sqrt{h^2+(L-\ell)^2}}\text{, where } 0<\Theta\le\frac{\pi}{2},$$

we can express the half-angle of bifurcation associated with $P_{\min}$ (see Eq. (E2.18-10)) as:

$$\cos\Theta = 2\left(\frac{R_2}{R_1}\right)^4 \tag{E2.18-11}$$

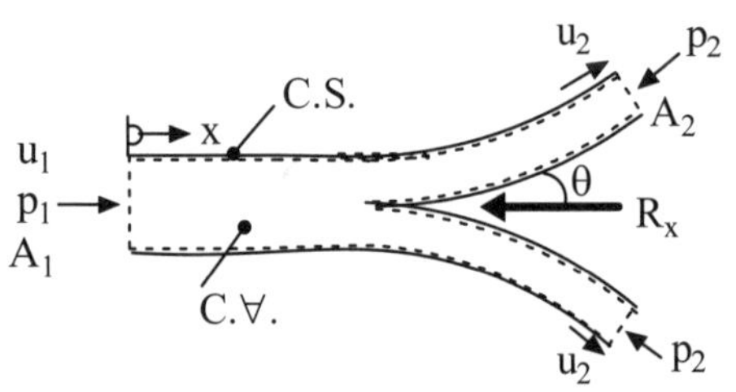

Case (ii): The net axial force exerted by the fluid flow due to net momentum change, neglecting frictional and gravitational effects, can be evaluated with a reduced form of Eq. (1.3.9), using averaged velocities:

$$\sum F_x = \int_{\text{c.s.}} u\,\rho\,\vec{v}\cdot d\vec{A} \tag{E2.18-12a}$$

i.e.,

$$p_1 A_1 - 2p_2 A_2 \cos\Theta - R_x = \rho u_1 A_1(-u_1) + 2\rho u_2^2 A_2 \cos\Theta \tag{E2.18-12b}$$

From Bernoulli along a representative streamline in the parent-daughter tube for frictionless flow:

$$p_1 + \frac{\rho}{2}u_1^2 = p_2 + \frac{\rho}{2}u_2^2 \tag{E.2.18-13}$$

From mass conservation,

$$u_1 A_1 = 2u_2 A_2 \tag{E.2.18-14}$$

Thus, eliminating p_2 and u_2 with Eqs. (E2.18-13 and 14), Eq. (E2.18-12) reads:

$$R_x = p_1 A_1 + \rho A_1 u_1^2 - \rho\frac{A_1^2}{2A_2}u_1^2\cos\Theta - 2A_2\left[p_1 + \frac{\rho}{2}u_1^2\left(1-\frac{A_1^2}{4A_2^2}\right)\right]\cos\Theta \tag{E2.18-15}$$

For the human abdominal aorta, the parameter ranges are: blood flow rate $500 \leq Q_1 \leq 2000\ mL/\min$, aortic neck diameter $20 \leq D_1 \leq 30mm$, and half-bifurcation angle $0 < \Theta \leq 80°$. As a base case, one could take $Q_1 = 1L/\min$, $p_1 = 120\,mmHg$, $\Theta = 30°$, $D_1 = 22mm$, and $D_2 = 12\,mm$ (see Graphs).

Graphs:

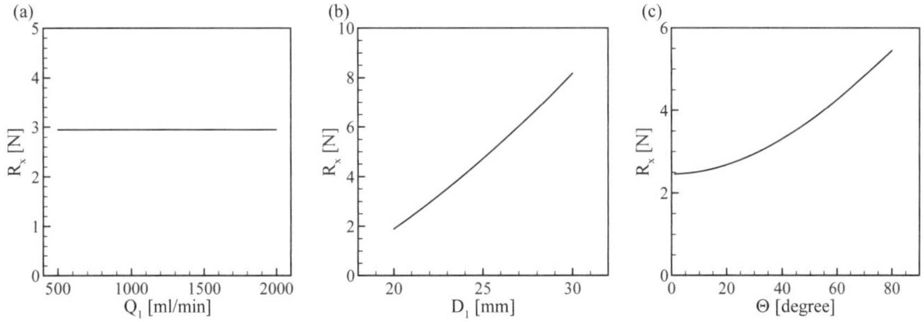

Comments:
The blood flow rate has a negligible effect on the axial force. In contrast, the axial force increases with increasing parent tube diameter and bifurcating angle because sudden constrictions, i.e., from a relatively large parent tube to small daughter tubes, as well as a large bifurcation angle cause higher net momentum changes, i.e., an elevated net pressure drop. It should be noted that frictional effects (neglected here) are indeed only minor contributors and that the computed stent-migration force range $2 \leq R_x \leq 8$ N matches in-depth analyses (see Sect. 5.4.2).

Example 2.19: Blood Flow in Noncircular Tubes

Arteries, when noncircular, may be elliptic in cross-sectional area. Assuming steady fully-developed flow of a Newtonian fluid, find expressions for the volumetric flow rate and the necessary pumping power.

<u>Sketch:</u>

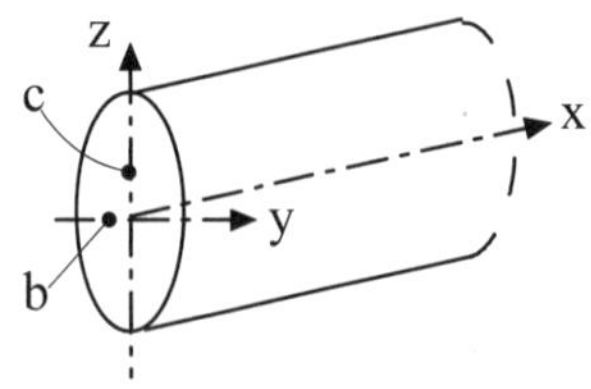

<u>Assumptions:</u>

- Steady laminar unidirectional flow
- Constant properties
- Rigid walls

<u>Concepts:</u>

- Reduced N-S equations
- Pumping power
  $P_{\text{dyn}} = F_{\text{pressure}} \cdot \bar{v}$
  where $F_p = \Delta p A$

<u>Solution:</u> For axial flow in a duct with an elliptic cross section,

$$\frac{y^2}{b^2}+\frac{z^2}{c^2}=1 \tag{E2.19-1}$$

the governing equation (see Eq. (1.3.65)) reads:

$$\frac{\partial^2 u}{\partial y^2}+\frac{\partial u}{\partial z^2}=\frac{1}{\mu}\frac{\Delta p}{L}=¢ \tag{E2.19-2}$$

subject to no-slip boundary condition and symmetry on y = z = 0.
The solution then is (Zamir, 2000; White 2003):

$$u(y,z)=\frac{\Delta p}{2\mu L}\frac{(bc)^2}{b^2+c^2}\left(\frac{y^2}{b^2}-\frac{z^2}{c^2}-1\right) \tag{E2.19-3}$$

The flow rate is:

$$Q=\frac{\pi\Delta p}{8\mu L}K^4=A\cdot\bar{v} \tag{E2.19-4}$$

where $K \equiv [2(bc)^3/(b^2+c^2)]^{1/4}$ and hence the pumping power can be expressed as:

$$P_{\text{dyn}}=\frac{8\mu L}{\pi K^4}Q^2 \tag{E2.19-5}$$

Clearly, when b = c = R, K = R and the Poiseuille flow results are recovered.

2.3 HOMEWORK PROBLEMS

<u>Notes:</u> Homework problems (HWPs) 2.1-2.24 are directly culled from Chapter II and their solutions should deepen the basic understanding of biofluid dynamics concepts.

"Reasonable system parameters" refers to the selection of physiologically correct values as given in BME handbooks (e.g., Moore & Zouridakis,

2003) and supplementary texts (e.g, Fung, 1997; Nichols & O'Rourke, 1998; Bird et al., 2002; Truskey et al., 2004, etc.).

2.1 Contrast static vs. dynamic equilibrium. Provide examples of "equilibrium assumptions/conditions" used to solve complex transport phenomena.

2.2 List the underlying assumptions as well as the pros & cons of "well-mixed compartment" models. Is the Reynolds Transport Theorem a lumped-parameter black-box reservoir or compartment model?

2.3 In reference to Fig.2.1.2(b), set up the equations to find $c_v(t)$ for a given Q(t) and $c_a(t)$ as well as $\forall_1$, R_1 and $\forall_2$, R_2.

2.4 Generalize the results of HWP2.3 to n compartments in parallel, where i=1,2,3 ... n.

2.5 Considering Example 2.2, select reasonable system parameters and inlet conditions and solve: (a) Equation (E2.2-3); and (b) Equations (E2.2-2a,b). Graph $c_v(t)$ as well as $c_B(t)$ and $c_T(t)$ and comment on the influence of key system parameters.

2.6 Consider the embedded membrane capillary depicted in Fig. 2.1.5. Solve the PDE for c(x,t) (see Eq. (2.1.5)), assuming reasonable "constants." Plot the species concentrations and comment.

2.7 Revisit HWP 2.6 and lift the assumptions of constant tissue concentration, i.e., $c_T=c_T(x,t)$, and uniform capillary flow, i.e., u=u(r,x); however, only Poiseuille flow u(r) and a very small membrane flux $V_M = ¢$ as a boundary condition are assumed (see Sect.1.3 and Example 2.7).

2.8 Redo Example 2.4 with a broader range of inlet conditions and system parameters. Graph the results and comment.

2.9 Develop a "compartment-in-series" model with $T_{in} \equiv T_A = 32°C$ and possible thermal boundary conditions to simulate body heat transfer (see Fig. 2.1.6b).

2.10 Discuss Equation (2.1.10) and provide a state-of-the-art review of advancements and applications of the bioheat transfer equation.

2.11 Reconsidering Example 2.5, which operational conditions and system parameters have to be reasonably changed to produce a nonlinear T(x), i.e., show the influence of bioheat convection? Simulate a case study.

2.12 Study the paper by Oka (1981) and related publications in order to derive Equations (E2.7-2) to (E2.7-5). Establish limiting parameter ranges for which Equations (E2.7-2) to (E2.7-4) are valid.

2.13 Find the species blood and tissue concentrations (see Example 2.8) for reasonable system parameters. Graph the simulation results and comment.

2.14 Concerning "joint lubrication and wear," provide an updated literature review, stressing theories of the three lubrication modes, parameter values and experimental/computational results (see Table 5.1.1).

2.15 Derive Equation (2.1.20), i.e., set up a mathematical model and describe necessary laboratory experiments.

2.16 Plot for a given L(t) (see Eq. (2.1.14)) the squeeze-film pressure p(θ,t), velocity $\partial h / \partial t$, and spacing h(θ,t). Which system parameters are most influential?

2.17 In very small vessels, such as the capillaries, red blood cell diameters may be significant compared to vessel diameter so that the fluid cannot be treated as being homogeneous (Sigma effect). In this case, the fluid may be approximated as a summation of finite cylindrical layers with thickness equal to the red blood cell diameter. Find the apparent viscosity of blood considering the Sigma effect.

2.18 Discuss the relationship among left ventricular volume, pressure, inflow, and outflow.

2.19 Similar to Example 2.16, now analyze the peristaltic pumping for two-dimensional flow in a circular tube of radius a. Find the velocity profile, volumetric flow rate, and overall pressure drop.

2.20 Windkessel theory (cf. Sect.2.2.3 and Example 2.13): For example, the aorta is modeled as an elastic chamber and the peripheral blood vessels as a representative rigid tube with resistance R. Thus, the inflow rate, or left ventricular ejection history, Q(t), can be expressed as:

$$Q(t) = C\frac{dp}{dt} + \frac{p}{R}$$

where C is the wall compliance.

(a) Interpret the biofluid mechanics meaning of each term and determine the dimensions of C and RC.

(b) Multiply the equation through by exp(t/RC), integrate from t=0 to t=t, and solve for p(t).

(c) Select a couple of Q(t)-functions, calculate p(t), plot p(t), and comment.

(d) How could the compliance be defined, measured, and/or calculated?

2.21 Pulsatile flow in elastic tubes: An improvement over the Windkessel Theory is 1-D inviscid flow where the pressure is p=p(x,t), i.e., describing a pressure wave, and the cross sectional area A=A(x,t):

$$\frac{\partial A}{\partial t} + \frac{\partial}{\partial x}(uA) = 0$$

$$\frac{\partial u}{\partial t} + u\frac{\partial u}{\partial x} = -\frac{1}{\rho}\frac{\partial p}{\partial x}$$

$$p - p_{ext} = p(A)$$

where p_{ext} is the pressure acting on the outside of the tube.

(a) Indicate the derivation of the first two equations.

(b) Assume u<<1 and hence $u\partial u/\partial x \approx 0$ as well as $A = \pi R^2$, where $2R = 2R_0 + \alpha p$, with α being the compliance constant of the tube. Show that the continuity equation, the simplified momentum equation, and the diameter-pressure relationship can be reduced to

$$\frac{\partial u}{\partial x} + \frac{\alpha}{R}\frac{\partial p}{\partial t} = 0$$

and

$$\frac{\partial u}{\partial t} + \frac{1}{\rho}\frac{\partial p}{\partial x} = 0$$

(c) Show that the last two equations can be combined to the wave equation:

$$\frac{\partial^2 p}{\partial x^2} - \frac{1}{c^2}\frac{\partial^2 p}{\partial t^2} = 0$$

where $c = R/(\alpha\rho)$.

(d) Postulate for the solutions of the equations in Section (b)

$$p = p_0 f(x-ct) + \hat{p}_0 g(x+ct)$$

and

$$u = u_0 f(x-ct) + \hat{u}_0 g(x+ct)$$

and show that "amplitudes"

$$p_0 = -\rho c u_0 \quad \text{and} \quad \hat{p}_0 = \rho c \hat{u}_0$$

Provide a physical meaning of these postulates/solutions.

(e) Look up the analytical solution to this problem and plot *p(x,t)* as well as the evolution of *u(x,t)* for suitable constants and initial/inlet conditions.

2.22 Fung (1990) outlines in "Biomechanics," Section 6.7, microcirculation as blood flow in a capillary sheet of, say, the alveolar region, where the sheet thickness

$$h = h(\Delta p = p_{blood} - p_{air}).$$

(a) Derive the Laplace equation describing $h=h(x,y)$ as

$$\left(\frac{\partial^2}{\partial x^2} + \frac{\partial^2}{\partial y^2}\right) h^4 = 0$$

where $h = h_0 + \alpha(p_b - p_a)$.

(b) Assuming that $h=h(x)$ only, derive an expression for the blood flow rate $\hat{Q} = hU$ per unit width and plot $h(x)$ for reasonable parameters.

2.23 Flow through a concentric annulus. Considering steady fully-developed flow of a Newtonian fluid in the annular space between two concentric cylinders, find expressions for the volumetric flow rate and friction factor. There is no slip at the inner ($r=b$) and outer radius ($r=a$).

2.24 Bifurcating vessels: (a) Draw the streamline of flow in a model of the abdominal aorta containing the left and right renal arteries (see Graph A);
(b) Draw the laminar, axial velocity profiles in a model of the carotid artery bifurcation (see Graph B);
(c) Considering a symmetric bifurcating airway (see Graph C), draw the secondary velocity pattern in the daughter tube during inspiration and that in the parent tube during expiration.

Graph A:

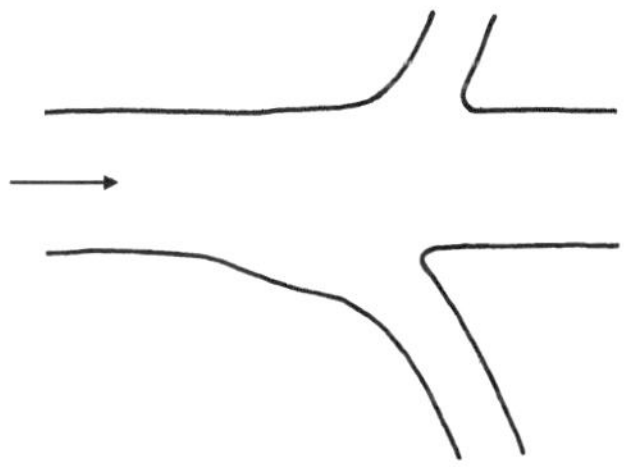

Graph B:

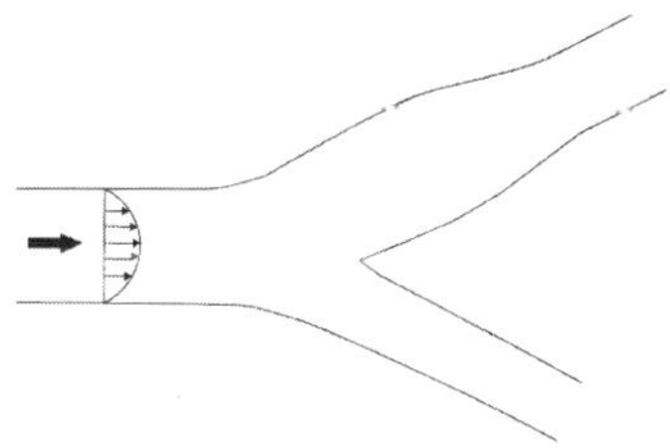

Graph C:

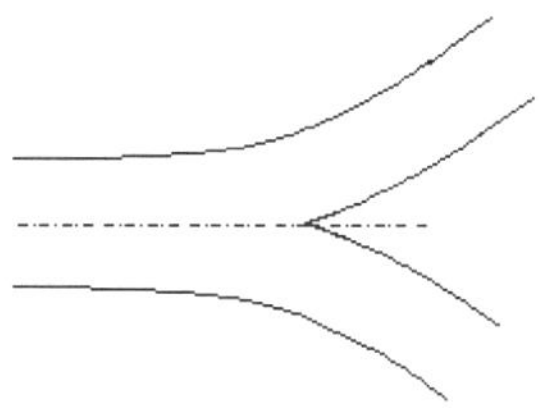

References

Bellhouse, B.J. and Talbot, L. (1969), J. Fluid Mech., 35:721-735.

Berger, S.A., Goldsmith, W., and Lewis, E.R. (eds.) (1996) "Introduction to Bioengineering," Oxford University Press, UK.

Bird, R. B., Stewart, W. E., and Lightfoot, E. N. (2002) "Transport Phenomena," John Wiley, New York, NY.

Buchanan, J.R. and Kleinstreuer, C. (1998), ASME J. Biomech. Eng., 120: 446-454.

Carlson, F.D. (1957) "Kinematic Studies of Mechanic Properties of Muscles" In: Tissue Elasticity, ed. J. W. Remington, Am. Physiol. Society, Washington, DC.

Carslaw, H.S. and Jaeger, J.C. (1959) "Conduction of Heat in Solids," Oxford University Press, London, UK.

Cebeci T. (2004) "Analysis of Turbulent Flows," Elsevier, Amsterdam, Oxford.

Charney, C. K. (1992) "Mathematical Models in Bioheat Transfer" In: Bioengineering Heat Transfer (Y.I. Cho, ed.) Advances in Heat Transfer, Vol. 22: 19-156.

Chen, M.M. and Holnes, K.R. (1980) Ann. N.Y. Acad. Sci., 335: 137.

Chhabra, R.P., Agarwal, L., and Sinha, N. K. (1999) Powder Technology, 101: 288-295.

Cooney, D.O. (1976) "Biomedical Engineering Principles," Dekker, New York.

Datta, A. K. (2002) "Biological and Bioenvironmental Heat and Mass Transfer," Marcel Dekker Inc, New York, NY.

Finlay, W.H. (2001) "The Mechanics of Inhaled Pharmaceutical Aerosols: An Introduction", Academic Press, London, UK.

Fung, Y. C. (1990) "Biomechanics: Motion, Flow, Stress, and Growth," Springer-Verlag, New York, NY

Fung, Y. C. (1997) "Biomechanics: Circulation," Springer, New York, NY.

Furley, M.J. (2003) in: "Biomechanics : Principles and Applications," Schneck, D.J. and Bronzino, J.D. (Eds.), CRC Press, Boca Raton, FL.

Greenberg, M. D. (1998) "Advanced Engineering Mathematics," Prentice Hall, Upper Saddle River, NJ.

Hoffman, J. D. (2001) "Numerical Methods for Engineers & Scientist," Marcel Dekker, Inc. New York.

Hu, X. and Weinbaum, S. (1999) Microvascular Research, 58:281-304.

Hyun, S., and Kleinstreuer, C. (2001) International Journal of Heat & Mass Transfer, 44: 2247-2260

Kamm, R.D. (2002) Annual Review of Fluid Mechanics, 34: 211-232.

Karch, R., Neumann, F., Neumann, M., and Schreiner, W. (1999) Computers in Biology and Medicine, 29: 19-38.

Kleinstreuer, C. (2003) "Two-Phase Flow: Theory and Applications," Taylor & Francis, New York, NY.

Mabuchi, K., Sakai, R., Ota, M., and Ujihira, M. (2004) Clinical Biomechanics, 19: 362-369.

Middleman, S. (1972) "Transport Phenomena in the Cardiovascular System," Wiley-Interscience, New York.

Milnor, W. R. (1989) "Hemodynamics," Williams & Wilkins, Baltimore, MD.

Moore, J. and Zouridakis, G. (2003) "Biomedical Technology and Devices Handbook", CRC Press, Boca Raton, FL.

Nichols, W. W., and O'Rourke, P. J. (1998) "McDonald's Blood Flow in Arteries: Theoretical, Experimental and Clinical Principles," Lea and Febiger, Philadelphia, PA.

Oka, S. (1981) "Cardiovascular Hemorheology," Cambridge University Press, New York.

Özisik, M. N. (1993) "Heat Conduction," Wiley-Interscience, New York, NY.

Pennes, H.H. (1948) J. Appl. Physiol., 1: 93-122.

Phillips, R.J., Armstrong, R.C., and Brown, R.A. (1992) Physics of Fluids A, 4:30-40.

Pollack, G. (2001) "Cells, Gels and the Engines of Life: A New, Unifying Approach to Cell Function," Ebner & Sons, Seattle, WA.

Pries, A.R., Secomb, T.W., Gessner, T., Sperandio, M.B., Gross, J.F., Gaehtgens, P. (1994). Circ. Res., 75:904-915.

Pries, A.R., Secomb, T.W., Jacobs, H., Sperandio, M., Osterloh, K., and Gaehtgens, P. (1997). Am. J. Physiol. Heart Circ. Physiol., 273: H2272-2279

Rosendahl, L. (2000) Applied Mathematical Modelling, 24:11-25.

Schmid-Schönbein, G.W., Lee, S.Y., and Sutton, D.W. (1989) Biorheology, 26: 215-227.

Schneck, D.J. and Bronzino, J.D., (Eds.) (2003) "Biomechanics - Principles and Applications", CRC Press, Boca Raton, FL.

Shapiro, A.H., Jaffrin, M.Y. and Weinberg, S.L. (1969) J Fluid Mech., 37: 799-825.

Spiegel, M.R. (1971) "Schaum's Outline of Theory and Problems of Advanced Mathematics for Engineers and Scientists," McGraw-Hill, New York, NY.

Truskey, G. A., Yuan, F., and Katz, D.F. (2004) "Transport Phenomena in Biological Systems," Pearson Prentice Hall, Upper Saddle River, NJ.

Weiss T. F. (1996) "Cellular Biophysics (Vol. 1 and 2)," The MIT Press, Cambridge, MA.

White, F. M. (2003) "Fluid Mechanics," McGraw-Hill, Boston, MA.

Wulff, W. (1974) IEEE Trans. Biomed. Eng., BME-21: 494.

Yang, W.J. (1989) "Biothermal-Fluid Sciences, Principles and Applications," Hemisphere, New York.

Zamir, M. (2000) "The Physics of Pulsatile Flow," AIP Press, Springer, New York.

Zhu, M., Weinbamn, S., and Lemmous, D.E. (1988) J. Biomech. Eng., 110: 74.

Zydney, A.L., and Colton, C.K. (1988) PhysicoChemical Hydrodynamics, 10: 77-96.

Chapter III

ANALYSES OF ARTERIAL DISEASES

Of special interest in biofluid dynamics are arterial diseases which originate inside the arterial wall and weaken the vessel wall, changing the local geometric configuration. Specifically, after decades of detrimental biochemical processes either "plaque" formation (i.e., atheroma protruding into the lumen) greatly reduces blood flow, or the locally weakened arterial wall dilates and forms a balloon-like vessel section (aneurysm), possibly leading to rupture. Biofluid dynamics applications at different stages of an arterial disease process may include:

(i) the simulation of transport and deposition of cells and biochemicals in blood vessels as well as abnormal stress-strain fields in the vessel walls, causing the disease;

(ii) an analysis of disease evidence in terms of atherosclerotic plaque formation or vessel dilation, generating locally geometric changes; and

(iii) the evaluation of remedies, e.g., optimal design and placement of implants such as bypass grafts or stents.

This chapter describes causes of partial or complete vessel occlusion, due to atherosclerosis, hyperplasia and/or thrombosis (Sect. 3.1), key factors in aneurysm formation, focusing on the abdominal aorta (Sect. 3.2), as well as treatment options, e.g., end-to-side bypass grafts (Sects. 3.1.4 and 3.1.5) and stent- grafts (Sects. 3.2.2 and 3.2.3).

3.1 VESSEL OCCLUSIONS

Vessel occlusion often occurs due to rapid thrombus formation in the vicinity of significant intimal thickening, called stenosis, which progressed over decades (see Figs. 3.1.1a, b). For example, as reviewed by Longest (2004), Davies (1994) describes two distinct patterns by which occlusive thrombosis may occur. Approximately one quarter of observed thrombi are superimposed on stenotic plaques, where the plaque itself does not undergo dramatic changes. This form of occlusive thrombosis typically occurs near a high-grade stenosis and is likely the result of endothelial denudation. In contrast, three quarters of occlusive thrombus are related to plaques that have undergone a deeper injury, typically extending into the core of the plaque, i.e., plaque tearing or disruption, called fissuring, ulceration, or rupture. Once the plaque cap has ruptured, a large amount of collagen is exposed to the blood providing a major stimulus for thrombi formation. The thrombus itself forms initially within the interior of the plaque, possibly advances to a degree that fills the remaining lumen, or is sheared off and suddenly fully occludes a smaller blood vessel downstream, leading to tissue ischemia and other major complications.

Atherosclerotic plaque exhibits a wide range of material properties that are associated with its multiple components and variable composition. Its mechanical properties can be classified as nonlinear and inelastic (see Salunke & Topoleski, 1997). Due to a wide range of plaque-cap structures and the long-term cyclic loading of the hemodynamic forces, definitive conditions for plaque-cap rupture have not been established. Nevertheless, for coronary arteries, it was observed that a majority of plaques that rupture and result in occlusive thrombosis are eccentrically situated and have a lipid pool that does not have an internal lattice of collagen supporting the cap (Richardson, et al., 1989). Such a structure will lead to a concentration of circumferential wall stress on the plaque cap in systole because of the inability of the pool to carry a load that is redistributed to the vessel wall. It has also been speculated that a factor contributing to the vulnerability of the plaque is infiltration of the cap tissue with macrophages (Salunke & Topoleski, 1997).

Clearly, a variety of outcomes are possible once the plaque has been disrupted, ranging from occlusive thrombosis to a healed fissure with no increase in stenosis. Clinically, it is important to know the outcome of a ruptured plaque cap. If a model of occlusion risk were available (see Sect. 3.3), it could be combined with a model for plaque-rupture risk to decide which patients are good candidates for surgical treatment and which patients can be managed medically. Currently, only stenosis severity is used, very often inappropriately, to determine the necessity for surgical intervention (Wootton & Ku, 1999). A nice multicomponent analysis of human carotid plaque is given by Tang et al. (2004).

3.1.1 Atherosclerotic Plaque Formation

As is well known, cardiovascular diseases, such as atherosclerosis, ischemia, hypertension, thrombosis, aneurysms, and stroke, continue to be major reasons for mortality and morbidity in the Western World. Although the causes are of genetic origin, accelerated by environmental factors, the basic disease mechanisms relate to abnormal or excessive cellular behavior:

- cell migration, replication, aggregation, and rupture;
- the production of vaso-active, growth-regulatory, inflammatory, degratory molecules; and
- the synthesis and organization of constituents of the extra cellular matrix (see Humphrey, 2002; Truskey et al., 2004, among others).

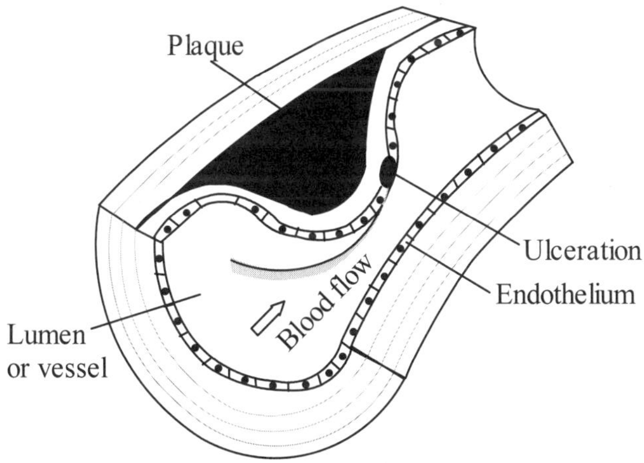

a) Schematic of "intimal thickening," i.e., stenotic development

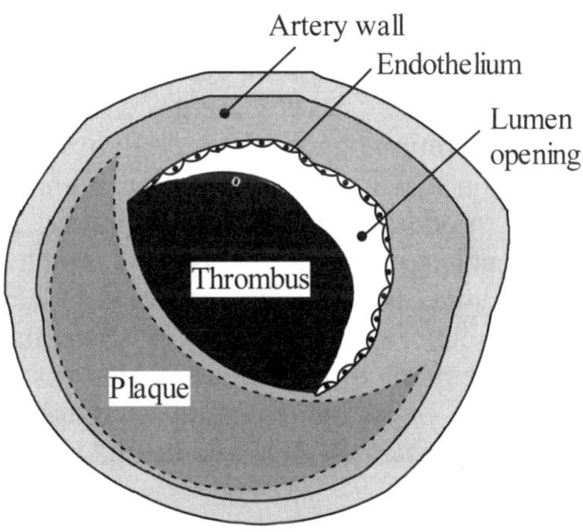

b) Partially occluded artery:
atherosclerotic plaque plus thrombus

Figure 3.1.1 Two examples of vessel occlusion.

3.1.1.1 A Particle-Hemodynamics Model

Many of these abnormal cellular functions may be triggered by locally "disturbed flow," i.e., complex aggravating blood flow patterns in curved or bifurcating, medium-to-large arteries. Specifically, high blood pressure, very low or very high wall shear stresses, and critical blood particle accumulation are the precursors of focal cell proliferation inside the arterial wall, endothelial cell dysfunction and elevated permeability of the endothelium, allowing for migration of low-density lipoprotein (LDL) and monocytes (WBC family) into the arterial wall, a process called atherosclerosis. Note, atherosclerosis is a common form of the more encompassing arteriosclerosis, i.e., the hardening of arteries. What follows is that monocytes gobble up oxidized LDL, turning into macrophages and leading to smooth muscle cell proliferation inside the arterial wall – labeled "intimal thickening" or stenotic development (see Fig. 3.1.1). Furthermore, the mechanics of the tissue (e.g., muscle, bone, etc.) surrounding a blood vessel may influence atherosclerotic plaque formation. For example, coronary arteries located on the heart surface develop atherosclerosis, while the intramural ones, i.e., those penetrating the heart muscle, do not. The same observation holds for vessels surrounded by the vertebrae. In any case, over long periods of time intimal thickening creates a progressive reduction of the cross sectional area at critical locations in medium to large arteries. In general, these stenotic developments may be the result of late-stage arteriosclerosis lesions as discussed, or due to a more rapid response to injury of the inner arterial wall, which is referred to as hyperplasia, or tissue overgrowth.

As mentioned, the slow process of vessel occlusion has two phases. The first stage is stenotic development, which may result from atherosclerotic plaque or intimal hyperplasia, both of which involve smooth muscle cell proliferation and lipid accumulation within the intima. This primary process of moderate lesion development may occasionally cumulate in vessel occlusion. Alternately, and much more frequently, thrombus formation superimposed on a stenotic development, or a thrombus (i.e., blood clot) arriving from an upstream vessel location, accounts for the ultimate mode of gradual or sudden vessel occlusion. This second phase leads to the acute ischemic syndromes of myocardial infraction, unstable angina, sudden ischemic death, and cerebral infarction.

3.1.1.2 A Pathway Model for Atherogenesis

Based on cell culture and in vivo studies over the past decade, a working model of atherogenesis has been developed. Initiation of atherosclerosis involves monocyte migration through a dysfunctional endothelium (Ross, 1999). Advanced intimal proliferation lesions of atherosclerosis may occur by at least two pathways as described by Ross (1986): Monocytes attach to the endothelium which may continue to secrete growth factors. Sub-endothelial migration of monocytes may lead to fatty-streak formation and release of growth factors. Fatty streaks may become directly converted to fibrous plaques through the release of growth factors from macrophages or endothelial cells or both. Macrophages may also stimulate or injure the overlying endothelium. In some cases, macrophages may lose their endothelial cover and platelet attachment may occur, providing three possible sources of growth factors, i.e., platelets, macrophages, and endothelial cells. Some of the smooth muscle cells in the proliferative lesion itself may form and secrete growth factors as well. As the atherosclerotic lesion progresses, there are gaps in the endothelial surface within which lipid-filled monocytes are impacted, and the presence of foam cells in peripheral blood suggests that some are emigrating from the intima. Platelets then adhere to the exposed connective-tissue matrix underlying the endothelial defects, and this is the stage in which a true thrombogenic dimension is involved in atherosclerosis. The attached platelets play a major role in plaque growth by stimulating smooth muscle cell migration and proliferation even though they may not significantly influence initiation.

3.1.2 Intimal Hyperplasia Development

Intimal hyperplasia is the rapid abnormal continued proliferation and overgrowth of smooth muscle cells in response to endothelial injury or dysfunction (Chervu and Moore, 1990). Endothelial injury or dysfunction often occurs at sites of bypass graft anastomoses, locations of balloon angioplasty, and stented vessels. At such locations, hemodynamic factors may be a primary or contributing source of endothelial dysfunction. Hyperplasia is often viewed as an accelerated form of atherosclerosis due to the similarities in the lesions. For example, Ross (1986) advanced the "response-to-injury" hypothesis to propose that intimal hyperplasia may be an early lesion on the pathway to atherosclerotic plaque. However, in similar sized plaques those developing due to hyperplasia tend to have a higher concentration of smooth muscle cells and a lower concentration of lipid accumulation than do atherosclerotic lesions which usually appear in low-shear regions (see Caro et al., 1971).

The work of Fry (1968) was the first to clearly demonstrate that acute endothelial changes could occur from extremely high wall shear stress, typically associated with local turbulence. Other sources concurred with and expanded on the observations of Fry (1968) by stating that endothelial cell modification can be induced by severe hemodynamic conditions or by surgical processes such as the creation of an anastomosis. Clowes et al. (1983) were first to clearly demonstrate that acute injury to the intima and media can produce hyperplasia. Subsequent studies have shown that such injury produces smooth muscle cell proliferation, which occurs at a rate proportional to the degree of the injury (Chervu and Moore, 1990). The growth of intimal hyperplasia may or may not subside when the endothelial layer is re-established. The injury model, then, does not require that endothelial cells be denuded, but that they be greatly disturbed by adverse hemodynamic conditions or a surgical procedure, resulting in smooth muscle cell proliferation. Platelets and other critical blood elements that adhere to the dysfunctional endothelium may be a major factor in stimulating this smooth muscle cell proliferation.

3.1.3 Thrombogenesis

Thrombosis is the formation of a blood clot, called a thrombus, inside an artery or vein. Thrombosis develops by the same mechanisms that control hemostasis, i.e., the clotting system which prevents blood loss in the event of vessel injury. A thrombus is primarily composed of platelets and red blood cells bound together by molecules in the cell membrane of the platelets, called membrane glycoproteins (GPs), by other proteins in the blood or inside the platelets, and by a network of polymerized plasma protein called fibrin (Colman et al., 1994). An arterial thrombus, as found in high shear regions, is composed primarily of platelets, with some fibrin and trapped red blood cells, which are typically found distal to the platelet-rich part of the thrombus. As mentioned, an arterial thrombus is usually found superimposed on an atherosclerotic plaque which has fissured (or ruptured) to expose subendothelium plaque components to the blood. Arterial thrombosis may also develop on artificial surfaces such as vascular grafts, heart valves, and stents, as well as in aneurysms or injured arteries.

Thrombi formed in the low shear environments, typical of the venous system, are composed mostly of red blood cells which are held together by a fibrin mesh, such as a fibrin tail. Low shear regions in which these thrombi are found include regions of flow stasis and recirculation which commonly occurs within branching blood vessels. The observable difference in thrombus composition due to environment is an indication that shear stress plays a primary role in thrombus formation.

A general view of the currently understood mechanisms of arterial thrombosis is provided in Fig. 3.1.2. In normal flow, platelets collide with the vessel wall due to both convective transport and the diffusive motion imposed by collisions with red blood cells. The converging flow field around a stenosis, flow stagnation points, and flow reattachment sites are all areas of high platelet-wall interactions. Some of these colliding platelets may adhere to the vessel wall, especially in the presence of a tissue factor, i.e., a chemical mediator present when the endothelium is injured. Platelets do not normally adhere to a healthy endothelium. Platelet deposition at dysfunctional sites increases with increasing shear rate; however, there is a limit to this relationship. As discussed the adhesion of a small number of platelet layers may significantly contribute to smooth muscle cell proliferation and intimal thickening. The platelet may be activated which makes adhesion, regulated by membrane glycoproteins and proteins in the vessel wall, more permanent. Activation also permits the aggregation of platelets, which may form a platelet plug that occludes the vessel, or may embolize (break free) totally or in part. Simultaneously, thrombin is produced at the injury site due to tissue-factor (TF) exposure. Thrombin causes coagulation of fibrin into a polymer network that may reinforce a platelet plug and incorporate red blood cells into the clot. During early hemostasis more thrombin is released as platelets are activated, accelerating the aggregation process; thus, a positive feedback mechanism is established. Healthy endothelial cells on the other hand trigger a negative feedback mechanism that limits the propagation of the clot to the region of injury, reducing the sustained rate of clotting and breaking down the clot. Other mechanisms which control thrombus growth rates, such as reduction by embolization, are poorly understood.

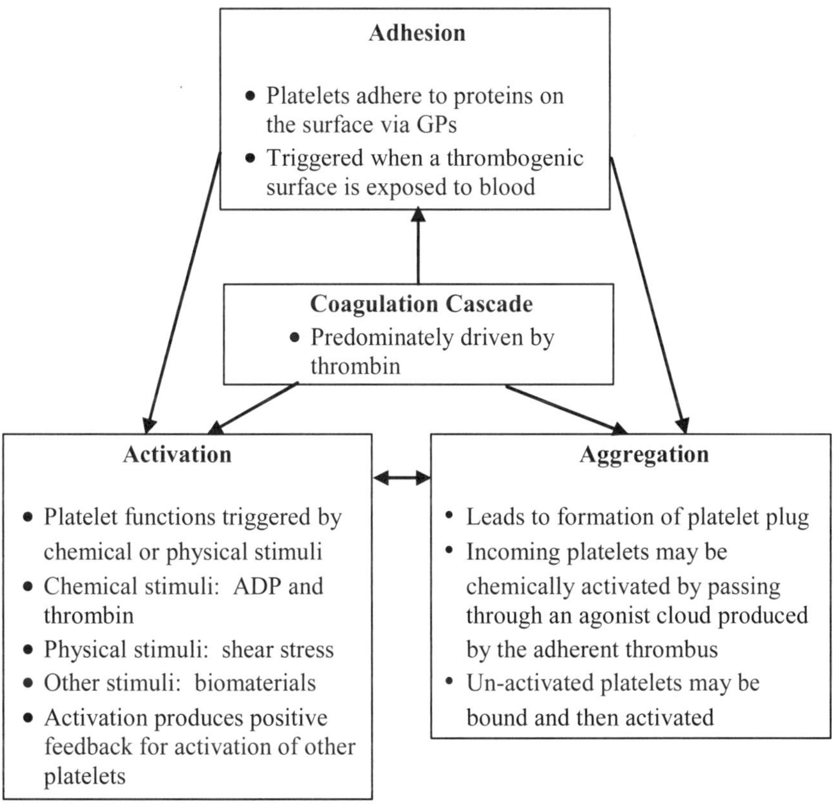

Figure 3.1.2 Simplified schematic of arterial thrombosis (Longest, 2004).

3.1.4 Particle-Hemodynamics

In this section, the equations governing Eulerian-Lagrangian particle-hemodynamics analysis are re-presented (see Sects. 1.3 and 1.4). The computational methods used to solve these equations are discussed by

Longest et al. (2004). A mathematical point-force model for blood particle motion that is valid in both far- and near-wall regions is employed.

Fluid flow equations and boundary conditions. The flow field equations describing transient laminar incompressible hemodynamics are the conservation of mass:

$$\nabla \cdot \vec{\mathbf{v}} = 0 \tag{3.1.1}$$

and momentum:

$$\frac{\partial \vec{\mathbf{v}}}{\partial t} + (\vec{\mathbf{v}} \cdot \nabla)\vec{\mathbf{v}} = \frac{1}{\rho}\left(-\nabla p + \nabla \cdot \vec{\vec{\tau}}\right) \tag{3.1.2a}$$

where $\vec{\mathbf{v}}$ is the velocity vector, p is the pressure, ρ is the fluid density, and the shear stress tensor is given as:

$$\vec{\vec{\tau}} = \eta(\dot{\gamma})\left[\nabla\vec{\mathbf{v}} + (\nabla\vec{\mathbf{v}})^T\right] \tag{3.1.2b}$$

The shear-rate-dependent absolute viscosity, η, is calculated with the transformed Quemada model of Buchanan and Kleinstreuer (1998):

$$\eta(\dot{\gamma}) = \left(\sqrt{\eta_\infty} + \frac{\sqrt{\tau_0}}{\sqrt{\lambda} + \sqrt{\dot{\gamma}}}\right)^2 \tag{3.1.2c}$$

The advantage of the Quemada model over, for example, the popular Casson model (see Sect.1.4.2) is the additional shear-rate modifier, λ, that allows for a better blood data fit in low ranges of the shear rate (i.e., $\dot{\gamma} \sim 10^{-2}\ s^{-1}$). Coefficients used in the original form of the Quemada model for a normal hematocrit (H) of 40% were taken from Cokelet (1987). The coefficients were transformed for Eq. (3.1.2c) using the methodology outlined in Buchanan et al. (2000), resulting in the empirical constants $\eta_\infty = 0.0309$ dyn·s·cm^{-2}, $\tau_0 = 0.07687$ dyn·cm^{-2}, and $\lambda = 0.047723$ s^{-1}. Using the above constants, the Quemada model predicts that the absolute viscosity is within 5% of the Newtonian limit, i.e., $\eta_\infty = 0.0309$ dyn·s·cm^{-2}, at a shear rate of 4,100 s^{-1}. Normal blood density has been assumed to be ρ = 1.054 g/cm^3. The local shear rate, $\dot{\gamma}$, used in Eq. (3.1.2c) is computed from the second scalar invariant of the rate of deformation tensor, i.e.,

$$\dot{\gamma}^2 = \left|II_{2\vec{\vec{D}}}\right| \tag{3.1.3}$$

where $\vec{\vec{D}}$ is the strain rate tensor such that

$$2\vec{\vec{D}} = \nabla\vec{\mathbf{v}} + (\nabla\vec{\mathbf{v}})^T \tag{3.1.4}$$

The second scalar invariant can be expressed as (Bird et al., 2002):

$$II_{2\vec{\vec{D}}} = \frac{1}{2}\left[\left(I_{2\vec{\vec{D}}}\right)^2 - trace\left(2\vec{\vec{D}}^2\right)\right] \tag{3.1.5a}$$

where the first scalar invariant is given by the trace of the rate-of-deformation tensor

$$I_{2\vec{\vec{D}}} = trace\left(2\vec{\vec{D}}\right) \tag{3.1.5b}$$

Expanding the shear rate, it can be expressed as

$$\dot{\gamma}^2 = | 4D_{11}D_{22} + 4D_{11}D_{33} + 4D_{22}D_{33} + 4D_{12}D_{21} + 4D_{13}D_{31} + 4D_{23}D_{32} | \tag{3.1.6}$$

A femoral bypass, as considered here, is the primary application site for most researchers. Following the work of Longest and Kleinstreuer (2003), inlet and boundary conditions, in addition to the "no-slip" wall condition, were selected in a manner to: (a) match physiological conditions as closely as possible with available data, and (b) facilitate numerical computation. To generate the time varying inlet velocity profiles, a transient Womersley solution was used (see Womersley, 1955 or Nichols and O'Rourke, 1998). Flow rate information over variable time-steps was used to compute a complex Fourier series approximation of the pressure gradient pulse (Buchanan, 2000). The transient velocity profiles were then computed using the Womersley solution for Newtonian flow. An extended inlet of approximately three diameters was added to allow non-Newtonian aspects of the velocity profile to develop fully. Flow field outlets were extended far downstream such that the velocity was normal to the outlet plane, i.e., fully developed flow profiles with no significant radial velocity component. For single outlet conditions, a pressure boundary patch was typically used. For multiple outlets, constant mass flow rate boundaries and flow division ratios were specified.

Flow rates through the femoral artery vary largely, depending on the quality of an individual's circulation system, location of the flow measurement, and conditions under which the measurements are made, e.g., resting or exercising. The various combinations of femoral diameter and flow rate can induce a wide range of Reynolds number waveforms. For example, Kleinstreuer et al. (1996) employed a "resting pulse" derived from femoral flow velocity measurements made in a dog (cf. Nichols and O'Rourke, 1998) and assumed a second "exercise" pulse for patients under active conditions. Steinman et al. (1993) made Doppler ultrasonography measurements of a mid-femoral flow waveform which are summarized in Table 3.1.1. Significant stenosis developments may lead to occlusions and a general loss of wall compliance. Furthermore, the inclusion of a bypass graft significantly alters waveform characteristics compared to measurements made in natural arteries. In general, waveforms that vary significantly from the standard femoral pulse are typically associated with both early and late graft failure (Inokuchi et al., 1982; Okadome et al., 1986, 1991). However, clinical evidence indicates that pulse severity alone does not necessarily indicate a predisposition for graft failure.

Table 3.1.1 Reynolds Number Waveforms for the Femoral Artery.

	Re_{max}	Re_{min}	Re_{mean}	f (beats/min)
Resting pluse	600	-67.8	113.5	60
Exercise pulse	480	295	356.5	100
Standard pulse (Steinman et al., 1993)	980	-375	125.0	60

Table 3.1.2 Characteristics of the Selected Type I Waveform.

	Q_{mean}	Re_{mean}	Re_{max}	Re_{min}	V_{mean}	V_{max}	V_{min}
Arterial conditions (Ø = 4 mm)	164.4 ml/hr	303.9	1054.	-5.9	21.8 cm/s	75.6 cm/s	-0.4 cm/s
Graft conditions (Ø = 6 mm)	164.4 ml/hr	202.6	702.7	-3.9	9.7 cm/s	33.6 cm/s	-0.2 cm/s
Graft conditions (Ø = 8 mm)	164.4 ml/hr	151.95	527.	-2.9	5.45 cm/s	18.9 cm/s	-0.1 cm/s

Note: All variables based on the Newtonian limiting viscosity of blood η = 0.031 dyn*s/cm^2 and ρ = 1.054 g/cm^3.

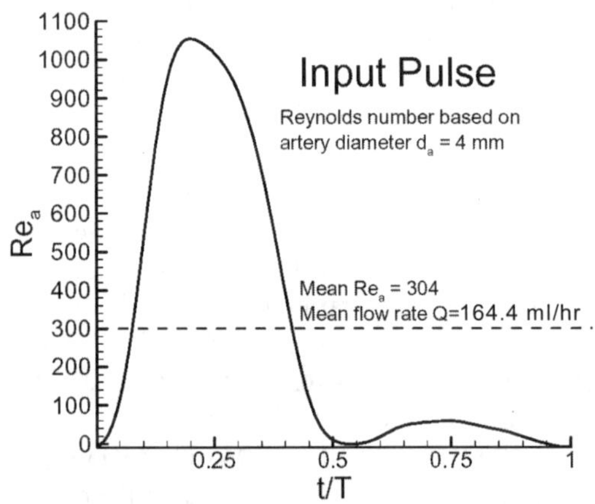

Figure 3.1.3 Selected femoral artery waveform (Type I) for numerical analysis.

To formulate a representative baseline pulse for numerical investigation of the femoro-popliteal bypass, a waveform should be selected due to frequency of clinical occurrence and the absence of extraneous complications. Therefore, a Type I pulse (see Table 3.1.2) has been selected as a representative basis for comparison and is illustrated in Fig. 3.1.3. This pulse is well within the range of measured mean flow-rates and falls within peak velocity and Reynolds numbers guidelines for non-stenosed grafts (cf. Table 3.1.2). Consistent with Type I waveforms, only a small amount of net retrograde flow is observed, which occurs around t/T = 0.5s. The resulting velocity profiles display a significant amount of retrograde flow in the near-wall region throughout diastole.

3.1.4.1 Equations of Particle Motion

Maxey and Riley (1983) have presented in detail the equation of motion for a small rigid sphere in a nonuniform flow field with creeping relative motion, i.e., Stokes flow. They began with the equation of motion for a spherical particle as

$$m_p \frac{d\mathrm{v}_i}{dt} = m_p\, g_i + \int_S \sigma_{ij}\, \hat{n}_j\, dS \tag{3.1.7}$$

where v_i is a component of the particle velocity, the surface integral is over the sphere, n is the unit outward normal, and the fluid stress tensor is given by

$$\sigma_{ij} = -\, p\, \delta_{ij} + \mu \left(\frac{\partial u_i}{\partial x_j} + \frac{\partial u_j}{\partial x_i} \right) \tag{3.1.8}$$

This derivation makes clear the fact that contributions to the particle motion equation include forces from both (a) the undisturbed flow field (although it may be nonuniform); and (b) the local disturbance flow which is created by the presence of the sphere. These forces are applied to the centroid of a discrete particle, which is referred to as a point-force approximation. In expanded form then, the equation of motion for a small, rigid, spherical particle subjected to a nonuniform transient three-dimensional flow field of a Newtonian fluid reads (Maxey, 1987):

$$m_p \frac{d\mathrm{v}_i}{dt} = \underbrace{m_f \frac{Du_i}{Dt}}_{term\ 1} - \underbrace{\frac{m_f}{2}\frac{d}{dt}(\mathrm{v}_i - u_i)}_{term\ 2} - \underbrace{6\pi a_p \mu(\mathrm{v}_i - u_i)}_{term\ 3} + \underbrace{(m_p - m_f) g}_{term\ 4}$$

$$- \underbrace{\frac{6\pi a_p^2\, \mu}{\sqrt{\pi \mathrm{v}}} \left(\int_0^t \frac{d}{d\sigma}(\mathrm{v}_i - ui) \frac{d\sigma}{\sqrt{t-\sigma}} + \frac{\mathrm{v}_i(0) - u_i(0)}{\sqrt{t}} \right)}_{term\ 5} + \underbrace{m_f \frac{d}{dt}\left(\frac{a^2\, \nabla^2\, u_i}{20} \right)}_{term\ 6}$$

$$+ \underbrace{\pi a_p^3\, \mu \nabla^2\, u_i}_{term\ 7} + \underbrace{a_p^4\, \mu \sqrt{\frac{\pi}{\mathrm{v}}} \left(\int_0^t \frac{d\left(\nabla^2\, u_i\right)}{d\sigma} \frac{d\sigma}{\sqrt{t-\sigma}} \quad \frac{\nabla^2\, u_i(0)}{\sqrt{t}} \right)}_{term\ 8}$$

(3.1.9)

where v_i is the particle velocity, u_i is the fluid velocity, and a_p is the particle radius. This equation requires that the particle Reynolds number approaches zero

$$\mathrm{Re}_p = \frac{2\,|\,u_i - \mathrm{v}_i\,|\,a_p}{\nu} << 1 \qquad (3.1.10a)$$

and the particle radius is sufficiently small

$$\frac{a_p}{L} << 1 \qquad (3.1.10b)$$

where L is a characteristic length of the flow field. Lift forces are assumed negligible due to the absence of particle rotation and the assumption of a small shear Reynolds number

$$\mathrm{Re}_g = \frac{(2a_p)^2}{\nu_c}\frac{dU}{dy} << 1 \qquad (3.1.11)$$

The first term in Eq. (3.1.9) is the result of pressure gradients that exists in the undisturbed non-uniform flow. The second term is the added mass, which is a transient term resulting from the linear acceleration of the fluid surrounding a particle. A small particle Reynolds number allows for the application of Stokes drag, term 3, which accounts for both shear and pressure contributions in a uniform linear non-accelerating flow field, and implies a small difference between the particle and material derivatives d/dt and D/Dt. The fifth term is the history integral, or Basset term, which is a transient consequence resulting from the delay in near-particle viscous effects as the relative velocity changes in a linear uniform flow field. The second term in the parentheses of the fifth component accounts for the effect of an initial relative velocity. Terms six through eight are the Faxen corrections to the added mass, the drag, and the history integral, respectively. These addenda account for the effects of nonuniform flow, such as swirl and curved particle paths; however, they do not account for or rely on particle rotation. Analytic solutions to the equation of particle motion (including the history term) are available by several techniques (Clift et al., 1978; Buchanan, 2000). For example, Michaelides and co-workers have applied Laplace transforms to arrive at an analytic expression of particle motion that is explicit (cf. Michaelides, 1997). The analysis of Buchanan (2000), following Coimbra and Rangel (1998), indicated that for a density ratio $\alpha = \rho_f / \rho_p$ < 8/5, the history term is only important for values of $D \equiv T / \tau_p$ < 100, where T is the pulse period and τ_p is the particle's relaxation time (or momentum response time), i.e.,

$$\tau_p = \frac{2\rho_p\, a_p^2}{9\mu_c} \qquad (3.1.12)$$

For blood flow in humans, the pulse period is $T \approx 1$ s and density ratios are very near unity, i.e., in plasma $\alpha = 0.983$ for platelets and $\alpha = 0.964$ for monocytes. Blood particle relaxation times are on the order of $\tau_p = 10^{-6}$ s resulting in a very large value of D, i.e., $D \approx 10^5$ to 10^6. Therefore, the history term may be neglected for the transport of platelets, monocytes, and red blood cells in plasma. Furthermore, Buchanan (2000) pointed out that the added mass term makes no significant contribution for liquid-solid multiphase flows or colloidal suspensions (such as blood) where the densities of the phases are very similar.

A reduced form of the Maxey equation (3.1.9) is the Basset-Boussinesq-Oseen (BBO) equation which features the first five terms of Eq. (3.1.9), as discussed by Crowe et al. (1998) and Kleinstreuer (2003), among others.

Finally, the resulting equation for blood particle motion, assuming only pressure and drag effects, is written as:

$$\frac{d\mathrm{v}_i}{dt} = \alpha \frac{Du_i}{Dt} + \frac{1}{\tau_p}(u_i - \mathrm{v}_i) \tag{3.1.13}$$

where v_i and u_i are the components of the particle and local fluid element velocity, respectively. If virtual mass effects are to be included, the above equation can be written

$$\frac{d\mathrm{v}_i}{dt} = \frac{3}{2}\beta\,\alpha \frac{Du_i}{Dt} + \frac{\beta}{\tau_p}(u_i - \mathrm{v}_i) \tag{3.1.14a}$$

where

$$\beta = \frac{1}{1 + \alpha / 2} := 0.693 \tag{3.1.14b}$$

It has been shown that the motion of a hardened red blood cell, as in the annular expansion of Karino and Goldsmith, can be modeled as a fluid element (cf. Longest & Kleinstreuer, 2003). However, both the drag and pressure terms make a significant contribution to the equation of particle motion due to variable flow fields. Furthermore, the continued collisions and hydrodynamic interaction between, and deformations of, the red blood cells in flowing blood induces a form of mixing, similar on a macroscopic scale to the intermolecular collisions that result in Brownian motion. For platelets and white blood cells, the movement induced by this mixing results in a persistent dispersion, and is often modeled with effective diffusion coefficients which are two to three orders of magnitude greater than those due to Brownian motion. This motion becomes a major mechanism for permitting cells and, possibly, some of the larger proteins found in the blood to interact with the wall. To provide an accurate model for lateral motion, particle dispersion coefficients should be a function of the local velocity gradient, local red blood cell concentration, and cell deformation. Unfortunately, models for a realistic red blood cell concentration field are rather complex; therefore, local variations in hematocrit will be ignored in defining particle dispersion coefficients.

3.1.4.2 Near-Wall Forces

Assumptions required by the equation of particle motion, as presented above, include a relatively small shear Reynolds number and negligible particle rotation. These assumptions are most valid for fluid-particle flow in a boundless media. However, as a particle approaches a fluid boundary, at which a no-slip condition holds, the shear forces and the rotation rate of the particle increase. In the near-wall environment, the enhanced shear field and the particle rotation may influence the particle's trajectory by generating lift and modifying the drag term. The near-wall forces to be considered for bio-particle transport include near-wall drag modifications (i.e., the lubrication forces), as well as lift and biochemical interactions.

When a particle is close to a solid boundary, Eq. (3.1.13) should be modified to include the effects of the wall on the trajectory. Consider a particle moving near a wall with both a normal and tangential velocity component with respect to the boundary. The Stokes drag term in Eq. (3.1.13) cannot accurately predict the wall-normal motion. Specifically, the form drag component in the Stokes drag approximation assumes that there is not a boundary in the direction of particle motion acting to increase the pressure in front of the particle. Hence, the Stokes drag approximation must be modified. Similarly, but to a lesser extent, the boundary influences the viscous drag on a particle moving tangential to a wall.

Cox and Brenner (1967) showed that for Stokes flow, the drag on a spherical particle of radius a_p moving normal to a wall is

$$F_n = -m_p \frac{1}{\tau_p}(\mathrm{v}_n - u_n) - m_p \frac{1}{\tau_p}(\mathrm{v}_n - u_n)\lambda_n \qquad (3.1.15)$$

where n is the wall-normal direction and h_p is the distance between the center of the particle and the wall. In the above representation, the first term is the traditional Stokes drag while the second accounts for the drag modification due to the presence of the wall. It is this second term that is often referred to as the lubrication force and can be viewed as an interaction force as derived by Crowe et al. (1998) for the case of two approaching spheres. Dahneke (1974) suggested the following fit to the results of Cox and Brenner (1967) for the lubrication coefficient at all particle wall separations:

$$\lambda_n = \frac{a_p}{\delta} \qquad (3.1.16)$$

where d is the distance between the particle surface and the wall, i.e.,

$$\delta = h_p - a_p \qquad (3.1.17)$$

For truly Stokesian flow, the lubrication coefficient may vary from zero to infinity in the limit, making it impossible for a particle to reach the wall in a finite amount of time. However, for particles moving through a gas, the continuum approximation breaks down when the gap between the surface of the particle and the wall becomes comparable with the molecular mean free path of the gas. Hence, the continuum model overpredicts the magnitude of the stress on the particle surface in the gap.

Similar to wall-normal motion, the drag coefficient in the tangential direction is also affected by the presence of the wall. Goldman et al. (1967a & b) showed that the drag force on a sphere near a wall in linear (uniform shear) flow is given by

$$F_t = -m_p \frac{1}{\tau_p}(\mathrm{v}_t - u_t) - m_p \frac{1}{\tau_p}(\mathrm{v}_t - u_t)\lambda_t \tag{3.1.18}$$

Goldman et al. (1967a & b) derived approximations for λ_t and illustrated that it was a function of shear rate, angular velocity, and the dimensionless ratio h_p / a_p. As the particle approaches a wall, λ_t diverges; however, unlike λ_t, it only diverges logarithmically.

The near-wall drag modifications given above, which were derived as asymptotic solutions in the limit of true Stokes flow, have proven to be reasonable estimates in comparison with experimental data (e.g., Young and Hanratty, 1991). Recently, Loth (2000) has reported near-wall drag modifications based on resolved-volume simulations of spherical particles for $\mathrm{Re_p} \ll 1$. In the context of Eqs. 3.1.15 & 3.1.18, the approximations are expressed as

$$\lambda_n = \frac{1.1}{\left(\dfrac{h_p}{a_p} - 1\right)} \tag{3.1.19a}$$

and

$$\lambda_t = 0.7 \frac{a_p}{h_p} \tag{3.1.19b}$$

Loth (2000) reports that these approximations provide a solution that is within 2% of simulated results.

In a strongly sheared flow, inertial lift forces are likely to be important. These enter the BBO-equation at a higher order in the particle Reynolds number than the terms in Eq. (3.1.9). There is a lack of rigorous mathematical justification for using the available results for lift forces (McLaughlin, 1994). Furthermore, all the existing theories of lift forces at small but finite particle Reynolds numbers are derived for steady laminar flows.

There are two causes of a composite lateral lift force on a spherical particle. The first is known as the Magnus lift force and develops due to rotation of the particle. The lift is created by a pressure differential between opposite lateral sides of the particle resulting from a velocity differential due to rotation. The sphere may rotate because of collisions with the wall, inter-particle collisions, or due to the enhanced near-wall shear field. Calculation of the Magnus effect requires direct knowledge of the particle rotation rate, which must be different from the local fluid element rotation rate, otherwise no lift is produced (cf. Crowe et al., 1998). Calculation of the rotation rate requires solving the angular momentum equation. Shear induced torque may be calculated directly if detailed knowledge of the particle's influence on the flow field is known. Collision-induced torques further complicate the calculation. Due to the difficulties associated with the calculation of particle spin, particularly in the presence of a moderate number of collisions, the Magnus force is usually neglected (Michaelides, 1997).

The second source of lift is the shear field of the fluid, which may cause lateral lift on the sphere even in the absence of rotation. A velocity gradient may give rise to sufficiently high and low pressures on opposite sides of a particle, which induces the Saffman lateral lift force. The basic Saffman lift force is relatively easy to compute (Crowe et al., 1998; Kleinstreuer, 2003), and has been corrected via an empirical correlation to allow for higher relative velocities (Mei, 1994).

Cherukat and McLaughlin (1994) derived an expression for Saffman-style lift which is applicable when the distance between the particle and wall is on the order of the particle radius. Assuming that the wall lies within the inner region of the sphere's disturbance flow, the Cherukat and McLaughlin (1994) lift approximation can be written as

$$F_{lift} = \rho_f a_p^2 u_s^2 \cdot \left(\frac{h_p}{a_p}, \frac{\dot{\gamma}\, a_p}{u_s} \right) \tag{3.1.20}$$

where u_s is taken to be the wall-tangent slip velocity

$$u_s = (\mathrm{v}_t - u_t) \tag{3.1.21}$$

Assuming the particle to be a freely rotating sphere, Cherukat and McLaughlin (1994) numerically integrated their asymptotic result and found it could be approximated as

$$\begin{aligned} I = & \left[1.7631 + 0.3561\kappa - 1.1837\kappa^2 + 0.845163\kappa^3\right] \\ & - \left[\frac{3.24139}{\kappa} + 2.6760 - 0.8248\kappa - 0.4616\kappa^2\right] \Lambda \\ & + \left[1.8081 + 0.8796\kappa - 1.9009\kappa^2 + 0.98149\kappa^3\right] \Lambda^2 \end{aligned} \tag{3.1.22}$$

where $\kappa = h_p / a_p$ and $\Lambda = \dot{\gamma}\, a / u_s$. For cases in which the slip velocity approaches zero, the velocity scale $\dot{\gamma}\, a_p$ should replace u_s in the above equations. Depending on the signs of $\dot{\gamma}$ and u_S, as well as the size of k, the near-wall lift may be directed toward or away from the wall. However, as a particle approaches the wall in the limit, the lift force inevitably acts to separate the particle from the wall. Application of the Cherukat & McLaughlin (1994) lift expression is based on the condition

$$a_p < h_p << \min\{L_s, L_G\} \tag{3.1.23a}$$

where the Stokes length scale is

$$L_S = \frac{\nu}{u_s} \tag{3.1.23b}$$

and the Saffman length scale is

$$L_G = \sqrt{\frac{\nu}{\dot{\gamma}}} \tag{3.1.23c}$$

For blood particle simulation, L_G is typically the limiting condition due to elevated shear fields and small particle response times. Assuming that Eq. (3.1.23a) implies that h_p should be one order of magnitude less than L_G, the near-wall lift expression is valid on the order of ten monocyte diameters.

For particle wall distances significantly greater than the particle radius, the near-wall lift expression of Cherukat and McLaughlin (1994) is no longer valid. To account for lift in wall bounded flow at all particle wall separations, Wang et al. (1997) suggested using a compilation of the available lift expressions, each valid in a band of wall separation distance. Lift expressions valid outside of the narrow near-wall region require significant axial lengths before a contribution to radial particle concentration is realized. In the context of bifurcating geometries, the region of interest is of insufficient length for significant radial motion away from the wall to occur as the result of lift. However, in the near-wall region, transverse motion on the order of a particle diameter becomes important with respect to particle deposition. Hence, for the case of blood particle deposition in bifurcating geometries, near-wall lift is the most critical. Beyond this region, insignificant particle lift will be assumed. Moreover, this is a necessary assumption owing to the fact that outside of the near-wall region collisional forces dominate lateral particle motion.

3.1.4.3 Hemodynamic Wall Parameters

As discussed in Sects. 3.1.1 and 3.1.2, endothelial cell injury or dysfunction and the interaction of critical blood particles with the vascular surface both play significant roles in the biophysical evolution of intimal thickening and thrombosis formation. Hemodynamic wall parameters are intended to identify sites where intimal thickening and thrombosis formation are likely, based on hemodynamic wall interactions.

Traditionally, identifying sites of endothelial cell dysfunction has been a primary purpose of the wall shear stress (WSS) based hemodynamic parameters as well as sites of excessive particle-wall interactions (Kleinstreuer et al., 2001). Hemodynamic wall parameters that are to be considered for (femoral) graft bypass and graft-end optimization are summarized in Fig. 3.1.4.

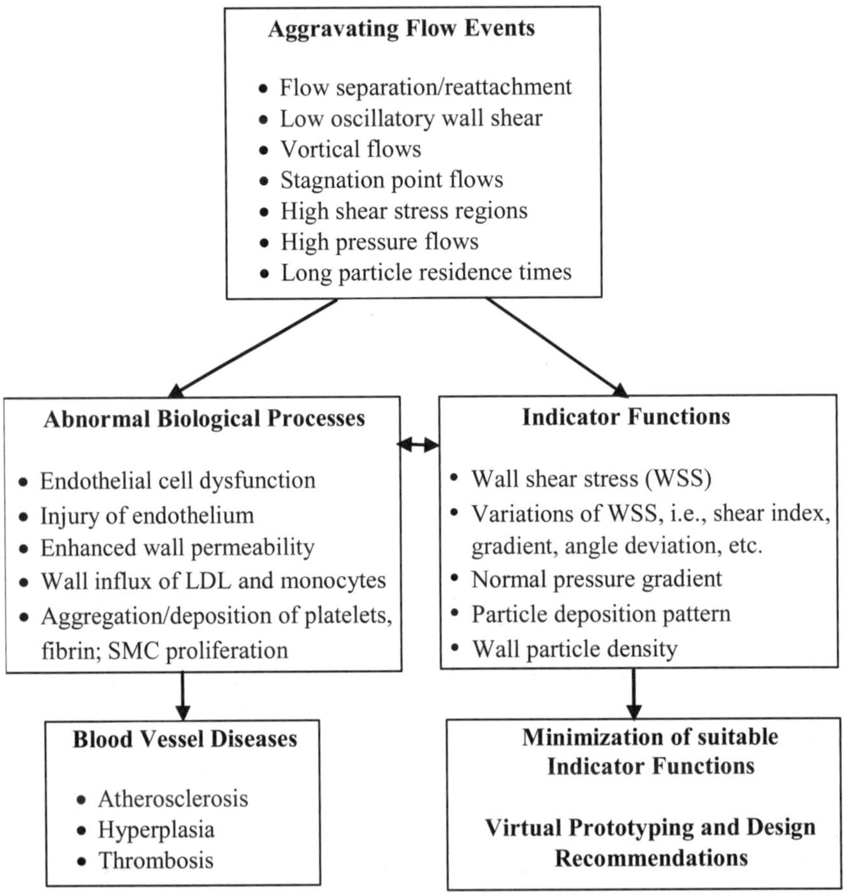

Figure 3.1.4 Flow events, biological processes, and methodology governing improved graft-end designs (Kleinstreuer, 2001).

Wall shear stress. It is widely held that the arterial wall is capable of changing its diameter in response to changes in WSS. For example, Zarins et al. (1987) demonstrated in vivo that the arterial wall, when subjected to as much as a tenfold increase in blood flow, gradually increased in diameter until a mean WSS of 15 dyn/cm^2 was restored. Furthermore, the response of endothelial cells to the time-averaged direction and magnitude of WSS in vitro has been well documented (e.g., Helmlinger et al., 1991).

The shear stress is a tensor within the flow field and reduces to a vector on a surface

$$\vec{\tau}_w = \hat{n} \cdot \vec{\vec{\tau}} \tag{3.1.24a}$$

where $\hat{n}$ is the local surface normal vector. The magnitude of the time-averaged wall shear stress vector can be written as

$$WSS = \left|\overline{\vec{\tau}_w}\right| = \left|\frac{1}{T}\int_0^T \vec{\tau}_w dt\right| \tag{3.1.24b}$$

Alternatively, the time-average of the wall shear stress magnitude can be written

$$\left|\overline{\vec{\tau}_w}\right| = \frac{1}{T}\int_0^T \left|\vec{\tau}_w\right| dt \tag{3.1.24c}$$

The oscillatory shear index monitors differences in these two time-averaged values.

Oscillatory shear index. Cyclic departure of the wall shear stress vector from its predominant axial alignment indicates flow disruption over time and is known as the oscillatory shear index, or OSI (Ku et al., 1985). The OSI, therefore, quantifies disturbed flow interaction with the wall and is formulated as (He and Ku, 1996)

$$OSI = \frac{1}{2}\left(1 - \frac{\left|\int_0^T \vec{\tau}_w \, dt\right|}{\int_0^T \left|\vec{\tau}_w\right| dt}\right) \tag{3.1.25}$$

where $\vec{\tau}_w$ is the instantaneous wall shear stress vector. The numerator of the shear stress fraction represents the magnitude of the time-averaged WSS while the denominator represents the time-average of the WSS magnitude. The shear stress fraction can vary from 1 to 0 which indicates no cyclic variation to 180-degree cyclic variation of the wall shear stress direction with time, respectively. The OSI can vary from 0 to 1/2, indicating the least and most severe temporal shear rate conditions, respectively.

Wall shear stress gradient. From a biological standpoint, endothelial cells have been shown to align themselves with the mean flow direction which corresponds to the local direction of the time-averaged wall shear stress. Surface coordinates that elucidate the interaction of instantaneous wall shear stress vectors and endothelial cells are defined as:

- m - temporal mean wall shear stress direction
- n -tangential to the surface and normal to m

• *l* - surface normal direction

The wall shear stress gradient can be obtained by calculating spatial derivatives of the wall shear stress vector $(\vec{\tau}_{w,i})$, which results in a nine component tensor

$$\nabla\bar{\tau}_w = \left(\frac{\partial}{\partial x}\hat{i} + \frac{\partial}{\partial y}\hat{j} + \frac{\partial}{\partial z}\hat{k}\right)\left(\tau_{w,x}\,\hat{i} + \tau_{w,y}\,\hat{j} + \tau_{w,z}\,\hat{k}\right)$$

$$= \begin{bmatrix} \frac{\partial\tau_{w,x}}{\partial x} & \frac{\partial\tau_{w,y}}{\partial y} & \frac{\partial\tau_{w,x}}{\partial z} \\ \frac{\partial\tau_{w,y}}{\partial x} & \frac{\partial\tau_{w,y}}{\partial y} & \frac{\partial\tau_{w,y}}{\partial z} \\ \frac{\partial\tau_{w,z}}{\partial x} & \frac{\partial\tau_{w,k}}{\partial y} & \frac{\partial\tau_{w,z}}{\partial z} \end{bmatrix} \quad (3.1.26)$$

The $\nabla\vec{\tau}_w$ tensor with respect to the xyz-coordinates (x_1,x_2,x_3) can be transformed to the *mnl*-coordinate system (s_1,s_2,s_3) by a standard component-wise tensor transformation

$$\frac{\partial\tau'_{w,i}}{\partial s_j} = a_{ik}\,a_{jl}\,\frac{\partial\tau_{w,k}}{\partial x_l} \quad (3.1.27a)$$

where the primes denote the *mnl*-coordinate system. The terms a_{ik} and a_{jl} represent the directional cosines of the *xyz*-coordinates rotated to the *mnl*-coordinates. The resultant tensor is

$$\nabla\vec{\tau}_w = \begin{bmatrix} \frac{\partial\tau_{w,m}}{\partial m} & \frac{\partial\tau_{w,m}}{\partial n} & \frac{\partial\tau_{w,m}}{\partial \ell} \\ \frac{\partial\tau_{w,n}}{\partial m} & \frac{\partial\tau_{w,n}}{\partial n} & \frac{\partial\tau_{w,n}}{\partial \ell} \\ \frac{\partial\tau_{w,\ell}}{\partial m} & \frac{\partial\tau_{w,\ell}}{\partial n} & \frac{\partial\tau_{w,\ell}}{\partial \ell} \end{bmatrix} \quad (3.1.27b)$$

Due to the coordinate system chosen, i.e., *mnl*, the components of the tensor affect the endothelial cell in normal and tangential directions. However, the components related to the *l*-directional wall shear stress and *l*-coordinate, i.e., all normal components and variations, are of no interest because the aggravating effects on the endothelium are caused by changes in surface, i.e., tangential, forces. Specifically,

$$\nabla \vec{\tau}_w := \begin{bmatrix} \dfrac{\partial \tau_{wm}}{\partial m} & \dfrac{\partial \tau_{w,m}}{\partial n} \\ \dfrac{\partial \tau_{w,n}}{\partial m} & \dfrac{\partial \tau_{w,n}}{\partial n} \end{bmatrix} \tag{3.1.27c}$$

The diagonal components $\partial \tau_{w,m} / \partial m$ and $\partial \tau_{w,n} / \partial n$ generate intracellular tension which causes widening and shrinking of cellular gaps. The off-diagonal components $\partial \tau_{w,m} / \partial n$ and $\partial \tau_{w,n} / \partial m$ cause relative movement of adjacent cells. Lei (1995) suggested that the normal components of the $\nabla \vec{\tau}_w$ tensor, i.e., those creating tension, are the most important ones with respect to intimal thickening due to atherosclerosis or hyperplasia. A scalar combination of the normal components can be written as

$$WSSG(t) = \left[\left(\frac{\partial \tau_m}{\partial m} \right)^2 + \left(\frac{\partial \tau_n}{\partial n} \right)^2 \right]^{1/2} \tag{3.1.28}$$

In order to employ the WSSG-concept to assess the impact of non-uniform hemodynamics, the absolute value of the local instantaneous wall shear stress gradient is either used directly or in its time-averaged dimensionless form. Specifically,

$$L_{SWSSG_{nd}} = \frac{1}{T} \frac{d_o}{\tau_o} \int_0^T WSSG(t) dt \tag{3.1.29}$$

in which d_o is a reference diameter and $\tau_o = 8\eta\, u_{mean} / d_o$ is the reference Poiseuille-type wall shear stress corresponding to the mean flow rate.

Wall shear stress angle deviation or gradient. Hyun et al. (2004) proposed that a measure of the angle between adjacent wall shear stress vectors would indicate regions of dysfunctional endothelial cells, and, hence, sites of intimal thickening. The wall shear stress angle deviation (WSSAD) for a control volume of area dA_i is formulated

$$WSSAD = \frac{1}{T} \int_0^T \left(\frac{1}{A} \int_S \arccos\left(\frac{\vec{\tau}_o \cdot \vec{\tau}_n}{|\vec{\tau}_o| \cdot |\vec{\tau}_n|} \right) dA_i \right) dt \tag{3.1.30}$$

where the surface stress vector at the location of interest is represented by $\vec{\tau}_o$, and $\vec{\tau}_n$ represents the four surrounding surface stress vectors. A similar, mesh independent parameter, the wall shear stress angle gradient, was devised for the publication of Longest and Kleinstreuer (2000). The rationale for this parameter, which has not appeared elsewhere, is provided in the following discussion.

Considering Eq. (3.1.30), the WSSAD can be used to evaluate single geometries and make comparisons among designs if and only if the inter-nodal distances between adjacent surface nodes are approximately equal. Grids should therefore be constructed to ensure that, in the region of interest, the inter-nodal distances at the wall are nearly equivalent. It is usually difficult to meet this requirement with the use of structured meshes for complex geometries.

A mesh independent wall shear stress directional parameter may be formulated using a gradient operation. For example, the WSSG scalar indicates how much $\tau_{w,m}$ and $\tau_{w,n}$ vary in the m and n directions, respectively. Similarly, the WSSAD reports spatial variation of the mean shear stress direction, i.e., variation of the m and n directions. If the WSSG components $\partial\tau_{w,m}/\partial m$ and $\partial\tau_{w,n}/\partial n$ were calculated as the WSSAD is, they would simply be the differences in $\tau_{w,m}$ and $\partial\tau_{w,n}$ values across one control volume in the m and n directions, respectively. In this formulation, the distance between the locations at which $\tau_{w,m}$ and $\tau_{w,n}$ values are sampled would be ignored resulting in a mesh dependent formulation. A mesh independent measure of change in $\tau_{w,m}$ and $\tau_{w,n}$ in the m and n directions is formed using the gradient operation (or directional derivatives). Similarly, to formulate a mesh independent measure of spatial variation in the WSS direction, the gradient operation can be used. The region of interest is a surface area segment defined by a central control-volume face and its four surrounding nodes. Using the central mean WSS vector as a reference, a scalar field of angular differences can be computed as with the WSSAD. In this formulation

$$\phi = \arccos\left(\frac{\vec{\tau}_o \cdot \vec{\tau}_n}{|\vec{\tau}_o|\cdot|\vec{\tau}_o|}\right) \tag{3.1.31a}$$

where, again, the stress vector at the location of interest is represented by $\vec{\tau}_o$ and $\vec{\tau}_n$ represents the surrounding stress vectors, i.e., vectors over each surface control-volume edge. The gradient of the scalar field of angular variations is then taken on an area-average basis assuming that each angular difference is constant for its respective control volume edge. The resulting wall shear stress angular gradient (WSSAG) is composed of three components, i.e,

$$\overrightarrow{WSSAG} = \frac{1}{A_i}\int_S \frac{\partial\phi}{\partial x}\, dA_i\, \hat{i} + \frac{1}{A_i}\int_S \frac{\partial\phi}{\partial y}\, dA_i\, \hat{j} + \frac{1}{A_i}\int_S \frac{\partial\phi}{\partial z}\, dA_i\, \hat{k} \tag{3.1.31b}$$

or

$$\overrightarrow{WSSAG} = \Theta_x \hat{i} + \Theta_y \hat{j} + \Theta_z \hat{k} \qquad (3.1.31c)$$

The gradient vector can then be transformed to the m and n coordinate system (such that the normal component l vanishes) to quantify specific directional changes in the surface coordinates. Alternately, the magnitude of the WSSAG vector is a coordinate independent scalar that can be computed on a time-average basis

$$WSSAG = \frac{1}{T} \int_0^T \left(\Theta_x^2 + \Theta_y^2 + \Theta_z^2 \right)^{1/2} dt \qquad (3.1.31d)$$

or, more succinctly,

$$WSSAG = \frac{1}{T} \int_0^T \left| \frac{1}{A_i} \int_S \nabla \phi \, dA_i \right| dt \qquad (3.1.32)$$

which represents the magnitude of the shear stress angle deviation over a distance. Due to the use of the gradient operation, this parameter is mesh independent even though the central vector was used to compute the scalar field of angles. This is because the mean shear stress vector is simply a reference, i.e., any other surface vector could be used for this purpose.

Lagrangian-based wall parameters. As discussed, blood particle depositions and stasis have been widely implicated with intimal thickening and thrombosis formation. In the absence of near-wall forces, it is often assumed that the initial point of particle wall contact represents a critical location of interest (cf. Hyun, 1998; Buchanan, 2000). However, the inclusion of near-wall forces coupled with the assumption of spherical particles and Stokes flow prevents particle wall contact. Blood particle stasis may then serve as a qualitative hemodynamic wall parameter. Ehrlich and Friedman (1977) examined the distribution of blood particle stasis in a Y-branch two-dimensional (2-D) model by tracking Lagrangian fluid elements assumed to be blood particles. For pulsatile axisymmetric stenotic flow, Kunov et al. (1996) tracked Lagrangian fluid elements which were assumed to be groups of activated platelets in a symmetric (2-D) stenosis and applied the quantitative concept of a local volumetric residence time which takes into account where particles accumulate and how long they remain there. For a similar stenotic configuration, Buchanan et al. (2000) generated maps of localized fluid element residence time for various input pulse frequencies.

This research hypothesizes that blood particle deposition is most likely in regions of near-wall particle stasis and/or elevated concentrations, coincident with regions of dysfunctional endothelial cells. Regions of enhanced particle-wall interaction may be quantified by the proposed non-dimensional near-wall residence time (NWRT) parameter. To derive a general model for the near-wall residence time (NWRT) and hence for the probability of blood particle deposition, the following factors have been considered:

1. Time that a particle spends in close proximity to an attached site
2. Availability of particles
3. Distance between the particle and the vascular surface
4. Bio-particle activation and surface reactivity (which may be proportional to flow induced surface dysfunction)
5. Independence of mesh size and particle loading

In conjunction with the correct particle dynamics equations (cf. Eqs. 3.1.13-3.1.22), a nondimensional near-wall residence time parameter (NWRT) is proposed. It indicates the probability of particle deposition, i.e., the larger the NWRT value, the higher the probability of particle deposition. Specifically,

$$NWRT = \frac{Q_{av}}{n_o V_{NW}} \sum_{p=1}^{n} \int_{path,p} \left(\frac{a_p}{h_p} \right)^S \frac{\vec{\mathbf{v}}}{\| \vec{\mathbf{v}} \|^2} \cdot d\vec{r} \qquad (3.1.33)$$

For a local near-wall region of volume V_{NW}, $\| \vec{\mathbf{v}} \|$ is the magnitude of the particle velocity, n is the number of activated blood particles that pass through the local volume, and $d\vec{r}$ is the local directional unit vector for integration along each trajectory. A near-wall layer of biophysical interaction is defined by h=50μm. The near-wall volume, $V_{NW} = A_{surf} \times h$, accounts for variations in near-wall control-volume size, such that the solution is practically mesh independent for a limiting value of local surface area, A_{surf}. Inclusion of the total number of particles, n_o, results in convergent profiles once a sufficient quantity of particles has been simulated. The average flow rate term Q_{av} is a physically relevant constant, which is included to form a dimensionless parameter. The distance between a particle center and the wall, h_p, is a key scaling (or probability) factor in the term $(a_p / hp)^s$, where a_p is the particle radius and, to match experimental results, s is set to unity for monocytes and a factor of two for platelets.

The NWRT concept is an approximate method which is particularly useful and necessary given that biophysical factors such as vessel surface roughness, actual blood particle shape, and nano-scale bond formations responsible for possible blood cell attachment, rolling, or re-suspension cannot feasibly be included in simulations involving relatively large-scale 3-D geometries. For the calculation of NWRT, at least 50,000 blood particles have been released randomly in both space and time over one pulse period at an upstream slice prior to the junction area. Simulations are continued until all particles have exited the flow field (see Sect.3.1.5).

3.1.5 Treatment Option: Femoral End-to-Side Graft Bypass

The treatment option and supportive equations discussed in Sect. 3.1.4 are now applied to a femoral bypass (see Fig. 3.1.5). Typically with blood vessel occlusion and age, the vessel wall deteriorates to such a degree that surgical removal of the atheromas is impossible and the diseased artery segment has to be bypassed. If a patient's sapheneous veins are not available as bypass tubes, synthetic grafts (i.e., ePTFE or Dacron) are used. Unfortunately, at the distal end where the graft meets the artery (see Fig. 3.1.5) restenosis may develop, especially in the toe region. Clearly, aggravating particle-hemodynamics factors (see Sects. 3.1.1 -3.1.4) play a significant role in restenosis. Thus, minimizing such wall parameters (see Fig. 3.1.4) and thereby generating amiable graft-to-artery geometries produces smooth laminar flow fields in the critical junction regions and, as a result, high patency rates.

Because the presence of the critical particles, i.e., a dilute suspension of micro-size monocytes and platelets, does not influence the blood flow, the particle trajectories can be evaluated after the velocity and pressure fields are known. Hence, as a first step, the flow field subject to suitable initial and boundary conditions was solved using CFX 4.4 (Ansys, Inc.; Canonberg, PA). This validated finite-volume program employs structured meshes where the complex flow domain is divided into somewhat uniform subregions, or blocks (see Figs. 3.1.6a,b).

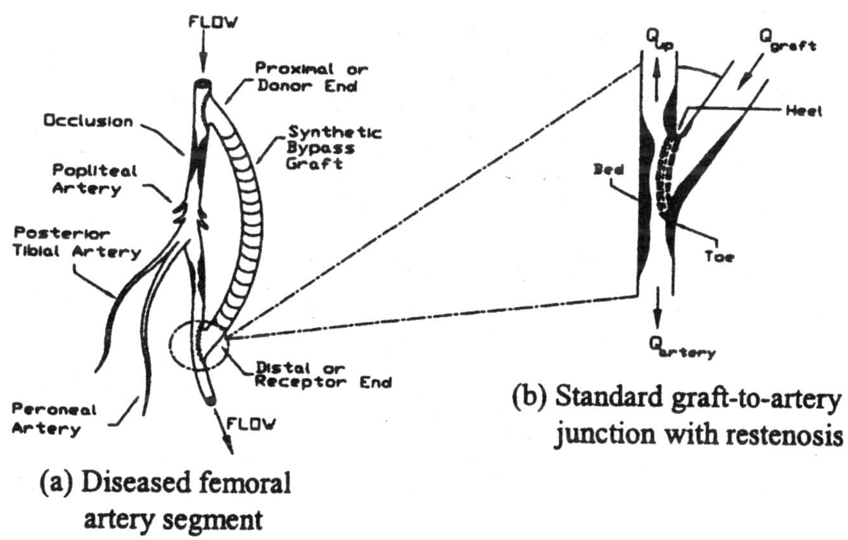

Figure 3.1.5. Example of occlusive diseases in branching blood vessels: (a) Femoral-to-popliteal bypass graft with (b) Restenosis in the region of the distal anastomosis (Kleinstreuer & Longest, 2003).

(a) Mid-plane computational mesh with block boundaries

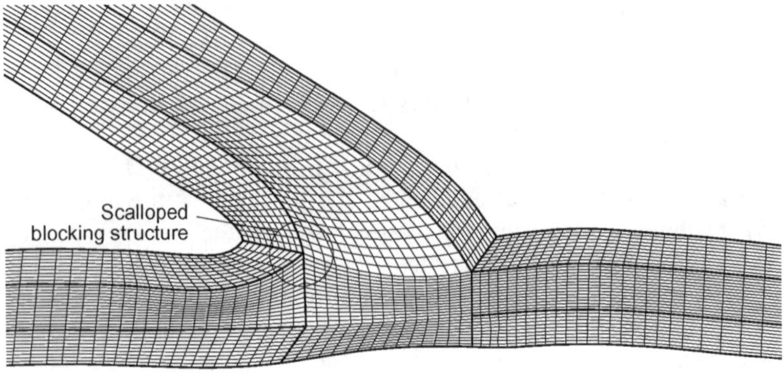

(b) Sample surface mesh

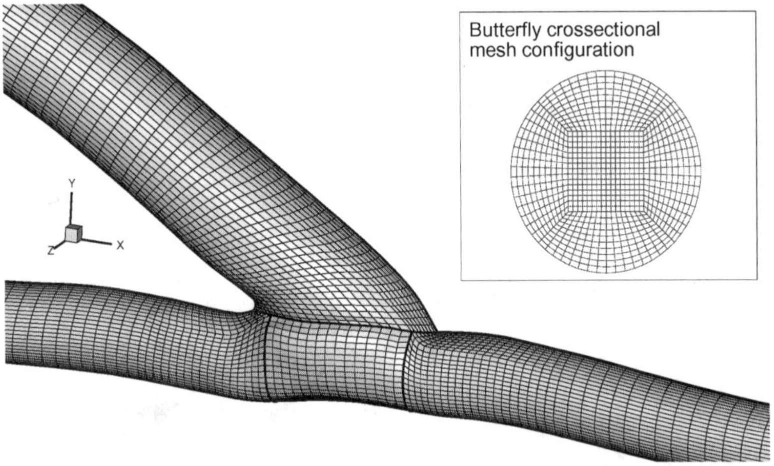

Figure 3.1.6 Computational mid-plane and whole domain mesh for a 1.5:1 graft-to-artery diameter ratio configuration.

3.1.5.1 Computational Fluid-Particle Dynamics Solution

The solution of the time-varying velocity field has been carried out with a validated finite-volume based algorithm (CFX 4.4) with user-Fortran programs added to account for the blood rheology, the input pulse, and the hemodynamic parameters. A structured, multiblock, body-fitted coordinate discretization scheme was employed. Parameters necessary for both temporal and mesh convergence were evaluated for a sample femoral-bypass configuration. These conditions provided a starting point for convergence validation in all subsequent configurations studied.

Solution of governing equations. The finite-volume solution relies on a modified form of the governing equations, i.e.,

$$\frac{\partial \rho \vec{\mathbf{v}}}{\partial t} + \nabla \cdot (\rho \vec{\mathbf{v}}\vec{\mathbf{v}} - \eta \nabla \vec{\mathbf{v}} u) = -\nabla p + \nabla \cdot \left(\eta (\nabla \vec{\mathbf{v}})^T\right) \qquad (3.1.34)$$

Integration of this equation for individual control or finite volumes results in

$$\int \frac{\partial \rho \vec{\mathbf{v}}}{\partial t} dV + \int \rho \vec{\mathbf{v}}\vec{\mathbf{v}} \cdot \vec{n} dA - \int \eta \nabla \vec{\mathbf{v}} \cdot \vec{n} dA = \int S dA \qquad (3.1.35)$$

where the pressure gradient and stress divergence have been included in the source term, S. With the exception of advection, all terms in Eq. (3.1.35) are discretized in space using second-order central differences. The advection terms are discretized using the higher-order upwind (HUW) differencing scheme of Thompson and Wilkes (1982). This upwind scheme was derived by integrating fluxes over the control-volume and is completely conservative and consistent with the control-volume formulation (Shyy et al., 1992). The coefficients of the convective terms were obtained using the Rhie-Chow (Rhie & Chow, 1983) interpolation formula which has been extended to include non-uniform grid corrections. Continuity is enforced by solving a pressure-correction equation to update the pressure and velocity fields. Specifically, the SIMPLEC algorithm of Van Doormal and Raithby (1984) was employed.

A fully implicit backward Euler scheme was used for temporal discretization. Higher-order schemes which allow for larger time steps, such as the Crank-Nicholson method, are available. However, variable time-step size and not the order of a scheme determine the accuracy of the solution. That is, the first-order scheme selected with sufficiently small time steps is capable of the accuracy achieved by higher-order schemes with larger time steps. The fully implicit backward Euler algorithm was chosen to rapidly generate a sufficiently large number of temporal velocity field solutions such that linear interpolation in time could be used for the integration of particle trajectories. The computational cost incurred by implementing this first-order method was reduced by a variable time-step routine for the solution of the velocity field.

The discretized transport equations are solved iteratively. There are two levels of iteration: (1) a linear inner iteration which solves for spatial coupling for a particular variable, and (2) a nonlinear outer iteration which solves for the coupling among variables. The outer iteration is a Picard scheme used to update the nonlinearities of the problem. Each variable is taken in sequence, regarding all other variables as fixed. The coefficients of the discretized equations are always reformed, using the most recently calculated values of the variables, before each inner iteration. For each outer iteration, the set of linearized difference equations is passed to a simultaneous linear equation solver which uses an iterative solution method. An exact solution is not required because this is just one step in the nonlinear outer iteration. With the exception of the pressure correction equation, all variables were solved for using an algebraic multigrid (AMG) method. This method solves the discretized equations on a series of coarsening meshes (cf. Lonsdale, 1993) chosen algebraically. The AMG method takes advantage of the properties of multigrid schemes which can converge in an ideal number of iterations for certain problems (Ferziger and Peric, 1999). However, rather than using nested grids which vary by a factor of two in mesh size to obtain approximate solutions on finer grids, the AMG method performs restriction and prolongation operations on the linearized matrix itself using nonsquare matrices to reduce the problem size and then performs smoothing operations. This "algebraic" method of prolongation and restriction is what gives the algorithm its name as well as its performance boost. The advantage of the AMG scheme is that it reaches a convergent solution monotonically and in fewer total (outer) iterations than the other schemes. The cost is that each iteration is more expensive in both CPU time and memory required, but the overall computational time is reduced. In general, for the transient three-dimensional problems solved, the AMG method has been used.

The typical convergence criterion for hemodynamic simulations is to halt the outer iteration procedure and advance the time-step when the global mass residual has been reduced from its original value by three orders of magnitude, in cgs units. However, Longest (1999) demonstrated that this criterion is often insufficient for variable viscosity conditions. Furthermore, the above convergence criterion often halts a solution process that is significantly reducing the mass residual. Therefore, monitoring the rate of both momentum and mass residual reduction, in addition to mass residual size, is recommended. To monitor convergence rate, a residual reduction factor (*rrf*) can be defined as

$$rrf = \frac{\text{avg. residual reduction over the last 5 outer iterations}}{\text{current residual magnitude}} \tag{3.1.36}$$

The inverse of the *rrf* indicates the number of outer iterations that would be required to drive the residual to zero, at the current rate of reduction. Appropriate values that resulted in a sufficiently convergent solution without excessive outer iterations were found to be $rrf_{mass} = 0.01$ and $rrf_{mom} = 0.06$. The composite convergence criterion is then:

(i) mass residual mass flow rate $\times 10^{-3}$

(ii) $rrf_{mass} \leq 0.01$

(iii) $rrf_{mom} = 0.06$

The mean mass flow rate divided by a factor of two was used in condition I when the time varying mass flow rate dropped below this value. To ensure that a converged solution had been reached, the above factors were reduced by an order of magnitude and results were compared. The stricter convergence criteria produced a negligible effect on both velocity and wall shear stress fields.

An adaptive time-stepping routine was implemented to maintain a highly efficient routine throughout the widely varying input pulse. Time-step size selection was based on the convergence rate of the previous time-step. Time-step size was either increased or decreased to maintain the total number of outer iterations around 75. This is a reasonable number of iterations considering typically under-relaxation factors for the AMG solver were 0.6 for momentum and 0.25 for viscosity. The resulting time-step size varied from 0.001 to 0.01 s and was found to sufficiently resolve Lagrangian particle tracks.

Blood particle properties and trajectory simulations. Human blood is a concentrated suspension of various types and shapes of deformable interacting cells in a complex aqueous solution. The platelets are very small oblate spheroids, or discoids (White, 1994) and constitute less than 1/800th of the cellular volume. Monocytes are a subclass of leukocytes, or white blood cells, that are also a dilute cellular constituent and are generally spherical in shape. More important to blood rheology are red cells which occupy about 45% of the blood volume and are capable of deforming their shape, in 0.006 seconds, under the influence of shear. The cellular contents of blood (including red cells, white cells, and platelets) are immersed in plasma, which displays Newtonian characteristics. Specific details of blood components are given in Table 3.1.3, repeated here for the reader's convenience. It is the red cells which, at the macroscopic level, determine the flow properties of blood, and which, at the microscopic level, determine the motions of the platelets and white cells.

Based on a dimensional analysis of blood particles, it has been shown that the interparticle hydrodynamic effects are significant and the effect of the red blood cells on the flow field cannot be ignored. However, due to a low Stokes number, the two-phase flow can be considered a multi-component mixture (homogeneous flow) with modified properties (see Kleinstreuer, 2003). Specifically, blood in sufficiently large vessels is typically modeled as a multi-component flow with a constant density influenced by components of the mixture and a variable viscosity dependent on shear rate and the mean hematocrit. The mixture density of whole blood is simply the sum of the bulk densities of the constituents. The viscosity of blood will be modeled with the semi-empirical Quemada formulation (Cokelet, 1987) which captures the red blood cell based shear thinning nature (cf. Eq. (3.1.2c)).

The low concentration and low Stokes number of monocytes and platelets allows for the motion of these cells to be computed in a one-way coupled manner, i.e., these particles do not affect flow field momentum. However, the continued collisions and hydrodynamic interactions with red blood cells results in a persistent lateral motion of platelets and monocytes.

Table 3.1.3 Properties of Blood Constituents.

Material	Red Blood Cell	Platelet	Leukoytes (all)	Plasma	Whole Blood
Density (g/cm^3)	1.09	1.069†	1.07-1.09	1.03	1.054
Viscosity ($dyn \cdot s/cm^2$)	--	--	--	0.014*	0.0309 **
Blood cell count ($\#/cm^3$)	5×10^9	3×10^8	7.0×10^6 (total)	--	--
Volume (μm^3)	88	5.17†,††	460	--	--
Size (μm)	7.7×2.8	3.0††	9.5	--	--
Volume fraction	0.42 – 0.46	1.9×10^{-3}	1.2×10^{-3} (total)	--	--
τ_p (s)	1.45×10^{-6}	1.12×10^{-6}	3.35×10^{-6}♦	--	--
A= ρ_p / ρ_p	0.967	0.983	0.964♦	--	--

† Representative mean value for unactivated platelet (Corash et al., 1977)
†† Platelet dimensions vary considerably
* at 37°C (may vary from 0.012 to 0.018)
** Newtonian limit at H = 40%
♦ Value for monocytes

In general, blood particle transport is characterized by highly pulsatile flow and large discrepancies in luminal and near-wall time scales. Hence, it was found necessary to adopt a highly flexible step-size adaptive integration routine. Particle tracking algorithms that are coupled to commercial flow field solvers often limit user access to step-size control parameters, if active step-size control is even implemented. Furthermore, commercial routines solve for the transient flow field as the particle trajectory is progressed; i.e., transient flow field solutions are usually not stored. Particles that remain in the flow domain for multiple pulse cycles and/or successive simulations within a single domain result in hundreds or even thousands of unnecessary flow field solution steps. To overcome the limitations of available commercial software, blood particle simulations have been computed in an effective 'off-line' manner using a separate Fortran 90 (f90) routine. This routine stores all velocity and geometry data in array format for all time-steps of one complete pulse cycle. It then repeatedly accesses and interpolates velocity field data for the calculation of particle trajectories over multiple pulses, in this one-way coupled formulation. Furthermore, the off-line f90 code provides full access to all adaptive step-size control parameters, allows for the specification of near-wall approximations, and has been effectively parallelized. The resulting algorithm is capable of computing 50,000 particles within a geometry of 90,000 control-volumes for approximately 10 pulse cycles within 22.5 hours, compared to approximately 200 hours required by typical commercial CFD packages. Furthermore, once the flow field solution has been generated and stored in array format, each additional group of 50,000 particles may be computed in 2.5 hours. The drawback of the f90 algorithm is the massive amount of storage required to accommodate the transient flow field solution.

3.1.5.2 Model Validation

Validation of the particle tracking algorithm applied in the vessel lumen away from wall boundaries has been established by comparison to the annular expansion results of Karino and Goldsmith (1977). The motion of an elliptical hardened red blood cell in an annular expansion under sinusoidal flow for a specific initialization position was reported as illustrated in the upper panels of Fig. 3.1.7. The lower panels of this figure illustrate the comparable numeric result, including drag and pressure gradient effects in the equation for particle motion. Indeed, the agreement is quantitatively very good (see Example 1.17 for details).

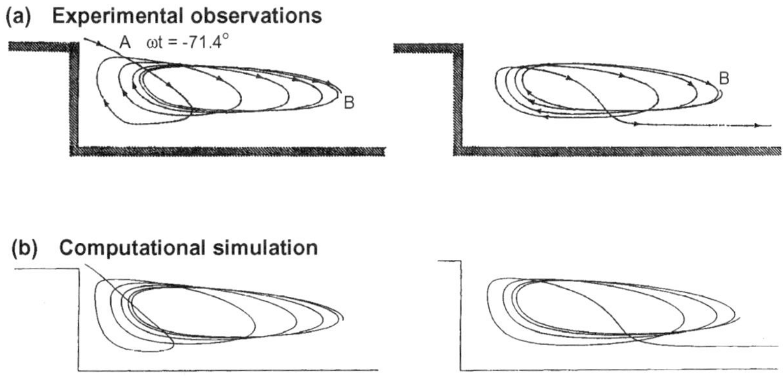

Figure 3.1.7 Comparison of: (a) experimental observation of a red blood cell trajectory (Karino and Goldsmith, 1977) (viewed in two stages); and (b) computational simulation of an idealized spherical particle trajectory.

The near-wall terms in the equation of particle motion (see Eq. (3.1.13)) are grounded in analytic derivations. Hence, their inclusion is fundamentally valid. Specifically, the interaction of blood particles with a responsive vascular surface is dependent on a variety of complex physico-biological mechanisms including particle pseudo-pod extension, receptor-ligand molecular binding, and active surface response. Hence, experimental studies reporting near-wall particle interaction and particle deposition cannot be used to evaluate the validity of the near-wall forces without the inclusion of a particle-wall deposition or wall-interaction model. For these reasons, the probabilistic NWRT model indicating possible particle deposition has been proposed (see Eq.(3.1.33)). Validation and applications of this model, which is inseparably coupled to the validity of the appropriate near-wall expression, are given in Longest et al. (2003, 2004).

3.1.5.3 Results for the Distal End-to-Side Femoral Bypass

Atherosclerosis may significantly occlude the femoral and popliteal arteries, more distal arteries, e.g., tibial and peroneal, or a combination of these vessels. In such cases, vascular bypass grafting is often the favored surgical option to restore blood flow to the lower extremities. Peripheral bypass grafts typically originate at the femoral artery (Fig. 3.1.8) with a proximal end-to-side anastomosis. In cases where the primary occlusive development resides in the lower femoral artery, the bypass typically terminates at the popliteal artery with an end-to-side distal anastomosis either above or below the knee, i.e., the femoropopliteal bypass. In cases where occlusive developments are more extensive, the distal bypass anastomosis may be positioned at the tibial or peroneal arteries, or at multiple sites. The use of an end-to-side anastomotic configuration is generally preferred for peripheral bypass grafting as it provides blood flow to terminal branches both proximal and distal to the anastomosis.

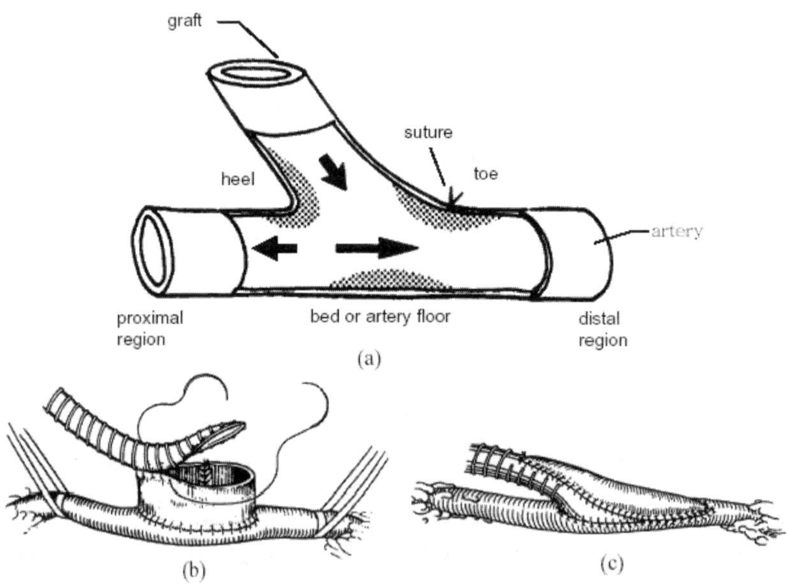

Figure 3.1.8 (a) Illustration of a conventional end-to-side distal anastomosis. Arrows indicate direction of blood flow and shading depicts typical sites of hyperplasia development (Sottiurai, 1999); (b) a vein-cuff (or Miller cuff) has been constructed between the synthetic graft and artery to potentially reduce DAIH (Miller et al., 1984); (c) A vein-patch (Taylor patch) has been constructed (Taylor et al., 1987).

Subsequent late-stage graft failures are predominately due to restenosis resulting from intimal hyperplasia and/or thrombosis at the distal anastomosis, i.e., distal anastomotic intimal hyperplasia (DAIH). Hyperplasia formation (see Sect. 3.1.2) has been reported to localize along the anastomotic suture-line, particularly at the graft toe and heel, and along the arterial floor opposite the graft as illustrated in Fig. 3.1.9 (Sottiurai et al., 1983; Sottiurai, 1999; Keynton et al., 2001; Loth et al., 2002). A common bypass failure scenario consists of a moderately occluded graft-to-artery anastomosis. Thrombosis (platelet adhesion, cell aggregation, and coagulation) may then occur in the region of constricted flow, resulting in sudden vessel occlusion.

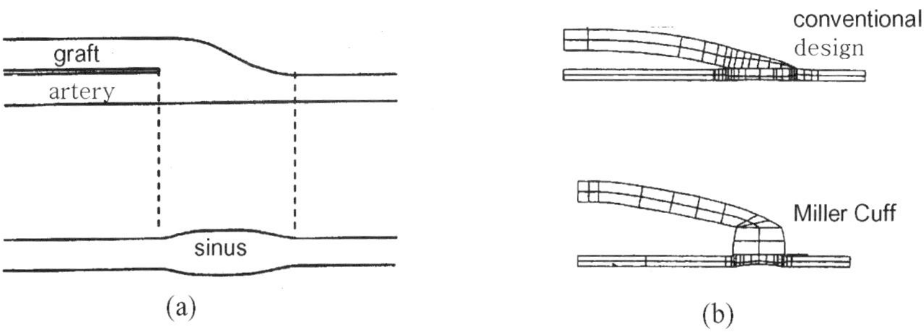

Figure 3.1.9 Realistic in vitro flow models from casts. (a) Model from canine iliofermoral grafts implemented by a variety of researchers, e.g., White et al. (1993) and Loth et al. (1997); (b) Models used by How et al. (2000) including a raised arterial floor in the region of the anastomosis.

While the pathological mechanisms responsible for the development of DAIH have not been fully elucidated, certain hemodynamic parameters have been qualitatively linked to stenotic development in arterial bypasses and other branching blood vessel configurations, both macroscopically and at the cellular level. These parameters include low mean and oscillatory wall shear stress, large spatial and temporal gradients in wall shear stress magnitude, and gradients in wall shear stress vector direction (see Eqs. (3.1.24)-(3.1.33) and Kleinstreuer et al., 2001). Other factors that have been implicated with DAIH formation include excessive intramural wall stress, graft-to-vessel compliance mismatch, blood particle stasis, and blood particle deposition (Kleinstreuer et al., 2001; Sottiurai, 1999; Ku, 1997).

To minimize DAIH occurrence, native saphenous vein is typically the conduit of choice for arterial bypass graft construction. Resulting long-term success (i.e., patency) rates have been reported as high as 80% at two years (Dalman and Taylor, 1993) and 49% at 4 years (Veith et al., 1986). However, the use of autologous saphenous vein is not an option in as many as 60% of patients requiring a lower-limb bypass (Pappas, 1998). In such cases, expanded polytetrafluoroethylene (ePTFE) conduits are typically implemented, which result in success rates similar to saphenous vein for applications where the distal anastomosis resides above the knee (Veith et al., 1986). For below-knee distal positioning, PTFE grafts typically perform poorly, with four-year patency rates of approximately 12%. To improve the success rates of below-knee applications, inter-positioned vein cuffs (Miller et al., 1984) and vein patches (Taylor et al., 1992) have been suggested and continue to show promise in clinical studies. For instance, Stonebridge et al. (1997) found a 52% success rate at 24 months for below-knee popliteal bypass grafts with inter-positioned vein cuffs (see Fig. 3.1.8). As surveyed by Longest (2004), the geometric, biofluids, and biomechanics parameters of interest in bypass graft design include:

- Graft-to-artery diameter ratio and graft angle, where, typically, $d_{graft} > 1.5 d_{artery}$ and $5° \leq \Theta \leq 45°$
- Inlet curvature, especially in the toe region
- Anastomosis shape and length
- Input waveform and maximum, mean, and minimum Reynolds numbers
- Outlet flow division, i.e., percentage of retrograde flow
- Graft-artery wall compliance and distensibilities.

The widely accepted view that hemodynamic factors initiate a biophysical cascade that is ultimately responsible for distal anastomotic intimal hyperplasia (DAIH) suggests that geometric modifications may significantly improve distal anastomotic performance. Considering the problematic below-the-knee distal femoral anastomosis clinical evidence indicates that inter-positioned vein cuffs and vein patches significantly extend graft function. A number of potential surgical, hemodynamic, and biological mechanisms have been associated with the improved patency of vein-supplemented designs (see Noori et al., 1999; Leuprecht et al., 2002); however, no single factor appears responsible. While clinical results appear promising, wide-spread implementation of vein-supplemented designs has been limited by the surgical rigor associated with constructing a cuff or patch as well as the presumed increased risk of surgical complications and technical errors. In order to expedite the surgical procedure and further mitigate inciting hemodynamic factors, several researchers have suggested the implementation of expanded grafts (Lei et al., 1996 & 1997; Harris and How, 1999; Longest and Kleinstreuer, 2000; Longest et al., 2000; cf. Fig. 3.1.4). As a result, expanded graft-end configurations are now commercially available (Fisher et al., 2002).

In an effort to improve distal anastomotic performance, Lei et al. (1996 & 1997) showed that large anastomotic flow areas, small continuously changing bifurcations angles, and smooth junction wall curvatures reduced hemodynamic wall parameters including the WSSG and OSI. They suggested an 'optimized' S-shaped configuration intended to minimize adverse 'disturbed flow' characteristics associated with DAIH (Fig. 3.10a). Following the lead of Lei et al. (1996), Harris and How (1999) proposed a pre-shaped PTFE cuff intended to eliminate the need for vein collar construction. While the design objective of Lei et al. (1996 &

1997) was to reduce disturbed flow patterns in the junction region, the pre-shaped cuff design of Harris and How (1999) was intended to increase vortical flow and wall shear stress (Figure 3.1.10c). Harris and How (1999) claimed the pre-shaped cuff eliminates regions of low WSS; however, no quantitative details in support of this argument have been provided. Longest and Kleinstreuer (2000) linearly combined several WSS-based hemodynamic wall parameters into one 'severity parameter' for analysis of an arteriovenous (AV) bypass configuration. Based on severity parameter mitigation, an 'optimized' AV distal anastomotic configuration was proposed (Fig. 3.1.10b). Indeed, clinical performance of anastomotic configurations designed to hemodynamically mitigate restenosis depends largely on the appropriateness of the wall parameters analyzed. While the pre-expanded designs of Lei et al. (1996) and Longest and Kleinstreuer (2000) successfully mitigated the hemodynamic parameters considered, e.g., the WSSG, recent findings strongly indicate multiple pathways for DAIH formation. For instance, expanded designs significantly reduce changes in WSS magnitude and direction by providing larger anastomotic areas to accommodate flow redirection. A downside of this strategy is lower WSS-values, which has been widely associated with DAIH development. Results of the current study indicate that significant interactions between critical blood particles and the vascular surface provide an extremely aggressive pathway for DAIH development. Furthermore, regions of low WSS and lumenal vortical patterns cannot effectively determine regions of micro-scale particle-wall interactions including adhesion. Therefore, analysis of WSS conditions and 'persistent vortex' formations may result in an insufficient model for DAIH formation as well as for graft design. Significant hood curvatures, which characterize the pre-expanded configurations (Fig. 3.1.9), are expected to result in elevated particle-wall interactions along the graft surface in the region of the anastomosis. For example, a low-angle configuration can be characterized by a continuous and relatively smooth hood curvature. Nevertheless, significant particle-wall interactions were observed along such a graft surface, which were consistent with the DAIH observations of Loth et al. (2002). Considering particle-wall interactions as a potential mechanism for DAIH formation, alternative anastomotic designs may be necessary. Thus, Longest & Kleinstreuer (2003) showed that the hemodynamic factor, expected to most aggressively elicit a localized hyperplasic response within the distal femoral anastomosis, is the composite NWRT model for platelets including surface reactivity and platelet activation conditions. Monocyte interactions with the vascular surface in regions of low-WSS also have the potential to incite DAIH formations. As supported by a number of studies at the cellular level, endothelial response to regions of significantly low WSS are considered a second pathway for DAIH development. Highly focal regions of significantly elevated WSSG and WSSAG values provide a third inciting mechanism for localized IH formation. Factors such as compliance mismatch, intramural stress and strain, surgical injury, and technical errors were not directly assessed in this study.

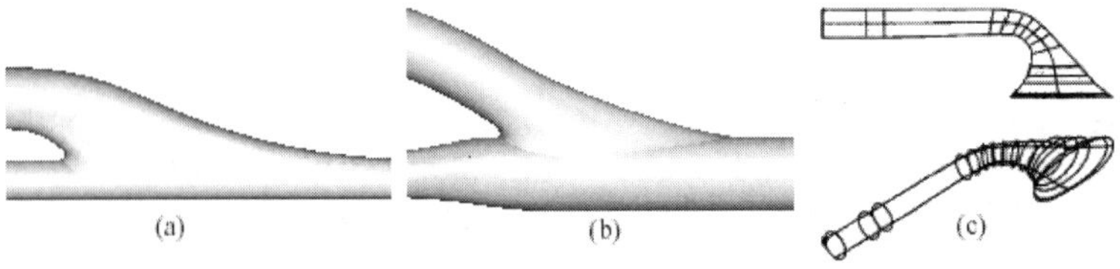

Figure 3.1.10 Expanded graft-end configurations intended to reduce DAIH formation: (a) S-shaped connector suggested by Lei et al. (1996, 1997); (b) 'new graft-end' design suggested by Longest and Kleinstreuer (2000) for arteriovenous access; (c) Distaflo™ graft proposed by Harris and How (1999) and marketed by IMPRA-Bard.

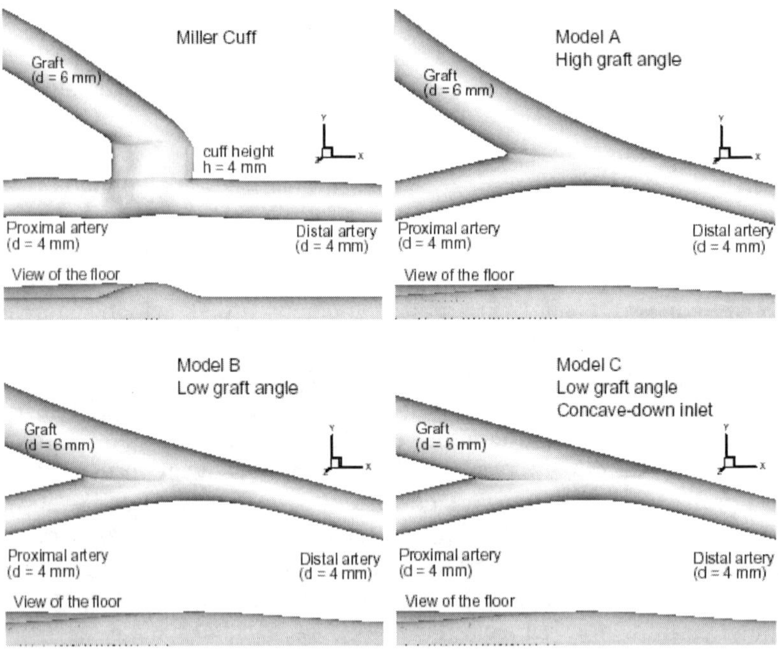

Figure 3.1.11 Geometric surface models including the currently implemented Miller cuff as well as virtually prototyped anastomotic configurations.

3.1.5.4 Novel System Designs and Discussion

Of interest are realistic and virtually prototyped distal anastomoses intended to reduce DAIH formations. Models include the clinically viable Miller cuff, as well as smooth unexpanded virtual prototypes with graft-to-artery diameter ratios of 1.5:1 (Fig. 3.1.11a). The Miller configuration has been constructed using a straight 45° graft-end cut and a 4 mm high venous cuff. In contrast, virtually prototyped models (Fig. 3.1.11b) implement a smoothly curved inlet-graft intended to gradually redirect the flow and potentially reduce disturbed hemodynamic conditions, particularly in the immediate junction area. Similarly, curvature of the recipient artery allows for a smooth transition, which is intended to redistribute IH away from regions critical to anastomotic flow delivery. Models A and B are characterized by high- and low-angle 'concave-up' grafts, whereas Model C represents a low-angle 'concave-down' configuration. In contrast to the previously expanded anastomotic configurations, utilizing both proximal and distal graft and arterial curvatures allows for smoothly connected unexpanded junctions intended to mitigate disturbed flow occurrence without reducing critical WSS-values and elevating particle residence times.

A transient Type I input pulse, consistent with post-surgical observations of the femoral bypass, has been implemented for all grafts (cf. Table 3.1.2).

While recent studies have suggested a variety of mechanisms potentially responsible for the improved clinical performance of the Miller cuff, current results indicate that hemodynamic conditions, including particle-wall interactions, may partially account for the redistribution of significant DAIH formations away from the critical arterial region. Nevertheless, the use of an autologous vein cuff is potentially the factor most responsible for the reduced hyperphlasic response in the region of the arterial suture-line (Kissin et al., 2000). Furthermore, the suturing of a vein segment to the artery facilitates the surgical procedure, resulting in a higher likelihood of a technically sound anastomosis.

Considering the virtually prototyped models, anatomic features consistent with venous anastomoses reduced the particle-hemodynamic potential for DAIH formations in locations critical to flow delivery. For instance, a concave-up graft inlet configuration (Fig. 3.1.11b) resulted in elevated particle velocities and WSS-values in the region of the anastomosis, which were observed to significantly reduce occurrences of the composite NWRT models for platelets and monocytes. Significant graft and arterial curvatures, as well as the application of an unexpanded design, reduced and redistributed WSS-based hemodynamic parameters and NWRT occurrence; however, the particle-hemodynamic potential for DAIH formation was not eliminated. Considering the proliferative nature of IH formations, it appears that eventual occlusion of the virtually prototyped configurations is expected.

In conclusion, the application of a multiple-pathway particle-hemodynamic model for IH in distal anastomotic design indicates that occlusive formations are an inevitable consequence of the un-physiological distal end-to-side anastomosis, particularly for the case of proximal outflow. Nevertheless, surgical benefits of the end-to-side distal anastomosis, such as ease of construction and the ability to deliver proximal outflow, ensure its continued implementation until a better alternative is proven. As such, results of this study suggest the implementation of concave-up graft inlets, relatively unexpanded anastomotic configurations, and arterial curvatures which moderate flow redirection. Graft and arterial curvatures can be clinically accommodated by the use of prefabricated external supports. Relatively new graft fasteners, such as vascular staples and clips (Segdi et al., 2001), may potentially reduce suture-line IH. Clinical testing will be necessary to determine if the unexpanded anastomotic design suggested can accommodate moderate IH formations, expected in the immediate junction and lateral wall regions, without significantly altering graft function.

Another important application of computational particle-hemodynamics is the prediction of geometrically optimal, biodegradable scaffolds on which complex arteries can be grown *in vitro*. Coupled to such an analysis is the evaluation of the stress field in those arteries under typical pulsatile loads, in order to prevent possible rupture at a later user stage. A key biomedical engineering example of fluid-structure interaction (FSI) dynamics is given in the next section.

3.2 ANEURYSMS

An aneurysm is an abnormal irreversible outward bulging of an artery, as found in a brain capillary, epicardial (i.e., coronary) artery or, most frequently, the abdominal aorta below the renal arteries (see Fig. 3.2.1). Possible causes include inherited genetic disorder, hypertension, atherosclerosis, life style (e.g., smoking, stress, no regular exercise, diet, etc.) as well as hemodynamic and biomechanical factors. Both brain aneurysms (BAs) and abdominal aortic aneurysms (AAAs) develop often over decades unnoticed. Specifically in about 40% of BA cases, people with unruptured aneurysms will experience peripheral vision deficits, speech complications, loss of balance, short-term memory difficulty, and/or sudden changes in behavior. Only 10% of AAA-patients have symptoms such as abdominal pain, back pain, fainting, and/or vomiting. However, when rupture suddenly occurs, BA-patients may experience the worst headache in their lives, nausea, vomiting, neck pain, blurred vision, eye pain, light sensitivity, loss of sensation, stroke symptoms, and/or death. The male-to-female ratio of ruptured aneurysms leading to death is 4:1 for AAA-patients (over 55) and about the same for BA-patients.

Aneurysms can be detected via angiography (i.e., injected contrast dye plus X-rays), ultrasonography, or computerized tomography (CT) scans. The latter, often augmented with angiography, can detect the aneurysm location and shape with precision. The two treatment options are either major open surgery or minimally-invasive endovascular aneurysm repair (EVAR). In EVAR, a catheter which is a small plastic tube containing an implant is inserted into the patient's femoral leg artery and navigated through the vascular system into the afflicted artery segment where the implant is deployed. For BA-patients, that implant is typically a platinum coil to be deposited into the brain aneurysm to fill out the cavity (Fig.3.2.1a), while for AAA-patients it is a bifurcating stent-graft which forms a new blood vessel and hence shields the weakened AAA-wall from the pulsatile flow (Fig. 3.2.1b).

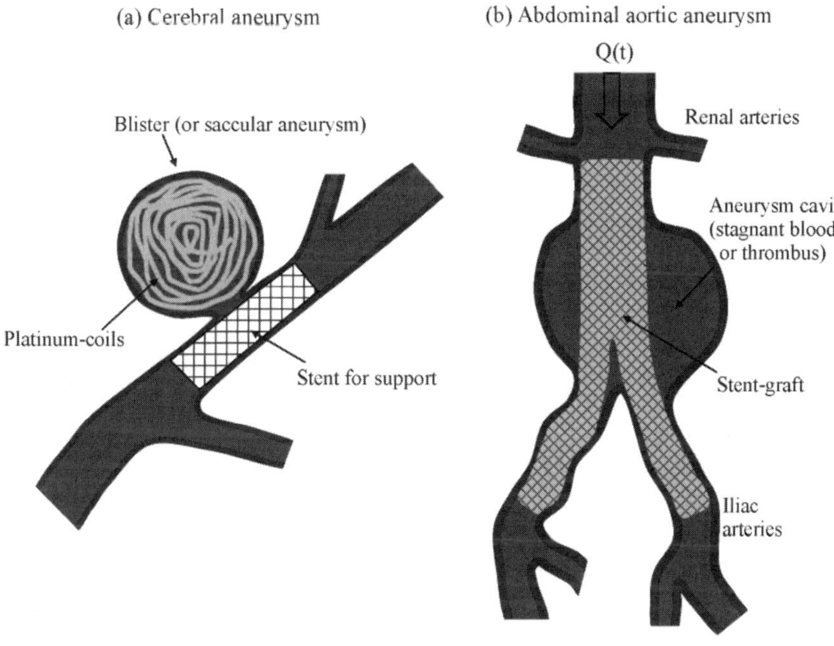

Figure 3.2.1 Schematics of aneurysms: (a) brain and (b) abdominal aorta.

3.2.1 Aortic Aneurysms

The overall mortality caused by sudden AAA-rupture is 90%, making it the 13th leading cause of death in the US, i.e., about 15,000 deaths every year. As mentioned, to prevent such a terminal event either open surgery or endovascular aneurysm repair (EVAR) is recommended when an AAA reaches a critical size, typically twice the aortic neck diameter. Table 3.2.1 compares the pros and cons of these two surgical interventions where the total cost for the modern procedure is much higher than for conventional aneurysm repair.

Table 3.2.1 Comparison between open surgery and EVAR (Greenhalgh et al., 2004).

Open surgery	**EVAR**
Suitable for almost any patient	Only for AAAs with "right anatomy"
Large abdominal incision	Small incision in groin
Average 6-day stay in hospital	Average 2-day stay in hospital
Full recovery after 6 weeks	Full recovery after 2 weeks
Morbidity: 29%	Morbidity: 18%
Blood transfusion: high	Blood transfusion: low
Mortality rate of selective surgery is 3.8%	Mortality rate 1.3 %
No long-term surveillance necessary	Long-term surveillance necessary
Average total cost: $12,500	Average total cost: $20,000

3.2.1.1 Mechanisms of AAA Development

The arterial wall mechanics and integrity are mainly determined by the matrix components of the wall. As alluded to in Sect. 2.2.3 and indicated in Fig. 3.2.2 these are predominantly elastin, collagen, and smooth-muscle cells. The distensible elastin is load-bearing at low pressures and responsible for the elastic recoil of the artery. Collagen is 1000 times stiffer and is load-bearing at high pressures, preventing over-dilatation and rupture of the vessel. Smooth muscle cells (SMCs) have the potential for contraction and relaxation with modulation of the wall mechanics. SMCs seem to be of minor importance in the abdominal aorta. Thus, the collagen-to-elastin ratio is the principal determinant of wall mechanics in the aorta. In healthy arteries, stiffness increases with age as a result of an increase in the collagen-to-elastin ratio in the wall, with men having stiffer arteries than women. Furthermore, these age- and gender-related differences are most pronounced in the aorta which may explain the 4:1 male-to-female ratio of AAA-mortality among the elderly. Thus, changes in composition and structure of the arterial wall will alter the wall mechanics. In any case, the increasing collagen-to-elastin ratio results in increased wall stiffness and decreased tensile strength. Aneurysm rupture occurs when the imposed stress exceeds the tensile strength of the wall. However, failure in the collagen matrix might not be detectable during normal pressure conditions and therefore the risk of vessel fragility cannot be predicted.

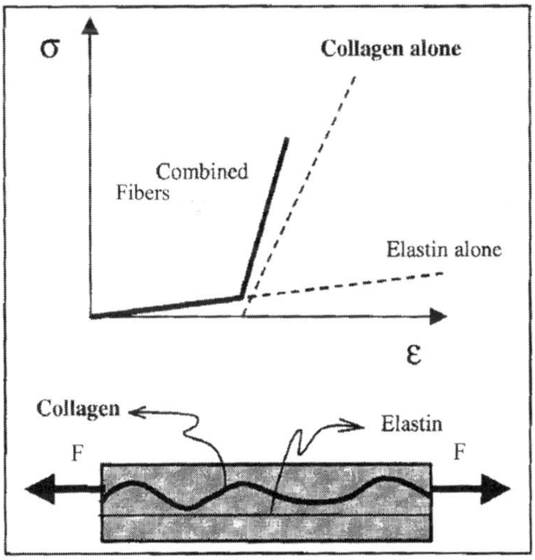

Figure 3.2.2 Comparison of elastin and collagen fibers in the arterial wall (after Raghavan 2002).

The stress-strain effect of the much higher collagen-to-elastin-ratios in AAAs is shown in Fig. 3.2.3, where the σ - ε curve of the AAA moves to the left considerably. The elastic modulus is much higher than that in the normal aorta, while its breaking stress decreases significantly.

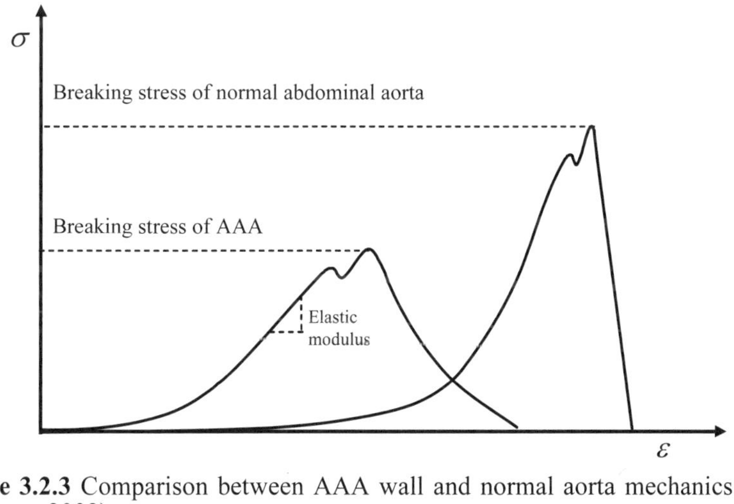

Figure 3.2.3 Comparison between AAA wall and normal aorta mechanics (after Raghavan 2002).

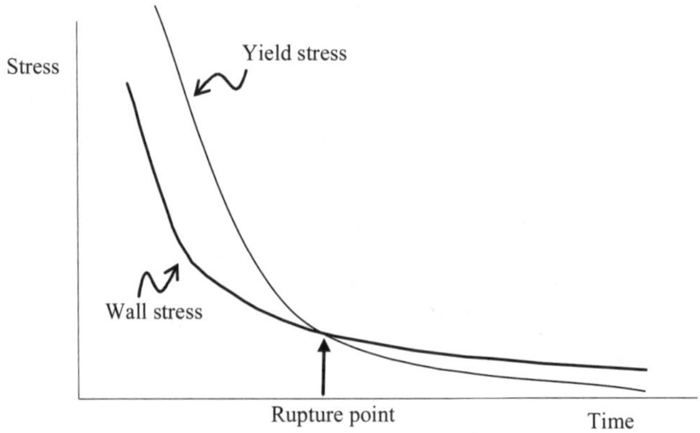

Figure 3.2.4 Effect of yield stress decrease on AAA rupture.

Enhanced wall stiffness is not necessarily advantageous for preventing AAA rupture; because, along with the increase of wall stiffness, the wall yield stress will accordingly decrease. For example, Young's modulus in an AAA wall may reach over four MPa, or almost three times that of the normal arterial wall whereas its yield stress is only 50% of the normal artery. Also, even though stiffness may become large with age, the yield stress of the wall will decrease significantly with respect to age. In summary, while a higher Young's modulus over time may reduce AAA-wall stress, the yield stress is possibly lower than the mechanical stress in the AAA wall, i.e., AAA rupture still may occur as the wall becomes stiffer (see Fig. 3.2.4).

3.2.1.2 AAA-Wall Stress and Rupture

The general consensus is that the peak wall stress is the best indicator of AAA rupture, although the maximum AAA diameter and growth rate as well as aortic neck asymmetry are very important as well and easier to measure. Generally, three stresses, i.e., longitudinal, circumferential, and Von Mises stresses, are considered in AAA rupture analyses. However, because of the complex mechanism of rupture and often complex AAA geometries, which stress actually causes AAA rupture may differ from case to case. Furthermore, how to define the critical values of a threshold stress and yield stress for different patients is rather complex, because stress measurements are not available *in vivo*, and software packages for fluid-structure analyses, such as Ansys-CFX Multiphysics, are not yet widely used (cf. Li & Kleinstreuer, 2005a).

Clearly, once the (local) mechanical stress exceeds the yield stress of the AAA wall, rupture may happen suddenly. Hence, the best and most accurate tool to determine AAA-rupture risk is the maximum wall stress (see Li & Kleinstreuer, 2005b). Computational stress analyses with CAT or

CT-scan based AAA models have been performed by several investigators, as discussed by Li (2005).

3.2.2 Treatment Option: Stent-graft Implants

In general, stents (or endografts) are expandable metal/plastic perforated tubes, wire-meshes, coils, etc. They are delivered in a catheter, i.e., a small plastic tube, through the patient's vascular system starting with the femoral artery, all the way to the afflicted site, e.g., the aortic aneurysm or brain aneurysm. The endovascular surgeon uses real-time X-ray technology, called fluoroscopic imaging, to visualize the patient's vascular system and to deploy the stent accurately.

For a brain aneurysm, which is typically a "blister," a thin platinum wire forming "coils" is deployed to fill out the bubble, called coil embolization or "coiling" (see Fig.3.2.1a). Thus, the blood flow into the aneurysm is blocked, a significant radial pressure drop occurs, and hence rupture of the artery may be prevented.

For an abdominal aortic aneurysm (AAA) a stent-graft (or endovascular graft, EVG) is employed, which consists typically of an NiTi-wire mesh embedded in synthetic graft material such as ePTFE or Dracon (see Fig.3.2.1b and Fig.5.4.6). There are three basic EVG configurations, i.e., simple tubular ones which are also used (without the graft sheath) for scaffolding coronary arteries, bifurcated ones making up the vast majority, and aorto-uni-iliac implants which have to be used when one iliac branch is highly stenosed. Once an EVG inside the catheter reaches its destination, it is deployed either via a self-expanding or balloon-expanding mechanism.

Clearly, if an EVG is correctly placed and no future complications occur, that implant functions like a new blood vessel, i.e., it restores normal blood flow and shields the weakened aneurysm wall from the pulsating pressure. However, major complications may occur because of stent-graft failure which implies improper implantation, material fracture or deterioration, endoleaks, and/or migration, i.e., axial EVG displacement over time. These events are often coupled and each has a multitude of underlying hemodynamics and biomechanics causes. For example, potentially most devastating is EVG migration, which occurs when the downward pulling force exceeds the EVG anchoring (or fixation) force. As a result, the aneurysm cavity may be again exposed to high blood pressure, potentially leading to aneurysm rupture. Thus, it is important to know the maximum migration force exerted by the net momentum change, i.e., pressure, inertia, and friction forces, on a representative EVG. This is a coupled fluid-structure-interaction (FSI) problem which can be solved numerically, given realistic flow input/output boundary conditions, geometries, and material properties.

3.2.3 Stented AAA-model Analysis

In order to demonstrate the benefits of a stent-graft implant, we consider laminar axisymmetric blood flow through a tubular stent-graft (or EVG) interacting with the stagnant blood in the cavity and the aneurysm wall (Fig. 3.2.5a). The EVG is assumed to be a uniform shell and self-expandable with the same diameter (say, 1.7cm) as the inner arterial wall. To simulate the effect of self-expansion and maintain the contact between the EVG and arterial wall, we applied a pre-stress in the neck of the EVG before the simulation began; thus, the EVG neck expands by 15% due to over-sizing together with the arterial wall to assume permanent contact. The maximum EVG fixation force is the key factor affecting migration. To study incipient EVG migration, a maximum migration force F, required to dislodge the EVG from the AAA neck, is calculated for different practical situations. The cavity is filled with stagnant fluid experiencing a time-dependent pressure as a result of the dynamic fluid-structure interactions between EVG, stagnant blood, and AAA wall.

Table 3.2.2 lists the structure parameter values used in the present simulations, of which Young's modulus is a key factor influencing the wall stress. Experimental data indicate that Young's modulus of an AAA is much higher than for a normal artery. Here, Young's modulus is assumed to be linearly changing with AAA diameter, i.e., from 1.2 to 2.7 MPa, as:

$$E = 1.2 + \frac{(2.7 - 1.2)(D_{AAA} - D_{neck})}{D_{AAA,\max} - D_{neck}} \tag{3.2.1}$$

The healthy artery section (neck) is incompressible with a Poisson ratio of 0.49, and the AAA wall is nearly incompressible with a Poisson ratio of 0.45. For a cylindrical NiTi-stent interwoven with graft material, no direct experimental data was available; thus, an equivalent parameter value supported by experiments was employed (Suzuki et al., 2001). The aneurysm wall thickness also plays an important role in wall deformation and stress. Considering the ballooning effect the wall thickness is assumed to be only 1.0-1.5 mm, changing from the neck to the maximum diameter site based on volume conservation.

A typical inflow velocity waveform is shown in Fig. 3.2.5b with a maximum Reynolds number (Re_{max}) of 1950 and average Reynolds number ($Re_{average}$) of 330. For the measured outlet pressure (see Fig. 3.2.5c), the peak and average pressures are 122 mmHg and 98.7 mmHg, respectively (Meter, 2000). The pulse period is chosen as T=1.2 s. The inlet velocity profile is uniform and fully developing in a tube of length ten times the neck diameter. It should be noted that the time-dependent Reynolds number will be somewhat different for the nonstented simulations as the internal diameter of the lumen changes slightly, i.e., $\Delta Re_m / Re_m < 5\%$.

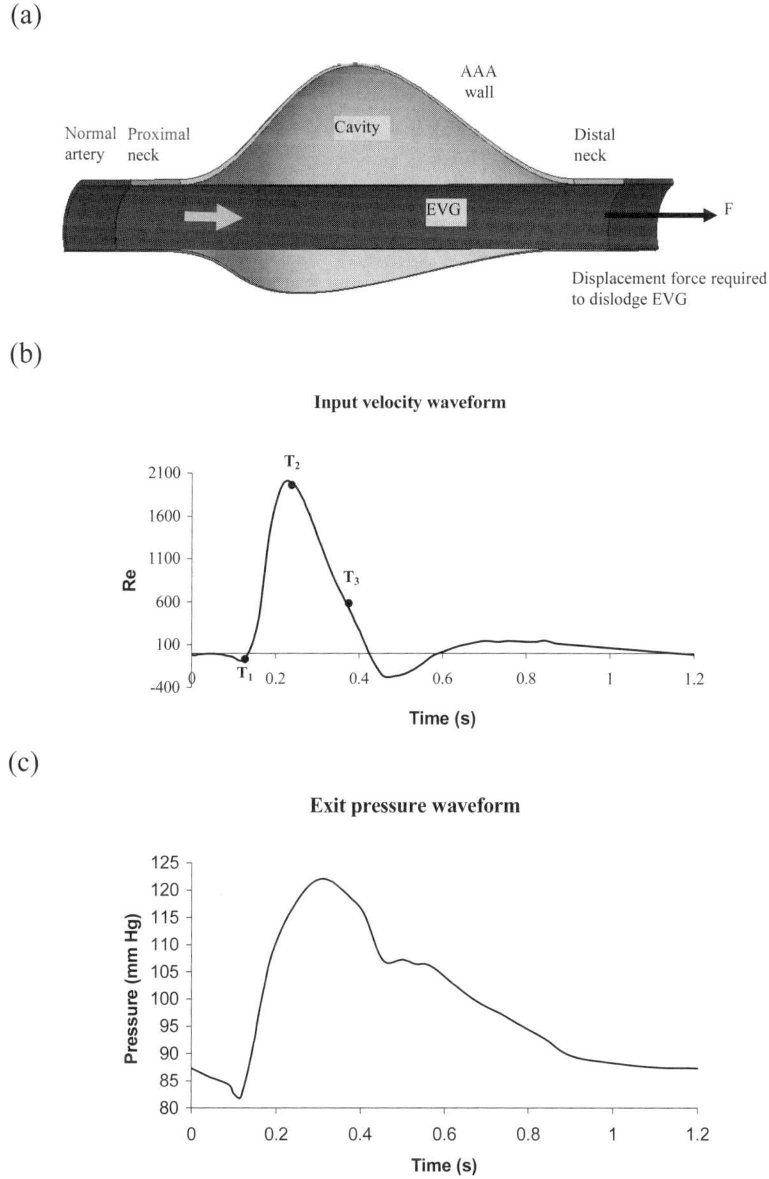

Figure 3.2.5 Abdominal aortic aneurysm model: (a) Schematic of stented AAA; (b) Input velocity waveform; and (c) Output pressure waveform.

Table 3.2.2 Parameters required for the stented AAA simulation.

Normal Artery		AAA Wall		EVG	
Wall thickness	1.5 mm	Wall thickness	1.0-1.5 mm	Equivalent thickness	0.2 mm
Inner diameter	17 mm	Maximum inner diameter	60 mm	Outer diameter	17 mm
Length	Inlet: 30 mm outlet: 30 mm	Length	80 mm	Length	120 mm
Young's modulus	1.0MPa	Young's modulus	1.2-2.7 MPa	Equivalent Young's modulus	19.8MPa
Poisson ratio	0.49	Poisson ratio	0.45	Equivalent Poisson ratio	0.27
Density	1.12 g/cm^3	Density	1.12 g/cm^3	Equivalent density	6.0 g /cm^3

The basic equations in tensor (or comma) notation, following Einstein's repeated index convention, are:

(Continuity) $u_{i.i} = 0$ (3.2.2)

(Momentum)

$$\rho \frac{\partial u_i}{\partial t} + \rho(u_j - \hat{u}_j)u_{i,j} = -p_{i,j} + \tau_{ij,j} + \rho f_i \quad \text{in } {}^F\Omega(t) \quad (3.2.3)$$

(Stress tensor) $\tau_{ij} = \eta \dot{\gamma}_{ij}$ (3.2.4)

(Non-Newtonian fluid model)
$$\eta = \frac{\eta_p}{\left[1 - \frac{1}{2}\left(\frac{k_0 + k_\infty \dot{\gamma}_r^{1/2}}{1 + \dot{\gamma}_r^{1/2}}\right) Ht\right]^2} \quad (3.2.5)$$

where u_i is the velocity vector, p_i is the pressure scalar, ρ is the fluid density, f_i is the body force at time t per unit mass, $\hat{u}_i$ is the wall displacement velocity at time t, ${}^F\Omega(t)$ is the moving spatial domain upon which the fluid is described, $\dot{\gamma}_{ij}$ is the shear rate tensor, η_p=0.014 dyn·sec/cm^2 is the plasma viscosity, $\dot{\gamma}_r = \dot{\gamma}/\dot{\gamma}_c$ is a relative shear rate with $\dot{\gamma}_c$ being defined by a "phenomenological kinetic model" (Buchanan et al., 2000), k_0=4.5862 is the lower limit Quemada viscosity constant,

k_∞=1.2917 is the upper limit Quemada viscosity constant, and Ht=40% is the hematocrit (see Buchanan et al., 2000).

3.2.3.1 Basic Structure Equations

The general governing equations for structure dynamics are:

$$\text{(Momentum)} \qquad \rho a_i = \sigma_{ij,j} + \rho f_i \quad \text{in } {}^S\Omega(t) \tag{3.2.6}$$

where a_i connects fluid flow dynamics with structure mechanics, i.e.,

$$a_i = \frac{d\hat{u}_i}{dt} \tag{3.2.7}$$

$$\text{(Equilibrium condition)} \quad \sigma_{ij} n_i = T_i \quad \text{on } {}^S\Gamma(t) \tag{3.2.8}$$

and

$$\text{(Constitutive)} \qquad \sigma_{ij} = D_{ijkl}\varepsilon_{kl} \quad \text{in } {}^S\Omega(t) \tag{3.2.9}$$

Here, ${}^S\Omega(t)$ is the structure domain at time t, n_i is the outward pointing normal on the wall surface ${}^S\Gamma(t)$, T_i is the surface traction vector at time t, ${}^S\Gamma(t)$ is the boundary of the structure domain, σ_{ij} is the mechanical stress tensor, D_{ijkl} is the Lagrangian elasticity tensor, and ε_{kl} is the strain tensor.

In order to analyze the stress distributions in both the endovascular graft and the arterial wall, the Von Mises stress, used as a material fracture criterion in complicated geometries, is employed. Specially,

$$\sigma_{Von\ Mises} = \frac{\sqrt{2}}{2}\sqrt{(\sigma_1 - \sigma_2)^2 + (\sigma_2 - \sigma_3)^2 + (\sigma_3 - \sigma_1)^2} \tag{3.2.10}$$

where σ_1, σ_2, and σ_3 are the three principal stresses (see Sect.1.5.2).

3.2.3.2 Numerical Method

Now, in order to find the maximum von Mises stresses in the AAA without and with a stent-graft, the coupled set of PDEs has to be solved numerically. For this particular problem, Li & Kleinstreuer (2005a) employed the finite-volume solver CFX 4.4 (Ansys, Inc.) to obtain the blood pressure distributions which, after integration, represent the loads on the walls. The force field due to the blood flow was then connected to ANSYS 7.1 (Ansys, Inc.) to obtain wall movement and wall stress distributions. In turn, wall motion changes the cross sectional area and hence influences the blood flow, i.e., velocity and pressure. Specifically, in order to satisfy mesh and solver compatibility, multinomial Lagrange interpolations for load and deformation between CFX and ANSYS were required. Concerning the flow solver, Lagrange interpolation was used to obtain the new domain boundaries and then the CFX-subroutine "Moving

Grid" was used for re-meshing. Concerning the structure solver, the flow pressure field was automatically interpolated to a pressure load-applying function in ANSYS. For fluid-structure interactions in the AAA cavity, a special finite element, Fuild79, can perform load and deformation transfer automatically. To connect CFX and ANSYS successfully, specific coupling routines were written in FORTRAN for CFX, APDL (ANSYS Parametric Design Language). APDL is a scripting language for ANSYS data and parameter operations, as well as UNIX scripts, so that the simulations with the two software packages could be synchronized. To reach mesh convergence, meshes for both fluid and solid domains were refined until grid independence of the results was achieved. A flow chart of the coupled fluid-structure interaction procedure is given in Fig. 3.2.6. In order to fulfill the FSI convergence between CFX and ANSYS, it needs several loops for each time step.

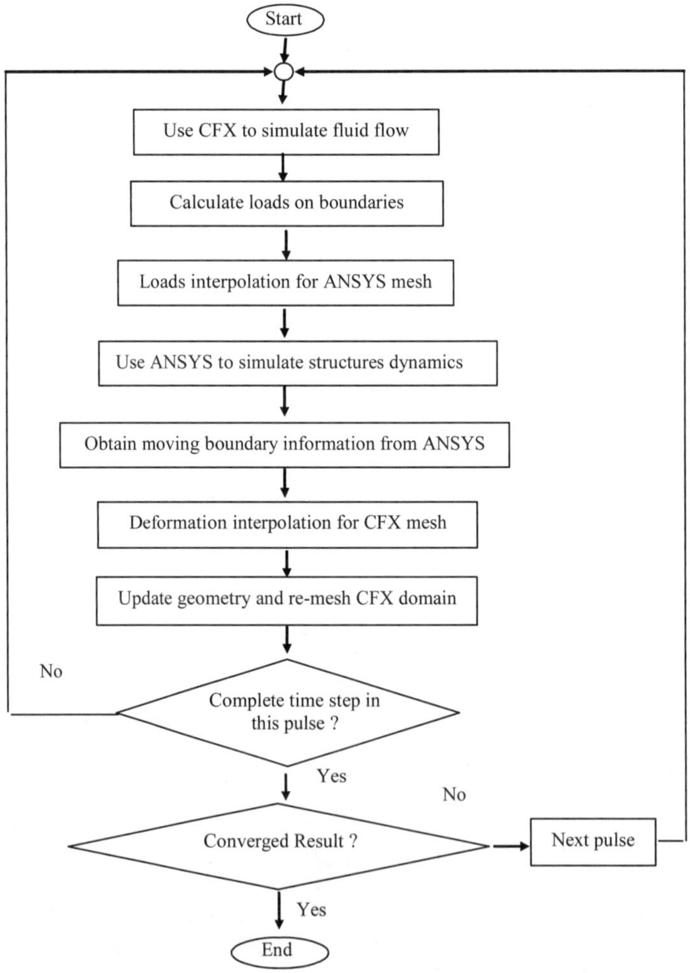

Figure 3.2.6 Flow chart for coupled CFX-ANSYS procedure.

3.2.3.3 Model Validations

In order to test the accuracy of the coupled CFX-ANSYS solver with user-supplied programs to simulate interacting flow and wall variables, several computer model validation studies were performed. For example, theoretical results for elastic cylindrical and spherical walls as well as clinical data, including diameter, pressure, and pulsatile wall motion for stented AAAs, were considered. The small differences between our present simulations and theoretical analyses as well as clinical observations are listed in Tables 3.2.3a and 3.2.3b.

Table 3.2.3a Comparison of results between simulations and theoretical analyses.

<table>
<tr><th colspan="2"></th><th>CFX-ANSYS simulation results</th><th>Theoretical results (Nichols et al., 1990)</th><th>Error</th></tr>
<tr><td rowspan="4">Straight Artery</td><td>Radial deformation (internal wall)</td><td>0.0212cm</td><td>0.0203cm</td><td>4.4%</td></tr>
<tr><td>Circumferential stress (internal wall)</td><td>0.119 MPa</td><td>0.115 MPa</td><td>3.4%</td></tr>
<tr><td>Radial strain (internal wall)</td><td>0.0209</td><td>0.0203</td><td>4.3%</td></tr>
<tr><td>Radial strain (external wall)</td><td>0.0161</td><td>0.0155</td><td>2.8%</td></tr>
<tr><td>AAA</td><td>Maximum circumferential stress</td><td>0.17 MPa</td><td>0.16 MPa (Laplace's equation)</td><td>5.8%</td></tr>
</table>

Table 3.2.3b Comparison of results between simulations and clinical observations.

	CFX-ANSYS Simulation results	**Clinical observations**
Von Mises stress drop	80%	75% (Sonesson et al., 2003)
Diameter decrease	10.2%	10% (Blankensteiju & Prinssen, 2002)
Maximum wall displacement	1.57 mm in non-stented AAA; 0.231 mm in stented AAA.	1 mm in nonstented AAA and 0.2 mm in stented AAA (Malina et al., 1998)

3.2.3.4 Results and Discussion

The results deal with the advantages of an endovascular graft (EVG) protecting the weakened arterial wall (Figs. 3.2.7-3.2.9) and aspects of the EVG-neck interface force in light of possible EVG migration and hence the specter of aneurysm rupture (Fig. 3.2.10). It should be noted that Figs. 3.2.8 and 3.2.9 were generated under pulsatile flow conditions (see Fig. 3.2.5), while for the others equivalent steady flow was assumed with P_{out}=120 mmHg and the associated Re=1200 because at this Reynolds number the peak systolic pressure is achieved. Although the wall-pressure distributions differ between steady and transient flows, the absolute pressure difference between the inlet and outlet for steady and pulsatile flows is very small, when compared to the (cyclic) cardiac pressure. The influence of flow patterns and hence wall stress effects on the structure are not as important as the cyclic pressure changes. It indicates that, under the same cyclic exit pressure, maximum deformation and wall stress values are very similar for steady and transient flows (see Table 3.2.4). Hence, for reasons of computational efficiency, equivalent steady flow was assumed to simulate aspects of the EVG effect and EVG-neck interface force. As mentioned, blood pressure is the main force to cause wall deformation and the exit pressure reaches its maximum value of 120 mmHg when Re = 1200, so that Re = 1200 was selected as a typical case.

Table 3.2.4 Comparison of EVG wall mechanical behavior between transient and steady state flow (Re = 1200).

	Maximum Von Mises stress (Pa)	Maximum deformation (mm)
Transient	6.80E+05	0.2225
Steady	6.77E+05	0.2216

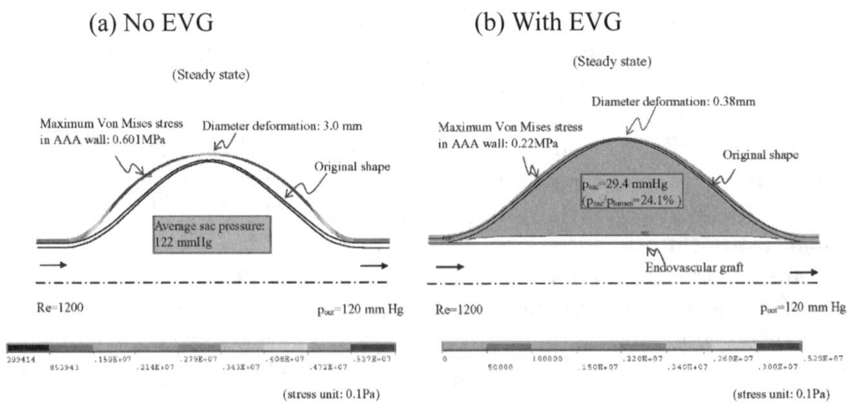

Figure 3.2.7 Comparison between stented and nonstented AAA.

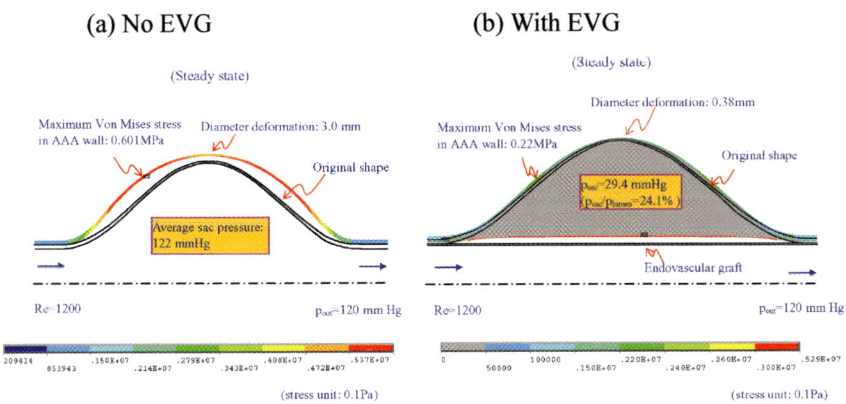

Fig. 3.2.7 Comparison between stented and nonstented AAA.

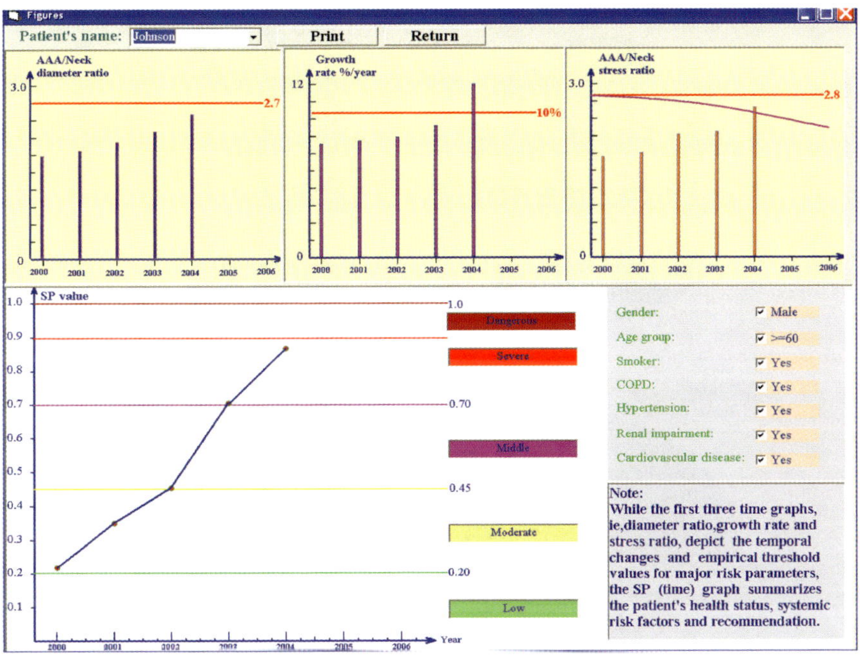

Fig. 3.3.9 Hypothetical example with time-dependent wall-stress threshold (Kleinstreuer & Li, 2005 and Kleinstreuer & Li, 2006).

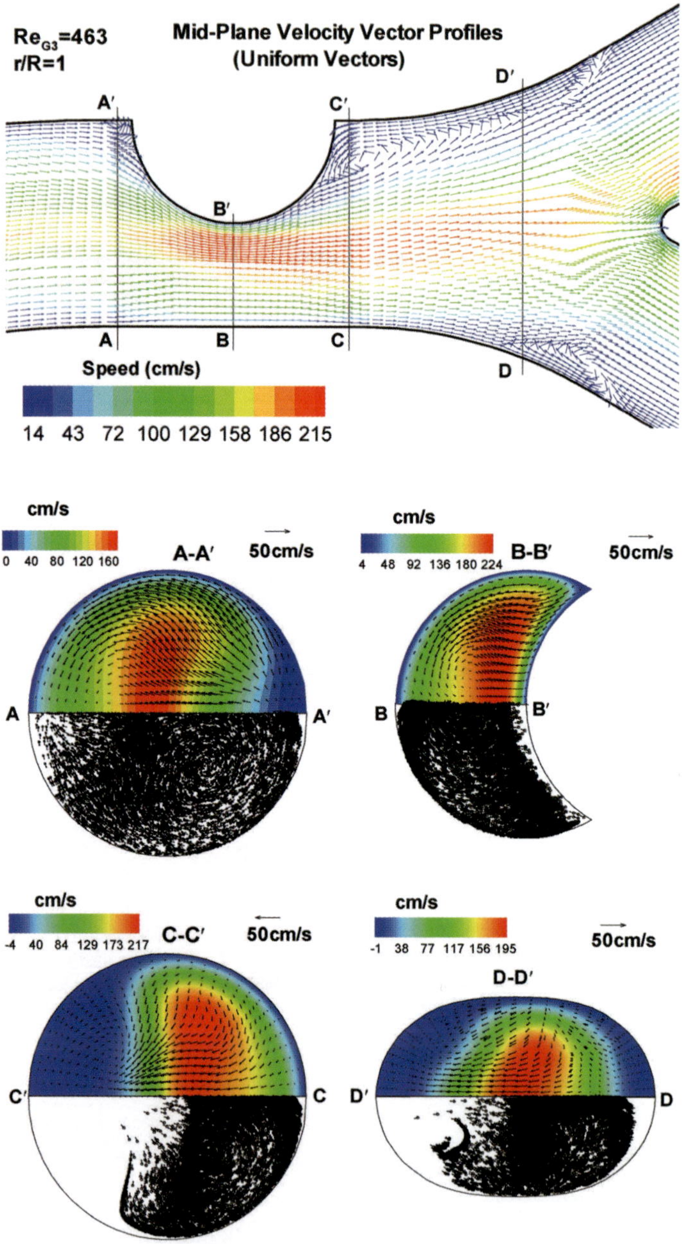

Fig. 4.1.6 Mid-plane velocity vectors, axial velocity contours, and secondary vectors (upper half of cross-sectional views) as well as particle distributions and flow directions (lower half of cross-sectional views) at selected cross sections near the tumor with r/R = 1.0 at Re_{G3} = 463 and St_{G3} = 0.12.

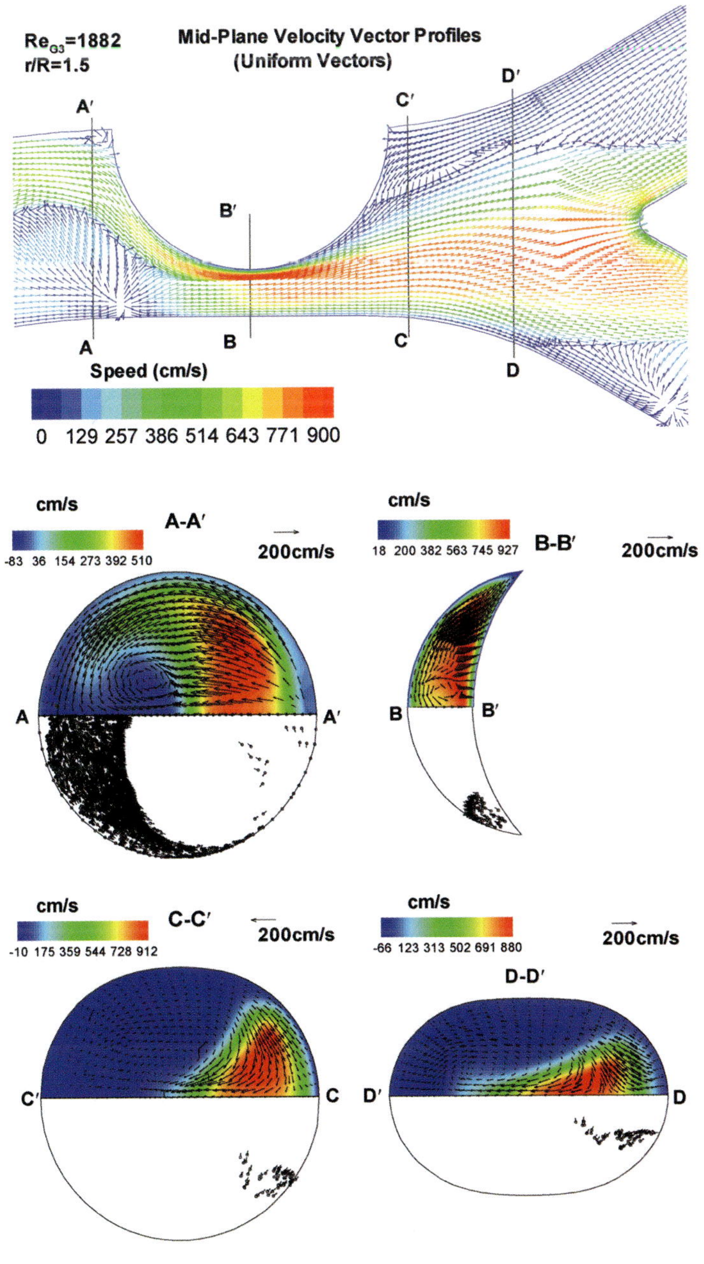

Fig. 4.1.7 Mid-plane velocity vectors, axial velocity contours, and secondary vectors (upper half of cross-sectional views) as well as particle distributions and flow directions (lower half of cross-sectional views) at selected cross sections near the tumor with r/R = 1.5 at Re_{G3} = 1882 and St_{G3} = 0.12.

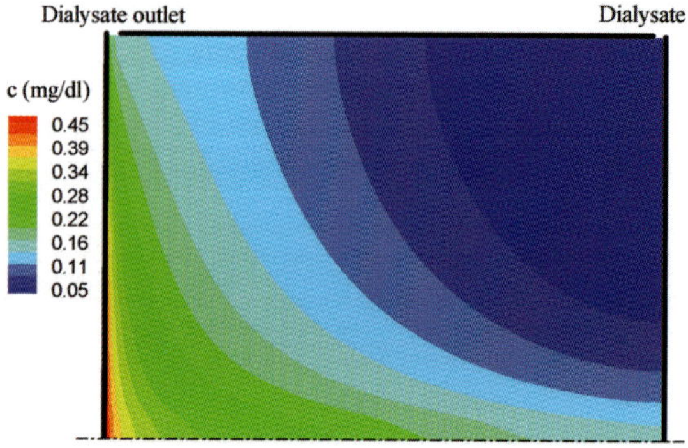

Fig. 4.2.8 Distribution of urea concentration in the shell side of an artificial kidney model.

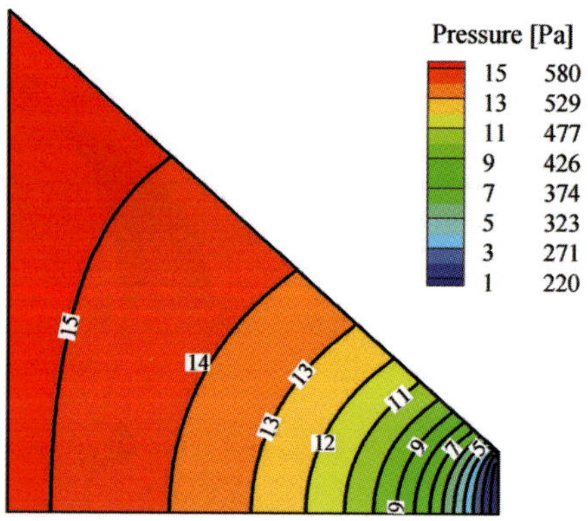

Fig. 4.3.3 Pressure distribution in liver acinus.

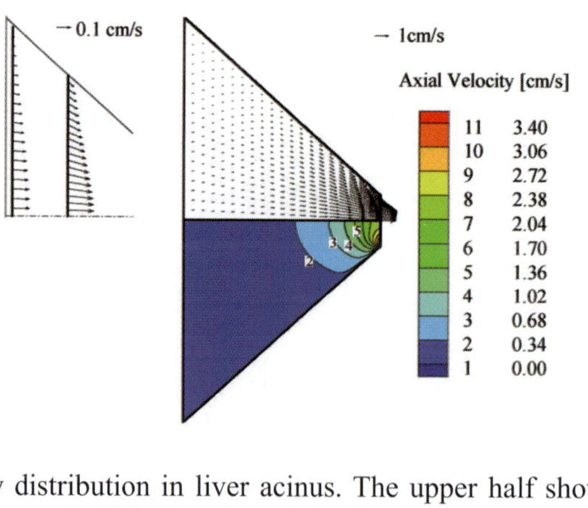

Fig. 4.3.4 Velocity distribution in liver acinus. The upper half shows the velocity vectors while the lower half depicts the axial velocity contours.

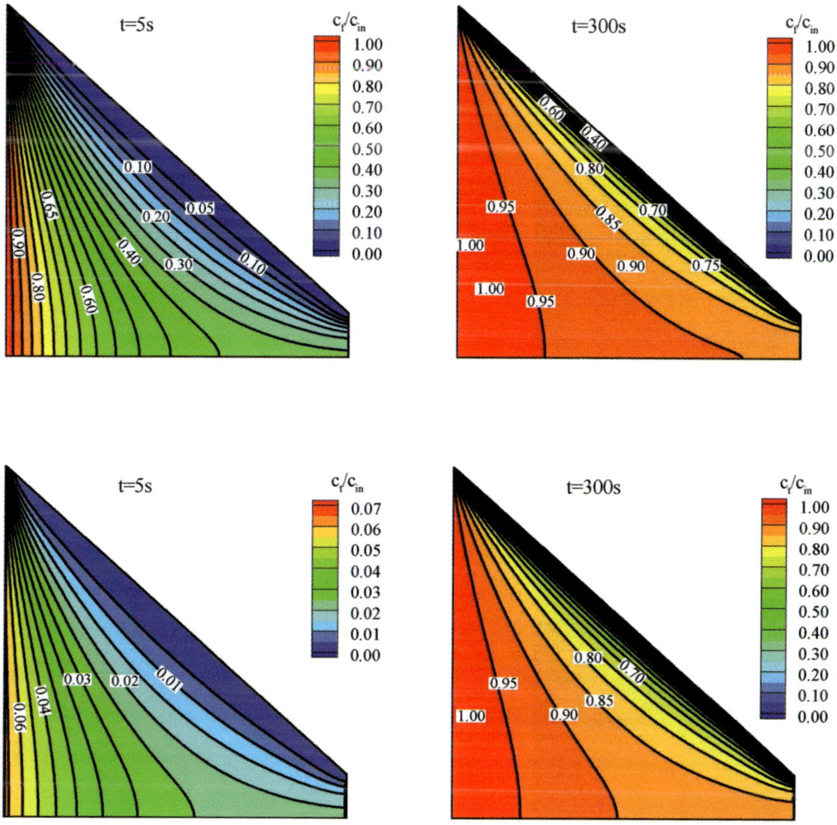

Fig. 4.3.5 Distributions of species concentration in the fluid and tissue for a step change in the inlet concentration at t = 5s and t = 300s.

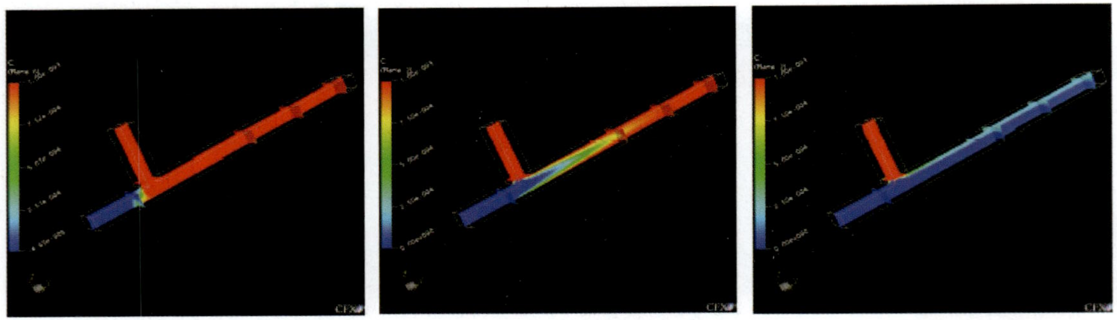

(a) (b) (c)

Fig. 5.2.7 Snapshots of concentration distributions during purging process: (a) 0.001 [sec]; (b) 0.017 [sec]; (c) 0.041 [sec].

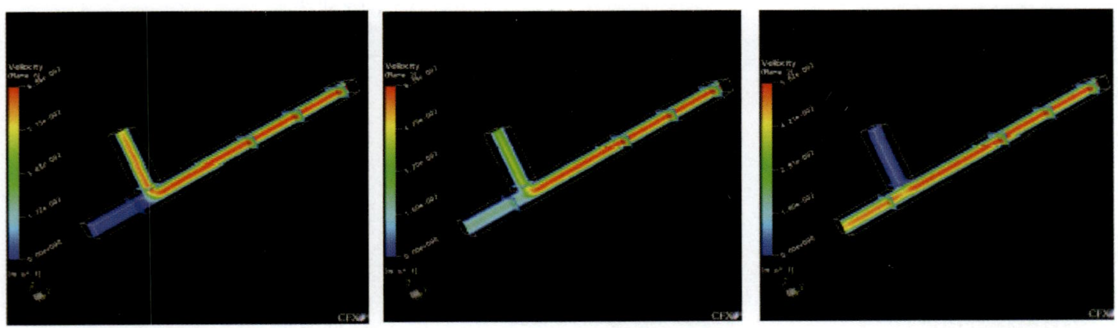

(a) (b) (c)

Fig. 5.2.8 The snapshots of velocity fields during purging process: (a) 0.001 [sec]; (b) 0.003 [sec]; (c) 0.009 [sec].

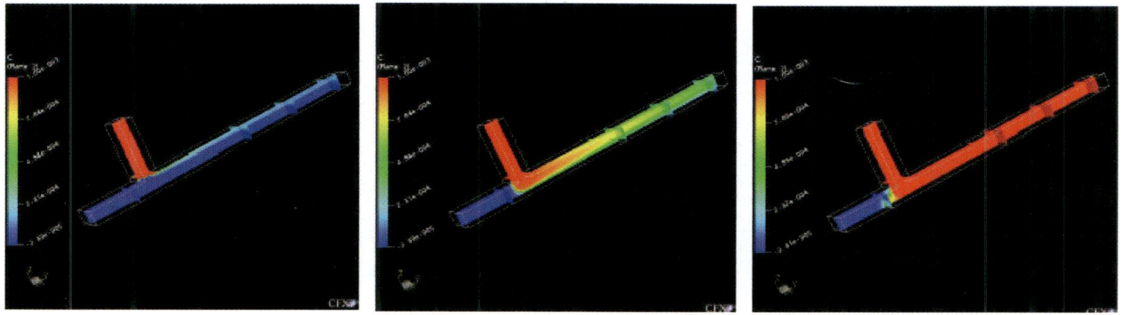

(a) (b) (c)

Fig. 5.2.9 The snapshots of concentration distributions during drug supply process: (a) 0 [sec]; (b) 0.05 [sec]; (c) 1 [sec].

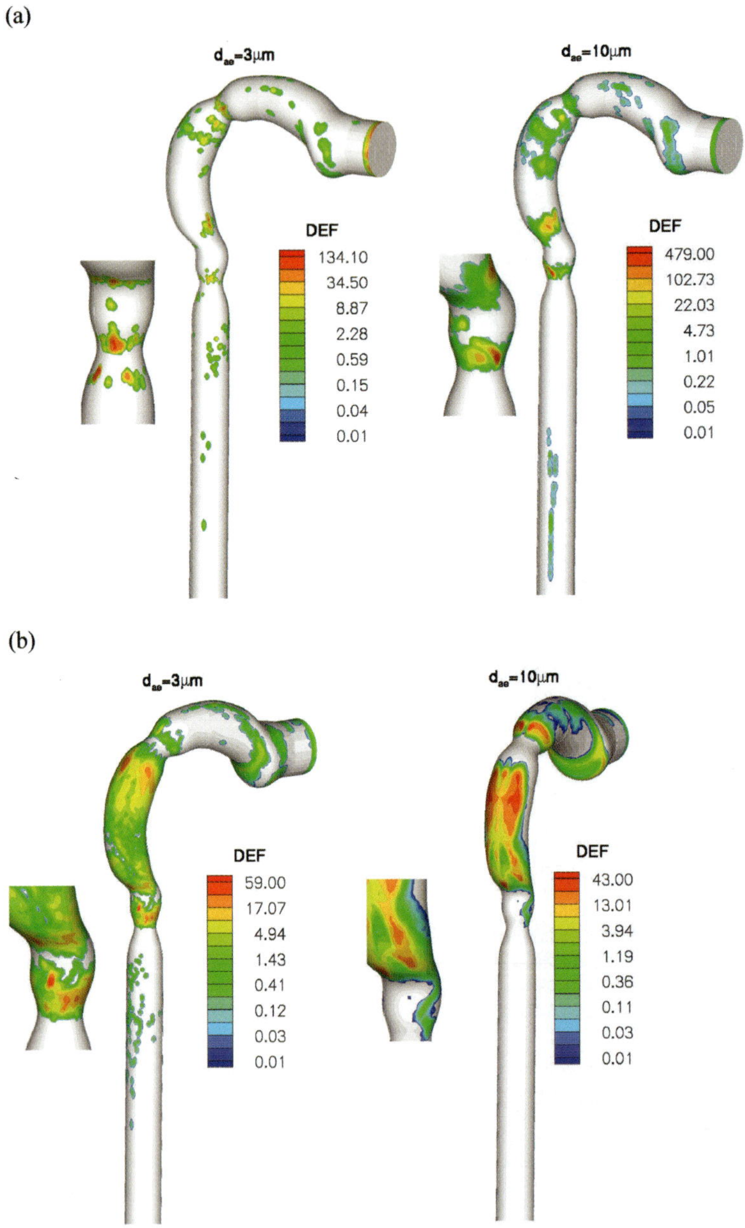

Fig. 5.3.4 3-D distributions of deposition enhancement factor (DEF) in the oral airway model for microparticles with: (a) Q_{in} = 15l/min; and (b) Q_{in} = 60 l/min.

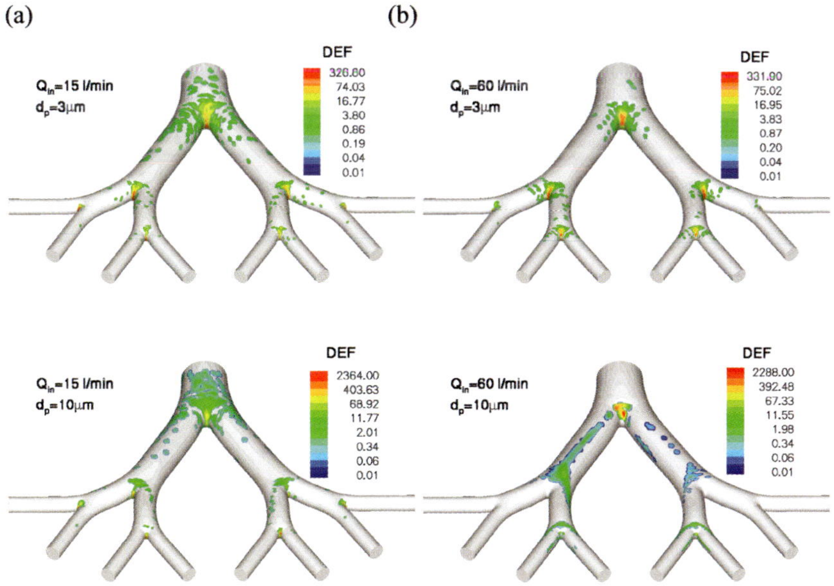

Fig. 5.3.5 3-D distributions of deposition enhancement factor (DEF) in the airways G0 to G3 for microparticles with: (a) Q_{in} = 15l/min; and (b) Q_{in} = 60 l/min.

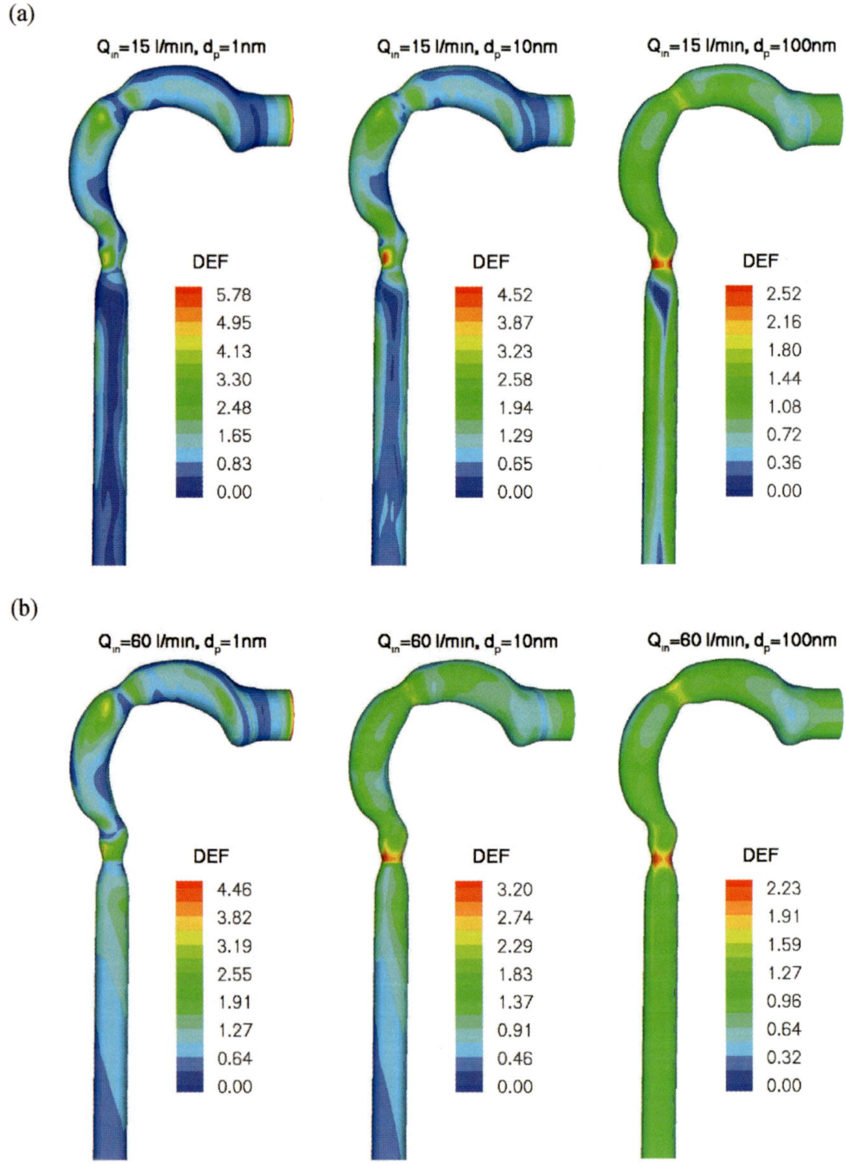

Fig. 5.3.10 3-D distributions of deposition enhancement factor (DEF) of nanoparticles in the oral airway model under steady inhalation with: (a) Q_{in} = 15 l/min; and (b) Q_{in} = 60 l/min.

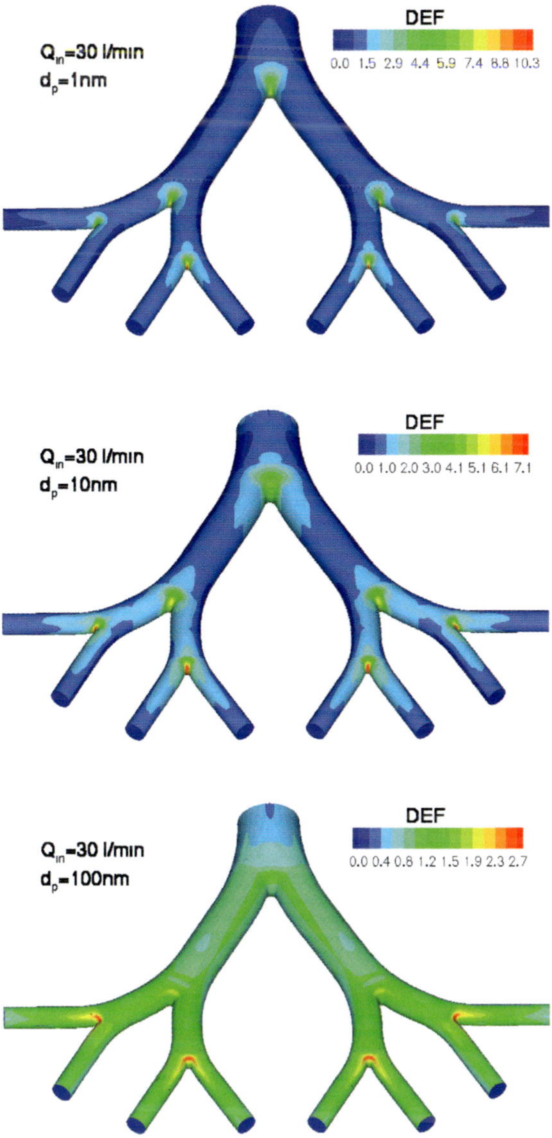

Fig. 5.3.11 3-D distributions of deposition enhancement factor (DEF) of nanoparticles in the bifurcation airway model under steady inhalation with Qin = 30 l/min.

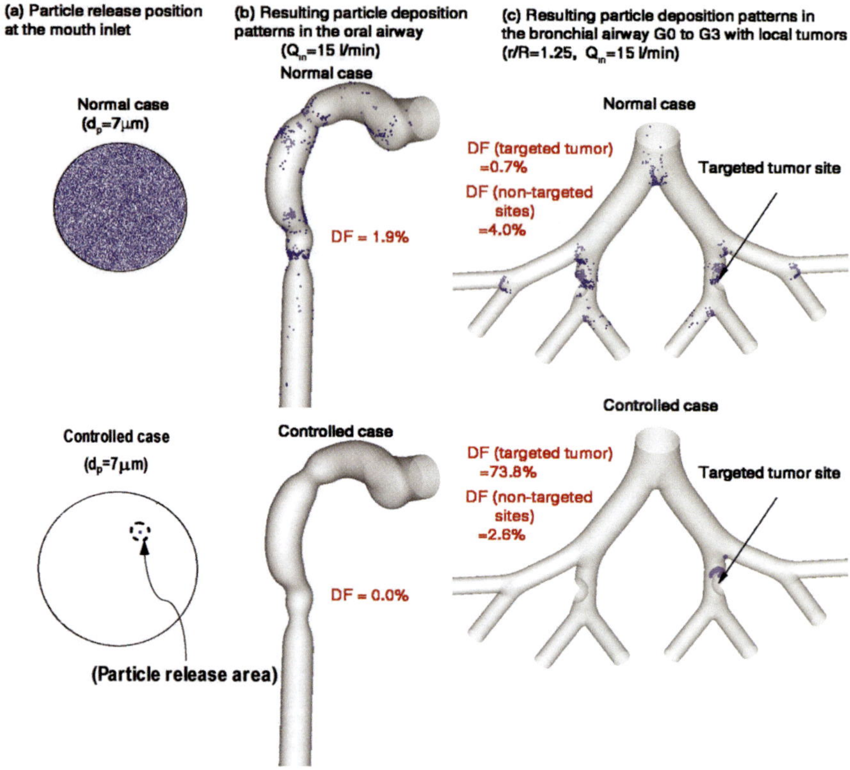

Fig. 5.3.15 Simulation-based drug delivery design for the upper airway starting from the mouth. A hemispherical tumor was placed in generation G2, comparing drug aerosol deposition fractions on the tumor surface for the normal case and controlled case.

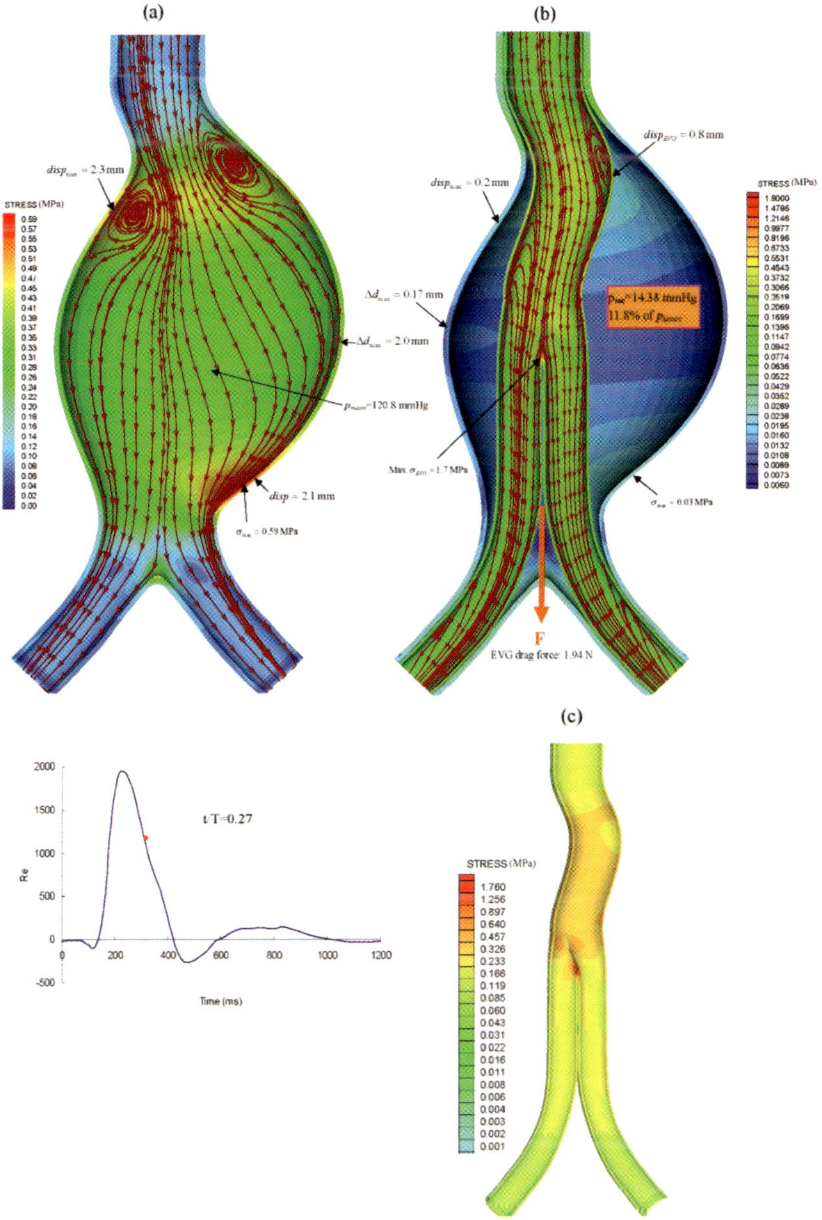

Fig. 5.4.10 Comparison between nonstented AAA and stented AAA (t/T = 0.27): (a) Nonstented AAA; (b) Stented AAA; (c) Wall stress distribution in EVG.

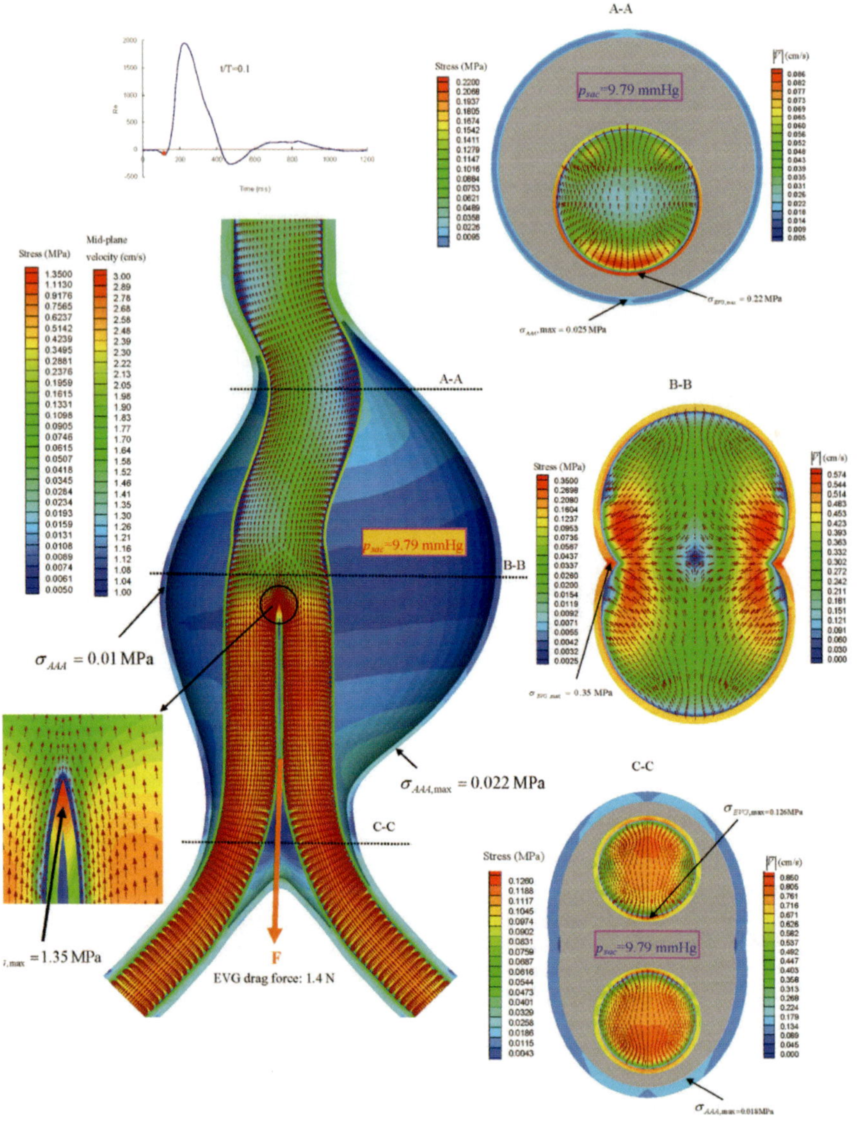

Fig. 5.4.11 Fluid-structure interactions of stented AAA (t/T = 0.1).

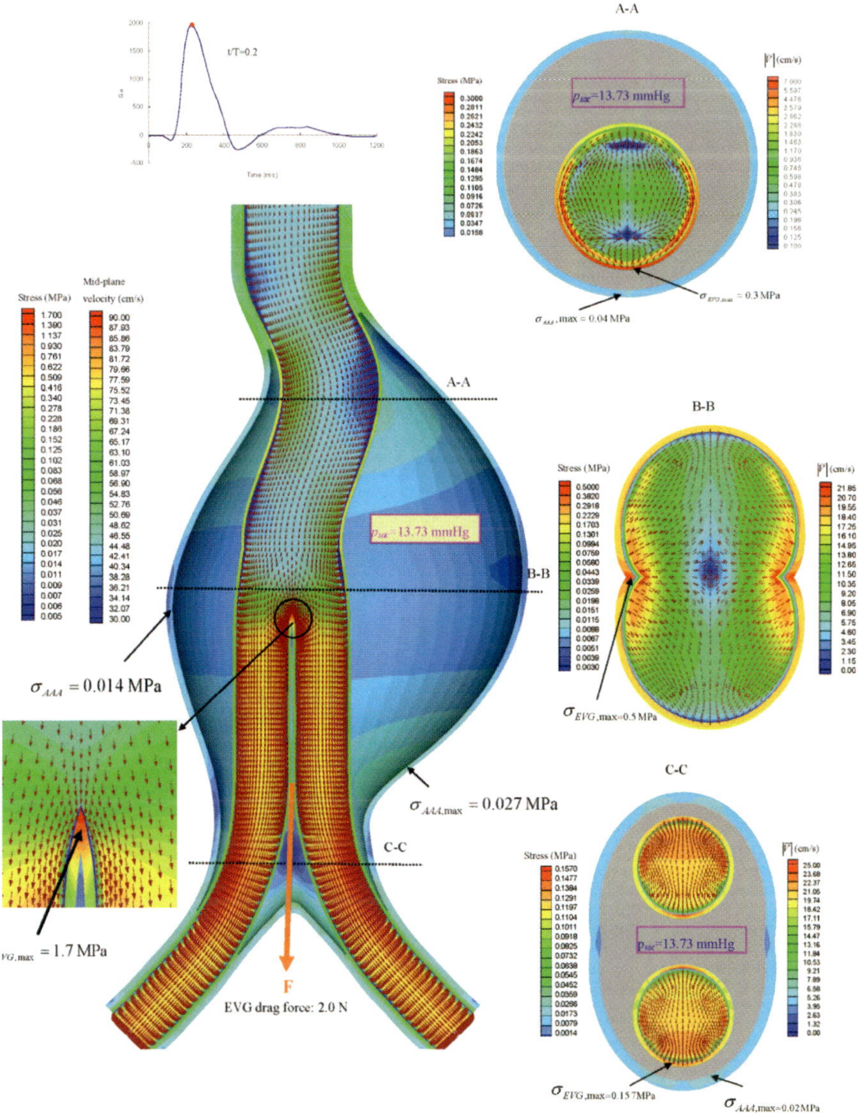

Fig. 5.4.12 Fluid-structure interactions of stented AAA (t/T = 0.2).

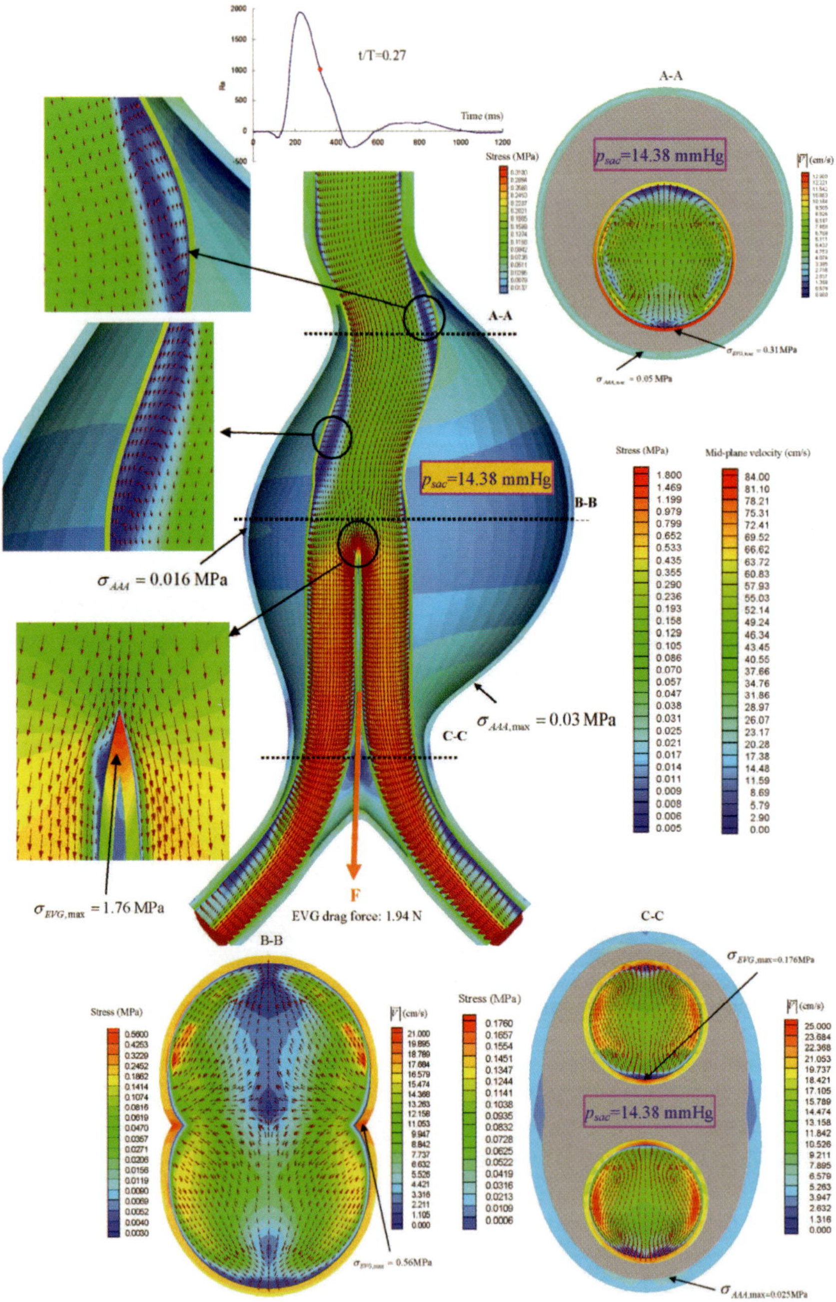

Fig. 5.4.13 Fluid-structure interactions of stented AAA (t/T = 0.27).

(a)

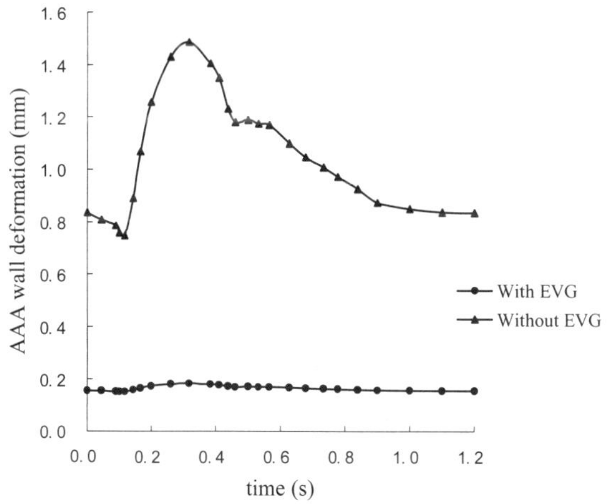

(b)

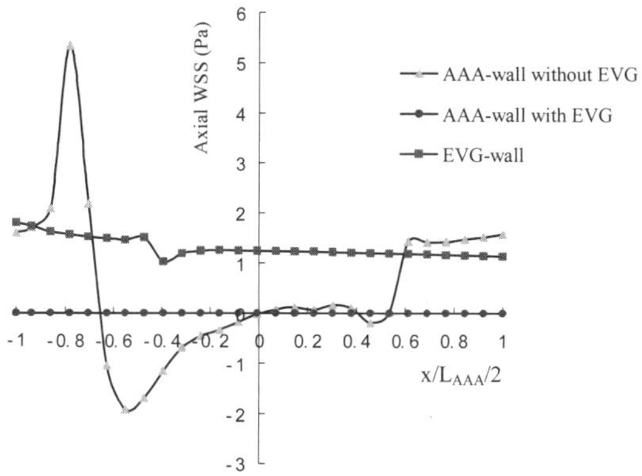

Figure 3.2.8 (a) Comparison of AAA wall deformation changes with time; and (b) WSS comparisons between stented and nonstented AAA models.

Comparing Figs. 3.2.7a and 3.2.7b, it can be deduced that the EVG is carrying most of the blood pressure load, i.e., the aneurysm expansion is now minimal, and the location of the maximum Von Mises stress moves from the AAA wall to the EVG. Specially, for the nonstented AAA, the maximum diameter change is up to 1.5 mm and the maximum Von Mises stress reaches 0.589 MPa in the AAA wall. After placement of the EVG, the change in maximum diameter is only 0.19 mm, while the maximum Von Mises stress in the AAA wall has decreased to 0.153 MPa. It was also found that the sac pressure was reduced from 122 mmHg to 29.4 mmHg, i.e., only 24.1% of the average lumen pressure. While in Fig. 3.2.7 the highest values in dmax -expansion and wall stress are displayed, Fig. 3.2.8a depicts temporal changes in $d_{AAA,\ max}$. It can be seen that an EVG may reduce the impact of pulsatile motion on the AAA wall significantly, i.e., the magnitude of wall deformation at the maximum diameter has decreased to only about 14% of that in the nonstented AAA. Furthermore, the wall motion is not sensitive to the pulsatile pressure wave anymore, which means that the influence of the blood pressure profile on potential AAA rupture is remarkably reduced.

In AAAs, high mechanical stresses and low wall shear stress (WSS) regions almost coincide. Such areas may be prone to thrombi formation; furthermore, strong sustained WSS changes may damage the endothelium (Kleinstreuer et al., 2001). Figure 3.2.8b indicates that an endovascular graft may reduce WSS-values in an AAA to zero, generating a near-uniform WSS distribution on the internal wall of the EVG. At present, there is no published clinical evidence regarding the role of WSS in the expansion and/or rupture of AAAs.

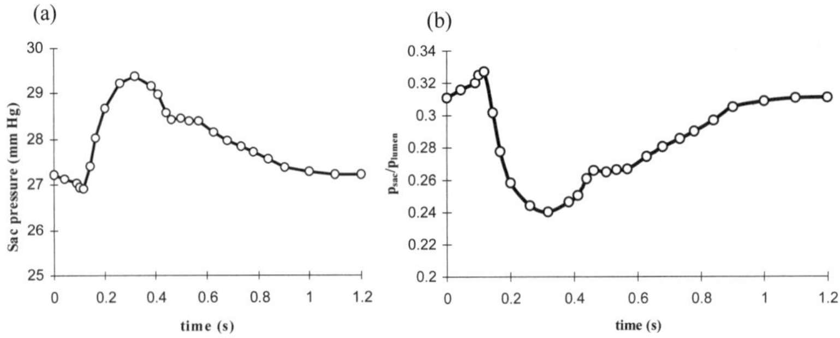

Figure 3.2.9 (a) Sac pressure changes with time; and (b) Ratio of sac pressure to lumen pressure changes with time.

In stented AAAs, even without endoleaks, the blood pressure in an AAA cavity is clinically not zero (Sonesson et al., 2003). Indeed, because the stagnant blood is incompressible, if only the EVG/AAA wall is flexible, the sac pressure can be generated by the moving boundary resulting from hydraulic effect and transfer to the surrounding wall. Our simulation shows that the cavity pressure changes between 26.9 to 29.4 mmHg (p_{sac}/p_{lumen} varying 24%-32.7%) during one cycle in the absence of endoleak in this study (Fig. 3.2.9a). When comparing Fig. 3.2.5c with Fig. 3.2.9a, it is evident that the sac pressure and the exit pressure waveform vary almost synchronously. The simulation results are in agreement with clinical observations and experimental results (cf. Sonesson et al., 2003 and Gawenda et al., 2004). Thus, under the conditions of fluid-structure interaction, the sac-pressure is nonzero in the absence of endoleaks and the stagnant blood can transmit pressure, thereby influencing the deformation of EVG and AAA walls.

Stent-graft movement may occur in AAAs if the EVG-neck fixation is insufficient to overcome the maximum force induced by the blood flow. In general, migration is triggered by momentum change, pressure, and wall shear stresses caused by the local blood flow, inappropriate aortic neck configurations (or tissue damage), and/or biomechanical degradation of the prosthetic material. EVG-migration is still a prevailing problem after endovascular aneurysm repair (EVAR). Migration may cause endoleaks, twisting and kinking, tortuosity, limb thrombus and occlusion, and ultimately EVG failure. It should be noted that momentum change and wall shear stress may be ignored in this straight tubular EVG when compared to the transient pressure load on a bifurcating EVG, which results in cyclic axial force changes. As mentioned, under the same inlet velocity and exit pressure conditions, the flexible wall deformations are very similar for both steady and transient flows. Thus, for the following migration analysis a steady luminal outlet pressure (p_{out}=120 mmHg) was assumed as the representative load for the coupled CFX-ANSYS simulations. Now, in order to calculate the maximum fixation force that the EVG can provide, a pulling force F was applied to the EVG (see Fig. 3.2.5a). This test force was increased gradually; once the EVG began to move, the value of the maximum pulling force F needed for EVG displacement was recorded, i.e., this pulling force is then equal to the maximum EVG fixation force.

Based on Coulomb's linear friction law, the friction coefficient and contact area determine the fixation force. Due to the lack of clinical test data for EVG-neck friction coefficients, a friction factor range of $0.1 \leq f \leq 1.0$ was selected based on the experiments of Lambert et al. (1999). It can be seen from Fig.3.2.10a that a higher friction coefficient increases the maximum EVG fixation force, reaching a maximum when f >0.8. Thus, although Coulomb's linear friction law is employed for each contact element, this correlation between the maximum fixation force and the friction coefficient is nonlinear for the entire EVG-AAA contact system. The reason is that the structure behavior of the EVG-neck contact system is nonlinear in light of the large displacements.

In addition to a sufficient friction coefficient, the aortic neck length and hence the contact area is important. Clinically, the proximal neck length should be greater than 13 mm. Figure 3.2.10b shows that the maximum fixation force is proportional to neck length. Therefore, if the patient has a long infra-renal neck, the best method to increase the fixation and to avoid EVG-migration is to utilize the extra length, if possible. Many patients have short infra-renal necks, and for them trans-renal fixation may be a promising choice.

Hypertension is very common in AAA patients. The simulation results show that high blood pressure may significantly decrease the maximum fixation force (Fig. 3.2.10c), partly because of "neck dilation" discussed below. If the blood pressure p is greater than 160 mmHg, the risk of migration is very high. It should be noted that the increased blood pressure can cause the contact to loosen but not to separate because the EVG is self expanding, i.e., a pre-stress on the EVG is applied which can produce 15% of over-sizing between EVG neck and arterial wall. Thus, if the arterial wall deformation caused by blood pressure is less than 15% of its diameter, the contact should remain. But the contact force will decrease accordingly which leads to the small force required to dislodge the EVG. As mentioned, for a bifurcated EVG, the pulsatile blood pressure may also cause a large longitudinal force on the EVG, which may lead to EVG migration. In conclusion, high blood pressure plays a very important role in EVG migration for both tubular and bifurcated EVGs.

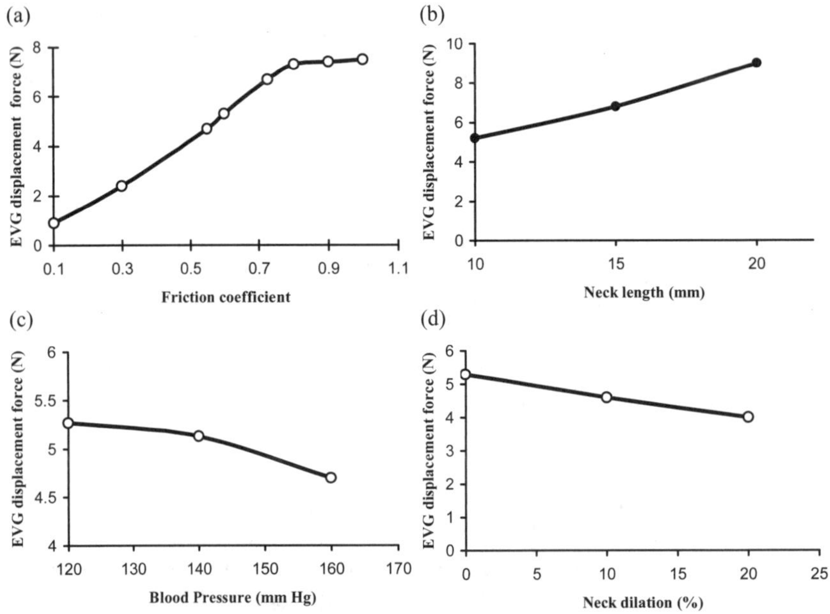

Figure 3.2.10 Relationship between migration force and: (a) friction coefficient; (b) neck length; (c) blood pressure; and (d) neck dilation.

"Neck dilation" implies that, due to biochemical processes in the wall tissue, the AAA neck diameter increases with time after EVG placement. As a result, the contact between EVG and AAA neck diminishes and the maximum fixation force that the EVG-neck can provide will be reduced. Neck dilation is a very common problem in EVG applications. In order to test this event, we increased the neck diameter by 0 to 20% of its original value to simulate neck dilation. Because of the pre-stress and blood pressure applied to the EVG, contact still remains between the AAA-neck and EVG, but the contact force is reduced (Fig. 3.2.10d) by up to 25%. Clinically, neck dilation is presently seen as the main reason for EVG migration. Researchers found that the neck diameter growth rate is about

0.5mm/year after open surgery. However, after EVAR, the neck diameter growth rate may reach 0.7-0.9 mm/year in the proximal region and 1.7-1.9mm/year in the distal region. Some researchers found that dilation also occurs in the super-renal artery without EVG placement. Therefore, dilation may cause EVG migration, but the cause of dilation may not only be due to the placement of an EVG in the neck; there are other, i.e., tissue-related, factors as well.

Based on clinical observations, neck shapes can be grouped into tubular, conical, reversed conical, barrel, and hourglass. The lengths of the neck segments are the same for all neck shapes where the geometries are shown in Fig.3.2.11. Clearly, the AAA region and all biomechanical parameters were identical for the five case studies. As expected, the tubular neck appears to be best for EVAR placement and neck fixation because the other four shapes are irregular and may decrease the contact area to a small ring. The hourglass shape is the worst and its maximum fixation force is only 10% of the straight neck configuration (Fig.3.2.11). Clinically, a neck with an hourglass or conical shape is not suitable for EVAR. Because different patients have different neck shapes, specific EVG designs for groups of patients may prevent migration.

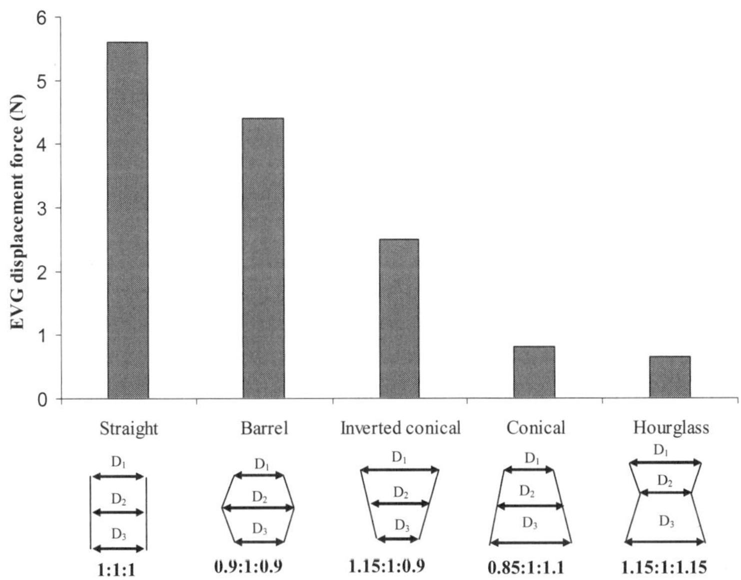

Figure 3.2.11 Influence of neck configuration on EVG migration.

3.2.3.5 Conclusions

The following conclusions can be drawn from this fluid-structure interaction analysis as applied to an axisymmetric stented AAA:

(1) An endovascular graft (EVG) can reduce pulsatile sac-pressure, wall motion, mechanical stress, and the maximum diameter in AAAs significantly, which may prevent AAA rupture effectively. The concentrated regions for low wall shear stress and high mechanical stress are eliminated.
(2) Sac-pressure (non-endoleak) may be caused by fluid-structure interactions between EVG, stagnant blood, and AAA wall. Both the sac pressure and the ratio of sac-pressure-to-lumen-pressure change with time.
(3) Simulation results indicate that the maximum EVG fixation force is determined by multiple factors, including EVG-neck friction coefficient, blood pressure, neck dilation, neck length, and neck shape.
(4) Coupled CFX-ANSYS simulations for blood flow and EVG-artery structure interactions are valid, predictive, and powerful for optimal surgical recommendations, improved EVG designs, and proper EVG placement.

It should be noted that in this study AAA asymmetry, intraluminal thrombus, iliac bifurcation, and angled neck were not considered. In patient-specific stented AAAs, these factors may significantly affect the fluid-structure interactions and hence possible EVG migration (see Sect. 5.4).

3.3 EXAMPLES OF COMPUTERIZED DISEASE MANAGEMENT

3.3.1 Introduction

Modern information technology is slowly advancing into the field of clinical medicine. While images of a patient's disease, generated via digital cameras, echocardiography, videos, ultrasound, MRI machines, and/or CT-scans, can be shared around the world for expert interpretation and second opinions (see Rangayyan, 2005), quantitative medical diagnostics and decision-making, using computer programs and CFD simulations, is rare. For some good reasons, there is a prevailing reluctance among surgeons to trust, for example, the assessment of disease severity based on nonlinear multivariable computer programs or to follow the recommendations for optimal surgical intervention based on elaborate computer model simulations and analyses. Quantitative medical diagnostic programs often require elaborate input data sets to be obtained from sophisticated CT-scan and/or MRI interpretations. Taking coronary bypass as an example, planned optimal surgical intervention requires realistic descriptions of the patient's coronary arteries in terms of geometry, hemodynamics, as well as biomechanics followed by fast and accurate solutions of such a fluid-structure interaction (FSI) problem.

Specifically, experimentally validated computational cardiovascular fluid dynamics analysis, based on model geometries from Magnetic Resonance Angiography (MRA), can provide the surgeon with quantitative guidelines for better surgery planning; for example, where to place a bypass graft, should a stent be implanted or not, which regions experience high pressure loads and hence wall stress concentrations, etc. A second example is the computational fluid-structure-interaction simulation of a (stented) aneurysm to provide quantitative information on the need for stent-graft insertion, suitable stent-graft selection and secure placement, possible stent-graft migration, endoleaks, etc. Other biomedical applications where CFD analyses play key roles include the respiratory system and device development. Clearly, experimental methods are sometimes impossible to use or fall short in predicting local deposition concentrations of inhaled toxic solid particles, droplets, and vapors, as needed in toxicology and health-effect studies. Furthermore, at all stages of the development of medical devices, e.g., cardio-pulmonary bypass systems, stent-grafts, hemodialysis machines, artificial valves and hearts, joint lubrication, nano-therapeutics implants, and drug-delivery devices, to name a few, computer modeling and virtual prototyping are essential. They can provide detailed physical insight and potentially optimal design solutions. References dealing with detailed computational fluid-particle dynamics of branching conduits include Zhang et al. (2005), Steinman et al. (2003), Kleinstreuer & Longest (2003), and Lohner et al. (2003), while Taylor & Draney (2004) reviewed experimental and computational methods in cardiovascular fluid mechanics applied to medical management. Section 3.2.3 illustrated simple applications of medical diagnostics and designs which are expanded and discussed in more detail in Chapter V.

Ultimately, high-resolution imaging of a patient's diseased system will allow for transient 3-D computer modeling of FSIs to simulate and analyze causes of the particular disease, illustrate the detrimental effects, and come up with an optimal treatment in form of drug therapy or surgical interventions, including device implantation and organ/tissue transplantations. As a first step, before such lofty goals can be achieved and hence computerized medical management will be widely accepted, much lower cost- and time-intensive solutions to patient-specific medical problems have to be demonstrated.

3.3.2 Image-File Conversion Steps

Computed tomography (CT) and magnetic resonance imaging (MRI) are most suitable for obtaining good pictures of bones (hard tissue) and soft tissue, respectively. Building 3D geometric surface models from CT and/or MRI scanned image files is being widely used in biomedical engineering and other disciplines. However, it is quite a challenging multi-step procedure, especially when the geometry of the object, e.g., an artery, organ, or tissue structure is complicated. Thus, the necessary procedural steps and associated software required can be summarized in flowchart form (see Fig. 3.3.1).

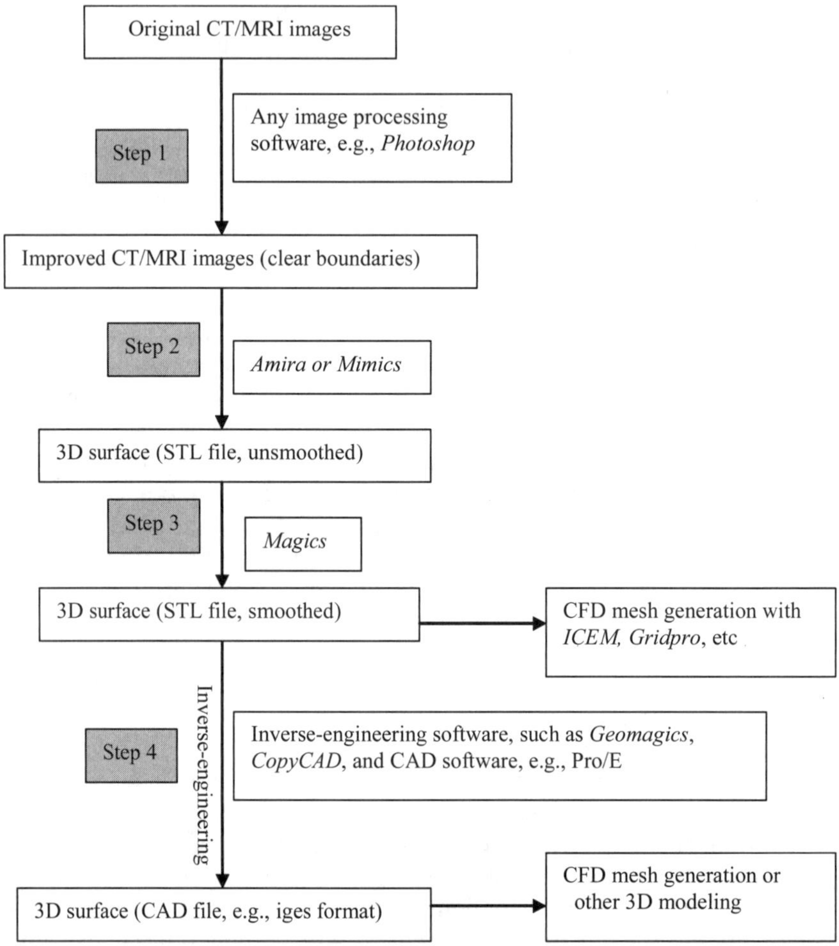

Fig. 3.3.1 Sequential multi-step procedure for image-file conversion to CFD Mesh.

Using the human nasal cavity as an example, each step is illustrated as follows. For simpler geometries, such as artery walls, Matlab Toolboxes may be employed to generate CAD-like geometry files from DiCom image files.

Step 1: Processing CT/MRI scanned images

Goal: *To obtain distinct grayscale-separated boundaries of the object.*

Software: *Photoshop (Adobe Systems Incorporated, San Jose, CA, USA).*

The original CT/MRI scanned data comes as a stack of images (Fig. 3.2.2). Those images are taken with a certain resolution, i.e., slice distances between each other from the start to the end of the organ. The images are gray-scaled to represent different organs, body fluids, or cavities based on the varying tissue densities. However, due to insufficient scanner resolution and unexpected tissue density changes, CT/MRI images often do not have well grayscale-separated boundaries of the organs. Such unclear boundaries will lead to distortion of organ surfaces which have to be constructed later. To eliminate this problem, it is recommended that CFD modelers examine each image with the help of research MDs or radiologists, and make necessary modifications using image processing software like Photoshop.

As indicated in Fig. 3.3.2, for the present case of a human nasal cavity there are 69 MRI slices at intervals of 1.5mm. The final image for each slice was carefully examined and manipulated to assure good grayscale-separated boundaries as indicated with three image examples given in Fig. 3.3.2.

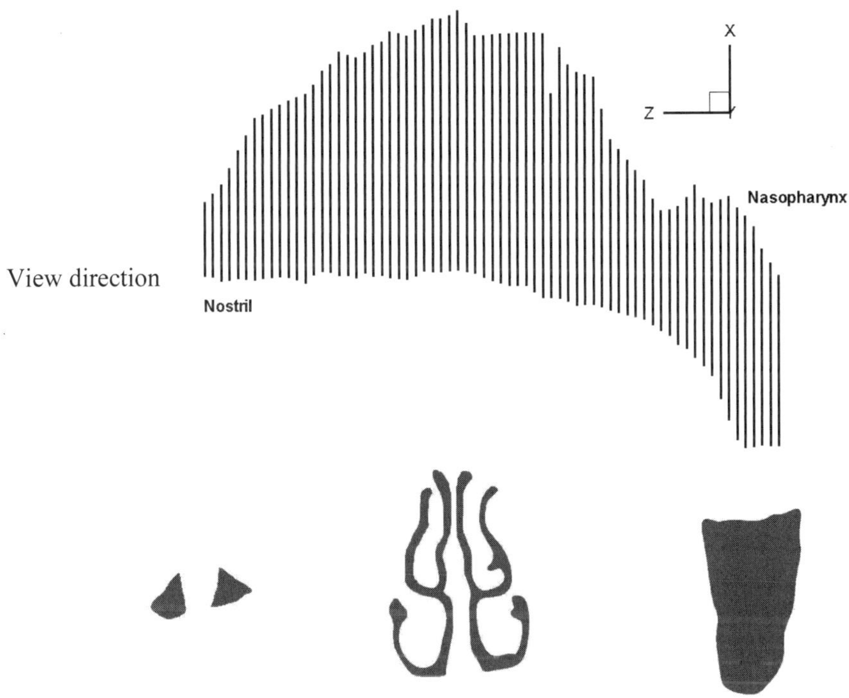

Fig. 3.3.2 MRI slices and local image contour.

Step 2: Constructing 3D surface from CT/MRI images

<u>Goal:</u> To obtain a 3D surface computer file from improved CT/MRI images

Software: Amira or Mimics

Amira (Mercury Computer Systems Inc, San Diego, CA) and Mimics (Materialise, Ann Arbor, MI) both are suitable for constructing 3D surfaces from CT/MRI images. They follow a very similar working procedure. For this particular case study of a human nasal cavity, Amira was used. Its robust automatic surface generation routine produces rough 3D surfaces (see Fig. 3.3.3) from a stack of images (see Fig. 3.3.2) in a minute. The computer file is in STL (stereolithography) format which is a widely used format in engineering, especially for rapid prototyping. The 3D STL surfaces are made up of three-sided "facets" or triangles (see Fig. 3.3.3). The direction of each facet is given defined with its normal unit vector. The smoothness parameter for a STL surface is often given as its maximum direction angle difference between any two neighboring facets, and it is called feature angle β.

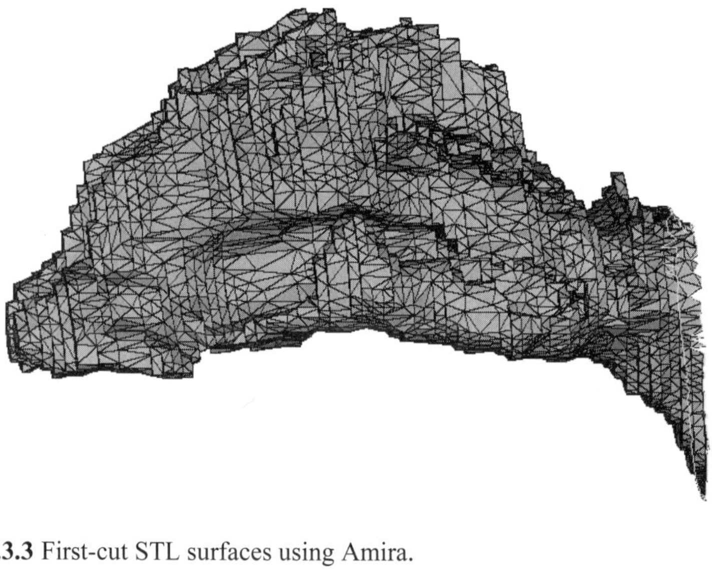

Fig. 3.3.3 First-cut STL surfaces using Amira.

Clearly, the first-cut 3D surface of the human nasal cavity produced by Amira is too coarse, with a maximum feature angle of $\beta = 160°$. This does not reflect a realistic human nasal cavity surface and is also unacceptable for CFD mesh generation.

Step 3: Smoothing surfaces

Goal: To smooth the 3D surfaces

Software: Magics

Magics (Materialise, Ann Arbor MI) was found to be the best software for 3D surface smoothing. While Magics has its own automatic smoothing scheme, the automatic smoothing function cannot accomplish the whole smoothing step because of mathematical limitations of auto-smoothing. Specifically, the perpetually running smoothing algorithm will not keep producing smoother surfaces, but will lead to distortion or shrinkage of surfaces. Hence, a proper approach was found to be a combination of auto-smoothing and user manual modification facet by facet. Such manual modification facet by facet is painstaking, but very effective. Magics graphically indicates which facets have a bad smoothness quality and provides very convenient operation to make improvements locally. Figure 3.3.4 shows the final 3D surface definitions in STL format after smoothing with Magics. Its maximum feature angle is only 35 degree, which is already within the range of good quality surfaces. Now the surface file is ready to be used in CFD mesh generators like ICEM or Gridpro.

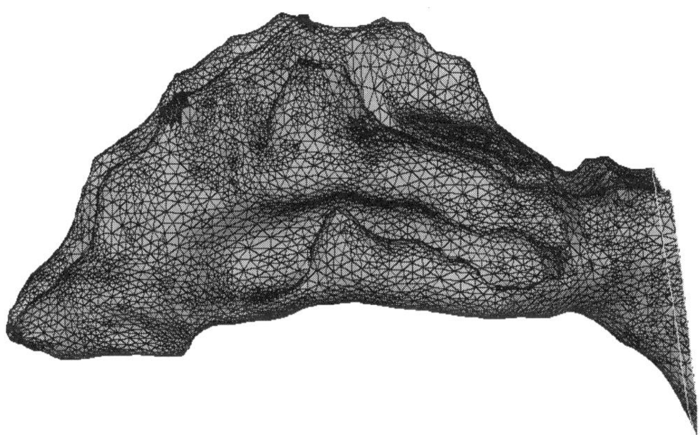

Fig. 3.3.4 Final 3D body surface of human nasal cavity (Shi, 2006).

Step 4: Converting STL format to CAD file format (Inverse-engineering)

<u>Goal:</u> *To convert the 3D surface STL file to the CAD file, e.g., iges format*

<u>Software:</u> *Imageware, CopyCAD, or Geomagic Studio*

The 3D surface in STL format is made up of a great number of small triangular facets. It is different from the 3D surface in CAD format which is described by Nonuniform Rational B-Splines (NURBS). The relationship between STL surface and NURBS surface is the analogy between data points and their curve-fitted connections. Thus, the transfer from STL to NURBS is a quite challenging topic, called inverse-engineering. Interestingly enough, conversion from NURBS to STL is a one second job.

Although the STL format file is acceptable for most mesh generators, a CAD format file is still highly desirable because the CAD file will provide much more room for further geometric processing. For example, if one wants to connect the human nasal cavity with oral airway models, it would be very inconvenient to construct the geometric linkage between the two models if any one is in the STL format.

Quite a few software packages, e.g., Imagware (UGS Inc., Plano, TX, USA), CopyCAD (Delcam Inc., Windsor, ON, Canada) and Geomagic Studio (Raindrop Geomagic Inc., Research Triangle Park, NC, USA) are capable to perform inverse-engineering. Figure 3.3.5 shows the 3D surface model in CAD (iges) format after completing inverse-engineering in Geomagic Studio. In this case, a certain airway length was added to the nasopharynx as an air outflow extension as well in preparation for future linkage to the oral airways.

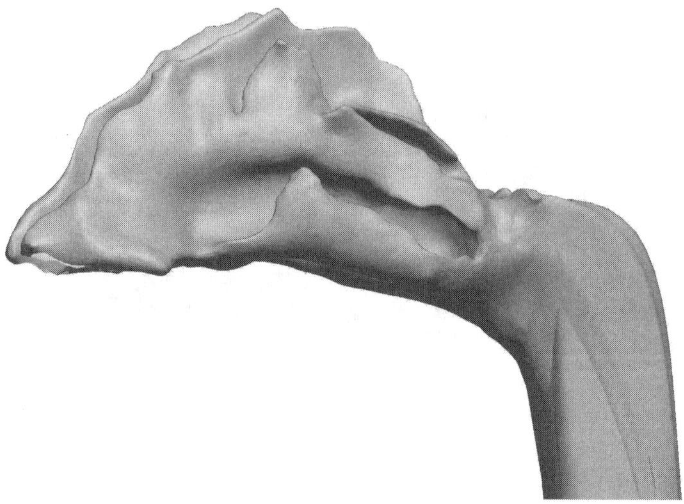

Fig. 3.3.5 3D CAD surface model of a human nasal cavity (Shi, 2006).

Step 5: CFD finite volume mesh generation

<u>Goal:</u> To generate a finite volume mesh for numerical analysis

<u>Software:</u> ICEM ((ICEM-CFD Engineering Inc., Berkeley, CA, USA)

ICEM has powerful functions to generate both structured and unstructured meshes for complicated biomedical 3D models. For the geometries with low-level complexity, e.g., arteries and lung airways, it is recommended to use a structured mesh. For the geometries with highly-irregular geometric features, e.g., human nasal cavity, generating tetrahedral elements with dense mesh resolution layers of prism elements would be more efficient and provides more flexibility. Figure 3.3.6 shows the final mesh for the human nasal cavities.

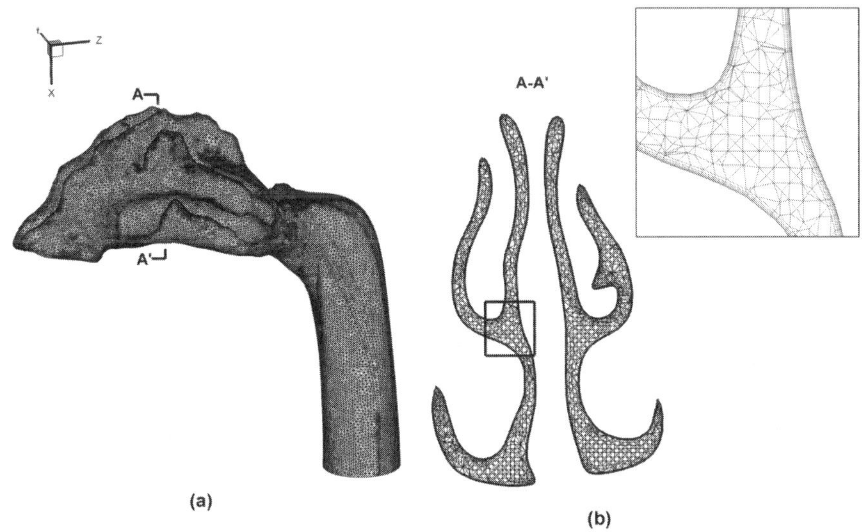

Fig. 3.3.6 Unstructured finite volume mesh (Shi, 2006).

3.3.3 A Stenosed Artery Model for Surgical Bypass Planning

As mentioned in Sect. 3.3.1, realistic computer models for medical planning are presently too slow and too expensive for routine use. Alternatively, algebraic global models or transient one-dimensional transport models have been proposed.

Examples include global lung models for the evaluation of toxic particle impact (Phalen & Oldham, 2001; Hofmann et al., 2002; Asgharian et al., 2004), pharmacokinetics modeling (Nestorov, 2003; Rodgers, 2005), for species pathway analysis, and 1-D models for vascular cardiovascular planning (Wan et al., 2002; Steele et al., 2003). Naturally, such dramatically simplified models require more empiricism in terms of input data sets, experimental correlations, and measured transfer functions. For illustration purposes the work of Wan et al. (2002) and Steele et al. (2003) is discussed, focusing on a 1-D computer model describing pulsatile flows in blood vessels, e.g., a straight human common carotid artery or a stenosed aorta with bypass graft. They incorporated an empirical nonlinear correlation for lumen pressure and local tube radius as well as minor loss coefficients for geometric non-uniformities, such as outflow resistance, junctions, and stenoses. The major conditions were Poiseuille-type flow, changing cross-sectional areas due to the elastic tube walls, and minor loss terms for the stenosis plus graft-to-artery junctions. The resulting flow equations in terms of Q(z,t) are (see Hughes & Lublieu, 1973):

$$\text{(continuity)} \qquad \frac{\partial A}{\partial t} + \frac{\partial Q}{\partial t} = 0 \tag{3.3.1}$$

$$\text{(z-momentum)} \frac{\partial Q}{\partial t} + \frac{\partial}{\partial z}\left[(1+\delta)\frac{Q^2}{A}\right] = -\frac{A}{\rho}\frac{\partial p}{\partial z} + \gamma\frac{\partial^2 Q}{\partial z^2} + h_e\frac{Q}{A} \tag{3.3.2}$$

where A=A(z, t) is the axisymmetric cross section and for Poiseuille-velocity profiles $\delta = 1/3$. The tube loss term is $h_e = K\left(\frac{Q}{2L}\right)$, where $K = \Delta p / \left[\rho / 2\left(\frac{Q}{A}\right)^2\right]$.

Clearly, selections of the "minor" loss coefficient are critical in providing accuracy to the 1-D modeling approach by realistically incorporating local obstructions (see Seeley & Young, 1976 for stenoses) and tube junctions. The inflow rate Q(z=0, t) is prescribed and at the outlets $Q(z = L, t) = p_L / R$, where $p_L = p_L(t)$ and R is the lumped downstream resistance.

The coupling between internal pressure p(z, t) and wall stress $\sigma_\theta(r,t)$ is given by Laplace's law for thin-walled tubes of base radius r_0 and wall thickness h, i.e.,

$$\sigma_\theta = \frac{r}{h}\Delta p = \frac{E}{1-\gamma}\frac{r - r_0}{r_0} \tag{3.3.3}$$

where $\Delta p = p - p_0$; $p(r = r_0) = p_0$; Poisson's ratio $\gamma = 1/4$ and E is Young's modulus. It is noted that $E[r_0(z)]$ is highly nonlinear for arteries, following the arterial volume compliance as:

$$\frac{\partial p}{\partial \forall} \approx \frac{3 r_0}{2hE} \tag{3.3.4}$$

Solving for the transmural pressure (see Eq.(3.3.3)) yields

$$p(z,t) = p_0 + \frac{4}{3}\frac{Eh}{r_0}\left(1 - \frac{r_0}{r}\right) \tag{3.3.5}$$

where $r_0 / r \equiv \sqrt{A_0(z) / A(z,t)}$. Olufson (1999) developed a curve-fitted correlation for arteries as:

$$\frac{Eh}{r_0(z)} = a + be^{cr_0} \tag{3.3.6}$$

where $a = 8.65 \times 10^5 g/(cm \cdot s^2)$, $b = 2 \times 10^7 g/(cm \cdot s^2)$, and $c = -22.53 cm^{-1}$.

1-D and 3-D simulation comparisons for various volume flow rates and pressure waves in a straight vessel via s are displayed in Fig. 3.3.7. The dimensions of this vessel were chosen to approximate the human common carotid artery, where the radius and length of the straight circular pipe were specified to be 0.4cm and 16cm, respectively. The results for 1-D finite element simulations were obtained by Wan et al. (2002) by solving Eqs. (3.3.1) and (3.3.2) with the prescribed flow rate at the inlet and resistance boundary conditions at the exit. The transient 3-D simulation results were achieved by solving continuity and momentum equations with the commercial finite volume software CFX. In the latter case, the inlet velocity profiles were determined by the analytic expression of transient developed flow in a straight tube, following the given input pulse expressed as Fourier series. The pressure at the outlet was specified as $p(t)=Q(t)R$ with R=3000 dynes·s/cm^5 . Three cycles were simulated to avoid start-up effects on the flow field. Clearly, the pressure wave at the inlet obtained from the 3-D simulation agrees very well with that from 1-D modeling with similar inlet flow waveform and outlet resistance boundary conditions. The minor damping of the propagating pressure waves are shown in both 1-D and 3-D simulations. The large lag of both volume flow rate and pressure wave at the outlet can be observed in the 1-D modeling, which perhaps is set up manually in order to exhibit the shifted wave because the same phenomenon is not shown in other 1-D simulations (e.g., Vignon and Taylor, 2004). Clearly, the lag of volume flow rate at the outlet does not appear in 3-D simulations, considering the mass balance at each time level.

Comparisons between 1-D and 3-D solutions for more complicated geometries, such as a porcine thoraco-thoraco aortic bypass graft, were also given by Wan et al. (2002). They showed that the peak differences for time-dependent flow rates were on the the order of 50% in the native aorta, but less than 10% in the bypass graft. In summary, the 1-D approximate method can be used to predict flow distributions and pressure waves in cardiovascular bypass grafts, which may be valuable for rapid evaluations of alternate treatment plans for patients with specific cardiovascular diseases. However, the 1-D approximation of blood flow ignores flow separation, secondary flows, and other complex, realistic flow phenomena. Thus, 3-D simulations are necessary for accurately analyzing realistic blood flow and associated particle dynamics as well as fluid-structure interactions in human blood vessels.

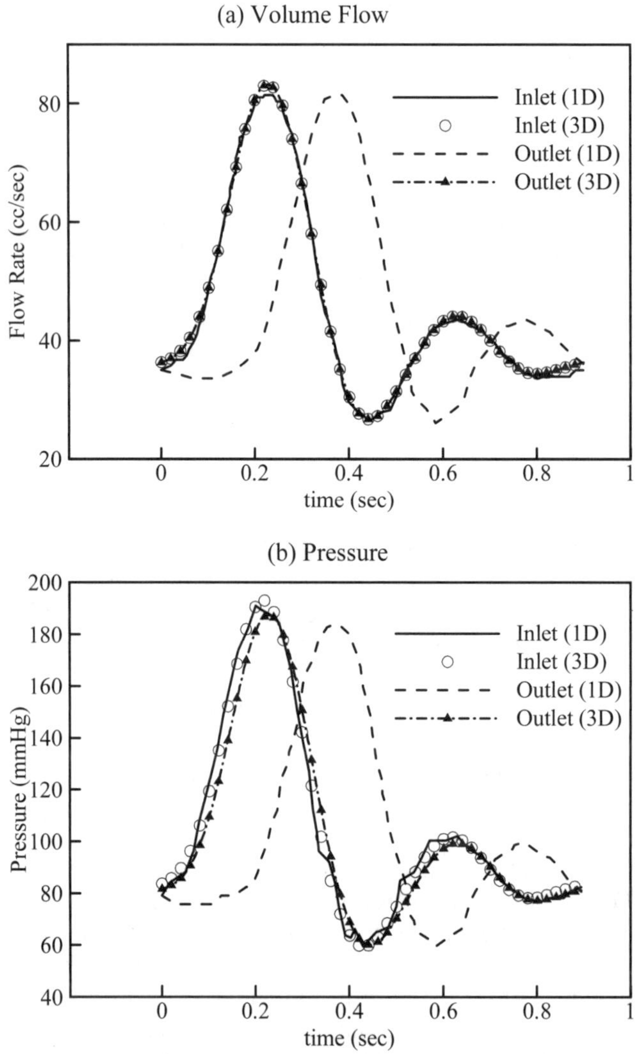

Fig. 3.3.7 Variations of volume flow rate and pressure waves in a straight circular pipe.

3.3.4 AAA-Rupture Prediction

Focusing on a representative abdominal aortic aneurysm (AAA), it is of interest to determine quantitatively whether it is necessary to repair AAAs via open surgery or endovascular aneurysm repair (EVAR). Hence to assess the risk of AAA rupture, key biomechanical factors and their threshold values, indicating possible AAA rupture, have been surveyed as summarized in Table 3.3.1 (Li, 2005).

Table 3.3.1 Parameters used in AAA rupture prediction (see Fig. 3.3.8).

Maximum diameter	Threshold: 5.5 cm (male); 5.0 cm (female)
Expansion rate	Threshold: 0.5cm/year
Mechanical stress	Threshold: 0.44MPa
Intra-luminal Thrombus	Threshold: $V_{ILT} / V_{AAA} = 0.62$
Diastolic pressure	Threshold: 90 mmHg and 105 mmHg as the middle and high risk levels for AAA rupture
Wall stiffness	Rupture threshold: stiffness begins to decrease
Asymmetry	High risk: asymmetry index $\beta = \frac{d_{AAA,\max}}{L_{AAA}} = 0.65$
Saccular index	High risk for AAA with large curvature

In order to estimate the actual wall stress, we start with the original Laplace equation, which correlates (blood) pressure with the average wall stress (see Sect. 1.5).

$$\sigma = \frac{pr}{ct} \tag{3.3.7}$$

where σ is the average wall stress, *p* is the pressure load, *r* is the radius, and *t* is the wall thickness, while *c* = *1* is for cylinders and *c* = *2* for spheres. Equation (3.3.7) greatly overestimates or underestimates actual aneurysm wall stresses because of the many underlying assumptions. In order to provide a useful, i.e., accurate and easy-to-calculate, predictor of the maximum wall stress in common AAAs (see Fig. 3.3.8), Eq. (3.3.7) was extended based on observed clinical evidence and computational analyses (Li & Kleinstreuer, 2005a). The functional form of

$$\sigma_{\max} = 0.006 \frac{(1-0.68a)e^{0.0123(0.85p_{sys}+19.5d_{AAA,\max})}}{t^{0.63}\,\beta^{0.125}} \quad \text{in [MPa]} \tag{3.3.8}$$

where $\sigma_{\max}$ is the maximum wall stress which occurs most frequently at a location two thirds of the maximum AAA diameter, the area ratio $\alpha = \frac{A_{ILT,\max}}{A_{AAA,\max}}$; the asymmetry index $\beta = \frac{l_p}{l_a}$, where l_p and l_a are the distances from the center point O to the posterior and the anterior (Fig.

3.3.8); p_{sys} is the systolic blood pressure (mmHg), $d_{AAA,\max}$ is the maximum AAA diameter (cm), and t is the wall thickness [mm] at the location of $d_{AAA,\max}$ [cm]. Specifically, $A_{AAA,\max}$ and $A_{ILT,\max}$ are the transverse areas of the AAA and intra-luminal thrombus (ILT) at the $d_{AAA,\max}$-location, respectively. For imaging techniques not providing area measurements, the transverse area can be approximately calculated as

$$A_{AAA,\max} = \frac{\pi d_{AAA,\max} H}{4} \tag{3.3.9a}$$

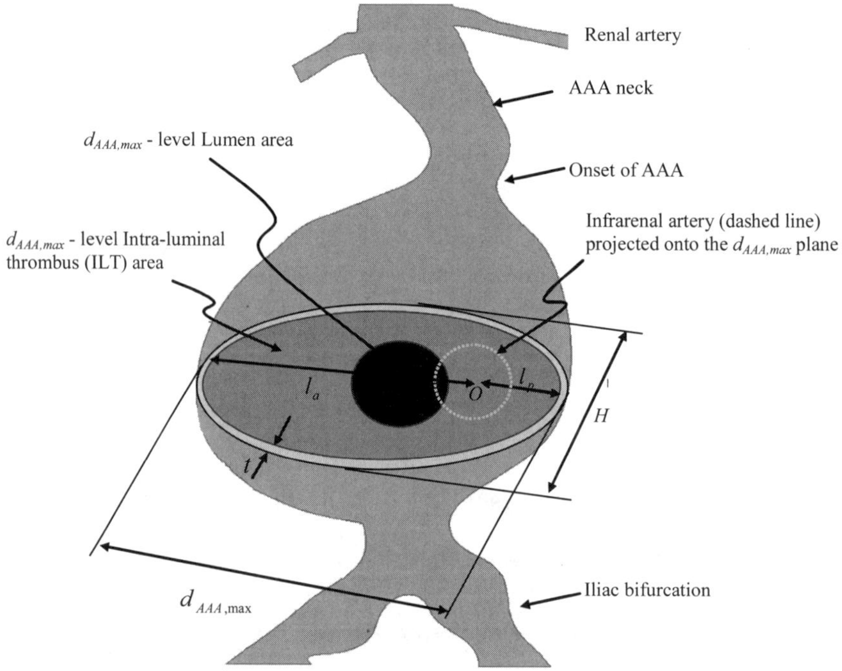

l_a is the distance from center point O to the AAA anterior side
l_p is the distance from center point O to the AAA posterior side

Figure 3.3.8 Geometric parameters of an AAA.

where H is the in-plane axis normal to the $d_{AAA,\max}$-measurement plane (see Fig. 3.3.8). The lumen area $A_{lumen,\max}$ may be calculated similarly; then, the ILT area is given as

$$A_{ILT,\max} = A_{AAA,\max} - A_{lumen,\max} \tag{3.3.9b}$$

If the wall thickness t is difficult to obtain from CT-scans, it may be approximated with a curve-fitted correlation

$$t = 3.9\left(\frac{d_{AAA,\max}}{2}\right)^{-0.2892} \quad \text{in [mm]}, \tag{3.3.9c}$$

where $d_{AAA,\max}$ is in [mm].

One can see that Eq. (3.3.8) not only represents the nonlinear correlation between wall stress and blood pressure, diameter, and wall thickness, but also correlates the effects of ILT and AAA asymmetry.

Notes:

- While Eq. (3.3.8) integrates the stress concentration effect caused by asymmetry to same extent, for seriously distorted geometries the evaluation of maximum wall stress magnitude (and location) is too complicated to predict with this simple equation.
- The ILT material property is assumed to be uniform. Even though the ILT close to the luminal surface is most newly formed and tends to be "older" away from the luminal surface toward the AAA wall, it was found that the use of mean ILT properties as opposed to the exact patient-specific parameters would result in a maximum error of only 5% in wall stress predictions.
- Residual stress is neglected because its magnitude is much smaller (3%) than the mechanical stress caused by the blood pressure in the artery wall.

In order to test the broader validity of the new wall stress equation, the average wall stresses were calculated, using data from ten different clinical and numerical AAA models, and the results compared with Eq. (3.3.8). As shown in Table 3.3.2, the maximum error using the new wall stress equation is 9.5%, whereas the Laplace equation generates a maximum error of 85.6%.

AAA-wall deterioration with time. Clearly, AAA-wall properties alter gradually with age, i.e., sustainable wall stress is a function of time. To predict AAA rupture more accurately, assessing time-dependent wall deterioration is very important.

Based on a literature review (see Li, 2005), the breaking (or tensile) strength is closely related to age, wall stiffness, i.e., collagen-to-elastin ratio, and ILT content. A nondimensionless model to estimate wall deterioration associated with time could be expressed as:

$$\frac{\sigma_{break,AAA}}{\sigma_{break,normal}} = \left(\frac{E_{p,AAA}}{E_{p,ref}}\right)^{a}\left(\frac{V_{ILT}}{V_{AAA}}\right)^{b} \tag{3.3.10}$$

where $\sigma_{break,AAA}$, $\sigma_{break,normal}$, $E_{p,AAA}$, $E_{p,ref}$, V_{ILT}, V_{AAA} are the breaking strength in the AAA wall and normal abdominal aorta, respectively, the pressure-strain elastic modulus, a reference pressure-strain elastic modulus, ILT volume, and AAA volume, where the exponents a and b may be deduced from clinical data sets.

Table 3.3.2 Comparisons with the new wall stress equation.

AAA model	p [mmHg]	$d_{AAA,max}$ [cm]	t [cm]	α	β	Max. Stress σ [MPa]	New wall stress equation (Eq. 3.2.2)		Laplace equation (Eq. 3.2.1) (cylinder)	
							Stress [MPa]	Error	Stress [MPa]	Error
Fillinger et al. (2003,2002)	120	6.7	0.19	0	0.4	0.32	0.335	4.7%	0.281	12.2%
	130	5.5	0.19	0	0.4	0.3	0.278	7.3	0.25	16.7%
Wang et al. (2002)	128	6.1	0.184	0.54	0.3	0.19	0.208	9.5%	0.282	48.4%
	155	6.4	0.175	0.3	0.9	0.35	0.343	2.0%	0.277	20.9%
Vorp et al. (1998)	120	6	0.15	0	0.3	0.33	0.34	3.0%	0.319	3.3%
Raghavan et al. (2000)	115	5.2	0.19	0	0.7	0.23	0.21	8.7%	0.209	9.1%
	188	5.5	0.19	0	0.9	0.43	0.46	6.9%	0.362	5.7%
Thurbrikar et al. (2000)	120	5.86	0.104	0	0.5	0.37	0.389	5.1%	0.449	21.4%
	120	5.86	0.158	0	0.5	0.28	0.299	6.7%	0.296	5.7%
Li & Kleinstreuer (2005)	120	5.0	0.05	0.15	1	0.43	0.412	4.2%	0.798	85.6%

AAA-rupture prediction. In order to quantify AAA-rupture prediction for individuals, key biomechanical parameters, most notably the wall stress (see Eqs. (3.3.8) and (3.3.10)), the ratio of maximum AAA diameter to normal aortic neck diameter, and the annual AAA-growth rate, have been calculated based on CT measurements and expressed in terms of a time-dependent severity parameter (SP).

$$SP = \sum_i \alpha_i \; P_i \tag{3.3.11}$$

where α_i are the weighting coefficients and P_i are the dimensionless biomechanics quantities (see Kleinstreuer & Li, 2006).

The severity parameter can vary between 0 and 1 (or 0 and 100%), where SP(t) < 0.2 indicates a low rupture risk, 0.2 SP(t) < 70 a moderate to average rupture risk, and where SP(t) > 0.70 surgical intervention should be considered. Figure 3.3.9 depicts a hypothetical sample SP(t) for patient "Johnson," supported by bar-graphs for the two commonly employed predictors, i.e., AAA-diameter ratio and AAA-wall stress ratio. The latter is shown in both constant and time-dependent modes.

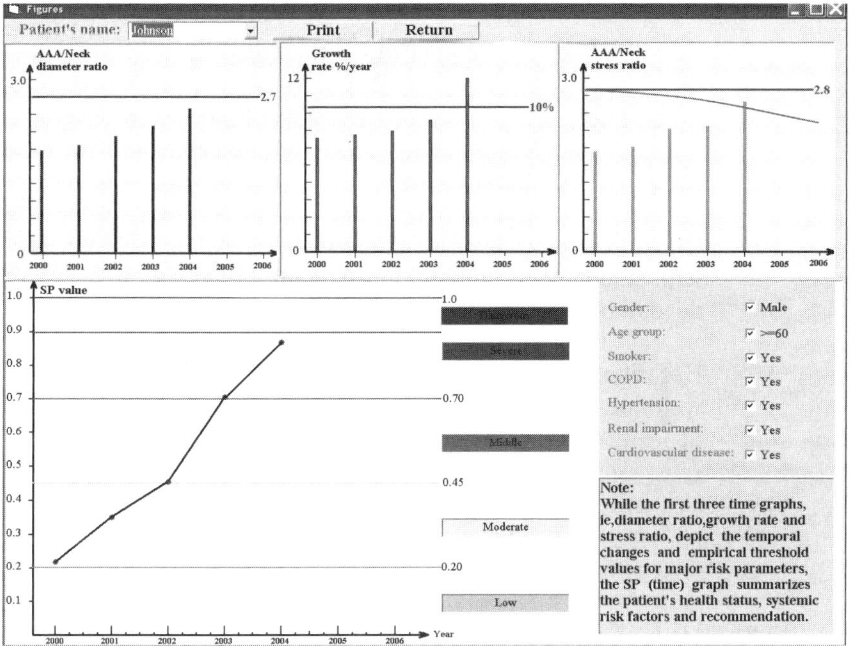

Fig.3.3.9 Hypothetical example with time-dependent wall-stress threshold (Kleinstreuer & Li, 2005 and Kleinstreuer & Li, 2006).

3.4 HOMEWORK PROBLEMS

Note: The homework assignment for Chapter III are more qualitative, i.e., literature reviews, critical assessments, and comparisons, rather than requiring quantitative problem solutions.

3.1 Describe and illustrate:
   (a) stenotic development, i.e., intimal thickening, leading to partial blood vessel occlusion
   (b) blood clot (or thrombus) formation leading to sudden downstream vessel occlusion
   (c) embolization

3.2 Discuss the causes/differences between "atherosclerosis" and "intimal hyperplasia" and give examples of their occurrences.

3.3 Graph Equations (3.1.2c) and (1.4.23) for different sets of model parameters (see Fig. 1.4.3).

3.4 The hematocrit, i.e., RBCs by volume, say, 40%, differs between people and changes in the blood vessels with location and time. Develop a conceptual model which takes into account $H = H(\vec{x}, t)$ for a proper blood rheology.

3.5 In conjunction with Section 1.4.3, define and discuss all Reynolds numbers, length scales, and time constants important in blood particle transport.

3.6 Update the discussion of "near-wall forces" (see Sect.3.1.4.2) with an emphasis on particle lift as important for micro-particle deposition.

3.7 Describe physically and mathematically blood-cell and vessel-wall interactions, such as particle lift, rolling, resuspension, deposition, attachment, and wall migration.

3.8 Critically review the pros & cons of hemodynamics wall parameters (see Sect.3.1.4.3) which are supposed to indicate the onset and progression of arterial diseases.

3.9 DAIH (distal anastomotic intimal hyperplasia) formation has been explained via two contrasting viewpoints: (a) aggravating multi-pathway particle-hemodynamics (see Sects. 3.1.1.2, 3.1.2 & 3.1.5.3); and (b) excessive, variable wall stress in critical junction areas. Discuss the merits of these two hypotheses.

3.10 Comment on the usefulness of the designs shown in Figs. 3.1.8-3.1.11 and suggest an "optimal" end-to-side configuration.

3.11 Questions concerning aneurysms:
(a) Who was the most famous person of the 20th century who died because of a sudden AAA rupture?
(b) Why do the platinum-coils in a saccular brain aneurysm reduce the local (radial) blood pressure?
(c) What are the differences and functions of stents in brain (or coronary) arteries with aneurysms and AAAs?
(d) Why is the total cost for EVAR almost twice as much as that for open surgery?

(e) What exactly causes aneurysm rupture and does this criterion change with time?
(f) In general, why is it recommended that old AAA-patients undergo EVAR while relatively young ones should opt for open surgery?

3.12 Discuss the make-up, properties, and functions of (a) SMCs; (b) elastin; and (c) collagen in arterial walls. What is their change in properties over time?

3.13 Write a brief report on imaging/scanning techniques, such as X-rays, ultrasound, MRI/MRA, CT/CAT, etc., as used for diagnostics, surgery, and computer modeling.

3.14 Mathematical AAA modeling questions:
(a) How are wall structure and fluid flow equations coupled?
(b) In computational fluid-structure interactions of AAAs, how exactly are wall movement and blood flow in changing conduits calculated over time?
(c) For low pressure loads with small variations Hooke's law (see Eq. (3.2.9)) may be applicable; however, stress-strain relations for arterial wall are nonlinear, in general. List in detail more realistic constitutive models, e.g., hyperelastic, viscoelastic, etc., and comment.
(d) How can Equation (3.3.11) be derived and is it the best indicator for possible AAA rupture?
(e) Figure 3.2.5c shows the outlet pressure waveform. Isn't the inlet pressure needed as well?
(f) What are the highest risks for EVAR vs. open surgery?

References

Asgharian B., Menache M.G., and Miller, F.J. (2004) Journal of Aerosol Medicine, 17: 213-224.

Bird, R. B., Stewart, W. E., and Lightfoot, E. N. (2002) "Transport Phenomena," John Wiley, New York, NY.

Blankensteijn, J.D., and Prinssen, M. (2002) Journal of Endovascular Therapy, 9:458-463

Buchanan, J. R. (2000) Computational Particle Hemodynamics in the Rabbit Abdominal Aorta. PhD Dissertation, Mechanical and Aerospace Engineering Department, North Carolina State University, Raleigh, NC.

Buchanan, J.R., and Kleinstreuer, C. (1998) Journal of Biomechanical Engineering, 120: 446-454.

Buchanan, J.R., Kleinstreuer, C., and Comer, J.K. (2000) Computers and Fluids, 29:695-724.

Caro, C.G., Fitz-Gerald, and Schroter, R.C. (1971) Proceedings of the Royal Society of London, Series B, 177:109-159.

Cherukat, P., and McLaughlin, J. B. (1994) Journal of Fluid Mechanics, 263:1-18.

Chervu, A., and Moore, W. S. (1990) Gynecology and Obstetrics, 171: 433-447.

Clift, R., Grace, J. R., and Weber, M. E. (1978) "Bubbles, Drops, and Particles," Academic Press, New York, NY.

Clowes, A. W., Reidy, M. A., and Clowes, M. M. (1983) Laboratory Investigation, 49: 327-332.

Coimbra, C. F. M., and Rangel, R. H. (1998) Journal of Fluid Mechanics, 370: 53-72.

Cokelet G. R. (1987) The rheology and tube flow of blood, In: Handbook of Bioengineering, R. Skalak and S. Chien (eds.), McGraw-Hill, New York, NY.

Colman, R. W., Marder, V. J., Salzman, E. W., and Hirsh, J. (1994) Overview of Hemostasis. In Hemostasis and Thrombosis. Colman, R. W., Hirsh, J., Marder, V. J., and Salzman, E. W. (eds.), Lipencott, Philadelphia, 3-18.

Corash, L., Tan, H., and Harvey, R. G. (1977) Blood, 49:71-87.

Cox, R. G., and Brenner, H. (1967) Chemical Engineering Sciences, 22: 1753-1777.

Crowe, C., Sommerfeld, M., and Tsuji, Y. (1998) "Multiphase Flows with Droplets and Particles," CRC Press, Boca Raton, FL.

Dahneke, B. (1974) Journal of Colloid and Interface Science, 48:520-522.

Dalman, R. L., and Taylor, L. M. Jr. (1993) Infrainguinal Revascularization Procedures. In: Basic Data Underlying Clinical Decision Making in Vascular Surgery, Porter, J. M., Taylor, L. M. Jr. (eds.), Quality Medical Publishing Inc., St. Louis, Missouri.

Davies, M. J. (1994) Mechanisms of thrombosis in atherosclerosis, In: Hemostasis and Thrombosis: Basic Principles and Clinical Practice, 3rd Ed., R. W. Colman, J. Hirsh, V. J. Marder, and E. W. Salzman (eds.), J. B. Lippincott Company, Philadelphia, 1224-1237.

Ehrlich, L. W., and Friedman M. H. (1977) Journal of Biomechanics, 10: 561-568.

Ferziger, J. H., and Peric, M. (1999) Computational Methods for Fluid Dynamics, 2nd ed., Berlin, Springer-Verlag.

Fisher, R. K., Kirkpatrick, U. J., How, T. V., Brennan, J. A., Gilling-Smith, G. L., and Harris, P. L. (2002) British Journal of Surgery, 89(4):514.

Fox, S. I. (1996) "Human Physiology," McGraw-Hill, Boston, MA.

Fry, D. L. (1968) Circulation Research, 22: 165-197.

Gawenda, M., Knez, P., Winter, S., Jaschke, G., Wassmer, G., Schmitz-Rixen, T., and Brunkwall, J. (2004) European Journal of Vascular and Endovascular Surgery, 27: 45-50.

Goldman, A. J., Cox, R. G., and Brenner, H. (1967a) Chemical Engineering Science, 22:637-651.

Goldman, A. J., Cox, R. G., and Brenner, H. (1967b) Chemical Engineering Science, 22: 653-660.

Greenhalgh, R.M., Brown, L.C., Kwong, G.P., Powell, J.T., Thompson, S.G. (2004) Lancet, 4:843-8.

Harris, P. L., and How, T. V. (1999) Critical Ischaemia, 9:20-26.

He, X., and Ku, D. N. (1996) Journal of Biomechanical Engineering, 118: 74-82.

Helmlinger, G., Geiger, R. V., Schreck, S., and Nerem, R. M. (1991) Journal of Biomechanical Engineering, 113:123-131.

Hofmann, W., Asgharian, B., and Winkler-Heil (2002) J. Aerosol Sci., 33: 219-235.

How, T. V., Rowe, C. S., Gilling-Smith, G. L., and Harris, L. (2000) Journal of Vascular Surgery, 31:1008-1017.

Hughes, T.J.R., and Lubliner, J. (1973) Mathematical Biosciences, 18:161-170.

Humphrey, J.D. (2002) "Cardiovascular Solid Mechanics: Cells, Tissues, and Organs," Springer-Verlag, NY.

Humphrey, J.D., and Delange, S.L. (2004) "An Introduction to Biomechanics," Springer, New York, NY.

Hyun, S. (1998) "Transient Particle-Hemodynamics Simulations in Three-Dimensional Carotid Artery Bifurcations," PhD Dissertation, Mechanical and Aerospace Engineering Department, North Carolina State University, Raleigh, NC.

Hyun, S., Kleinstreuer, C., Longest, P.W., and Chen, C. (2004) ASME J. Biomechanical Engineering, 126:188-195.

Inokuchi, K., Kusaba, A., Kamori, M., Kina, M., and Okadome, K. (1982) Surgery, 92:1006-1015.

Karino, T., and Goldsmith, H. L, (1977) Philosophical Transactions of the Royal Society of London, Series B, 279: 413-445.

Keynton, R. S., Evancho, M. M., Sims, R. L., Rodway, N. V., Gobin, A., and Rittgers, S. E. (2001) Journal of Biomechanical Engineering, 123: 464-473.

Kissin, M., Kansal, N., Pappas, P. J., DeFouw, D. O., Duran, W. N., and Hobson, R. W. (2000) Journal of Vascular Surgery, 31: 69-83.

Kleinstreuer, C. (2003) "Two-Phase Flow: Theory and Applications," Taylor & Francis, New York, NY.

Kleinstreuer, C., Hyun, S., Buchanan, J. R., Longest, P. W., Archie, J. P., Truskey, G. A. (2001) Critical Reviews in Biomedical Engineering, 29: 1-64.

Kleinstreuer, C., Lei, M., and Archie, J. P. (1996) ASME J. Biomechanical Engineering, 118:506-510.

Kleinstreuer, C. and Li, Z. (2005) "Patent Application for Computerized Medical Management Software," North Carolina State University, Raleigh, NC

Kleinstreuer, C., and Li, Z. (2006) "Analysis and Computer Program for Rupture-risk Prediction of Abdominal Aortic Aneurysms," Biomedical Engineering Online.

Kleinstreuer, C., and Longest, P.W. (2003), "Particle-hemodynamics Analyses of End-to-side Anastomoses: A Computational Comparison Study," In: Vascular Grafts - Experiment and Modeling, A. Tura (ed.), Advances in Fluid Mechanics Series; WIT Press, Ashurst, UK.

Ku, D. N. (1997) Annual Review of Fluid Mechanics, 29: 399-434.

Ku, D. N., Giddens, D. P., Zarins, C. K., and Glagov, S. (1985) Arteriosclerosis, 5(3):293-302.

Kunov, M. J., Steinman, D. A., and Ethier, C. R. (1996) J. Biomechanical Engineering, 118: 158-164.

Lambert, A.W., Williams, D.J., Budd, J.S., and Horrocks, M. (1999) European Journal of Endovascular Surgery, 17: 60-65.

Lei, M. (1995) "Computational fluid dynamics and optimal design of bifurcating blood vessels," PhD Dissertation, Mechanical and Aerospace Engineering Department, North Carolina State University, Raleigh, NC.

Lei, M., Kleinstreuer, C., and Archie, J. P. (1996) Journal of Biomechanics, 29:1605-1614.

Lei, M., Kleinstreuer, C., and Archie, J. P. (1997) ASME J. Biomechanical Engineering, 119:343-348.

Leuprecht, A., Perktold, K., Prosi, M., Berk, T., Trubel, W., and Schima, H. (2002) Journal of Biomechanics, 35:225-236.

Li, Z., (2005) "Computational Analysis and Simulations of Fluid-structure Interactions for Stented Abdominal Aortic Aneurysms," PhD Dissertation, North Carolina State University, Raleigh, NC.

Li, Z., and Kleinstreuer, C. (2005a) Annals of Biomedical Engineering, 33: 209-213.

Li, Z., and Kleinstreuer, C. (2005b) ASME J. Biomech. Engineering, 127: 662-671.

Lohner, R., Cebral, J., Soto, O., Yim, P., and Burgess, J.E. (2003) Int. J. Numerical Methods in Fluids, 43: 637-650.

Longest, P. W. (1999) "Computational Hemodynamic Simulations and Comparison Study of Arteriovenous-access Grafts for Hemodialysis," MS Thesis, Mechanical and Aerospace Engineering Department, North Carolina State University, Raleigh, NC.

Longest, P.W. (2004) "Computational Analyses of Transient Particle Hemodynamics with Applications to Femoral Bypass Graft Designs," Ph.D. Dissertation, MAE Dept., North Carolina State University, Raleigh, NC.

Longest, P.W., and Kleinstreuer, C. (2003) Medical Engineering and Physics, 25: 843-858.

Longest, P.W., Kleinstreuer, C., and Buchanan, J.R. (2004) Computers and Fluids, 33: 577-601.

Longest, P. W. and Kleinstreuer, C. (2000) Journal of Medical Engineering and Technology, 24: 102-110.

Longest, P. W., Kleinstreuer, C., and Andreotti, P. J. (2000) Critical Reviews in Biomedical Engineering, 28:141-147.

Lonsdale, R. D. (1993) International Journal of Numerical Methods for Heat Transfer and Fluid Flow, 3:3-14.

Loth, E. (2000) Progress in Energy and Combustion Science, 26:161-223.

Loth, F., Jones, S., Zarins, C. K., Giddens, D. P., Nassar, R. F., Glagov, S., and Bassiouny, H. S. (2002) Journal of Biomechanical Engineering, 124:44-51.

Loth, F., Jones, S., Giddens, D., Bassiouny, H., Glagov, S., and Zarins, C. (1997) ASME J. Biomechanical Engineering, 119:187-194.

Malina M., Lánne T., Ivancev K., Lindblad B., and Brunkwall J. (1998) Journal of Vascular Sugery, 27:624-31.

Maxey, M. R. (1987) Physics of Fluids, 30:1579-1582.

Maxey, M. R., and J. J. Riley (1983) Physics of Fluids, 26: 883-889.

McLaughlin, J. B. (1994) International Journal of Multiphase Flow, 20: 211-232.

Mei, R. (1994) Journal of Fluid Mechanics, 270: 133-174.

Meter, O. (2000) Annals of Biomedical Engineering, 28: 1281-99.

Michaelides, E. E. (1997) Journal of Fluids Engineering, 119: 233-247.

Miller, J. H., Foreman, R. K., Ferguson, L., and Fads, I., (1984) Aust. N. Z. J. Surgery, 54:283-285.

Nestorov I. (2003) Clinical Pharmacokinetics, 42: 883-908.

Nichols, W. W., and O'Rourke, P. J. (1998) "McDonald's Blood Flow in Arteries: Theoretical, Experimental and Clinical Principles," Lea and Febiger, Philadelphia, PA.

Noori, N., Scherer, R., Perktold, K., Czerny, M., Karner, G., Trubel, W., Polterauer, P., Schima, H. (1999) Eur. J. Endovasc. Surg., 18:191-200.

Olufsen, M. (1999) American Journal of Physiology, 276: 257-268.

Okadome, K., Onohara, T., Yamamura, S., and Sugimachi, K. (1991) Annals of Vascular Surgery, 5: 413-418.

Pappas, P. J., Hobson, R. W., Meyers, M. G., Jamil, Z., Lee, B. C., Silva, M. B., Goldberg, M. C., and Padberg, F. T. (1998) Cardiovascular Surgery, 6: 19-26.

Phalen R.F., and Oldham M.J. (2001) Respiration Physiology, 128:119-130.

Raghavan M (2002) "Lecture notes on cardiovascular bio-solid mechanics section," Dept. of Biomechanical Engineering, University of Iowa.

Rangayyan, R.M. (2005) "Biomedical Image Analysis," CRC Press, Boca Raton, FL.

Rhie, C. M., and Chow, W. L. (1983) AIAA Journal, 21: 1527-1532.

Richardson, P. D., Davies, M. J., and Born, G. V. R. (1989) Lancet, 2:941.

Rodgers, T., Leahy, D., and Rowland, M. (2005) Journal of Pharmaceutical Sciences, 94: 1259-1276.

Ross, R. (1986) New England Journal of Medicine, 314:488-500.

Ross, R. (1999) New England Journal of Medicine, 340:115-126.

Salunke, N. V., and Topoleski, L. D. T. (1997) Critical Reviews in Biomedical Engineering, 25:243-285.

B. D. Seeley, and D. F. Young (1976) Journal of Biomechanics, 9:439-448.

Shi, H. (2006) "Analyses and Simulations of Inhaled Particle Uptake and Droplet Spray Dynamics in Representative Human Nasal Airways," PhD Dissertation, North Carolina State University, Raleigh, NC.

Shyy, W., Thakur, S., and Wright, J. (1992) AIAA Journal, 30: 923-931.

Sonesson B., Resch T., Lánne T., and Ivancev, K. (1998) Journal of Vascular Surgery, 28: 889-894.

Sottiurai, V.S., Yao, J.S.T., Flinn, W.R., and Batson, R.C. (1983) Surgery, 93:809-817.

Sottiurai, V. S. (1999) International Journal of Angiology, 8:1-10.

Steele, B.N., Wan, J., Ku, J.P., Hughes, T.J.R., and Taylor, C.A. (2003) IEEE Transactions on Biomedical Engineering, 50:649-656.

Steinman, D.A., Vinh, B., Ethier, C.R., Ojha, M., Cobbold, R.S.C., and Johnston, K.W. (1993) Journal of Biomechanical Engineering, 114: 112-118.

Steinman, D. A., Milner, J. S., Norley, C. J., Lownie, S. P., and Holdsworth, D.W. (2003) Am. J. Neuroradiology, 24:559-566.

Stonebridge, P.A., Prescott, R.J., and Ruckley, C.V. (1997) Journal of Vascular Surgery, 26: 543-550.

Suzuki, K., Ishiguchi, T., Kawatsu, S., Iwai, H., Maruyama, K., and Ishigaki, T. (2001) Cardiovascular and Interventional Radiology, 24:94-98.

Tang, D., Yang, C., Zheng, J., Woodard, P.K., Sicard, G.A., Saffitz, J.E., and Yuan, C. (2004) Annals of Biomedical Engineering, 32(7): 947-960.

Taylor, C.A., and Draney, M.T. (2004) Annual Review of Fluid Mechanics, 36: 197-231.

Taylor, R. S., McFarland, R. J., and Cox, M. I. (1987) Eur. J. Vasc. Surg., 1:335.

Taylor, R.S., Loh, A., McFarland, R.J., Cox, M., and Chester, I.F. (1992) British Journal of Surgery, 79: 348-354.

Thompson, C. P., and Wiles, N. S. (1982) "Experiments with higher-order finite difference formulae," Harwell, UK: AERE Harwell, Report No. AERE-R 10493.

Truskey, G. A., Yuan, F., and Katz, D.F. (2004) "Transport Phenomena in Biological Systems," Pearson Prentice Hall, Upper Saddle River, NJ.

Van Doormal, J.P., and Raithby, G.D. (1984) Numerical Heat Transfer, 7: 147-163.

Veith, F.J., Gupta, S.K., Ascer, E., White-Flores, S., Samson, R.H., Scher, L.A., and Towne, J.B. et al. (1986) Journal of Vascular Surgery, 3: 104-114.

Vignon, I., and Taylor, C.A. (2004) Wave Motion, 39: 361-374.

Wan, J., Steele, B.N., Spicer, S.A., Strohband, S., Feijoo, G.R., Hughes, T.J.R., and Taylor, C.A. (2002) Computer Methods in Biomechanics and Biomedical Engineering, 5:195-206.

Wang, Q., Squires, K.D., Chen, M., and McLaughlin, J.B. (1997) Int. J. Multiphase Flow, 23:749-763.

White, J. G. (1994) Anatomy and structural organization of the platelet, in: Hemostasis and Thrombosis, R. W. Colman, J. Hirsh, V. J. Marder, and E. W. Salzman (eds.), Philadelphia, J. B.Lippencott, pp.397-413.

White, S. S., Zarins, C. K., Giddens, D. P., Bassiouny, H., Loth, F., Jones, S. A., and Glagov, S. (1993) ASME J. Biomechanical Engineering, 115: 104-111.

Wootton, D. M., and Ku, D. N. (1999) Annual Reviews of Biomedical Engineering, 10: 299-329.

Womersley, J. R. (1955) J. Physiology, 127:553-563.

Young, J. B., and Hanratty, T. J. (1991) AIChE Journal, 37: 1529-1536.

Zarins, C. K., Zatina, M. A., Giddens, D. P., Ku, D. N., and Glagov, S. (1987) Journal of Vascular Surgery, 5: 413-420.

Zegdi, R., Lajos, P., Ponzio, O., Bruneval, P., and Fabiani, J.-N. (2001)

ASAIO Journal, 47:329-332.
Zhang, Z., Kleinstreuer, C., Donohue, J.F. and Kim, C.S. (2005). J. Aerosol Science, 36: 211-233.

Chapter IV

BIOFLUID MECHANICS OF ORGAN SYSTEMS

Organs are groupings of primary tissues, i.e., muscle, connective, nervous, epithelial, and fibrous, into anatomical and functional units. In terms of surface area, the largest organ is the skin which has various functions provided by all five primary tissues. Interactive organs that perform primary functions can be grouped into systems (Table 4.1.1).

The first three, i.e., circulatory, respiratory, and regulatory systems, are of special interest in biofluid mechanics. They interact to provide proper blood flow and blood constituents. While Chapters II and III dealt with various aspects of the circulatory system, this chapter provides some background information on the respiratory system, including pulmonary problems, as well as the urinary and regulatory systems, i.e., the kidneys and the liver, respectively. Nevertheless, some intriguing convective and diffusive transport phenomena occur in the other organ systems as well (see Fox, 1996; Enderle et al., 2000; Truskey et al., 2004, among others). *Clearly, a strong knowledge base and elevated skill level of Chapter I and Appendix A material are essential for dealing with the research projects discussed in this chapter.*

As already alluded to in Chapter II, the transport of fluids and species from one domain to another across membranes of capillaries and tubules takes place on the microscale, due to:

- driving forces such as hydrostatic (blood) pressure and concentration gradients leading to osmotic pressure and/or diffusion (see Sect. 1.1.2);
- large local surface areas, often enhanced by microvilli;
- elevated membrane/wall permeability because of controlled openings via carrier-protons and ion-channels (Sect.1.1.2).

The specific examples discussed in this chapter include $CO_2 - O_2$ gas exchange in the lung, plasma-water filtration, salt absorption, and waste-product discharge in the kidneys, as well as regulation of the blood's chemical composition in the liver.

Table 4.1.1 List of body systems, major organs, and their functions.

System	Major Organs	Primary Functions
Circulatory	Heart, blood vessels, lymphatic vessels	Movement of blood and lymph
Respiratory	Lungs, airways	$O_2 - CO_2$ gas exchange w/blood
Regulatory	Liver, kidneys, ureter	Regulation of blood composition, pressure, and clearance
Immune	Bone marrow, lymphoid organs	Defense of the body against invading pathogens
Integumentary (covering)	Skin, hair, nails	Protection, thermoregulation
Skeletal	Bones, cartilages	Movement and support
Endocrine (ductless glands)	Hormone-secreting glands, such as the pituitary, thyroid, and adrenals	Secretion of regulatory molecules called hormones
Nervous	Brain, spinal cord, nerves	Regulation of other body systems
Muscular	Skeletal muscles	Movements of the skeleton
Digestive	Mouth, stomach, intestine, liver, gallbladder, pancreas	Breakdown of food into molecules that enter the body
Reproductive	Gonads, external genitalia, associated glands and ducts	Continuation of the human species

After Fox (1996)

4.1 THE LUNGS

The respiratory system consists of the conducting zone, i.e., the upper airways, trachea plus bronchial tree, and the respiratory zone, i.e., the bronchioles and alveolar sacs, where the $O_2 - CO_2$ gas exchange takes place. The upper airways include the nose/mouth, pharynx, and larynx (Fig. 4.1.1). The nose with its two cavities separated by the nasal septum contains three airflow passages on each side, i.e., the main passage with the narrow olfactory region as well as the two meatuses. They are created by

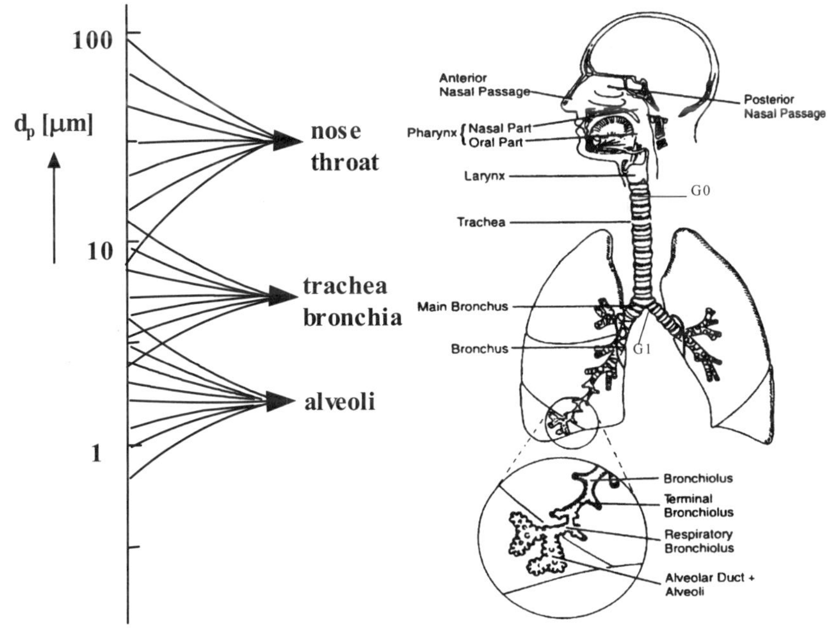

Fig.4.1.1 Human respiratory system and likely locations of microparticle deposition.

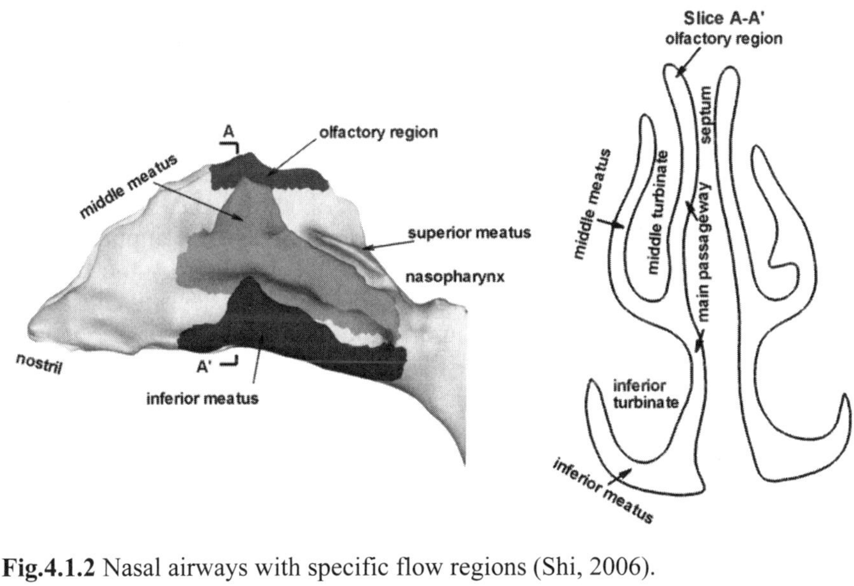

Fig.4.1.2 Nasal airways with specific flow regions (Shi, 2006).

the projection of turbinates from the lateral wall (Fig. 4.1.2). Mucous layers on the passage walls condition the inhaled air and trap fine/ultrafine particles and microorganisms. The turbinates are bones with warm mucous-membrane surfaces that also contain sensitive nerves that detect odors or induce sneezing to expel irritating particles. The pharynx (or throat) is a muscular section where air flows from the nose to the larynx/ trachea and food slides from the mouth to the esophagus/stomach. The regulator is the epiglottis which closes during swallowing and is open during breathing. The larynx (or voice box) is a cartilaginous structure connecting the pharynx and trachea. It mainly contains the vocal cords which produce sound because of vibration when air passes through the opening, called the glottis. The trachea (or windpipe) is a smooth-muscle tube supported by incomplete, i.e., C-shaped, cartilaginous rings. After the airway constriction in the larynx, airflow in the trachea and below may become turbulent at elevated breathing rates. The trachea divides at the carina to form the left and right bronchi (or branches) leading to the lungs. Specifically, these bronchi bifurcate into three secondary bronchi in the right lung and into two in the left lung. This portion of the respiratory tract is also lined with mucous membranes where the cilia (i.e., fine hairs) sweep mucus and embedded particles toward the pharynx for clearing. Further repeat-bifurcations of the bronchi rapidly reduce the local airflow rate and geometric dimensions until the terminal bronchioles are reached which become the alveolar ducts. Clearly, even when the airflow in the trachea is turbulent, say, during exercise, continuous airflow rate division quickly reduces the local Reynolds number all the way to $Re < 1$ in the alveolar region. At the end of the alveolar ducts millions of small sacs (the alveoli) are clustered, surrounded by capillary beds connected to both the pulmonary artery and the pulmonary vein. The thin-walled alveoli consist of a layer of epithelial cells for the $O_2 - CO_2$ exchange, surfactant production to control the sac stability via variable surface tension, and scavenging of foreign material, such as bacteria.

In addition to the upper airways and lungs, the respiratory system also assures for proper breathing in the thorax (or chest) and the abdomen, where the two are separated by a diaphragm. Neural inputs to and feedback from these pulmonary structures are controlled by the nervous system. Specifically, on inspiration the respiratory muscles and chest wall, together with the downward moving diaphragm act like a "pump" creating a negative-pressure field (i.e., a partial vacuum) by enlarging the thoracic space. On expiration, the respiratory muscles relax, the diaphragm moves from its flat position upwards again, and the lung deflates (Pedley, 1977). The rate and depth of ventilation, i.e., air movement in and out of the respiratory tract, is neurologically controlled to maintain the required CO_2/O_2-levels and, in conjunction with the kidneys, a normal body fluid pH of 7.4. In addition to ventilation and subsequent gas diffusion in the alveolar region, blood perfusion via the bronchial and pulmonary circulation supplies nutrients and oxygen to the lungs. Specifically, the bronchial arteries, branching from the thoracic aorta and intercostal arteries, perfuse the trachea and bronchi as well as the lungs, tissue, nerves, and outer layers of the pulmonary arteries and veins. The pulmonary circulation is a low-pressure system where venous blood is transported from the right ventricle to the right and left lungs (see Fig.1.1.2 and Sect.2.2). In the pulmonary capillary bed the gas exchange takes place and then O_2-rich blood is returned through the pulmonary veins to the left atrium.

Clearly, the lungs are very elastic, due to a high content of elastin and collagen fibers. They expand nonlinearly under transpulmonary pressure to a point. This distensibility is called compliance which relates changes in lung volume to net pressure changes, i.e., $C=\Delta\forall/\Delta p$, where $C=C(\forall)$ as indicated in Fig. 4.1.3.

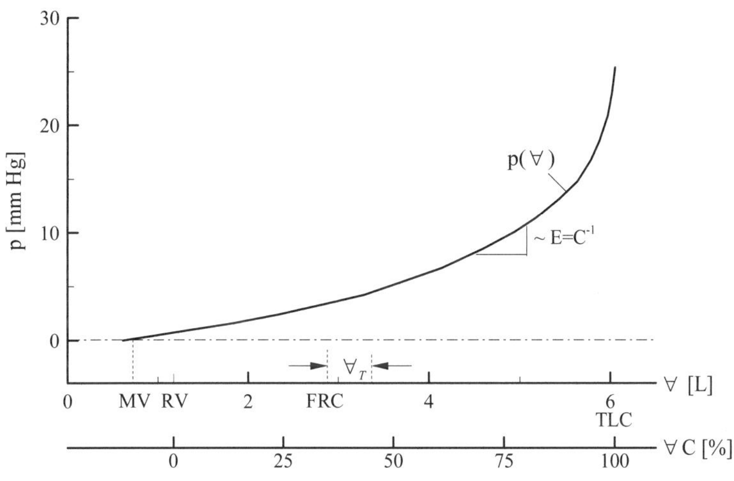

After Staub (1991)

Fig. 4.1.3 Pressure-volume curve of the lung. Note: MV ≙ minimal volume; RV ≙ residual volume at minimally inflated lung, FRC ≙ functional residual capacity (FRC ≈ 3 liters), i.e., air volume at the start of tidal inhalation; V_T ≙ tidal volume, i.e., net average volume during periodic (or tidal) breathing as needed for metabolic requirements; TLC ≙ total lung capacity (TLC ≈ 6 liters), i.e., air-volume in maximally inflated lung; VC ≙ vital capacity, i.e., the largest inhaled volume (VC = TLC-RV ≈ 4 liters); and IC ≙ inspiratory capacity (IC = TLC – FRC ≈ 3 liters).

Lung compliance and elasticity are inversely proportional, i.e., $E = 1/C$. The inflation limit (see Fig. 4.1.3) is largely due to the inelasticity, i.e., elastic resistance, of the collagen fibers when they are stretched.

The net result of fluid secretion (active Cl^- transport) and absorption (active transport of Na^+) is a thin aqueous film in the alveoli, generating a surface tension σ [F/L]. In turn, this creates a pressure in the alveolar sacs of radii r, which could be estimated from fluid statics as $p = 2\,\sigma/r$. This implies that smaller alveoli (without film surfactant) with higher pressure levels would reduce in size, even collapse, and drive air into the larger sacs. However, a surface-active agent (surfactant), produced by alveolar cells, gradually lowers the surface tension as the alveoli get smaller during expiration and hence prevents the sacs from collapsing (see Grotberg, 1994).

Two-phase flow in the lung is best illustrated considering the important roles surface tension and surfactants play in lung inflation and function of newborns. While normal infants have developed a fully functioning surfactant system assisting in inflating the fluid-filled lungs at birth, premature babies lack this surfactant system and hence exhibit very high surface-tension levels. As a result, premature infants cannot inflate their lungs and breathe on their own, thus requiring a mechanical ventilator for oxygen supply (see Ghadiali et al., 2002; Ghadiali & Gaver, 2003, among others). When analyzing the fluid dynamics and physiology of collapsible respiratory airways, one should recall that pulmonary surfactants exhibit non-uniform (bulk) concentrations and nonlinear surface tension properties (Morris et al., 2001).

Air bubble progression and surface film dependence in lung-airway opening. The steady opening of a liquid-filled airway can be modeled as the progression of a semi-infinite bubble in a rigid axisymmetric 2-D conduit. In letting the channel move to the left with constant velocity U, the air bubble is stationary and generates a liquid film $\delta_f = \delta_f(\mu, \gamma, U)$. This problem and its solution was provided by Prof. S. Ghadiali (Lehigh University).

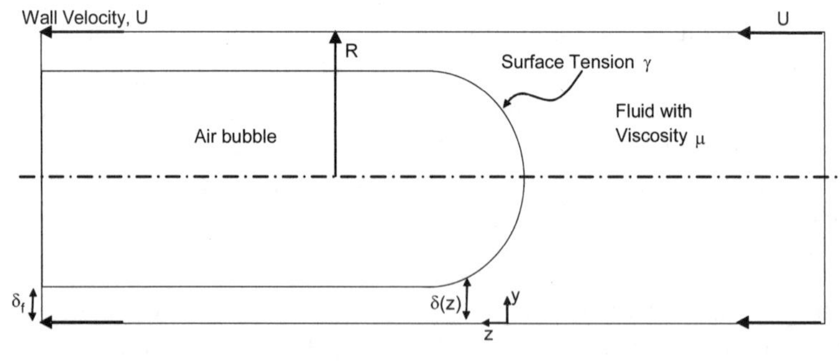

Fig.4.1.4 System schematic.

Assuming steady, incompressible, low-Reynolds-number flow with no gravity effect, the Navier-Stokes equations in the thin film reduce to:

$$\frac{\partial p}{\partial z} = \mu \frac{\partial^2 u}{\partial y^2} \tag{4.1.1}$$

The Young-Laplace Equation, which is a force-balance relationship, can be used to express the fluid pressure p in terms of the surface tension γ

$$-p = \gamma\kappa = \gamma \frac{\partial^2 \delta}{\partial z^2} \tag{4.1.2}$$

Here κ is the interfacial curvature and differentiating Eq. (4.1.2), we can write with Eq.(4.1.1):

$$\gamma \frac{\partial^3 \delta}{\partial z^3} = -\mu \frac{\partial^2 u}{\partial y^2} \tag{4.1.3}$$

The boundary conditions for this problem are u(y=0) = U and du/dy=0 at y=δ(z). Note that we are assuming a constant surface tension at the air-liquid interface and no Marangoni stresses. Integration of Eq. (4.1.3) with these boundary conditions yields:

$$u(y) = U - \frac{\gamma}{\mu}\frac{\partial^3 \delta}{\partial z^3}\left[\frac{y^2}{2} - \delta y\right] \tag{4.1.4}$$

We can calculate the flow rate through a vertical section in the thin film as:

$$Q = \int_0^{\delta(z)} u(y)dy = U\delta + \frac{\gamma}{\mu}\frac{\partial^3 \delta}{\partial z^3}\frac{\delta^3}{3} \tag{4.1.5}$$

However, as $z \to \infty$, the flow becomes uniform and Q=Uδ_f. This yields the following governing equation for the film thickness

$$\delta^3 \frac{\partial^3 \delta}{\partial z^3} + \frac{3\mu U}{\gamma}\delta = \frac{3\mu U}{\gamma}\delta_f \tag{4.1.6}$$

Using the following scales $z = \zeta\delta_f(3Ca)^{-1/3}$ and $\delta = \eta\delta_f$, where Ca is the capillary number defined as Ca=μU/γ, the dimensionless governing equation is:

$$\eta \frac{\partial^3 \eta}{\partial \zeta^3} = 1 - \eta \tag{4.1.7}$$

For this third-order differential equation, three boundary conditions have to be specified plus a 4th condition in order to determine the film thickness, δ_f.

- The first condition is that as $\zeta \to \infty$, $\eta \to 1$.

- In the neighborhood of η=1, Eq. (4.1.7) reduces to $d^3\eta/d\zeta^3=1-\eta$. The particular solution to this equation can be written as: $\eta=1+A\exp(-\zeta)+B\exp(\zeta)$.
- However, η cannot grow exponentially, so B=0. The coefficient A may be chosen arbitrarily because the solution should be invariant to a shift in the origin.
- The 4th and last condition is that there should be a smooth transition in interfacial curvature from the thin film region to a region near the bubble cap. The assumption is that the curvature in the bubble-cap region is constant and therefore at some value of ζ, i.e.,

$$\frac{\partial^2 \eta}{\partial \zeta^2} = \alpha \tag{4.1.8}$$

  where α is a constant that is related to the curvature in the bubble-cap region. One can perform a stepwise numerical integration of Eq. (4.1.7) by starting at a large value of ζ and decreasing ζ until $d^2\eta/d\zeta^2$ is constant. This integration yields a value of α=0.643.

By writing Eq. (4.1.8) in dimensional form and setting the curvature for a projected two-dimensional spherical bubble to $d^2\delta/dz^2 = 1/R$, where R is the channel radius, we obtain:

$$\frac{\delta_f}{R} = 0.643(3Ca)^{2/3} \tag{4.1.9}$$

This expression holds for low capillary numbers and has been validated with experiments (Bretherton, 1961).

4.1.1 Respiratory Tract Geometry

In order to simulate and analyze airflow as well as inhaled particle transport and deposition in the lung, realistic geometries of the respiratory system are needed. However, there are several obstacles which have to be addressed before computer modeling or laboratory testing can begin. Clearly, the system morphology is very complex, ranging from intricate nasal airways to the pharynx, larynx, and trachea of the upper airways, and then progressing through 23 generations of the lungs, ending in the alveolar regions with millions of alveoli of different shapes and diameters of $\mathcal{O}(100\mu\text{m})$. Furthermore, the geometries of the smaller airways and alveoli change nonlinearly with time as a natural part of breathing. Finally, there is the problem of subject variability, i.e., each individual has a different respiratory tract morphology which changes with age and may become locally obstructed because of pulmonary diseases.

In general, the lungs consist of a series of bifurcating tubes, where each bifurcation leads to a new lung "generation" numbered 1-23 (see Fig. 4.1.1 and Table 4.1.2). As the tubes get smaller and shorter, they end via the terminal bronchioli and alveolar ducts into the millions of sacs with a

Table 4.1.2 Dimensions of the Weibel A lung geometry (Weibel, 1963) scaled to a 3 ltr lung volume, and using a volume of 10^{-5} ml per alveoli, is compared to the symmetric lung geometry used by Finlay et al. (2000). The underlines in the table indicate the border between the alveolar and tracheo-bronchial regions in the models (after Finlay et al., 2000).

Generation	Finlay et al. model length (cm)	Scaled Weibel length (cm)	Finlay et al. model diameter (cm)	Scaled Weibel diameter (cm)
0 (trachea)	12.456	10.26	1.81	1.539
1	3.614	4.07	1.414	1.043
2	2.862	1.624	1.115	0.71
3	2.281	0.65	0.885	0.479
4	1.78	1.086	0.706	0.385
5	1.126	0.915	0.565	0.299
6	0.897	0.769	0.454	0.239
7	0.828	0.65	0.364	0.197
8	0.745	0.547	0.286	0.159
9	0.653	0.462	0.218	0.132
10	0.555	0.393	0.162	0.111
11	0.454	0.333	0.121	0.093
12	0.357	0.282	0.092	0.081
13	0.277	0.231	0.073	0.07
14	<u>0.219</u>	0.197	<u>0.061</u>	0.063
15	0.134	0.171	0.049	0.056
16	0.109	<u>0.141</u>	0.048	<u>0.051</u>
17	0.091	0.121	0.039	0.046
18	0.081	0.1	0.037	0.043
19	0.068	0.085	0.035	0.04
20	0.068	0.071	0.033	0.038
21	0.068	0.06	0.03	0.037
22	0.065	0.05	0.028	0.035
23	0.073	0.043	0.024	0.035

total surface area for rapid diffusion of nearly 100 m^2. The first and still widely used mapping of the human lung was done by Weibel (1963), known as the symmetric planar Weibel A model. In reality, the airways are asymmetric, out-of-plane, and distensible. Thus, improvements were published by Horsfield & Cummings, 1968; Phalen et al., 1978; Weibel, 1991; and Finlay et al., 2000, among others. Table 4.1.2 contrasts the traditional Weibel A model with the updated geometries of Finlay et al. (2000). It should be noted that the conducting airways, i.e., extrathoracic airways plus tracheobronchial region, have an air volume of only about 150 ml, while the alveolar region encompasses 2-6 liters. Typical inhalation flow rates range from 10 to 60 l/min for adults.

4.1.2 Pulmonary Disorders and Treatment Options

Lung diseases can drastically change the airflow mechanics because of local airway constrictions (e.g., tumors and bronchitis) and destruction of alveolar tissue (i.e., emphysema). Bronchitis and emphysema, the two most common causes of respiratory failure, are together called chronic obstructive pulmonary disease (COPD), which is the fifth leading cause of death in the US. Another pulmonary disorder because of a genetic defect in the Caucasian population is cystic fibrosis. The end effect is that the airway fluid in the alveolar region becomes excessively viscous and difficult to clear. A modern approach to combat lung and other diseases starts with the inhalation of targeted drug aerosols. For example, the characteristics of drug aerosols, types of inhalers and aerosol dynamics are discussed in Hickey (2004) and by Finlay (2001), among others. In order to illustrate some of the difficulties encountered when targeting drug aerosols, computer simulation results of effective drug aerosol deposition on a bronchial tumor are presented next.

Fluid-particle dynamics surrounding a hemi-spherical lung tumor. Presently, lung cancer causes 28% of all cancer deaths, where bronchogenic carcinoma accounts for more than 90% of all lung tumors. Next to surgery and radiation therapy, chemotherapy with (aggressive) drug aerosols is a viable treatment option.

As a special case of local airway obstruction, we consider a hemispherical lung tumor of radius r situated in the second branch of generation 5 (G5.2) with radius R (see Fig. 4.1.5). In varying the tumor radius in the range of $0.2 \leq r/R \leq 2.0$, the branch area can be blocked anywhere from 2% to 100%. Inspired by the clinical observation that the effect of inhaled drug-aerosol treatment diminished as the tumor grew beyond a certain size, Kleinstreuer & Zhang (2003) analyzed the airflow pattern and local micro-particle depositions for different inlet conditions at G3 for this Weibel A model (see Table 4.1.2).

Breathing patterns are pulsatile in nature. Although air flow and particle transport can be considered as quasi-steady under normal breathing conditions, particle deposition data for cyclic flow cannot be simply evaluated assuming steady flow at the mean flow rate given by the average Reynolds number, Re_{mean}. However, Zhang et al. (2002) have found a matching flow rate, i.e., an inlet Reynolds number $Re_{match} \approx 0.50(Re_{mean} + Re_{max})$ representing the inhalation cycle, which generates very similar particle deposition efficiency values (< 5% error) and deposition patterns as in equivalent pulsatile flows. Hence, with the matching Reynolds and Stokes numbers representing resting and moderate exercise breathing (see Table 4.1.3), we expect that the computationally efficient constant-flow deposition data duplicate the cyclic-flow scenarios.

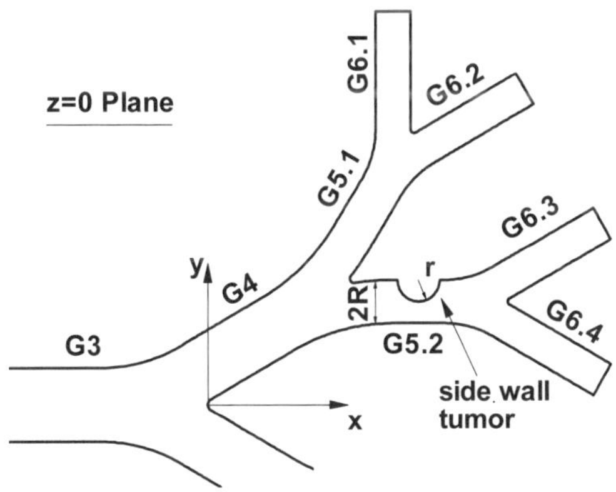

Fig.4.1.5 Schematic of a symmetric triple bifurcation airway model (generations G3-G6) with a hemispherical tumor.

Assuming steady, laminar incompressible air flow, monodispersed non-interacting spherical micron-size particles in smooth rigid impermeable conduits, the describing equations are:

(Continuity)

$$\nabla \cdot \vec{\mathbf{v}} = 0 \tag{4.1.10}$$

(Momentum)

$$(\vec{\mathbf{v}} \cdot \nabla)\, \vec{\mathbf{v}} = -\frac{1}{\rho} \nabla p + \nabla \cdot [\nu(\nabla \vec{\mathbf{v}} + (\nabla \vec{\mathbf{v}})^{tr})] \tag{4.1.11}$$

where $\vec{\mathbf{v}}$ is the fluid velocity vector; ρ is the fluid density; p is the pressure; and ν is the fluid kinematic viscosity.

The motion of liquid or solid particles suspended in air is governed by Newton's Second Law (see Sect.1.4.3):

$$m_p \frac{d^2 \vec{\mathbf{x}}_p}{dt^2} = \sum \vec{\mathbf{F}}_p \tag{4.1.12}$$

where m_p is the mass of a single particle, x_p is the displacement of the particle, and $\sum \vec{\mathbf{F}}_p$ are the forces acting on the particle. Considering micron-size particles, large particle-to-air density ratios, nonrotating spheres, and relatively low air shear stress variations, most of the known particle forces other than the drag force can be discounted using order of magnitude arguments (Kleinstreuer, 2003). First, these particles are relatively large, so that Brownian motion can be neglected. Second, the particulate material considered is far denser than air, causing terms that depend on the density ratio, such as pressure force, virtual mass effect, and Basset force, to be very small. Third, the lift forces in the present Stokes flow limit $(Re_p \leq 1)$ are negligible because of a lack of measurable particle spin (Magnus lift) and the laminar, low-level fluid shear fields (Saffman lift). For near-wall particle drag modifications, see Longest & Kleinstreuer (2003). Finally, the effects of gravity cannot be considered because gravitational settling is only important for large particle sizes and low flow rates which occur at the lower airways (say, from generation G9 on). More specifically, the resulting simplified particle equation of motion, i.e., Eqs. (4.1.12) and (4.1.13), which produced excellent particle deposition results when compared with measurements (Zhang et al., 2002), will allow elucidation of the physical processes at work. Hence, the drag can be considered as the dominant point force, and it may be stated as:

Table 4.1.3 Representative respiration data and particle characteristics with matching Reynolds numbers and Stokes numbers.

Physical State		Resting	Moderate exercise
Respiration rate for inhalation (Q_{in}, l/min)		15	60
Tidal volume (ml)		500	1153
Time ratio of inspiratory phase (t_{in}/t_{total})		0.47	0.474
Breathing frequency (cycles/min)		14	24.5
Mean Reynolds number [a] at G3		388	1586
Matching Reynolds number at G3		463	1882
Particle diameter (d_p, μm)		3 –7	3 –7
Particle density (ρ_p, kg/m^3)		540-6480	600-5000
Range of mean Stokes number [b] at G3		0.017-0.102	0.017-0.102
Matching Stokes number range at G3		0.02 – 0.12	0.02 – 0.12
Representative particle concentrations	Normal particle inlet	4.5×10^{-5}	1.25×10^{-5}
	Controlled inlet stream	5.63×10^{-4}	1.41×10^{-4}

[a] Reynolds number $Re=UD/\upsilon$, where U is the mean velocity and D is the tube diameter

[b] Stokes number $St=\rho_p d_p^2 U/(18\mu D)$

$$\sum \vec{\mathbf{F}}_{\mathbf{p}} = \vec{\mathbf{F}}_{Drag} = \frac{1}{8}\pi d_p^2 \rho C_{DP}(\vec{\mathbf{v}} - \vec{\mathbf{v}}_p)\,|\,\vec{\mathbf{v}} - \vec{\mathbf{v}}_p\,| \qquad (4.1.13a)$$

where d_p and $\vec{v}_p$ are the particle diameter and velocity, respectively; C_{DP} is the drag coefficient given as (Clift et al., 1978):

$$C_{DP} = C_D / C_{slip} \qquad (4.1.13b)$$

where

$$C_D = \begin{cases} 24/\mathrm{Re}_p, & \text{for } 0.0 < \mathrm{Re}_p < 1.0 \\ 24/\mathrm{Re}_p^{0.646} & \text{for } 1.0 < \mathrm{Re}_p \le 400 \end{cases} \qquad (4.1.14a,b)$$

and

$$C_{slip} = 1 + \frac{2\lambda_m}{d_p}\left[1.142 + 0.058\exp\left(-0.999\frac{d_p}{2\lambda_m}\right)\right] \qquad (4.1.15)$$

Here, λ_m is the mean free path in air. The local particle Reynolds number is $\mathrm{Re}_p = \rho\,|\,\vec{\mathbf{v}} - \vec{\mathbf{v}}_p\,|\,d_p/\mu$, where μ is the dynamic viscosity of the fluid. For the drag force, the Faxen correction was also ignored because the particle diameter is much less than the characteristic length associated with the air flow field velocity distribution. Integration of Eqs. (4.1.12, 4.1.13) for each non-interacting particle yields

$$\frac{d\vec{\mathbf{x}}_p}{dt} = \vec{\mathbf{v}}_p \qquad (4.1.16a)$$

subject to the initial conditions

$$\vec{\mathbf{x}}_p(t_0) = (x_0, y_0, z_0) \qquad (4.1.16b)$$

For the steady inhalation phase a parabolic fluid velocity profile with random, parabolically distributed monodispersed particles was specified at the inlet. Such a type of inlet particle distribution was verified to be suitable for experimental analyses and data comparisons. The initial particle velocities were set equal to that of the fluid and one-way coupling was assumed between the air and particle flow fields due to the dilute particle suspension, i.e., low mass loading ratios. The boundary conditions for the governing equations include symmetry with respect to the plane of the bifurcation (i.e., z = 0; cf. Fig. 4.1.5), and no fluid slip at the rigid impermeable walls. Particle deposition and withdrawal occur when the particle's center comes within a radius from the wall, which mimics the fact that solid particles or liquid droplets "deposit" as long as they touch the moist epithelial or mucus layer. A uniform pressure condition was employed at the outlets.

The numerical solution of the fluid flow equations was carried out with a user-enhanced finite-volume based program CFX4.3 (see Kleinstreuer & Zhang, 2003 for details).

The present computational fluid-particle dynamics (CFPD) model has been validated with various experimental data sets for single as well as double bifurcations under steady or cyclic flow conditions in terms of velocity profiles, particle deposition patterns, and deposition efficiencies (Comer et al., 2000, 2001; Zhang et al., 2002). The good agreements between experiments and theory instilled confidence that the present computer simulation model is sufficiently accurate to analyze fluid-particle dynamics and aerosol deposition in a four-generation (triple bifurcation) lung model segment with local tumors.

The tumor obstructing the G5-airway (see Fig. 4.1.5) further amplifies the asymmetry of the flow field between branches G5.2 and G5.1. The branch flow rate ratio declines with tumor size, because of the reduction of the airway lumen for both inhalation conditions. Due to the relatively high Peclet numbers present, air and particle flow fields upstream of small to medium-size tumors are almost the same; hence, the graphical results are only depicted for the tumor area as well as the downstream region. For example, Figs. 4.1.6 and 4.1.7 show airflow fields and particle transport in the vicinity of the tumor for different inlet flow rates and tumor sizes. For a given cross section, the particle distribution was captured just when these particles passed through this section. Both mid-plane velocity vector plots demonstrate at various degrees the basic phenomena of stagnation point flow after A-A′, accelerating flow in B-B′, and recirculating flow in C-C′. Specifically, very large recirculation zones are formed at the back of the tumor at low inhalation rates (cf. Fig. 4.1.6) and high flow rates with an expanding cross section downstream of the large tumor (cf. Fig. 4.1.6). Such flow separation-and-reattachment events are a function of the local geometric features, downstream geometric transitions, and the local Reynolds number. In cross section A-A′ of Fig. 4.1.5, the maximum velocity has slightly shifted back towards the center of the tube due to the upstream secondary motion, which pushes high-velocity fluid to the low-velocity region. The secondary flow depicted in Slice A-A′ will move the particles from the outside (A′) to the inside (A), and as a result two distinct particle swirls appear. However, the particle concentration is quite uniform for this case, and the upstream effect of the tumor is small. It is expected that the particles near the tumor side (A′) may impact the tumor directly and deposit on the tumor. However, some particles may follow the streamlines and bypass the tumor so that the particle concentration seems to be higher at cross section B-B′. Particles deposit on the tumor surface and avoid entrainment into the recirculation zone so that there are no particles near the outside of the tube (C′), directly downstream of the particle occlusion. At cross section D-D′, some particles move from the inside (D) to the outside (D′) carried by secondary flows; these particles may convect downstream into generation G6.3.

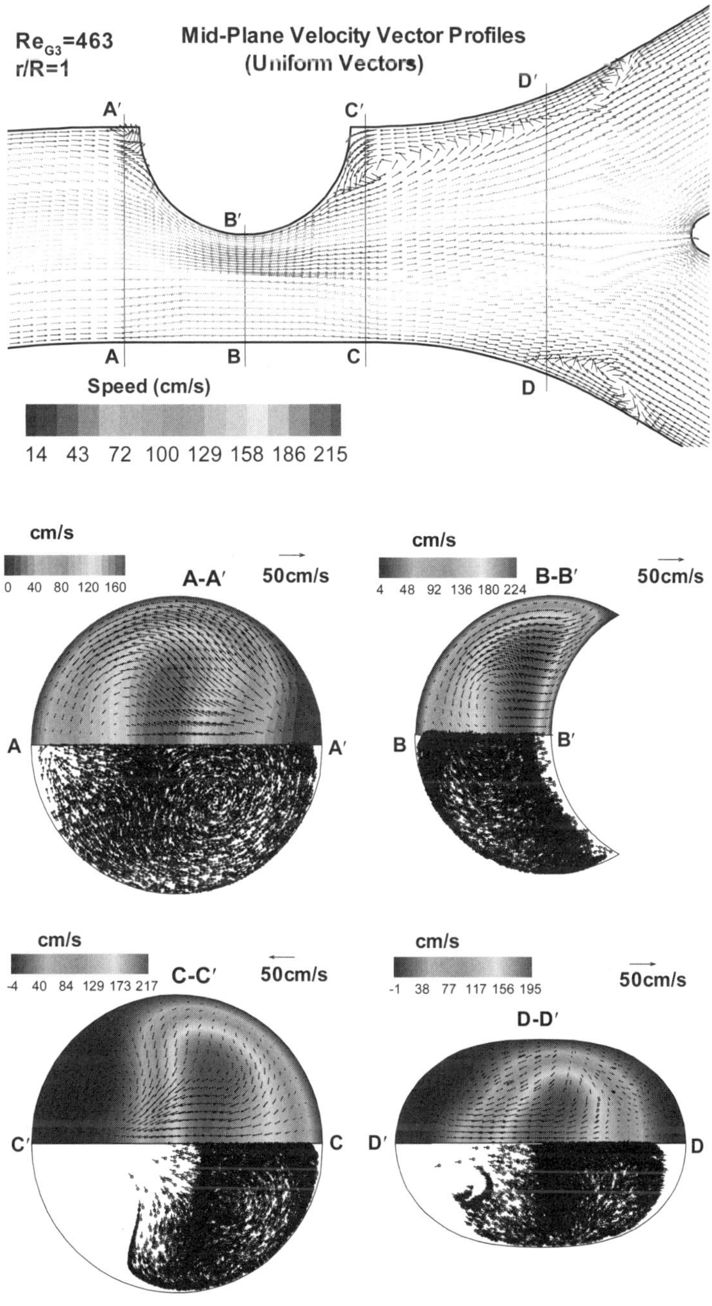

Fig. 4.1.6 Mid-plane velocity vectors, axial velocity contours, and secondary vectors (upper half of cross-sectional views) as well as particle distributions and flow directions (lower half of cross-sectional views) at selected cross sections near the tumor with r/R=1.0 at Re_{G3}=463 and St_{G3}=0.12.

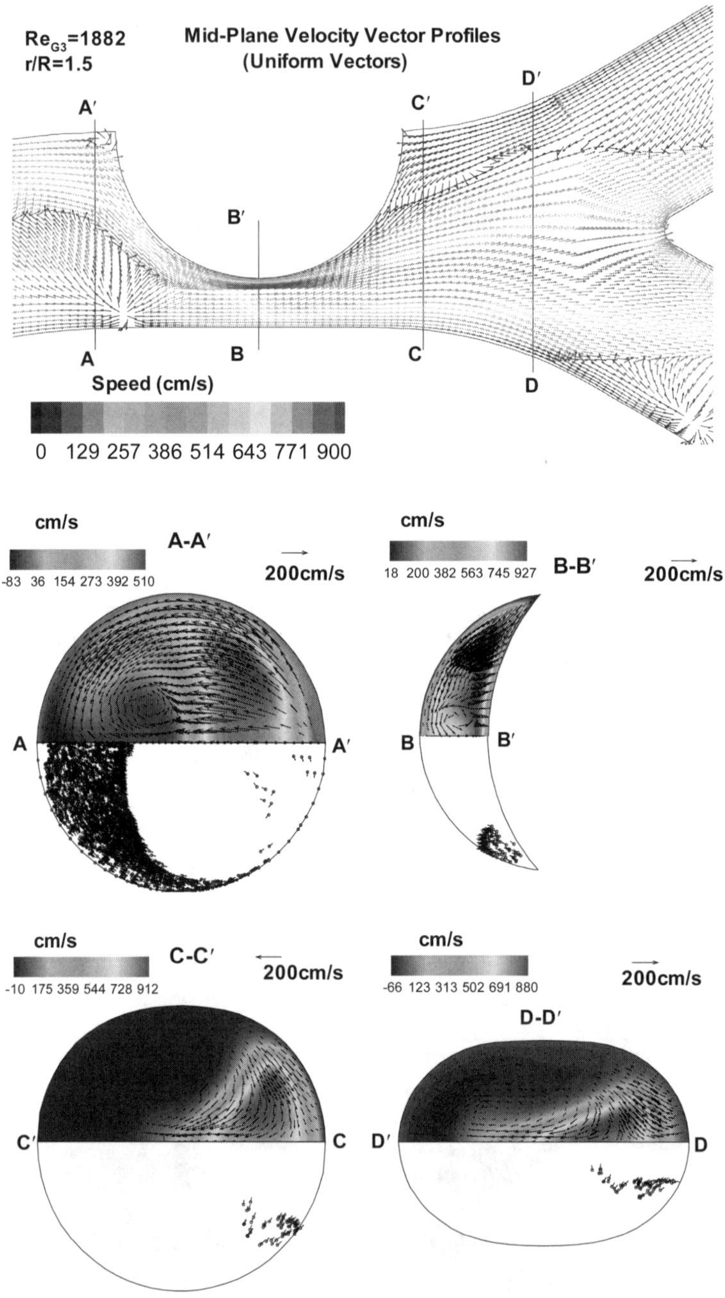

Fig. 4.1.7 Mid-plane velocity vectors, axial velocity contours, and secondary vectors (upper half of cross-sectional views) as well as particle distributions and flow directions (lower half of cross-sectional views) at selected cross sections near the tumor with r/R=1.5 at Re_{G3}=1882 and St_{G3}=0.12.

For the large-size tumor (r/R=1.5) under moderate exercise breathing condition (i.e., $Re_{match} = Re_{G3} = 1882$), the situation changes somewhat because the high inlet Reynolds number has a measurable effect on the flow field approaching the obstruction, i.e., it may not only influence the flow fields downstream but also to a small extent upstream (Fig. 4.1.7). As shown in Fig. 4.1.7 stagnation point flow and a recirculation zone are generated in front of the tumor. At the same time, the recirculation zone behind the tumor extends further into G6.3 when compared to the medium-size tumor case (r/R=1, Fig. 4.1.6). The flow ratios between branch G5.1 and branch G5.2 become larger for larger tumors present at branch G5.2 due to the blockage effect. At cross section A-A′, the maximum axial velocity zone has shifted to the outside (A′) due to the blockage effect. The strong secondary flow drives particles from the outside (A′) to the inside (A), so that only a few particles exist near the outside of the tube (A′). In addition, the number of particles at cross section A-A′ is obviously smaller than that for the medium-size tumor case (r/R=1) because of the decreasing flow rate. Since most particles near the tube center impact on the tumor surface, the number of particles at sections B-B′, C-C′, and D-D′ is very small, which indicates that very low particle depositions can be expected at the downstream generations G6.3 and G6.4. The flow fields near the second carinal ridge and in generations G5.1, G6.1, and G6.2 change significantly for the large-size tumor case as well.

Particle deposition on the tumor surface for different $(Re-St)_{match}$ pairs may be influenced by the local occlusion (i.e., tumor size and location), as well as inlet velocity profile and particle distribution. In general, particle deposition, occurring mainly along the front surface of the tumor due to impaction, increases with increasing Stokes number and Reynolds number. However, attributed to the secondary flow and blockage effect of the obstruction, some particles deposit along the top/bottom and inside wall of the tube in the geometric transition zone. This is demonstrated for the case r/R=1.25, where most particles land on the front surface of the tumor and some deposit on the tubular wall. Clearly, particles landing on the tumor surface are most desirable for targeted drug aerosol delivery. For this relatively large-tumor-size case, very few particles enter and deposit at the downstream bifurcations because of the tumor blockage effect.

In order to properly evaluate the overall concentration of particles appearing on a tumor surface, the deposition fraction (DF), i.e., mass of deposited particles to total particle mass entering the inlet tube (here G3), is employed. When plotting the DF as a function of tumor radius for different $(Re-St)_{match}$ pairs, a critical tumor size with maximum DF-values can be identified (Fig. 4.1.8). For the present system and most inlet Re-St combinations, the critical tumor size is r/R≈1.25, i.e., about half of the daughter branch G5.2 is occluded by this type of obstruction. The occurrence of this phenomenon can be attributed to the variations of branch air/particle flow rates as well as local flow fields with tumor size. Although the presence of small-to-medium size tumors (e.g., 0<r/R<1.25) results in a reduction of flow rate and particle number entering the diseased branch, the likelihood of particle deposition increases with tumor growth due to inertial impaction. Hence, DF-values continue to increase with tumor size. However, when the particle number associated with the flow rate approaching the tumor branch decreases drastically in the cases of large-size tumors, i.e., 1.25<r/R≤2.0, the number of deposited particles, i.e., DF, decreases although the probability of particle impaction is higher. The effect of tumor size may vary somewhat with the inlet Reynolds and Stokes numbers as indicated in Fig. 4.1.8. In addition, there are still a few particles depositing at the tumor site if the branch is completely blocked by the tumor. This can be explained by the existence of a low positive flow near the inside wall which moves particles to the inner tube wall or tumor site. Table 4.1.4 summarizes the maximum deposition fractions, where for $(r/R)_{critical}$ $DF_{\max} \approx 11\%$. Clearly, targeted drug aerosol delivery is necessary (see Sect. 5.3.2).

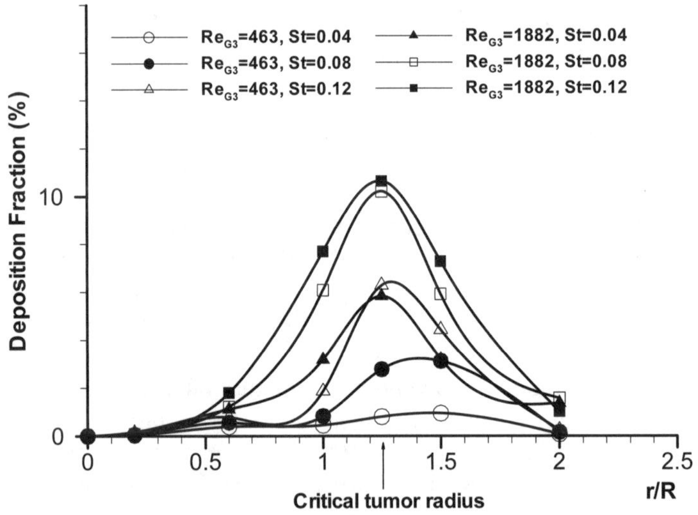

Fig. 4.1.8 Variations of DF-values on the tumor surface vs tumor size.

Table 4.1.4 Maximum DF-values on variable tumor surface (Re=1882; St=0.12).

Tumor size: r/R	0.2	0.6	1.0	1.25	1.5	2.0
Maximum DF (%)	0.1	1.8	7.7	10.5	7.2	1.3

Assuming symmetrically bifurcating rigid lung airways, hemispherical tumors, and constant-diameter non-interacting micron-size particles, findings of the present study can be summarized as follows:

1. As expected, the presence of a singular tumor in bronchial airways changes the flow distributions among different branches. While medium-size tumors generally affect the nearby and downstream flow fields, large-size tumors may change the flow features upstream and in the sister branches. The effects of tumors on the flow characteristics in the bronchial airways are related to the inlet flow rate (i.e., breathing condition) as well. Smooth inhalation wave forms and dilute aerosol uptake can be accurately simulated with matching Re-St pairs.

2. Particle deposition on the tumor surface may be influenced by the local occlusion (tumor size), as well as upstream flow and particle distributions as a function of Reynolds number, Stokes number, tumor locations, and tumor size. (The influence of tumor shape would be important as well.) The particle deposition fraction (DF) on a tumor increases significantly when switching breathing patterns from resting to moderate exercise. When other conditions (e.g., tumor location, inlet Reynolds numbers, and Stokes numbers) are fixed, DFs on the tumor surface increase first and then decrease with the tumor size. Maximum DF-values are generated for a hemispherical tumor size where the lumen area of the afflicted tube is approximately half-way occluded by the tumor.

3. Separately it was shown that the release positions for aerosols which land on the tumor surface concentrate for high Re-St pairs in a specific area of the inlet plane, though this zone may vary somewhat with the particle release conditions and tumor size. Hence, controlled particle release could dramatically increase the deposition of drug aerosol at desired lung target sites and at the same time decrease the influence of aggressive drugs on the healthy airway epithelium (see Sect. 5.3.2).

4.2 THE KIDNEYS

The kidneys are an amazingly efficient pair of filters and regulators. In fact, every 40 minutes or so the entire blood volume (≈5-6 liters in adults) is cleared of the blood's waste products, the blood's pH is regulated to stay at 7.4, and the blood pressure is controlled at a normal arterial mean of 100 mmHg. Thus, the kidneys filter out and discharge via the urinary tract biochemicals as well as water to clean the blood and to achieve constant osmolarity, i.e., homeostasis of the plasma concentration. Secretion of the anti-diuretic hormone (ADH) regulates the correct water balance in the body. The driving forces for blood filtration, water reabsorption, and secretion of ADH as well as toxins and ions include pressure differentials, concentration gradients, and solute differences in solutions. Transport of fluid (i.e., water), cells, proteins, and molecules occurs selectively across tubular walls which act like semi-permeable membranes (see Sect. 1.1.2). Such membranes feature variable pores with micro-channel sizes where the solute transport can be passive, e.g., diffusion, osmosis, and ultrafiltration, or active, i.e., against a concentration gradient and hence requiring metabolic energy.

4.2.1 Kidney Structure and Functions

Each adult kidney weighs about 160 grams and is about the size of a fist (see Fig. 4.2.1a). The renal capsule covers the medulla which houses 8 to 15 conical renal pyramids. Each renal pyramid consists of "nephrons," i.e., tubules plus capillary beds, the functional unit of the kidney that is responsible for the formation of urine (Fig. 4.2.1b). Specifically, filtration takes place in the capillary bed (or glomerulus) which is covered by the Bowman capsule. Water and dissolved solutes (minus proteins, blood cells, and platelets) pass from the blood plasma, supplied by the renal artery, through rather large pores (i.e., 200-500Å) of the glomerular capillaries to the inside of the capsule and the lumina of the nephron tubules. It is a net filtration pressure (p = 10 mmHg), i.e., hydrostatic blood pressure minus osmotic pressure, which generates across large capillary surface areas a high volume of ultra-filtrate, called the glomerular filtration rate (GFR≈ 180 liters/day). Approximately 99% of the filtered water (plus salt, NaCl) is immediately returned to the vascular system via reabsorption to maintain a normal water balance and hence normal blood osmolarity and blood pressure (see Fig. 4.2.2). The filtrated water passes through filtration slits into Bowman's capsule, called ultrafiltrate, which then migrates into the lumen of the proximal convoluted (i.e., coiled) tubules (see Fig. 4.2.1b). The tubular walls have millions of microvilli (tiny fingerlike projections of wall-cell membranes) which form large effective surface areas for re-absorption of water, salt, and other molecules into the surrounding capillaries, as needed by the body. Leaving the renal cortex, the fluid descends the "Loop of Henle" (see medulla region in Fig. 4.2.1b) where most of the filtered water is reabsorbed via osmosis into the capillaries. While the descending limb is passively permeable to water, the ascending limb is impermeable to water but actively transports Na^+, followed passively by Cl^-. The net effect is that salt is extruded into the surrounding tissue fluid so that water is drawn out of the descending limb by the osmotic pressure and collected by the capillaries. From the loop of Henle, the remaining water flows via the distal convoluted tubules to the collecting duct. Because 99% of the filtered water has to be reabsorbed, an anti-diuretic hormone (ADH, or anti water-loss hormone) triggers formation of water channels in the cell membranes of the collecting ducts, which in conjunction with the osmotic pressure secures sufficient water re-absorption. Thus, proper ADH release and its concentration control collecting-duct permeability and hence normal water balance, osmolarity, and ultimately urine discharge. The fluid (now called urine) in the collecting ducts, located in the renal cortex and medulla region, is then drained into the renal pelvis (see Fig. 4.2.1a) from which it leaves the kidney via the ureter. The ureter terminates in the urinary bladder to which the urethra is connected.

a) Kidney midplane

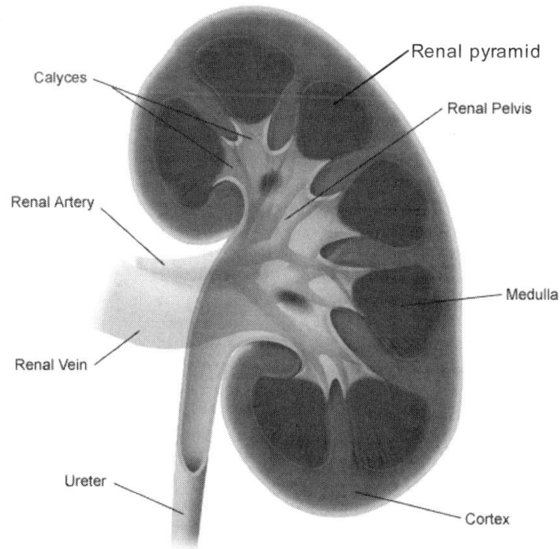

(b) Blood flow and plasma clearance

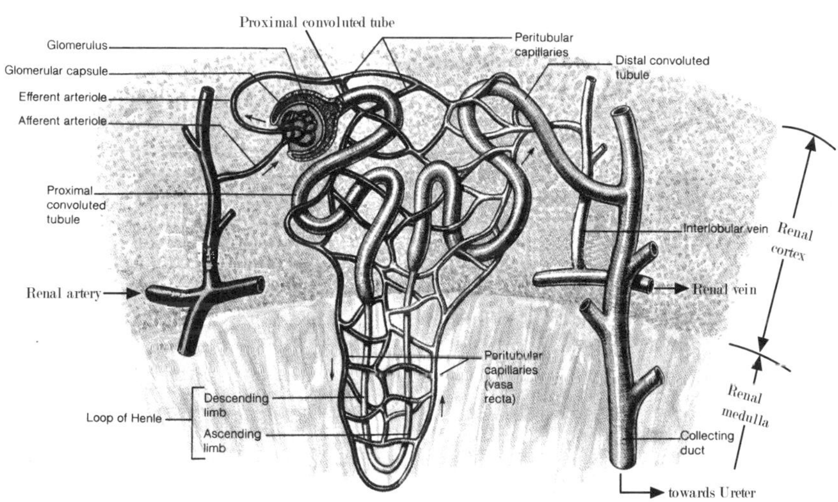

Fig. 4.2.1 Kidney structure and nephron function (after Fox, 1996).

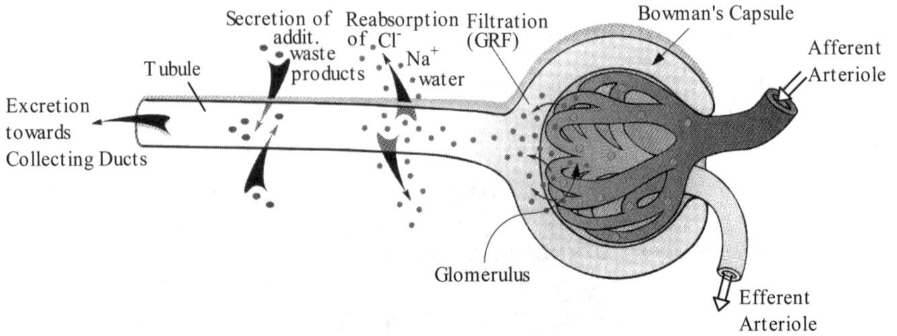

Fig. 4.2.2 Microscale kidney function: Filtration, re-absorption, and secretion in a nephron (after Fox, 1996).

4.2.2 Fluid Flow and Mass Transfer in an Artificial Kidney Model

As discussed in Sect. 4.2.1, one of the kidney's key functions is the removal of waste products from the blood. Renal insufficiency, due to destroyed nephrons or reduced plasma filtration, can cause hypertension and high urea concentrations in the blood, called uremia, accompanied by elevated plasma H^+- and K^+-ions which may lead to uremia coma. Uremia patients (350k in the US alone) have to be often connected to dialysis machines, known as artificial kidney machines, to maintain homeostasis. Hemodialysis is the separation of unwanted molecules on the basis of size, e.g., urea and other wastes, to diffuse easily through an artificial semi-permeable membrane. Thus, just as it occurs across the glomerular capillaries, plasma is cleansed of just these wastes (e.g., urea, creatinine potassium, fluid, etc.) as they migrate in a dialyzer from the blood across the porous membrane into the dialyzing fluid because of a pressure difference, called ultrafiltration. However, re-absorption of water and needed species is not possible with dialysis machines (see Fig. 4.2.3a). The dialysis solution, called dialysate, washes the wastes and extra fluid out of the kidney machine. It contains chemicals which bind to the waste products and has to be prescribed for every patient to be most effective.

These machines are basically convective-diffusive countercurrent mass exchangers. Typically, inside a cylindrical housing (or shell) of $D\approx 3$-4 cm and L = 20 – 30 cm are several thousands of hollow fibers (see Fig. 4.2.3b). A hollow fiber is a long tiny tube of inner diameter $d\approx 200\ \mu m$ and semi-permeable membrane thickness $t = 15\text{-}50\ \mu m$ (Liao et al., 2003). People with kidney problems often have to undergo hemodialysis three times per week, i.e., proper vascular access to withdraw blood and return it is very important. Longest & Kleinstreuer (2000) analyzed access-point problems and discussed solutions mitigating stenotic developments.

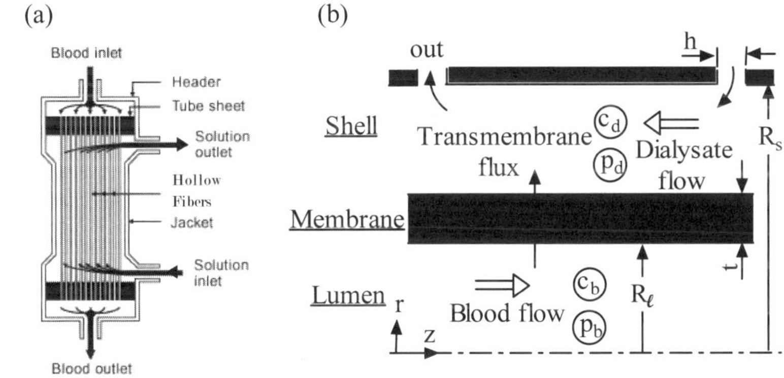

Fig. 4.2.3 Hemodialysis: (a) A typical hollow fiber dialyzer (kidney machine); and (b) Single hollow fiber model.

Kidney machine model. Focusing on a bundle of hollow fibers (say, n = 12,000) as the key element of a dialyzer (see Figs. 4.2.3a,b), three flow processes have to be considered, i.e., lumen-side, trans-membrane, and shell-side flows. Following Liao et al. (2003), assumptions and approaches for fluid flow and mass transfer in this representative dialyzer unit are:

- Steady 2-D laminar dilute suspension flow in the filter lumen, using the reduced Navier-Stokes and mass transfer equations (see Sects. 1.3.1 and 1.3.2).
- Steady radial trans-membrane mass fluxes for solution and solute, employing the Kedem-Katchalsky equations (see Sect. 1.1.2).
- Steady 2-D dialysate, i.e., shell-side, flow through a porous medium formed by these numerous hollow fibers, using Darcy's law (see Sec. 1.4.4).

Lumenal Blood Flow. The incoming blood with its waste products is considered to be Newtonian, entering axisymmetric and fully developed each filter lumen at a low Reynolds number. Thus, the governing equations and boundary conditions read:

Continuity:
$$\frac{1}{r}\frac{\partial(r\,\mathrm{v}_r)}{\partial r}+\frac{\partial \mathrm{v}_z}{\partial z}=0 \tag{4.2.1}$$

z-Momentum:
$$\mathrm{v}_r\frac{\partial \mathrm{v}_z}{\partial r}+\mathrm{v}_z\frac{\partial \mathrm{v}_z}{\partial z}=-\frac{1}{\rho}\frac{\partial p}{\partial z}+\nu\left[\frac{1}{r}\frac{\partial}{\partial r}\left(r\frac{\partial \mathrm{v}_z}{\partial r}\right)+\frac{\partial^2 \mathrm{v}_z}{\partial z^2}\right] \tag{4.2.2a}$$

r-Momentum:
$$\mathrm{v}_r\frac{\partial \mathrm{v}_r}{\partial r}+\mathrm{v}_z\frac{\partial \mathrm{v}_r}{\partial z}=-\frac{1}{\rho}\frac{\partial p}{\partial r}+\nu\left[\frac{\partial}{\partial r}\left(\frac{1}{r}\frac{\partial}{\partial r}(r\mathrm{v}_r)\right)+\frac{\partial^2 \mathrm{v}_r}{\partial z^2}\right] \tag{4.2.2b}$$

Mass Transfer: $$\frac{v_r}{r}\frac{\partial(rc)}{\partial r} + v_z\frac{\partial c}{\partial z} = D\left(\frac{\partial^2 c}{\partial r^2} + \frac{\partial^2 c}{\partial z^2}\right) \quad (4.2.3)$$

Boundary Conditions:

- Inlet (z = 0) $v_z(r) = v_{max}\left[1 - \left(\frac{r}{R}\right)^2\right]$; $v_{max} = \frac{2Q_{total}}{\pi n R^2}$

  $v_r = 0$ and $C = C_{b,0}$

- Outlet (z = L) p = 0
- Centerline (r = 0) $\frac{\partial v_z}{\partial r} = 0, \frac{\partial c}{\partial r} = 0, v_r = 0$
- Membrane surface (r = R) $v_z = 0$; $v_r = J_v$ (see Eq. (4.2.4));

  and $J_s = -D\frac{\partial c}{\partial r} + cJ_v$

Membrane Mass Transfer. Because of fluid pressure and species concentration differentials between lumen area and shell side both solution and solute cross the lumen wall, i.e., the semi-permeable membrane (see Fig. 4.10b). Specifically, we have:

Solution Flux: $$J_v = L_M(p_b - p_d) - \sigma RTL_M(c_b - C_d) \quad (4.2.4)$$

Solute Flux: $$J_s = \overline{C}_M(1-\sigma)J_v + D_M(C_b - C_d) \quad (4.2.5)$$

Note that the dimension of the solution flux $[L^3/(L^2T)]$ is that of a velocity [L/T], while that of the species flux is $[M/(L^2T)]$. The hydraulic membrane permeability L_M, the reflection coefficient, s, which indicates the solute fraction blocked by the membrane, and diffusive permeability D_M are intrinsic membrane properties and have to be measured (see Table 4.2.1). The average solute concentration inside the membrane is $\overline{C}_M[M/L^3]$.

Table 4.2.1 Model parameter values.

Hollow Fiber	R = 100 μm; L=20.32 cm; t = 15 μm
Shell-side:	N = 12,000 hollow fibers; R_s = 1.7 cm; $k_r = 4.16 \times 10^{-11}$ m^2; $k_z = 1.29 \times 10^{-9}$ m^2
Membrane-Urea:	$L_M = 1.15 \times 10^{-10}$ m/s/Pa; $D_M = 9.76 \times 10^{-6}$ m^2/s; σ = 0; D = 1.81×10^{-9} m^2/s

Dialysate Flow. Assuming radially-directed uniform inlet and outlet velocity profiles of the dialysate through outer-shell areas $A_{in} = A_{out} = 2\pi R_s h$, the "porous medium" flow equations read:

Continuity: $$\frac{1}{r}\frac{\partial(r\,v)}{\partial r} + \frac{\partial u}{\partial z} = S_v \tag{4.2.6}$$

z-Momentum: $$u = -\frac{1}{\mu} k_z \frac{\partial p}{\partial z} \tag{4.2.7a}$$

r-Momentum: $$v = -\frac{1}{\mu} k_r \frac{\partial p}{\partial r} \tag{4.2.7b}$$

Mass Transfer: $$v\frac{\partial c}{\partial r} + u\frac{\partial c}{\partial z} = S_s \tag{4.2.8}$$

Boundary Conditions:

- Inlet $u = 0$ and $v = -\dfrac{Q_d}{2\pi R_s h}$
- Outlet $p = 0$
- Walls $u = v = 0$

Note that the source terms S_v and S_s for solution and solvent, respectively, can be neglected under the assumption of low ultrafiltration and back filtration rates. Liao et al. (2003) suggested correlations to estimate these source terms when trans-membrane fluxes are significant.

Focusing on urea clearance <test solute> and using a cellulose triacetate membrane, the system parameter values, as given by Liao et al. (2003), are summarized in Table 4.2.1.

Of interest are the velocity, pressure, and urea concentration distribution at the shell-side as well as the urea clearance [ml/min] from the solution (blood) for different flow rates, say, $200 \le Q_d \le 1000$ ml/min and $100 \le Q_b \le 500$ ml/min. Maximizing waste product clearance is of course the goal of optimal dialyzer design and operation. Specifically,

$$\kappa \equiv (Q_b C_b /_{in} - Q_b C_b /_{out}) / C_{b,in} \tag{4.2.9}$$

Typical lumen inlet values are $Q_{b,in} = 120$ ml/min and $C_{b,in} = 100$ mg/dL. The flow chart for the sequential calculation steps is given in Fig. 4.2.4.

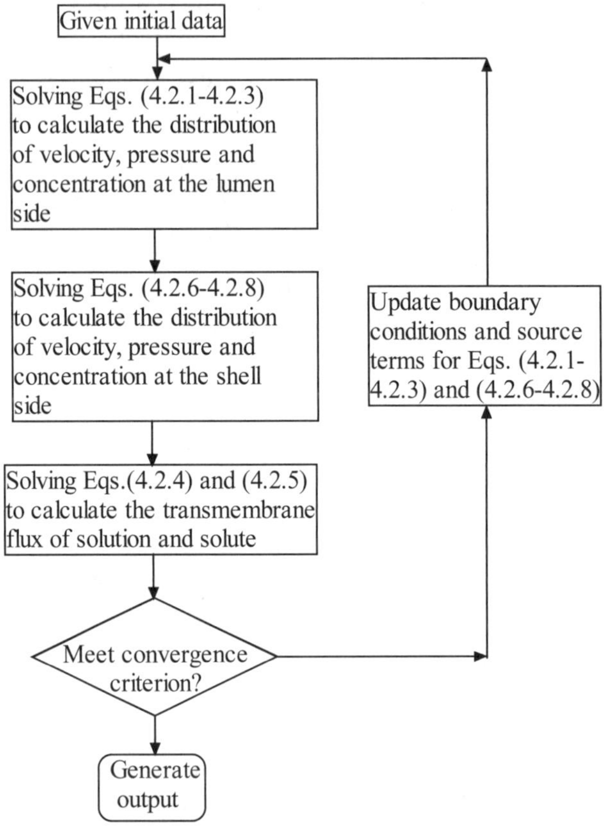

Fig. 4.2.4 Computational flow chart for urea clearance in a kidney machine model.

Results and Discussion. Using the commercial finite volume code CFX4 (Ansys, Inc.), the coupled continuity, momentum, and mass transfer equations in both lumen and shell sides (i.e., Eqs. 4.2.1-4.2.3 and 4.2.6-4.2.8) were solved numerically. Then, the solution and solute fluxes were updated according to the Kedem-Katchalsky equations following the simulation steps (see Fig. 4.2.4).

Figure 4.2.5 depicts the variations of the urea clearance rate (see Eq. 4.2.9) as a function of dialysate and blood flow rates. Clearly, the urea clearance increases with an increasing dialysate or blood flow rate due to the enhanced mass fluxes.

Figures 4.2.6-4.2.8 show the velocity, pressure, and urea concentration distributions at the shell side for flow rates Q_b=400 ml/min, Q_d=400 ml/min and an inlet urea concentration at the lumen $C_{b,in}$=70mg/dL. Because of the much larger resistance in the radial direction ($k_{rr} \ll k_{zz}$), the axial flow is dominant in most areas of the dialysate side, as shown in Fig. 4.2.6. The radial flow is only significant near the inlet and outlet due to inlet/outlet effects. Moreover, flow becomes weak near the centerline below both inlet and outlet. The pressure distribution is non-uniform near the dialystate inlet and outlet and becomes uniform in mid-region due to the interaction between isotropic permeability and entrance/outlet effects (see Fig. 4.2.7).

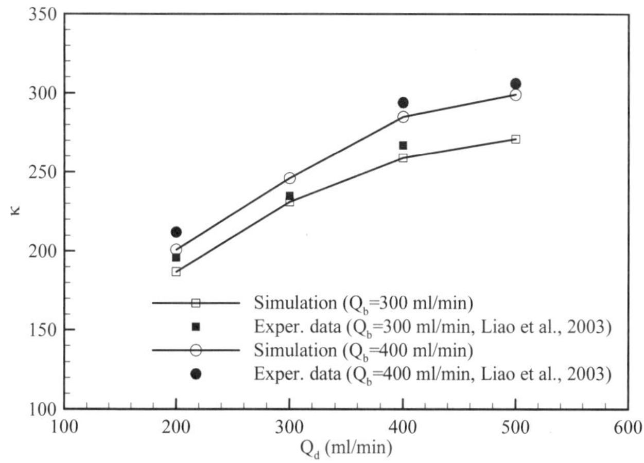

Fig. 4.2.5 Computational and measured urea clearance rates in an artificial kidney model.

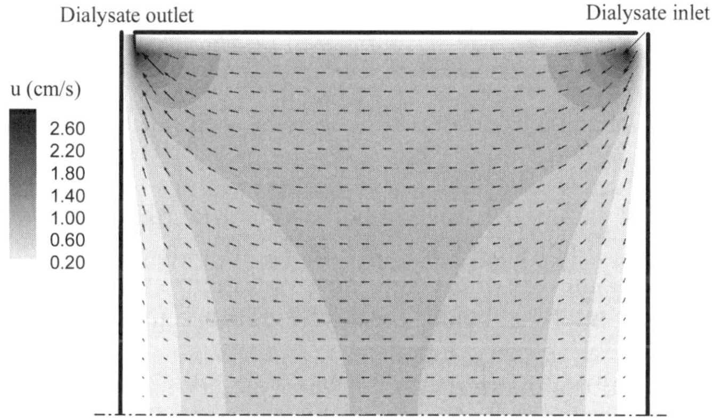

Fig. 4.2.6 Velocity vectors and axial velocity contours at the shell side of an artificial kidney model (see Fig.4.2.3b).

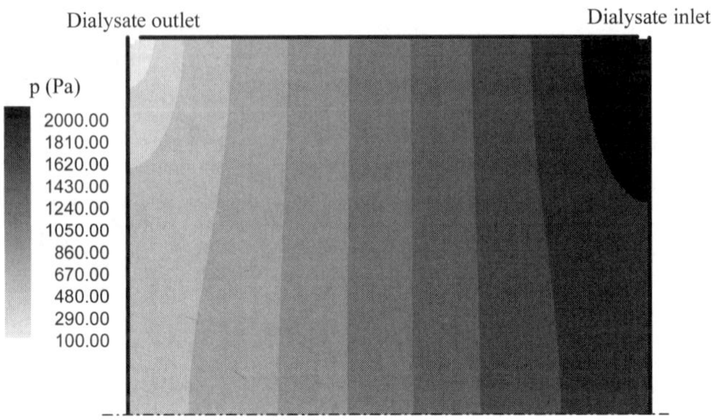

Fig. 4.2.7 Pressure distributions in the shell side of an artificial kidney model.

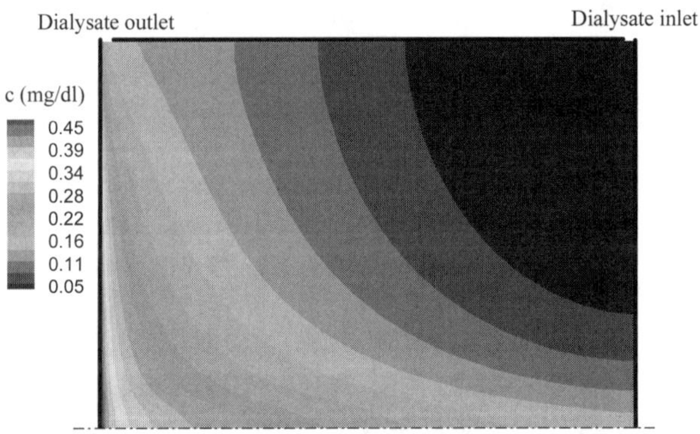

Fig. 4.2.8 Distribution of urea concentration in the shell side of an artificial kidney model.

Associated with both axial and radial velocities, the urea concentration in the shell is also non-uniform (see Fig. 4.2.8). The concentration in the center is larger than that in the periphery and the maximum concentration appears at the centerline below the outlet because the solute can remain longer in these low-velocity regions.

A more uniform distributed dialysate velocity and concentration may contribute to a higher performance of an artifical kidney. Therefore, effective biofluid analyses are very helpful to evaluate the urea clearance rate and optimally design the artificial kidney in order to achieve high clearance of urenary solutes of toxins.

4.3 THE LIVER

The liver, which is the largest internal organ, regulates the chemical composition of the blood (Fig. 4.3.1a). Specifically, it can remove hormones, drugs, and other compounds via different biochemical processes (see Table 4.3.1). Clearly, compared with all other organs, the liver has the largest variety of functions.

Table 4.3.1 Liver functions (after Fox, 1996).

Category	Processes
Detoxication of Blood	• Digestion of toxins by Kupffer cells • Chemical alteration of biologically active molecules (hormones and drugs) • Production of urea, uric acid, and other molecules that are less toxic than parent compounds • Excretion of molecules in bile
Carbohydrate Metabolism	• Conversion of blood glucose to glycogen and fat • Production of glucose from liver glycogen and from other molecules (amino acids, lactic acid) by gluconeogenesis • Secretion of glucose into the blood
Lipid Metabolism	• Synthesis of triglyceride and cholesterol • Excretion of cholesterol in bile • Production of ketone bodies from fatty acids
Protein Synthesis	• Production of albumin, a major component of the plasma proteins • Production of plasma transport proteins • Production of clotting factors (fibrinogen, prothrombin, and others)
Secretion of Bile	• Synthesis of bile salts • Conjugation and excretion of bile pigment (bilirubin)

(a) The liver's central role in regulating blood chemical composition

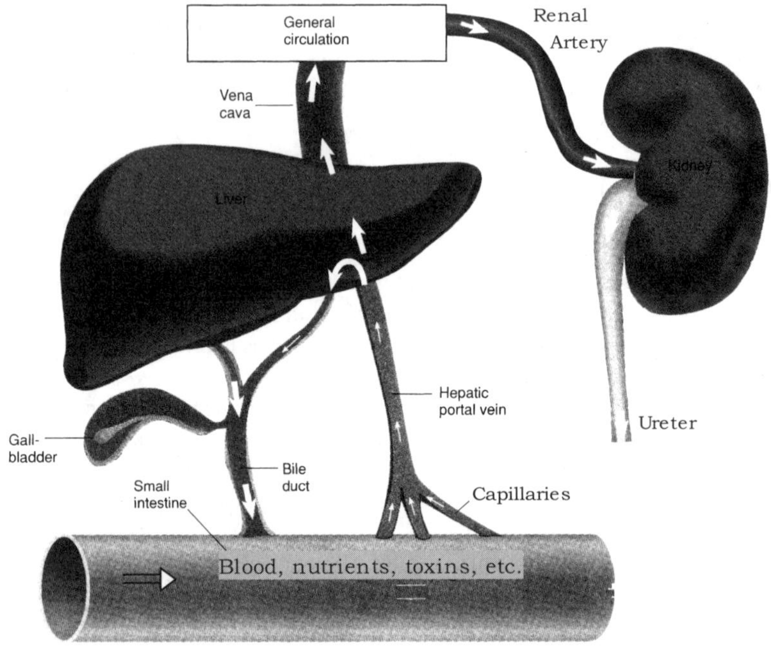

(b) Blood and bile flow in a liver lobule

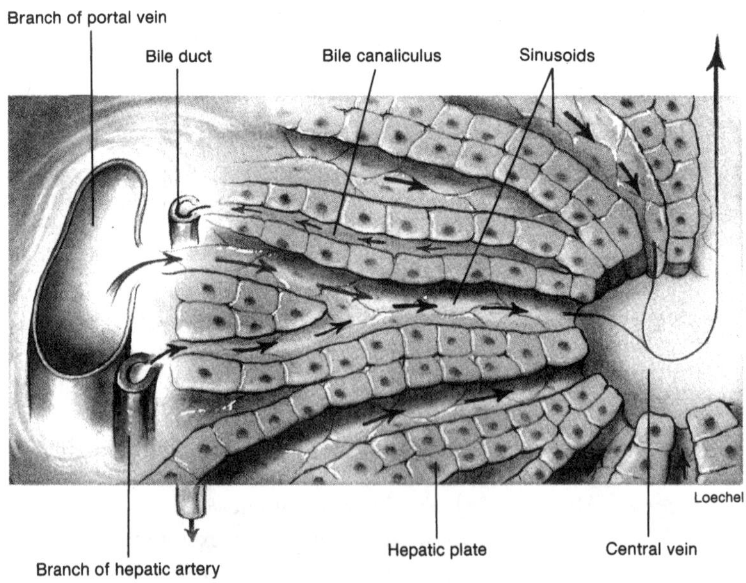

Fig. 4.3.1 Blood and bile circulation in the liver (after Fox, 1996).

4.3.1 Liver Structure and Functions

Apart from a patch where it is connected to the diaphragm, the liver is covered entirely by a thin, double-layered membrane that reduces friction against other organs. Based on surface appearance, it can be divided into major folded lobes (see Fig. 4.3.1a). Surprisingly, there are only two types of cells forming the structure of the liver. The liver cells, called hepatocytes, form hepatic plates that are one to two cell-layers thick. These plates are separated by microchannels, called sinusoids (see Fig. 4.3.1b). The channel walls are lined with Kupffer cells which, like white blood cells, can "devour" toxins, bacteria, etc. As indicated in Fig. 4.3.1a, nutrients in the blood from the digestive system are carried via the hepatic partial vein to capillaries in the liver, as it receives arterial blood via the hepatic artery. The hepatic plates are arranged into active units called liver lobules, each with a central vein which collects the arterial/venous blood mixture flowing through the sinusoids (Fig. 4.3.1b). The central veins of all the lobules converge and deliver the blood to the inferior vena cava. In contrast, bile is produced by the hepatocytes and secreted into microchannels, called bile canaliculi, which empty into the bile duct (see Fig. 4.3.1).

An example where liver and kidney function together is the conversion, removal, and discharge of ammonia, a highly toxic chemical produced in the liver from amino acids and the action of bacteria in the intestine. The liver has the necessary enzymes to convert ammonia into less toxic urea molecules, which are secreted into the blood and carried to the kidneys, where they are excreted into the ureter (see Sect. 4.2 and Fig. 4.3.1a).

4.3.2 Fluid Flow and Mass Transfer in a Liver Model

As discussed in Sect. 4.3.1, one of the liver's key functions is effective uptake, storage, transport, and distribution of chemical substances to the blood and bile. The rate at which the liver uptake and transport various chemical species, such as a drug, is of importance in studying liver physiology and pharmacology as well as treating liver diseases. Mass transport of chemical species between blood and the hepatocytes (liver cells) may occur at the level of the hepatic acinus. Hence, simulating the fluid flow and mass transfer in the hepatic acinus is useful in evaluating the uptake and transport of chemical species in the liver. As shown in Fig 4.3.2a, the hepatic acinus consists of the bile duct, terminal portal venule, terminal hepatic arteries, sinusoids, and terminal hepatic venule.

Hepatic acinus model. Focusing on a hepatic acinus as the key element of the liver (see Figs. 4.3.2 a,b), three transport processes have to be considered, i.e., fluid flow in the acinus, as well as mass transport in blood and tissue and across the membrane. Following Lee & Rubinsky (1989), the assumptions and approaches for fluid flow and mass transfer in this representative acinus model are:

- Steady, laminar, incompressible porous flow in the acinus with sinusoids;
- Transient diffusion-convection in the blood vessels and surrounding tissue;
- Simple diffusion of chemical species across the cell membrane, neglecting any metabolism.

(a) Hepatic acinus (after Gumucio, 1983) (b) Schmetic of hepatic acinus model

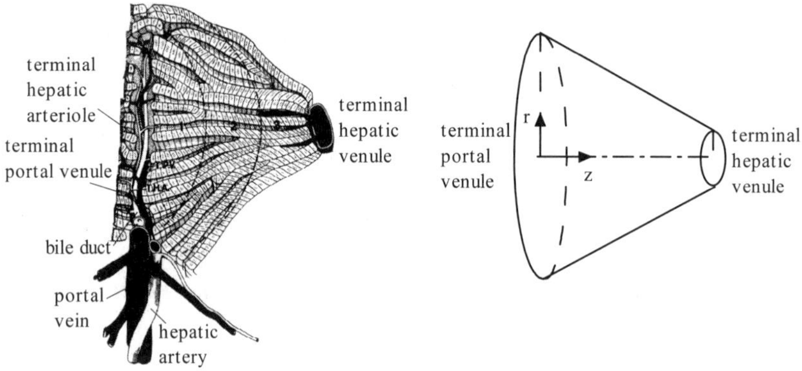

Fig. 4.3.2 Hepatic acinus.

Fluid Flow. The incoming blood with chemical species is considered to be Newtonian, entering the sinusoids at a low Reynolds number (Re < 1). Thus, the "porous medium" flow equations read:

Continuity:
$$\frac{1}{r}\frac{\partial(r\mathrm{v}_r)}{\partial r}+\frac{\partial \mathrm{v}_z}{\partial z}=0 \tag{4.3.1}$$

z-momentum:
$$\mathrm{v}_z=-\frac{1}{\mu}k_z\frac{\partial p}{\partial z} \tag{4.3.2a}$$

r-momentum:
$$\mathrm{v}_r=-\frac{1}{\mu}k_r\frac{\partial p}{\partial r} \tag{4.3.2b}$$

Boundary conditions:

Inlet $\quad p=p_1$ (4.3.2c)

Outlet $\quad p=p_0$ (4.3.2d)

Walls $\quad \mathrm{v}_r=\mathrm{v}_z=0$ (4.3.2e)

Mass transfer in blood and tissue. Considering the macroscopic mass diffusion in a control volume which contains blood vessels and surrounding tissue, the convection-diffusion equation for chemical species reads:

$$\frac{\partial}{\partial t}\left(c_f + \frac{V_t}{V_f}c_t\right) + \mathrm{v}_r\frac{\partial c_f}{\partial r} + \mathrm{v}_z\frac{\partial c_z}{\partial z} = D\left[\frac{1}{r}\frac{\partial}{\partial r}\left(r\frac{\partial c_f}{\partial r}\right) + \frac{\partial^2 c_f}{\partial z^2}\right] \tag{4.3.3a}$$

where c_f and c_t are the concentration in the blood and tissue, respectively; V_t and V_f are volume of tissue and sinusoids, respectively.

Initial condition: $c_f(r,z,t=0) = c_0$ (4.3.3b)

Inlet: $c_f = c_{in}$ (4.3.3c)

Outlet and walls: $\frac{\partial c}{\partial n} = 0$ (4.3.3d)

Mass transfer across the cell membrane. The simple diffusion of chemical species transfer across the cell membrane from the fluid is governed by Fick's law, i.e.,

$$\frac{\partial c_f(r,z,t)}{\partial t} = \frac{P_s S_A}{V_t}\left(c_f(r,z,t) - c_t(r,z,t)\right) \tag{4.3.4}$$

where P_s is the species permeability across the cell and S_A is the surface area of the tissue.

Numerical method and model parameters. The commercial finite volume software CFX4 was employed to solve the fluid flow and chemical species distribution in the three-dimensional axisymmetric geometric representation of the liver acinus. The steady fluid flow equations (Eqs. 4.3.1 and 4.3.2) were solved first to obtain the flow fields. Once the velocity fields were obtained, the transient species transport in the fluid (i.e., Eq.4.3.3) was also solved with CFX4. For a certain time instant, the net change in the fluid chemical species concentration exchanging with the tissue, i.c., $\frac{\partial}{\partial t}\left(\frac{V_t}{V_f}c_t\right)$, is treated as a source term in Eq. (4.3.3). After the species concentration in the fluid, i.e., c_f, was obtained by solving Eq. (4.3.3) at a certain time, concentration in the tissue (i.e., c_t) was updated with Eq. (4.3.4) by forward time differencing. The time step used was small enough so that a converged solution could be achieved. The parameter values used for this simulation are given in Table 4.3.2.

Table 4.3.2 Parameter values (adapted from Lee & Rubinsky, 1989).

Parameter	Value
Pressure portal vein, p_1	6 cm H_2O
Pressure hepatic vein, p_0	2 cm H_2O
Porosity, ε	0.11
Length of sinusoid, L_s	0.03 cm
Diameter of sinusoid, d_s	7 μm
Surface area/volume tissue, S_A / V_t	556 cm^{-1}
Volume tissue/volume fluid, V_t / V_f	8
Permeability in r-direction, k_r	$4.3\times10^{-6}cm^3g^{-1}s$
Permeability in z-direction, k_z	$2.6\times10^{-6}cm^3g^{-1}s$
Membrane permeability, P_s	4.8×10^{-5}cm/s
Kinematic viscosity, ν	0.0272 cm^2/s
Diffusion coefficient, D_s	6.7×10^{-7} cm^2/s

Pressure and velocity fields. Figure 4.3.3 shows a pressure distribution in the acinus assuming a constant permeability. Due to symmetry, only half of the cone is displayed. Clearly, the pressure gradient becomes steeper close to the hepatic venule due to the contraction of the flow domain starting from the terminal portal venule. The pressure distribution in the radial direction is also non-uniform because of the geometry effect.

Figure 4.3.4 depicts the fluid velocity distribution in terms of velocity vectors and axial velocity contours. Again, the velocity increases rapidly from the inlet to the outlet due to the contraction. In most parts of the acinus, the velocity at the center is larger than that at the periphery because of the wall viscous effects and larger flow resistance in the radial direction (i.e., $k_r<k_z$). However, the "squeeze" of the contracting wall pushes the fluid downwards; hence, the maximum velocity may shift a bit from the center near the outlet.

In summary, the pressure and velocity distributions within the acinus are non-uniform and functions of position. In addition, the converging flow path and increasing cross-sectional area of the sinusoids towards the hepatic vein would change the hydraulic permeability, which may also have a significant effect on pressure and flow distributions in the liver acinus (Lee & Rubinsky, 1989).

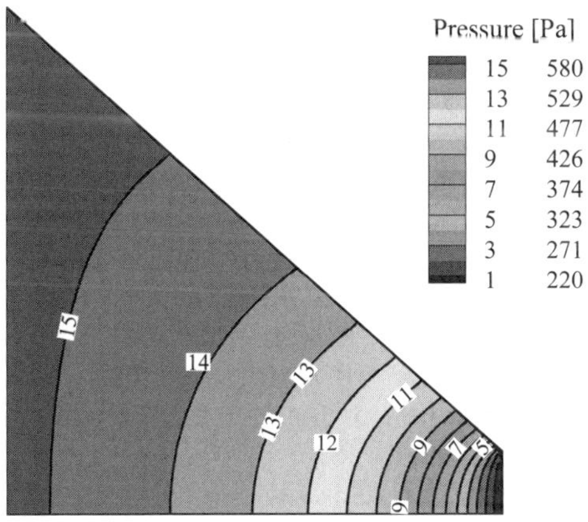

Fig.4.3.3 Pressure distribution in liver acinus.

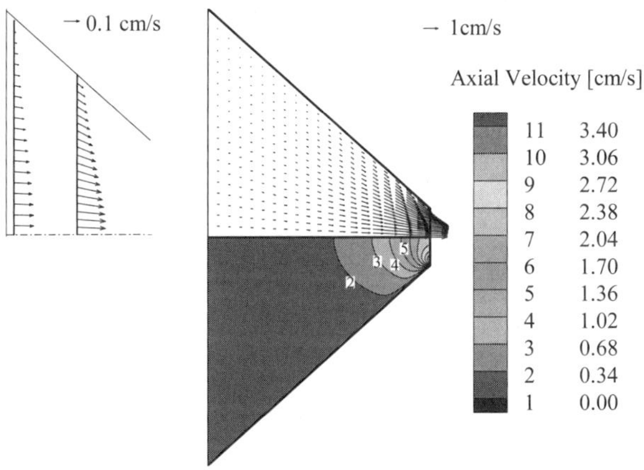

Fig.4.3.4 Velocity distribution in liver acinus**.** The upper half shows the velocity vectors while the lower half depicts the axial velocity contours.

Transient mass transfer with a step change in inlet chemical species concentration. The transient responses of the fluid and tissue concentration in the liver acinus to a step change in the inlet species concentration have also been simulated. Figure 4.3.5 shows both the species concentration distributions in the fluid and tissue after 5s and 300s. As expected, the fluid concentration at the center changes more rapidly than that at the periphery with a more rapid fluid flow. Associated with the low-velocity region, a distinct low concentration zone appears near the wall. At the initial stage of the transport process (e.g., t=5s), the species concentration in the tissue is much lower than that in the fluid, which indicates a delay for chemical species transferring from the fluid to the tissue. After 300s, the chemical species has already convected and diffused into most regions of the fluid and tissue within the acinus. The concentration variations as a function of time and position can be observed more clearly from Figs. 4.3.6-4.3.8. As shown in Figs. 4.3.6 and 4.3.7, both the fluid and tissue concentrations along the centerline and wall initially have a rapid decay and then tend to be flat. The average fluid concentration at the outlet (i.e., effluent concentration) is also plotted in Fig. 4.3.8 as a function of time. The effluent concentration is only a synthesis of the different fluid concentrations at the outlet, which cannot reflect the multi-dimensional, transient mass transfer processes in the liver (Lee & Rubinsky, 1989). Therefore, the comprehensive biofluid dynamics simulations are very helpful to analyze the hepatic uptake process of various chemical substances.

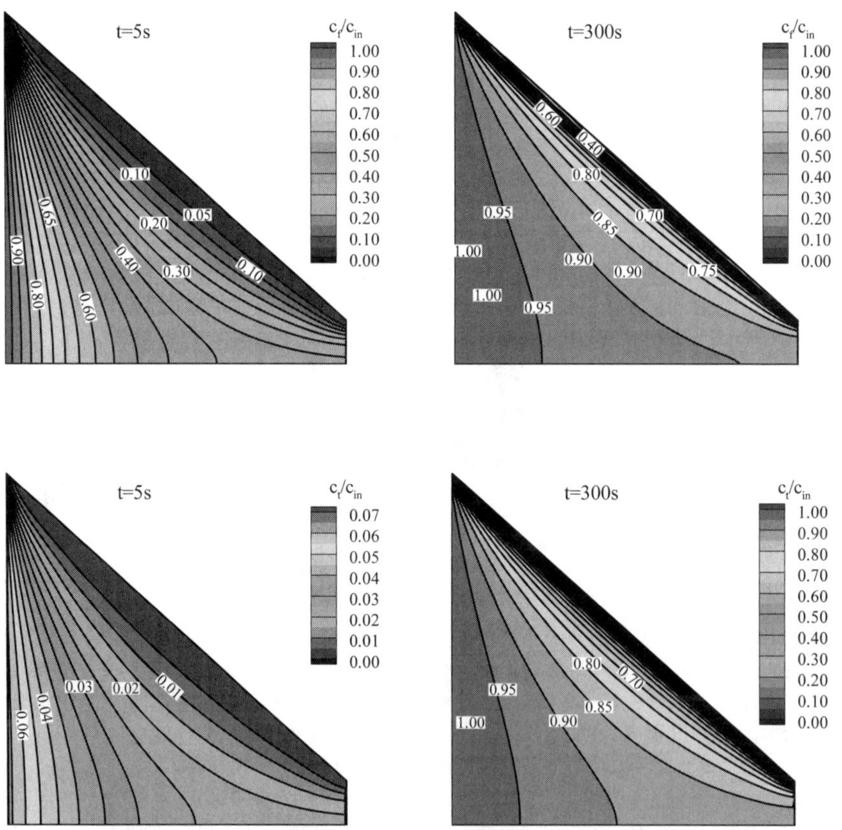

Fig.4.3.5 Distributions of species concentration in the fluid and tissue for a step change in the inlet concentration at t=5s and t=300s.

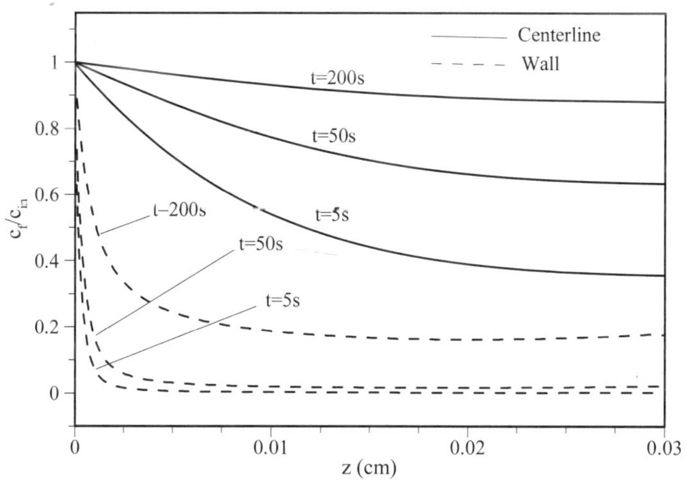

Fig. 4.3.6 Variations of fluid concentration as a function of time and position.

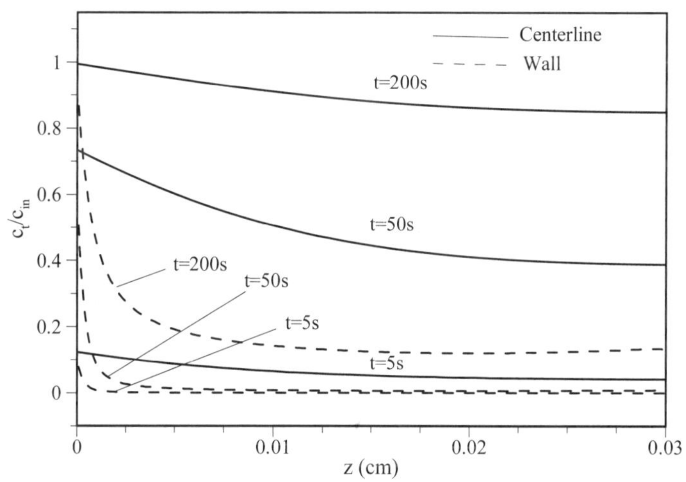

Fig. 4.3.7 Variations of tissue concentration as a function of time and position.

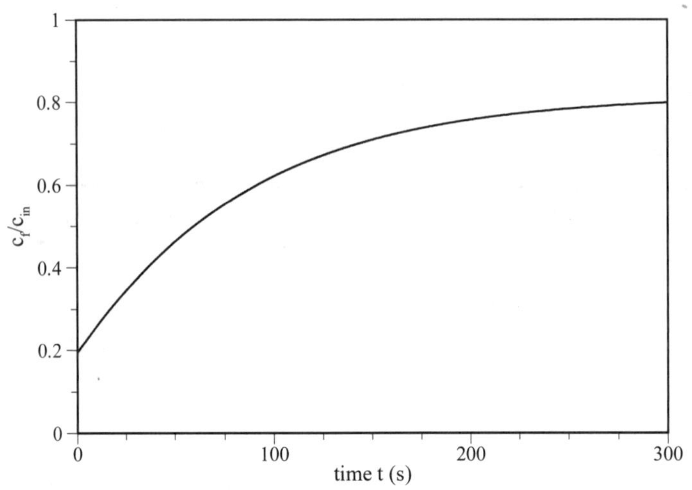

Fig. 4.3.8 Variations of mean fluid concentration at the outlet as a function of time.

4.4 HOMEWORK PROBLEMS

<u>Note:</u> As the Chapter III assignments, the following HWPs are somewhat open-ended, requiring literature reviews, creative thinking, and perhaps the use of basic software, such as MATLAB or studentFluent <www.fluent.com>.

4.1 Air and water-like liquids (e.g., blood, lymph, urine, etc.) are the body's biofluids. Design a systematic biofluid flow chart showing individual pathways and biofluid exchanges. Depict oxygen as well as particle transport, uptake, and migration.

4.2 Considering particle size, $1\text{nm} \le d_p \le 100\mu\text{m}$, inhalation flow rate, $10 \le Q \le 60$ l/min, and tidal volume, $0.5 \le \forall_T \le 2$ l, i.e., net air-volume inhaled, plot inhaled particle depositions as a function of volumetric lung region, $0 \le \forall_T \le 0.5$ l, i.e., along a pathway from mouth/nose to terminal alveoli.

4.3 In extension of HWP4.2, what are the particle transport and deposition mechanisms as a function of particle size range and inhalation flow rate?

4.4 Focusing on lung compliance and elasticity, obtain experimental values from the literature for, say, pressure-radius correlations of different regions of the human respiratory system. If available, categorize data sets according to age groups, i.e., children, adults, and seniors.

4.5 Effect of Surfactant on Bubble Progression and Film Thickness. The presence of surfactant in the pulmonary and respiratory systems is well known to reduce the surface tension, γ, which according to the Young-Laplace equation will reduce the pressures required to displace fluid with air bubbles. Although this reduction in pressure is desirable in order to protect the delicate cells and tissues that line respiratory airways, the presence of surfactant can also have a profound impact on the thickness of fluid left behind in the thin film. Surfactant transport will be governed by the following convection/ diffusion relationship

$$\frac{\partial \Gamma}{\partial t} + \nabla_s \cdot (\Gamma u_s) = D\nabla_s^2 \Gamma + J_{bulk} \tag{H4.1}$$

Here, Γ is the concentration of surfactant molecules absorbed onto the air-liquid interface, u_s is the velocity at the air-liquid interface, and J_{bulk} is the flux of surfactant to the interface from the bulk solution. For an insoluble surfactant $J_{bulk}=0$. Assuming steady-state conditions and that convection dominates diffusion (i.e., Pe=UR/D>0), Eq. (H4.1) reduces to

$$\nabla_s \cdot (\Gamma u_s) = 0 \tag{H4.2}$$

From continuity, $\nabla_s \cdot u_s = 0$, and the governing equation for Γ is $u_s \nabla_s \cdot \Gamma = 0$. Since surfactant molecules may be convected along the interface, a gradient in surface concentration may develop and therefore $\nabla_s \cdot \Gamma$ may be nonzero. Under these conditions the velocity at the interface, u_s, is zero and the interface is acting like an incompressible thin solid membrane at which the interfacial velocity drops to zero in a reference frame fixed to the bubble. Derive an expression for δ_f/R under these conditions. Using this relationship discuss the impact of surfactant on the film thickness at both high and low capillary numbers.

<u>Hint:</u> The boundary condition at the interface is now $u_s(y=\delta(z))=0$.

4.6 Provide a critical literature review on muco-ciliary transport in the human lung. As always (see Table 5.1.1) state the latest theoretical models, computer simulations, experimental findings, and applications to particle uptake and clearance as well as pathological scenarios.

4.7 "Subject variability" is a major problem in biomechanical engineering research. Comment on approaches to obtain and model human respiratory tract geometries representative of different age groups and pathological cases.

4.8 Figure 4.1.8 indicates that a high G3-inlet Reynolds number combined with a relatively large Stokes number generates a maximum deposition fraction for a tumor of critical size r/R=1.25. In contrast, low Re-St pair values produce hardly any particle deposition.
(a) Provide fluid-particle dynamics explanations for these two scenarios.
(b) If more realistically the inlet would be the mouth rather than generation 3, would the outcome be trend-wise the same?

4.9 Develop a compartmental model (see Sect.2.1.1) of the kidney from the inflow/outflow micro-scale to the nephron micro-scale. Estimate the ranges of Reynolds numbers and Schmidt numbers.

4.10 In continuation of HWP 4.9, collect typical kidney values for water flow, blood flow, species concentrations (i.e., waste products), transfer functions, and reaction coefficients in order to set up a lumped parameter model of the kidney.

4.11 In continuation of HWP 4.10, solve a set of first-order rate equations for a simplified kidney model.

4.12 Develop a compartmental model of the liver from the inflow/outflow micro-scale to the acinus micro-scale. Estimate the ranges of Reynolds numbers and Schmidt numbers.

4.13 In continuation of HWP 4.12, collect typical liver values for blood flow, species concentrations, transfer functions, and reaction coefficients in order to set up a lumped parameter model of the kidney.

4.14 In continuation of HWP 4.13, solve a set of first-order rate equations for a simplified liver model.

References

Bretherton, F.P. (1961) J. Fluid Mech., 10: 166-188.

Clift, R., Grace, J. R., and Weber, M. E. (1978) "Bubbles, Drops, and Particles," Academic Press, New York, NY.

Comer, J.K, Kleinstreuer, C., Hyun, S., and Kim, C. S. (2000) ASME Journal of Biomechanical Engineering, 122: 152-158.

Comer, J. K., Kleinstreuer, C., and Zhang, Z. (2001) Journal of Fluid Mechanics, 435: 25-54.

Cumucio, J.J. (1983) Am. J. Physiol., 244: G578-582.

Enderle, J.D., Blanchard, S.M., and Bronzino, J.D. (2000) "Introduction to Biomedical Engineering," Academic Press, San Diego.

Finlay, W.H. (2001) "The Mechanics of Inhaled Pharmaceutical Aerosols: An Introduction," Academic Press, London, UK.

Finlay, W.H., Lange, C.F., King, M., and Speert, D. (2000) Am. J. Resp. Crit. Care Med., 161:91-97.

Fox, S. I. (1996) "Human Physiology," McGraw-Hill, Boston, MA.

Ghadiali, S.N., Banks, J., and Swarts, J.D. (2002) J. Appl. Phys., 93: 1007-1014.

Ghadiali, S.N., and Gaver, D.P.III. (2003) J. Fluid Mech, 478: 165-196.

Grotberg, J.B. (1994) Annual Review of Fluid Mechanics, 26: 529-571.

Hickey, A.J. (2004) "Summary of Common Approaches to Pharmaceutical Aerosol Administration," In: Pharmaceutical Inhalation Aerosol Technology, Hickey A.J. (ed.), Marcel Dekker, Inc., New York, NY, pp.385-421.

Horsfield, K., and Cumming, G. (1968) J. Appl. Physiol., 24:373-383.

Kleinstreuer, C. (2003) "Two-Phase Flow: Theory and Applications," Taylor & Francis, New York, NY.

Kleinstreuer, C., and Zhang, Z. (2003) Journal of Biomechanical Engineering, 125: 197-206.

Lee, C.Y.C., and Rubinsky, B. (1989) Int. J. Heat Mass Transfer, 32:2421-2434.

Liao, Z., Poh, C.K., Huang, Z., Hardy, P.A., Clark, W.R., and Gao, D. (2003) Journal of Biomechanical Engineering, 125:472-480.

Longest, P.W., and Kleinstreuer, C. (2000) J. Medical Eng. & Tech., 24: 102-110.

Longest, P.W. and Kleinstreuer, C. (2003) Medical Engineering and Physics, 25: 843-858.

Morris, J., Ingenito, E.P., Mark, L., Kamm, R.D., and Johnson, M. (2001) J. Biomech Eng, 123: 106-113.

Pedley, T. J. (1977) Ann. Rev. Fluid Mech., 9: 229-274.

Phalen, R.F., Yeh, H.C., Schum, G.M., and Raabe, O.G. (1978) Anatom. Rec., 190:167-176.

Shi, H. (2006) "Analyses and Simulations of Inhaled Particle Uptake and Droplet Spray Dynamics in Representative Human Nasal Airways," PhD Dissertation, North Carolina State University, Raleigh, NC.

Staub, N.C. (1991) "Basic Respiratory Physiology," Churchill Livingstone, New York, NY.

Truskey, G. A., Yuan, F., and Katz, D.F. (2004) "Transport Phenomena in Biological Systems," Pearson Prentice Hall, Upper Saddle River, NJ.

Weibel, E. R. (1963) "Morphometry of the Human Lung," Academic Press, New York, NY.

Weibel, E.R. (1991) in "The Lung: Scientific Foundations," Crystall, R.G., West, J.B. et al. (eds.), Raven Press, New York, NY.

Zhang, Z., Kleinstreuer, C., and Kim, C.S. (2002) J. Aerosol Sci., 33: 257-281.

Chapter V

CASE STUDIES IN BIOFLUID DYNAMICS

Mathematical modeling and validated computer simulations can lead to new physical insight to complex systems. Equally important for an engineer is employing this computational tool for virtual prototyping of, say, improved surgical procedures or a new drug delivery device or a new implant. Such outcome relies heavily on the researcher's interdisciplinary knowledge base and multifaceted skills. Zeroing in on actual biofluid dynamics systems, this chapter provides a more in-depth extension of the previous two chapters with respect to mathematical modeling and computer simulation techniques. All three applications (Sects. 5.2 to 5.4) are drawn from results of the author's research group. *The type of case studies selected is rather immaterial as long as the projects are representative and insightful.* Clearly, computational fluid-particle dynamics and fluid-structure interaction analyses play major roles in the development of most mechanical, electromechanical, chemical, and pharmaceutical systems, as well as in the understanding of transfer processes in the human body.

Quite often, complex systems have to be decomposed or reduced before they can be analyzed in more detail and a solution can be sought. Especially optimization of a dynamic, nonlinear system requires the decomposition of the problem into "simpler" interconnected subsystems. Hence, decomposition might yield reduction in dimensionality and conceptual simplification, which lead to more flexible and more realistic system modeling, the so-called hierarchical, multilevel modeling approach for complex systems. Examples include atmospheric, human-body, ecologic, oceanographic, socioeconomic, manufacturing, or space systems on a large scale, and photochemical smog formation, flame front analysis, bio-MEMS transport phenomena and design, capillary blood flow, or enzymatic fermentation on a very small scale.

Sufficient and reliable measurements are most important for preparing input data sets and for checking the predictive capabilities of the model. Usually, such data sets are incomplete or simply not available and alternative ways have to be found. Occasionally, a few definite values of key input parameters may be replaced by appropriate value ranges. This is done with parametric sensitivity analyses where the effects of input data variations on the dependent variables are recorded. A thorough computer model validation with measured data sets is imperative and hence, because of this desirably strong interaction between theoreticians and experimentalists, a word on data reduction and analysis is appropriate. For example, in regression analysis, given data sets are evaluated (usually with a statistical package) for the relative significance of all terms in a postulated equation. The most important terms are retained and their coefficients are evaluated using the least-square routine. The resulting equation, a lumped parameter model, possesses the best continuous fit of the discrete data sets. Such a rather simple (empirical) modeling approach is widely employed for comparison purposes, in parameter estimations, and input/output model designs.

5.1 PREREQUISITES FOR MODELING AND SIMULATING

In Sect. 1.2.2, we discussed the two frameworks in which flow fields are traditionally described and mentioned that the Eulerian viewpoint is preferred by engineers to derive and apply the conservation laws (Sect. 1.3). Indeed, many fluid-particle dynamics problems are solved with a coupled Eulerian (fluid flow) and Lagrangian (particle motion) approach (cf. Sects. 1.4.3 and 5.3.1). In this section, we elaborate and extend the previous ideas of flow model development and consider modeling techniques for the simulation of more complex biofluid dynamics systems. Again, the problems or systems under consideration are restricted to:

(i) Continuum mechanics: velocity, stress, and deformation rates are continuous point functions, ignoring any local molecular effects.

(ii) Deterministic processes: random phenomena are either time-smoothed and represented by deterministic functions or random perturbation terms are added to simulate particle inlet distributions, turbulent particle transport, etc.

(iii) The Eulerian flow description is used for the carrier fluid only, or for both phases in case of two-fluid modeling, i.e., the velocity field is expressed as $\vec{\mathbf{v}} = \vec{\mathbf{v}}(\vec{\mathbf{x}}, t)$. Occasionally, the Lagrangian description is employed for particle trajectories based on Newton's 2nd law, where $\vec{\mathbf{v}}_{particle} = \vec{\mathbf{v}}_p\left[t, \vec{\mathbf{r}}_p(t)\right]$ and $\vec{\mathbf{r}}_p(t)$ is the particle position vector.

Mathematical modeling has matured over the years to a discipline in its own right as manifested in the number of specialized journals issued, and annual conferences held in this field. Computational fluid dynamics (CFD) is now established as a predictive design tool in aerospace, automobile, chemical process, drug, electronics, manufacturing, and nuclear industries. Indeed, CFD analyses are most useful and cost effective for today's engineering problems, which are nonlinear, three-dimensional, and transient. Although we are predominantly interested in the development and use of appropriate models for the understanding and prediction of complex transport phenomena in the human body and medical devices, a specialized group of researchers in the field of "mathematical modeling and computer simulation" focuses on the development of new modeling methodologies. Such methodologies may include generalizing theories and error analyses, ways of parameter estimation, as well as the development of new robust numerical methods, automatic adaptive mesh generation schemes, nonlinear optimization techniques, computer algorithms and tera-scale computing. Finally, special software is needed for the graphic presentation of results in either color and three dimensions or as fluid-particle flow visualizations on a CD-R or on websites (e.g., http://www.mae.ncsu.edu/research/ck_CFPDlab/index.html). Thus, specialists provide scientists and engineers engaged in R&D with new tools to represent and study real-world system processes. However, in many instances, time pressure, limited availability of trained people, low budgets, or insufficient information may restrict a thorough system analysis, mathematical modeling, and computer simulation. As a result, complex engineering systems are in reality too often described by very idealized models, utilizing algebraic correlations or linearized steady-state differential equations for problem solutions. Nevertheless, today's worldwide competition and increasing cost pressures coupled with a high demand for quality products require the use of computer hardware and software that accurately predict transport phenomena and system optimizations and hence prevent problems in product design and development. In fact, sophisticated computer tools have raised CFD models to such complex levels that they sometimes outstrip our ability to obtain experimental data with which to evaluate them.

In summary, Chapter V provides physical insight to some representative biofluids applications. To begin with, the important aspect of problem understanding and system conceptualization as a first step in modeling is discussed in Sect. 5.1.1. Modeling approaches and different stages in model development are outlined in Sect. 5.1.2. Section 5.1.3 contains a brief discussion of (1) important aspects in the development of system-specific modeling equations from the general field equations, and (2) proven software for computer simulations in fluid mechanics and heat transfer. The next three sections represent an updated version of Sect. 5.1 in Kleinstreuer (1997).

5.1.1 Problem Recognition and System Conceptualization

Before starting with the theoretical analysis of any research project, two basic aspects have to be considered:

(i) There has to be sufficient information on the particular project available so that all problem aspects are fully recognized, and the main objectives can be achieved in light of the available resources.

(ii) A preliminary assessment has to show that it is cost effective and manageable to develop an accurate and predictive mathematical model that will have the potential to lead toward new discoveries or fundamental advances in the knowledge base and/or technology base; if necessary, complex-system decomposition has to be considered.

System Identification

- Problem recognition and understanding of research project
- Statements of specific objectives and expected results
- Advantages and cost-efficiency of developing/using a math model
- Conceptualization and design of modeling framework
- Acquisition of data sets for model parameters and model validation

System Modeling

- Selection of dependent and independent variables
- Selection of model type and modeling approach
- Determination of *basic* equations plus initial and/or boundary conditions (cf. "deductive" approach)
- List of assumptions and postulates for derivation of modeling equations
- Development of submodels to gain closure

System Solution

- Selection of appropriate solution technique(s) and/or appropriate software package
- Identification of appropriate computer platform (e.g., desk-top, workstation, supercomputer, etc.)
- Validation of computer simulation model (e.g., experimental data sets, flow visualization, exact solutions)
- Display of model results, interpretation, and recommendations
- Use of model for design, analysis, process development, and decision making

Figure 5.1.1 Considerations and steps in model development.

As a first step toward system identification (Fig. 5.1.1), a preliminary literature review together with field observations and/or experimental data sets may help to understand a given project, i.e., the issues in question, the important transport phenomena, and the scope of the investigation. Sources and tools for literature search are textbooks, governmental reports, refereed journal articles, and conference proceedings, using library resources including machine searches, employing, for example, the citation indices "Web of Science" or "Medline," the World Wide Web on the INTERNET, etc. The resulting review of the relevant background material should contain information on the state-of-the-art in the area of interest, the data base needed for parameter values and model validation, as well as the potentially successful modeling approach and solution technique to be employed. At this point it is a valuable exercise to briefly describe the specific system to be analyzed (and how it fits into the larger picture, if appropriate), to state the objectives and anticipated results, and to outline the research plan as concisely as possible (cf. first two blocks of Fig. 5.1.1). Such a short document is similar to a preproposal, which should include a time-activity schedule as well as a preliminary budget.

When the research objectives and project significance are known and when the system boundaries, the dependent and independent variables plus the associated transfer processes are determined, the *system identification* step has been completed. As indicated in Fig. 5.1.1, this first step sets the stage for the subsequent phase of project development (cf. Sects. 5.1.1 and 5.1.2). Although some corrections are possible at a later point in time, a major redefinition of the system may imply a heavy loss in time and money.

5.1.2 Types of Models and Modeling Approaches

Different types of models and related modeling approaches are discussed in this section, together with generic preliminary steps such as literature review and decomposition of complex systems.

Basic definitions. Models may be *classified* as

- verbal (i.e., a concept, hypothesis, or theory);
- physical (e.g., laboratory bench-scale, electric/electronic analog, pilot plant, etc.); or
- mathematical (i.e., deterministic, empirical, stochastic; continuous, molecular discrete; analytical, numerical, etc.).

One could define a *model* as a (mathematical) representation of the real process; the actual operation of the model, e.g., the computer program, is the *simulation*. Another definition for *system simulation* is representation of the system's behavior by moving it from state to state in accordance with well-defined operating rules and subsequent observation of its dynamic performance, using computers. Alternatively, some authors categorized modeling of a physical or biochemical process as direct analysis, reconstruction, or identification. In *direct analysis*, the goal is to determine the output for a given set of input and system parameters. In turn, *reconstruction* implies determination of a system's input parameters; this is also called "*inverse problem modeling*." The identification problem is finding the system parameters when the system's input and output are given. In two-phase flow modeling we typically proceed with a direct analysis.

Design is akin to modeling since it involves the manipulation of elements representing physical systems. Software for computer-aided design (CAD), engineering (CAE), or manufacturing (CAM) can be used to develop a plan (blueprint) from which a device, a piece of machinery, a vehicle, a plant, or a whole city can be built and operated. In contrast, *mathematical modeling* focuses on the analysis of transfer processes in order to improve existing systems, i.e., modeling assists in designing new systems. Interesting hybrids are *expert systems* where advances in artificial intelligence are combined with computer simulation models in order to make "optimal" decisions and to manage a system, a plant, or a project most efficiently. *Experimentation* is a controlled observation of either the real system (field data acquisition) or a physical model (laboratory measurements), which should precede mathematical model development or should operate in tandem.

Basic physical understanding has to be acquired *before* mathematical descriptions can be postulated. Specifically, literature reviews and experimental data are generally necessary for problem recognition, model input, calibration, and verification. A guide for reviewing background literature is given in Table 5.1.1. In summary, experimentation, design, modeling, and simulation are iterative approaches or tools that may lead to a detailed description of a given system, the accurate prediction of its behavior, and, ultimately, improved designs or perhaps new products.

Physical laws describing natural processes in terms of physical quantities have to be dimensionally homogeneous; in other words, the units of each term on both sides of an equation have to be the same. A standard operation in mathematical modeling is to *nondimensionalize* the governing equations in order to generate dimensionless groups on which the solution must depend. In addition, *dimensional analysis* (cf. Buckingham Pi Theorem) can produce groupings of parameters that determine the system's dynamics and transfer processes. Similarly, *scale analysis* (cf. Sect. 1.5.5 and Kleinstreuer, 1997) derive dimensionless groups as well as functional dependences of key dependent variables.

A firm notion of *equilibrium* is important when selecting proper auxiliary equations for a complete system representation. An equation of state relating pressure and density or a parameter equation for the net absorption of a constituent may serve as examples. A system is in a state of equilibrium if no macroscopic changes with respect to time occur. Hence, all energy potentials are balanced and uniform throughout the domain of interest. Thermodynamic equilibrium is the most stringent type, comprising mechanical, thermal, chemical, electrostatic, as well as phase equilibrium. Hence, there is no macroscopic energy, matter, or charge flow within a

Table 5.1.1 Patterns for Review of Mathematical Modeling Publications and Indirectly a Guide for Writing Project Reports or Journal Articles.

Items to be Considered	Steps to be Taken
REFERENCES:	Reference reviewed publications and three to four key articles or reports; create special data files for program input and model verification. Distinguish between theoretical (analytical/numerical), experimental, and clinical papers.
ABSTRACT:	Review basic features of the project: (i) problem solved and new results gained, including new physical insight; (ii) type of mathematical model used (deterministic, stochastic); (iii) approach taken (lumped parameter, distributed parameter); (iv) how the problem was solved (solution technique and type of software used); and (v) outcome of clinical studies or laboratory observations, if applicable.
ASSUMPTIONS & APPROACH:	Make a list of all stated as well as implicitly used simplifications and reflect on their significance. Address approach selection, fluid/solid properties, flow regimes, time-dependence, dimensionality (see Sects. 1.2 and 1.2.3); solution method or lab set-up.
GOVERNING EQUATIONS:	Examine the basic equations and constitutive equations, including closure models as well as initial and boundary conditions; check final set of modeling equations.
INPUT DATA:	List data required for model input (geometry, ambient conditions, equation parameters and coefficients (I.C., B.C., etc.); reflect on possible sources for data acquisition (e.g., analytic solution, laboratory or field measurements); check on value ranges of dimensionless groups.
SOLUTION TECHNIQUE:	Review merits of employed transformations and solution techniques (i.e., analytic, asymptotic, approximate, numerical). Check if the mesh resolution is adequate. Check if mass and momentum residuals are small, say, 10^{-3} or less. Examine "model validation," i.e., comparisons with experimental data, resolution, completeness, etc. State computer requirements (e.g., mesh size, CPU-time, etc.) and type of computer used.
RESULTS & DISCUSSION:	Review results critically; check the used data. Comment on model validation, output, and display. Do the results provide *new* physical insight? What are the useful applications?
LIMITATIONS:	State area(s) of applications w/degree of accuracy and reliability; list major limitations.
CONCLUSIONS:	Summarize the salient discoveries and discuss their meaning/importance.

system in thermodynamic equilibrium, though the molecules are free to produce such flows. For example, equilibrium between the liquid and vapor phases occurs not necessarily when all changes cease, but when these molecular, i.e., microscopic, changes just balance each other, so that the macroscopic, i.e., the gross or average properties, remain unchanged. As seen in this broader context, equilibrium is actually a dynamic process on a molecular scale, even though it is treated as a static condition in macroscopic terms. The conservation principles are normally written in the form of partial differential equations expressing the balance of momentum, mass, and energy at any point in space and time. The Navier-Stokes equations, for example, express the condition of equilibrium, namely that for each fluid particle there is equilibrium between body forces, surface forces, and inertia forces. This is a reasonable assumption for fluid flows at normal densities; flows exhibiting shock waves, certain chemical reactions, rapid thermodynamic changes, and/or sorption processes require additional (nonequilibrium) terms for their mathematical description.

Modeling approaches. Model characteristics and modeling approach are directly linked. For example, a lumped parameter approach is taken when the gross (macroscopic or integral) behavior of a system is of interest. Distributed parameter models are used when the engineering variables at each discrete point in the system have to be evaluated (microscopic or differential approach). Statistical analysis, for example via stochastic modeling, becomes necessary when due to the complexities of random processes/phenomena only certain probabilities and definite trends of key variables can be obtained. Other characteristics of a given system that requires special modeling approaches and solution techniques may include the following:

(1) *Coupling*: A large or complex system can often be decomposed into simpler subsystems that are weakly coupled, i.e., certain modeling equations can be solved independently. For example, dilute particle suspension or forced convection heat transfer problems can usually be solved in computing the velocity field *independently*, i.e., one-way coupling. In general, coupled systems of equations have to be solved simultaneously, which requires higher computer resources and might cause (numerical) error magnification. Stability problems may turn up when coupled nonlinear equations representing system components with feedback and time delay have to be solved numerically. Packaged simulation languages avoid these problems by feeding the difference equations to the equation solver (Runge-Kutta, Gear, etc.) in the correct sequential order.

(2) *Linearity*: Linearity of a system is examined by comparing the response of a system to input. If a system response is directly proportional to an input, for all inputs, then the system is linear. For linear differential equations the principle of superposition can be employed via a product solution.

An equation is nonlinear when the exponent of the dependent variable (or its derivative) is not equal to unity or if dependent variables form products or appear in arguments of any transcendents. Nonlinear algebraic or ordinary differential equations without singularities can be solved with standard software packages (Newton-Raphson, Piccard or Runge-Kutta, Adams-Bashforth algorithms). Solutions of nonlinear partial differential equations, e.g., the Navier-Stokes equations, may cause (stability) problems when the nonlinear advection terms are dominant (cf. Roache, 1976).

(3) *Coefficient Type*: Coefficients of (differential) equations describing a real system are in most cases variable. Elaborate submodels are sometimes necessary to reflect the functional changes of coefficients with independent and principal variables.

(4) *Homogeneity*: Representation of systems with "forcing functions" results in inhomogeneous equations.

In summary, the main difficulties encountered in the development and validation of (deterministic) computer simulation models for advanced research projects are

- system decomposition;
- mesh generation and adaptation;
- acquisition of data sets for program input and model validation;
- derivation of the (correct and complete) modeling equations plus boundary conditions;
- selection of an efficient solution algorithm; and
- identification of sufficient computing resources.

These topics and others are further discussed in the next section and in the subsequent case studies.

5.1.3 Mathematical Representation and System Simulation

Mathematical modeling steps. The core of any mathematical model is the correct or at least adequate representation of all important transport and conversion phenomena occurring in the given (real-world) system. After selecting a suitable modeling approach (cf. Sect. 5.1.2), this requires the derivation of the system-oriented governing equations plus boundary conditions and necessary submodels for closure (cf. Fig. 5.1.1). To accomplish that, the basic equations (cf. Appendix B) have to be tailored to problem-oriented equations based on justifying assumptions and resulting postulates. In addition, associated initial and/or boundary conditions have to be defined. The derivation of modeling equations is a delicate process requiring mathematical skills and sometimes "good engineering feeling," especially in two-phase flow and biofluid mechanics. Specifically:

- Select a coordinate system that simplifies the boundary conditions and problem solution.
- Determine significant transport/conversion phenomena and evaluate relevant dimensionless groups; exclude negligible

processes based on physical insight gained from literature reviews, successful solutions to *similar* problems, etc.

- Check the significance of certain terms in the governing equations by comparing their value range with observations and measurements; perform a *scale analysis* if sufficient information is available; if necessary, do a *relative order of magnitude analysis* (ROMA) for each term of the basic equations.
- Subdivide the system domain into simpler regions or decouple transfer processes.
- If possible, *transform* the governing equations plus boundary conditions in order to reduce their mathematical complexity or to make them more tractable for a special solution method.

In case suitable sets of *differential* equations cannot be directly deduced from the basic equations in terms of fluxes (cf. Appendix B), the control volume approach should be employed as a starting point to derive the system-oriented equations representing energy, material, or force balances in *integral* form (cf. Sect. 1.3).

Computer simulation. Back-of-the-envelope estimates, scale analysis, and rough approximate calculations are most valuable to quickly identify/assess a technical problem at hand. However, many theoretical and industrial problems in fluid mechanics and heat transfer require experimental analyses and/or computer solutions. Once the governing equations, auxiliary relationships, and boundary conditions have been obtained, a suitable *numerical method* has to be selected in order to program the problem-oriented modeling equations and to simulate the system. All numerical schemes require the subdivision or discretization of the computational domain into a finite difference grid, finite element mesh, or a control volume mesh. The gradients (i.e., flux terms) of the governing equations are approximated by finite difference ratios between nodal points or at volume surfaces as in the finite difference or control volume method, respectively. Alternatively, approximation or shape functions for all dependent variables in association with each finite element are selected, integrated, and assembled, leading again to a coupled system of (nonlinear) algebraic equations. In any case, there are basically three choices for advancing from the mathematical modeling stage to a computer simulation:

(1) Develop your own numerical code based on a proven finite difference, finite element, or control volume method (cf. Patankar, 1980; Bradshaw & Cebeci, 1984; Cuvelier et al., 1986; Minkowycz et al., 1988; Hirsch, 1990a,b; Pepper & Heinrich, 1992; Reddy & Gartling, 1994; Tannehill et al., 1997; Roache, 1998; Ferziger & Peric, 1999; Cebeci, 2004). Numerous software routines are available at computer center libraries to aid in

executing specific tasks, e.g., matrix conversion, solution of elemental ODEs, as well as parabolic and elliptic PDEs (cf. Press et al., 1992, 1996; and IMSL-software).

(2) Most universities and R&D companies maintain licenses for multipurpose mathematical software such as Maple, Mathcad, Mathematica, Matlab, FlexPDE, studentFluent, etc., with which most of the previous chapter problems and some course projects can be solved and the results efficiently plotted.

(3) A more powerful but also cost-intensive option is the utilization of a general purpose software package for fluid flow, heat transfer, two-phase flow, and/or fluid-structure interaction problems, such as CFD-TWOPHASE, FLUENT/FIDAP, FIRE, ANSYS-CFX, CFD-LIB, Star CD, etc. Flexible and efficient mesh generators, fast and accurate equation solvers, realistic subroutines for turbulence, two-phase flow, non-Newtonian fluids, etc., vivid postprocessors, and *broad allowance for user-supplied programs* are the hallmarks of good commercial packages (see App. C).

(4) Emerging numerical codes for solving complex biofluid flow and bioparticle transport problems are the lattice Boltzmann method (Succi, 2001), immersed boundary methods (e.g., Peskin, 1977; Li, 1997, among others), and particle movers (see Loth, 2000).

With respect to items (1) and (3), the following aspects should be kept in mind. Computer codes have to account for the problem's unique characteristics, such as complex (3-D) geometries, time dependence, coupled momentum, heat and mass transfer, moving boundaries, turbulent flows including transitional effects, particle dynamics, coupled nonlinear (unknown) interfacial conditions, chemical reactions, fluid rheology, fluid-solid interactions, etc. Local oscillations and "false diffusion" have to be avoided. Efficient and accurate mesh generation is usually the *limiting module* in CFD code applications. Hence, a number of mesh generators, such as GridPro, ICEM-CFD, I-DEAS, and PATRAN, have been developed that may act as pre-processors for most commercial packages. Conventional meshes are structured where as a first step the flow domain is subdivided into connected blocks; alternatively, for more complex geometries, unstructured meshes are more suitable. If available, the hardware's parallel-processing capabilities should be exploited. In many cases, locally adaptive mesh refinement techniques are more advantageous than implementing elaborate higher-order algorithms. Unstructured meshes allow for more flexibility accommodating complex geometries than structured meshes. Commercial software packages should allow for user-supplied programs and exceptional alternatives to "default."

Generally, transient three-dimensional projects can be very CPU time intensive when using a *finite element* method. *Finite difference* methods, although faster, are too awkward and possibly inaccurate, requiring elaborate transformations to tackle geometrically complex problems. Control- or finite-volume-based programs are presently most popular. Supercomputers are certainly not the answer for many CFD problems because they become less and less accessible for most researchers for economic reasons. A practical solution is the use of vectorized numerical codes on high-end workstations with several parallel processors.

Model validation. Ideally, several complete sets of reliable experimental data should be available to compare advanced simulation results with measurements. Data sets for velocity, pressure, and temperature profiles are even more desirable than values for integral or lumped parameters such as skin friction coefficients, flow rates, mass transfer rates, as well as average or local dimensionless numbers (e.g., *Nu, Sh, Cp*, etc.). Profiles of the dependent variables for different input conditions allow a very detailed comparison.

If measured data sets are not available, the computer simulation model may be checked against analytical results for a simplified test case. Field or laboratory data sets have to be carefully scrutinized before they can be used. Information should be available on the experimental setup, the process conditions, what has been measured and what has been kept constant, and the *uncertainty estimates*. As indicated earlier, a joint theoretical/experimental venture is the best approach for solving complex problems accurately because it avoids the costly generation of incomplete, i.e., often useless or at least superficial data sets (cf. Fig. 5.1.2). Such an interaction is also helpful in producing plots that display the results most efficiently, i.e., insightful, concise, and usable. Figure 5.1.2 also indicates the growing problem of data processing as a result of DNS (cf. Joseph, 2001).

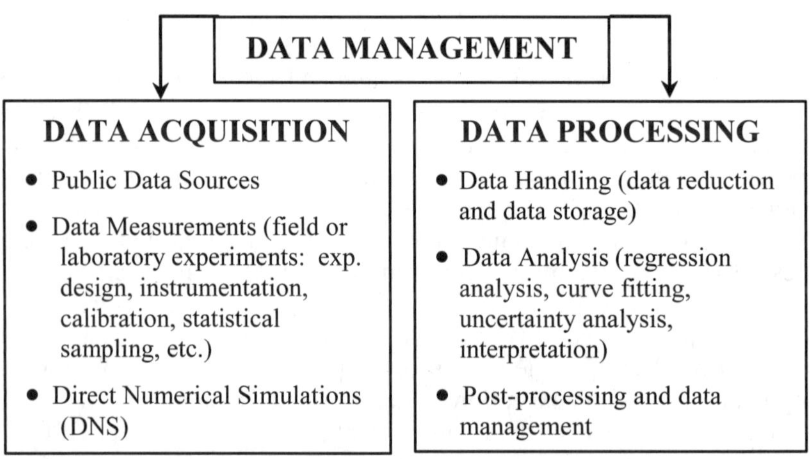

Figure 5.1.2 Aspects of data acquisition and processing (after Vemuri, 1978).

Display of results. Occasionally, computer simulation results are beautifully plotted, employing sophisticated graphics routines, but the display is somewhat void of insightful content. Considerations for the preparation of meaningful graphics, include the following:

- Derivation of dimensionless groups for plotting, using scale analysis or nondimensionalized parameters of the governing equations.
- Use the same coordinates for your graphs as used in benchmark publications to allow for easy comparison.
- Plotting of differential parameters, such as velocity and shear stress, as color-coded vectors as well as integral parameters in order to display the full range of resolution the simulator is capable of exploring; however, what counts is not the quantity but the quality of figures, i.e., those that provide new physical insight.

After the computer simulation model has been validated, novel results and the enhanced understanding of the project should be clearly displayed. *Parametric sensitivity analyses* and unique ways of plotting the research results, using *scale analysis*, help to achieve these goals. A potentially powerful display of CFD results is animated movies, i.e., computer-generated flow visualization employing postprocessor software such as WAVEFRONT, AVS, EnSight, or FIELDVIEW.

Finally, the graphs or videos have to be interpreted and unique features have to be explained rather than describing the "ups and downs" of single or multiple curves. Conclusive statements summarizing the physical insight and new understanding as well as potential design applications of the research should round out the project work.

Virtual prototyping. After the design-and-analysis phase of a given system or desired product has been completed, the phase of optimal-prototype-development starts. This is accomplished, via *virtual prototyping*, i.e., on the computer, rather than via actual physical prototype construction. Nowadays, not only machine parts or molecular structures of components but entire automobiles, transportation systems, medical procedures, and consumer product lines are first being designed, developed, and prototyped in *virtual reality*, before they are built. Clearly, the goal is to optimize a key objective function subject to several sets of constraints. This can be done via guided trial and error analyses or more formally with nonlinear optimization packages such as VMCON (cf. Crane et al., 1980).

The following computational two-phase flow case studies demonstrate the use of basic principles and may convey physical insight by means of a comparison study and a design project employing the “mixture flow” approach (Sect. 5.2). Particle transport and interacting media studies, employing the "separate flow" approach (see Sect.1.4.3), are discussed in Sects. 5.3 and 5.4.

Report writing. It is extremely important to write-up the research work in a professional report. Such a project report forms the basis for interdepartmental information exchange, M.S./Ph.D. theses, conference papers, journal articles, research work of others, etc. A good report should feature the following items:

- ***Abstract*** (i.e., summary of the essentials of the project report)
- ***Introduction*** (i.e., problem statement with project goals and relevance; literature review and novel contributions)
- ***Theory*** (i.e., system sketch, assumptions, concepts/approach used; governing equations plus derivations, boundary conditions and necessary submodels; solution method employed and model validation)
- ***Results and Discussion*** (i.e., depiction of the results in terms of figures, graphs, and tables; discussion of new physical insight and solutions of <biomedical> engineering problems)
- ***Conclusions*** (i.e., summary of salient points and their implications)
- ***References***.

5.2 NANODRUG DELIVERY IN MICROCHANNELS

Part of the more advanced endeavors in biomedical engineering is the development of devices for controlled nanofluid flow in microchannels. Specifically, it is of interest to deliver rapidly nano-size drugs dispersed in a liquid at predetermined, near-uniform concentrations to fixed micro-channel endpoints. Examples include devices for compound selection toward new drug development, testing of target-cell responses to stimuli, and dosed delivery of therapeutic drugs. In order to design, build, and further miniaturize such drug delivery systems, theoretical and technological challenges in microfluidics, microphysics/chemistry, and microfabrication as well as biosensing, actuation, and control have to be met. Other micro-system applications in medicine and biology include accurate, fast, non-invasive diagnostics, neural prosthetics, and minimally-invasive surgery, or micro-robotics (Dario et al., 2000; Whitesids & Strook, 2001; Stone & Kim, 2001, among others). Looking ahead, a new design paradigm is necessary for "nanofluidics," requiring theoretical solutions from statistical or molecular dynamics because the basic continuum assumption does not hold for such relatively large Knudsen numbers (see Sect. 1.2.1 and Li, 2004).

Focusing here naturally on microfluidic systems, i.e., intermittent or continuous laminar flows of liquid mixtures in channels of $L \geq 10$ μm, basic differences when compared to macro-channel flows are discussed in

Sect. 5.2.1. An application of dissolved drug delivery in multiple micro-channels is illustrated in Sect. 5.2.2.

5.2.1 Flow in Microchannels

Microfluidics refers to micro-scale gas or liquid flow configurations, minute fluid dosimetry, and/or devices such as MEMS (micro-electro-mechanical-systems) which include microchannels, motors, nozzles, valves, pumps, amplifiers, actuators, sorters, mixers, probes/sensors, turbomachines, etc. Characteristically, surface-to-volume ratios of MEMS are high, e.g., 10^6 m^{-1} for a one-micrometer device. Small length scales approaching the fluid's mean-free-path, say, for certain gases, may invalidate the continuum mechanics assumption. Fluid flow phenomena usually of no concern on the macro-scale may become very important in microchannel flow. Examples for liquid flow include fluid slip for polymeric liquids at stationary solid walls, lower apparent viscosity, higher viscous dissipation, surface roughness effects, interactive fluid-wall forces, channel entrance effects, dominance of diffusion and laminar flow, lower Reynolds number transition to turbulence, formation of gas bubbles, and surface tension effects. As measured in the laboratory, there are even variations in the pressure gradient and friction factor for fully-developed flow, leading to flow rates different than those obtained for comparable macro-channel flows.

One of the basic tasks of lab-on-a-chip devices (or bio-MEMS) is micro-mixing, which is largely due to diffusion and laminar convection. Typically, $\mathrm{Re}_{D_h} < 100$, where the hydraulic diameter $D_h = O(10\mu\mathrm{m})$. Ways of micro-mixing include the following:

- Chaotic Mixer I, where helical grooves in the walls of microtubes fold liquid streams into themselves.
- Chaotic Mixer II, where transverse impinging jets perturb the main flow, causing chaotic advection.
- Static Micromixer, where several bifurcating channels connect on two layers.
- Microplume Mixer, where an array of vertical jets inject a second fluid into a shallow container with a primary moving fluid.
- Acoustic Streaming, where ultrasonic flextural channel-plate waves cause mixing and net convection.

Whatever the task of bio-MEMS, e.g., mixing, separating, drug or fluid dosimetric, drug targeting, reactive flow production, etc., the choice of *driving force* is important. While micro-channel flow could be driven in the traditional way of a net pressure force, problems with miniaturized pumping systems and the small force requirements may suggest unconventional driving forces such as electrokinetics (Li, 2004), surface tension and phase change (Probstein, 2003), thermo-phoresis, etc.

5.2.1.1 Numerical Solution Techniques

As discussed at the micro-scale level, let alone transport phenomena on the nano-scale, certain wall effects, short-range forces, and molecular characteristics of a given fluid, not important on the macro-scale, may have to be taken into account. The guiding dimensionless group is the Knudsen number, i.e.,

$$Kn = \frac{\lambda}{L} \tag{5.2.1}$$

where, for example, L = $\rho/|\partial\rho/\partial y|$ to calculate the *local* Kn-number, or $L = l_{device}$ for the *global* Kn-number (see Fig. 1.2.1).

Molecular dynamics (all Knudsen numbers). Applications to microfluidics, employing molecular dynamics, are presently quite impossible (Binder et al., 2004). The typical time scale to simulate a system's dynamics ranges from 0.1 s to 10^3 s. Hence, for a chemically realistic molecular dynamics simulation, with a necessary time step in the range of 1 fs (= 1×10^{-15} s) to 10 fs, the task is quite taxing, considering that one would have to simulate over a span of 10^{15} time steps which is many orders of magnitude more than what is feasible nowadays.

Molecular dynamics (MD) is a computer simulation technique where the time evolution of a set of interacting atoms/molecules is followed by integrating their equations of motion. In molecular dynamics we follow Newton's law, i.e., for each atom i in a system consisting of N particles

$$F_i = m_i a_i \tag{5.2.2}$$

Here, m_i is the particle mass, a_i its acceleration, and F_i the force acting upon it, due to the interactions with other molecules. Therefore, in contrast to the Monte Carlo method, molecular dynamics is a deterministic technique: given an initial set of positions and velocities, the subsequent time evolution is in principle completely determined. In more pictorial terms, i.e., in virtual reality, atoms/molecules will "move" around, bumping into each other, oscillating in waves in concert with their neighbors, perhaps evaporating away from the system if there is a free surface, etc. A computer would have to calculate a trajectory in a 6N-dimensional phase space (3N positions and 3N momenta). However, such trajectories are usually not particularly relevant by themselves. Thus, molecular dynamics is connected with statistical mechanics to keep solutions manageable. Like the Monte Carlo method (see next section) it is a way to obtain a set of particle-path configurations distributed according to some statistical distribution functions, or statistical ensembles. An example is the micro-canonical ensemble, corresponding to a probability density in phase space where the total energy is a constant E:

$$\delta(H(\Gamma) - E) \tag{5.2.3}$$

Here, H(Γ) is the Hamiltonian, and Γ represents the set of positions and momenta; δ is the Dirac function, selecting only those states which have a specific energy E. Another example is the canonical ensemble, where the temperature T is constant and the probability density is the Boltzmann function

$$\exp(-H(\Gamma)/\kappa_B T) \tag{5.2.4}$$

According to statistical physics, physical quantities are represented by averages over configurations distributed according to a certain statistical ensemble. A trajectory obtained by molecular dynamics (see Eq. (5.2.2)) provides such a set of configurations. Therefore, a measurement of a physical quantity by simulation is simply obtained as an arithmetic average of the various instantaneous values assumed by that quantity during the numerical MD simulation run. Statistical physics is the link between the microscopic behavior and thermodynamics. In the limit of very long simulation times (and high N-values), one could expect the phase-space to be fully sampled, and in that limit this averaging process would yield the thermodynamic properties. In practice, the runs are always of finite length, and one should exert caution to estimate when the sampling may be good (i.e., the "system at equilibrium") or not. In this way, MD simulations can be used to measure thermodynamic properties and therefore evaluate, say, the phase diagram of a specific material. Beyond this "traditional" use, MD is nowadays also used for other purposes, such as studies of nonequilibrium processes, and as an efficient tool for optimization of structures overcoming local energy minima, e.g., simulated annealing.

Direct Simulation Monte Carlo ($0.1 \le Kn \le 10$). Direct Simulation Monte Carlo (DSMC) method was introduced by G.A. Bird in 1978. *Direct* simulation implies that physical phenomena are modeled using kinetic theory, and *Monte Carlo* implies that the scheme adopts random numbers. The use of random numbers is the distinguishing feature of a Monte Carlo procedure, and the essentially probabilistic nature of, say, a gas flow at the molecular level makes it an obvious subject for a simulation approach based directly on the physics of the individual molecular interactions (Bird, 1994). Using each simulated molecule to represent a large number of actual molecules reduces computational requirements to a manageable level but introduces statistical error which is inversely proportional to the square root of the total number N of simulated particles. The simplicity of the algorithm allows for straightforward incorporation of higher-order physical models and for application to complex geometries.

The direct-simulation method is similar to the molecular-dynamics method in that a large number of simulated molecules are followed simultaneously. The essential difference is that the intermolecular collisions are dealt with on a probabilistic rather than a deterministic basis. This requires the assumption of *molecular chaos* and restricts the method to *dilute gas flows* in which the mean spacing between the molecules is large in comparison with the molecular diameter. When the ratio of the mean molecular spacing δ to the molecular diameter d is large (δ/d), the fluid becomes *dilute*. The DSMC method may be used for dilute fluids when the ratio of the mean-free-path to the molecular diameter is at least $\lambda/d \geq 10$. DSMC is a statistical method that can be used very efficiently in the transition regime, where $\mathcal{O}(0.1) < Kn(= \lambda/L, \text{ or } \delta/L) < \mathcal{O}(10)$. DSMC remains valid, though extremely expensive to use, for much lower values of Kn.

Another computational approximation involves the time interval over which molecular motions and collisions are *uncoupled.* The effect of this step is negligible if the global timestep is smaller than the mean collision time. Particle motions are modeled deterministically, while the collisions are treated statistically.

The primary drawback of using DSMC is its cost. Whereas idealized calculations are possible on small computers, significant computational resources are required to simulate somewhat realistic flows. A sample application of Couette flow with Argonne gas can be viewed at www.mae.ncsu.edu/research/ck_CFPDlab/index.html.

Concerning microfluidics applications, the Navier-Stokes equations are valid for low Kn-number flows at high enough densities. DSMC can be used for high-Kn-number flows, and it is still possible to use DSMC for the regimes which overlap with Navier-Stokes, where Kn ~ 0.1 and δ/d ~ 10; but, the computations become quite expensive. Molecular dynamics, however, becomes expensive for low-density problems because the computational time step is limited by the time scales of the interaction potential and the method of solution. In the regime of parameter space in which $Kn^{-1} = L/\delta \sim 100$, there are few particles per volume element and, therefore, few collisions. When this is the case, large fluctuations occur in the calculations of the mean properties of the system, such as density, pressure, and temperature. A DSMC calculation would require many ensembles to get good statistics.

There is an important physical problem for which the limit of $L/\delta \sim 100$ becomes a serious practical issue. Currently, basic experiments and practical microdevices involve flows characterized by high Kn numbers and very low velocities. In a typical case, for example, the channel dimensions are 1.2 μm high by 5 μm wide by 3000 μm long. The undisturbed helium gas pressure and temperature might be 1 atm and 298 K, respectively, and the inflow velocity might be ~ 20 cm/s. In this case, Kn $\approx$0.13 and DSMC is a possible solution method.

In general, DSMC cannot be used for reasons involving both the number of time steps and the resolution required. Because of the relatively high number density at 1 atm, the mean collision time is very small (approximately 10^{-10} s). In DSMC, the computational time step must be less than the mean collision time. The particle transit time, which is the time that a flow with speed 20 cm/s takes to pass through the channel, is

10^{-2}s. Therefore, to reach the steady-state flow condition requires 10^8 time steps. A fully three-dimensional simulation of these experiments would also be impractical because of the large number of computational cells required (the limit of memory). One condition that must be satisfied during a DSMC procedure is that the smallest dimension of the computational cells must not be greater than one half of the mean-free-path. For the microchannel described above, a minimum of 500,000 cells would be required. Considering that each cell contains a certain number of particles (at least 20), a total of 10^7 particles would be required. Such a calculation is certainly beyond the capabilities of current computers. There is also the problem that statistical scatter due to small perturbations increases as the flow velocity becomes very subsonic, that is, when $V_f << V_{th}$, where V_f and V_{th} are flow and mean molecular thermal velocities, respectively. In the example above, the thermal speed is approximately 1000 m/sec. With 0.2 m/sec flow, the statistical noise is four orders of magnitude greater than the required signal. The solution of microchannel flow at 0.2 m/s would require over 10^8 samples in a time averaged steady flow because the statistical fluctuations decrease with the square root of the sample size. A general method that will solve problems for flows in the low-speed, high Kn-number regime is still needed.

Direct numerical simulation (Kn≤0.1). Direct numerical simulation of solid-liquid flows is a way of solving the initial value problem for the motion of particles in fluids exactly, without approximation. The particles are moved by Newton's laws under the action of hydrodynamics forces computed from the numerical solution of the fluid equations. One must simultaneously integrate the Navier-Stokes equations (governing the motion of the fluid) and the equations of rigid-body motion (governing the motion of the particles). *These equations are coupled through the no-slip condition at the particle boundaries, and through the hydrodynamic forces and torques that appear in the equations of rigid-body motion.* These hydrodynamic forces and torques must be those arising from the computed motion of the fluid, and so are not known in advance, but only as the integration proceeds. It is crucial that no approximation of these forces and torques be made other than that due to the numerical discretization itself so that the overall simulation will yield a solution of the exact coupled initial value problem up to the numerical truncation error. Joseph (2002) clearly mentioned their goal to be doing direct numerical solutions with *many thousands* of particles in three dimensions, with *large volume fractions*, for various kinds of suspensions and slurries. Particle tracking methods take into account the particles and the fluid motion to understand particulate flow. Particle tracking methods move the particles by Newton's equations for rigid-bodies using forces that are modeled from single particle analysis or from empirical correlation rather than from forces which are obtained by direct computation from the fluid motion. Joseph (2002) mentioned that the approximate methods include simulations based on potential flow, Stokes flow, and point-particle approximations so that they all simplify the computation by ignoring some possible important effects like *viscosity* and *wakes* in the case of potential flow, *inertial forces* which produce lateral migration and across-the-stream orientations in the case of Stokes flow, and the effect of *stagnation* and *separation* points in the case of point-particle approximations.

Approximate Methods ($Kn \leq 0.5$). In micro-scales the characteristic dimension of the flow conduits are comparable to the mean-free-path of the gas media they operate in. Under such conditions the "continuum hypothesis" may break down. Constitutive laws that determine stress tensors and heat flux vectors for continuum flows have to be modified in order to incorporate the rarefaction effects. The very well known "no-slip" boundary conditions for velocity and temperature of the fluid on the walls are subject to modifications in order to incorporate the reduction of momentum and energy exchange of the molecules with the surroundings. Deviation from continuum hypothesis is identified by the Knudsen number, which is the ratio of the mean-free-path of the molecules to a characteristic length scale, i.e., $Kn = \lambda/L$. This length scale L should be chosen in order to include the gradients of density, velocity, and temperature within the flow domain. For example for external flows boundary layer thickness, and for internal fully developed flows channel half thickness should be used as the characteristic length scale. According to the Knudsen number the flow can be divided into various regimes. These are: continuum, slip, transition, and free-molecular flow regimes.

A discrete particle or molecular based model is the Boltzmann equation. The Boltzmann equation is an integro-differential equation and solutions of this equation are limited. The continuum based models are described by the Navier-Stokes equations. A special case is the Euler equation corresponding to the inviscid continuum limit, which shows a singular limit since the fluid is assumed to be inviscid and nonconducting. Euler flow corresponds to $Kn = 0.0$. The Navier-Stokes equations can be derived from the Boltzmann equation using the Chapman-Enskog expansion (Liboff, 1998). When the Knudsen number is larger than 0.1, the Navier-Stokes equations break down and a higher level of approximation is obtained by carrying second order terms (in Kn) in the Chapman-Enskog expansion. A special form of such an equation is called the Burnett equations, for which the solution requires second-order accurate slip boundary conditions in Kn. The Burnett equations and consistent second-order slip boundary conditions is subject to some controversy and a better way of solving high-Knudsen number flows is through molecular based direct simulation techniques such as the Direct Simulation Monte Carlo method (DSMC). Microflows are typically in the slip and early transitional flow regimes. Therefore, Navier-Stokes equations with appropriate slip boundary conditions govern these gas flows (Beskok and Karniadakis, 1994).

Another powerful numerical approach for simulating biofluid flow systems is the lattice Boltzmann method (LBM), especially when the flow domain is complex and parallel processing is an option (see Succi, 2001). For example, focusing on alveolar flow, Li and Kleinstreuer (2006) demonstrated the influence of the geometric structure on airflow patterns in the human alveolar region.

5.2.1.2 Microchannel Flow Effects

For Newtonian liquids, the slip velocity is negligible in microchannels. However, surface phenomena, i.e., surface roughness effect and viscous dissipation effect which were assumed to be unimportant for liquid flows in macro-channels, turn out to be significant (Koo and Kleinstreuer, 2003). Specifically, to simulate the surface roughness effect, a Porous Medium Layer (PML) model was introduced (Kleinstreuer and Koo, 2004). The PML model (see Fig. 5.2.1) complements the Navier-Stokes equations by adding not only the Darcy term and Brinkman term, the latter representing the frictional resistance force which is proportional to the mean flow speed, but also the Forchheimer term which is caused by the form drag and is proportional to the square of the flow velocity.

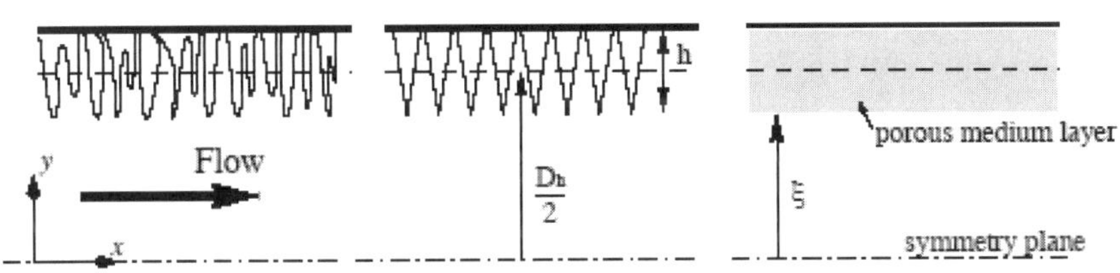

Figure 5.2.1: Porous medium layer equivalent to surface roughness and simple microchannel geometry: a) real surface roughness; b) homogeneous distribution of identical roughness elements; and c) mid-plane view of conduit with idealized roughness layer, or porous medium layer (PML).

In general, the use of the Navier-Stokes equations appears to be appropriate for micro-channel flows of liquids as long as $l_{system} \geq 0.1\,[\mu m]$ for conduits filled with, say, water at standard conditions. This condition is based on the global Knudsen number for liquids, Kn $=(\lambda_{IM}/l_{system})$, which is very small, where the intermolecular length for water molecules is $\lambda_{IM} = 3\,\text{Å}$ (Kleinstreuer, 2003). Nevertheless, as the system size decreases, surface phenomena, such as near-wall forces and relative roughness, become more and more important, and hence the Navier-Stokes equations have to be augmented as shown below, starting with a review of the friction factor for liquid flow in macro- vs. micro-conduits.

Friction factor analysis. Based on the continuum mechanics assumption, the continuity equation and equation of motion (or the Navier-Stokes equations for constant-fluid-property cases) are being solved in the conventional theory approach. With the resulting information, the friction factors for various internal and external flow systems can be calculated. The friction factor is defined as the ratio between the wall shear stress and the dynamic pressure, i.e.,

$$f = \frac{\tau_\omega}{\frac{1}{2}\rho\bar{u}^2} \tag{5.2.5a}$$

where $\bar{u}$ is the mean fluid velocity. The friction factor defined in Eq. (5.2.5a) is called the Fanning friction factor, or skin-friction coefficient. There is another commonly used friction factor called the Darcy friction factor, which is four times greater than the Fanning friction factor (Kleinstreuer, 1997):

$$f_D = \frac{4\tau_\omega}{\rho\bar{u}^2} = 4f \tag{5.2.5b}$$

Friction factor deviations can be expressed as percentage change

$$\Delta f^* = \frac{f_{observed} - f_{base}}{f_{base}} \tag{5.2.5c}$$

where f_{base} (see Eqs. (3.1a), (3.2), (3.3a) or (3.4)) is the theoretical friction factor value and $f_{observed}$ is the experimental or computational micro-channel value.

Friction factor relationships for turbulent fully-developed flow through macro-channels are typically expressed as a function of the Reynolds number and relative roughness height. For example, Colebrook combined the surface roughness and Reynolds number effects in an interpolation formula, expressed as (White, 1999):

$$\frac{1}{f_{D,turb}^{1/2}} = -2.0\log\left(\frac{\varepsilon/d}{3.7} + \frac{2.51}{\mathrm{Re}_{D_h} f_{D,turb}^{1/2}}\right) \tag{5.2.6}$$

where $Re_{D_h} = \frac{\overline{U}D_h}{\nu}$ and D_h is the hydraulic diameter. For laminar pipe flow, $Re_D < 2,300$, the friction factor is given by

$$f_{D,lam} = \frac{C_f}{Re_D} \tag{5.2.7a}$$

or

$$f_{D,lam} Re_D = C_f \tag{5.2.7b}$$

where $C_f = 64$ for circular tubes, and for rectangular channels, C_f will have different values that are determined by the aspect ratio b/a, knowing the hydraulic diameter (White, 1991).

The friction factor in the entrance region of any conduit is always larger than that for fully-developed flow; the difference increases with the Reynolds number for a given channel length and decreases as the channel length increases. Shah and London (1978) suggested an experimentally validated formula for the laminar friction factor, which is valid within 2.4 percent for many macro-scale duct shapes (White, 1991):

$$f_{app} Re \approx \frac{3.44}{\sqrt{\zeta}} + \frac{f_p Re + K_\infty / 4\zeta - 3.44/\sqrt{\zeta})}{1 + c/\zeta^2} \tag{5.2.8}$$

with $Re_D = \dfrac{U_0 D_h}{\nu}$ where $f_p Re = 14.23$, $K_\infty = 1.43$, c = 0.00029 and z is a Graetz-type variable defined as $\zeta = (x/D_h)/Re_{D_h}$ for channels, where x is the distance from the inlet.

Nanofluid viscosity. The mixture of a suspension of nanoparticles (e.g., a drug) and a carrier fluid (e.g., an aqueous solution) is called a nanofluid. As the nanodrug concentration α_d in the mixture increases, the effective dynamic viscosity changes. For dilute particle suspensions, i.e., $\alpha_d < 0.1$, according to Einstein:

$$\mu_{eff} = \mu_0 (1 + 2.5\alpha_d) \tag{5.2.9}$$

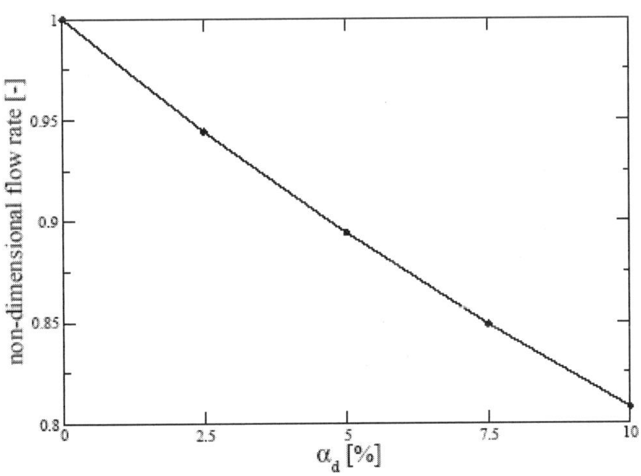

Fig. 5.2.2 Effect of particle concentration on mass flow rate supplied to the delivery channel.

The impact of Eq. (5.2.9) on the mass flow rate of the drug supply side is shown in Fig. 5.2.2. The nondimensional flow rate represents the ratio of the inlet drug flow rate with particles to that without particles. The pressure at the drug inlet was kept at 95 [Pa], and 60 [Pa] for the purging flow inlet. The mass flow rate decreases linearly with particle concentration α_d and decreased to 80 [%] compared to the pure liquid case, which means the mean flow velocity decreases correspondingly.

Surface roughness effect. The effect of surface roughness on flow characteristics was investigated by Mala and Li (1999) and Qu et al. (2000), who proposed a roughness-viscous model (RVM). It assumes that the roughness-viscosity is a function of y, the distance from the wall. The viscosity introduced by surface roughness has a finite value at the wall and zero at the channel center, and is proportional to the Reynolds number. Sharp (2001) claimed that the RVM does not fully explain friction factor trends at high Reynolds numbers referring to their experimental results. In this study, surface roughness effects are investigated with a thin porous medium layer (PML) on the walls for which the Brinkman-Forchheimer extended Darcy equation holds (Fig. 5.2.1). Thus, the effects of surface roughness on the friction factor, pressure gradient, and flow structures in microchannel flow are elucidated. Generally, the PML results show that surface roughness may increase or decrease the friction factor, depending upon the roughness-equal-PML permeability, i.e., if the Darcy number, $Da = \kappa / D_h^2$ approaches ∞, then the PM resistance vanishes and when $Da \rightarrow 0$, the PML acts like a solid coating, reducing the cross-sectional flow area. The PML, or surface roughness, effects on the pressure gradient and friction factor are analyzed using a steady fully-developed flow model. Typical relative surface roughness values, which depend on the wall material and machining process, are listed in Table 5.2.1. Generally, micromachining results in higher roughness than etching processes, and microchannels made out of stainless steel have higher values of surface roughness than those of silicon. A typical roughness value for etching processes was found to be 20 [nm], which is negligible in most microchannels where $D_h > 10\,[\mu\text{m}]$. However, considering flow in channels with $D_h \leq 10\,[\mu\text{m}]$, the effect of surface roughness should be taken into account.

The governing equations for steady incompressible non-isothermal mixture flow in microchannels with porous wall layers can be expressed as

$$\nabla \cdot \vec{u} = 0 \tag{5.2.9a}$$

$$(\vec{u} \cdot \nabla)\vec{u} = -\alpha\vec{S} - \alpha\nabla p + \nu_{eff}\nabla \cdot (\nabla\vec{u} + \nabla\vec{u}^T) \tag{5.2.9b}$$

$$(\vec{u}\cdot\nabla)T=\nabla\cdot((\alpha_{eff})\cdot(\nabla T))+\frac{\eta_{eff}}{\rho c_p}\Phi \tag{5.2.9c}$$

and

$$\nabla\cdot(\vec{u}\alpha_d)=\nabla\cdot\left[\left(D+\frac{\nu_T}{Sc}\right)\nabla\alpha_d\right] \tag{5.2.9d}$$

where p is the pressure, η_{eff} is the effective viscosity, ν_{eff} is the effective kinematic viscosity, which is the sum of molecular kinematic viscosity and turbulent viscosity, $\vec{S}$ is the source term which becomes the resistance vector for flow in a porous medium layer (see Sect. 3.3.1), Φ is the viscous dissipation function, $\dot{q}$ is a possible heat source/sink, α_d is the particle concentration, D is the particle diffusion coefficient, ν_T is the turbulent viscosity, and Sc is the Schmidt number.

Table 5.2.1 Typical values of relative surface roughness ($h/D_h \times 100[\%]$).

Author	Relative surface roughness [%]	Material
Pfhaler et al. (1991)	~ 1 [%]	Silicon
Peng et al. (1994)	~ 0.6 – 1 [%]	Silicon
Mala and Li (1999)	~ 3.5 [%]	Stainless steel, fused silica
Papautsky et al. (1999)	~ 2 [%]	Silicon
Xu et al. (1999, 2000)	~ 1 – 1.7 [%]	Aluminum, silicon

On the right-hand side of Eq. (5.2.9b), the first term captures the augmented surface forces induced by the scale-down of the system, and the second term is the pressure force acting on the fluid.

In case of turbulence effects, the low-Reynolds number k-ω model (Wilcox, 1998; Zhang and Kleinstreuer, 2003) could be used. The LRN k-ω model equations are:

$$u_j\frac{\partial k}{\partial x_j}=\tau_{ij}\frac{\partial u_i}{\partial x_j}-\beta^*k\omega+\frac{\partial}{\partial x_j}\left[(\nu+\sigma_k\nu_T)\frac{\partial k}{\partial x_j}\right] \tag{5.2.10a}$$

and

$$u_j\frac{\partial \omega}{\partial x_j}=\alpha\frac{\omega}{k}\tau_{ij}\frac{\partial u_i}{\partial x_i}-\beta^*\omega^2+\frac{\partial}{\partial x_j}\left[(\nu+\sigma_\omega\nu_T)\frac{\partial \omega}{\partial x_j}\right] \tag{5.2.10b}$$

where the turbulent viscosity is given as $\nu_T=c_\mu f_\mu k/\omega$, and the function f_μ is defined as

$$f_\mu = \exp[-3.4/(1+R_T/50)^2]; \qquad R_T = \rho k/(\mu\omega) \qquad (5.2.10c,d)$$

and

$$C_\mu = 0.09,\ \alpha = 0.511,\ \beta = 0.8333,\ \beta^* = 1,\ \sigma_k = \sigma_\omega = 0.5 \quad (5.2.11a\text{-}e)$$

The source term in the momentum equations, Eq. (5.2.9b), simulating the resistance in the porous medium, is the resistance vector, $\vec{R}$, which can be represented as (Bear, 1972):

$$\vec{R} = (R_C + R_F\,|\vec{u}\,|^\beta)\vec{u} \qquad (5.2.12)$$

where R_C is a resistance constant, R_F is the resistance speed factor, and b is a resistance speed power (usually 1.0). The quadratic drag term (R_{Fu^2}) in Eq. (5.2.12) is dominant for relatively high-Reynolds number flows (Nield and Bejan, 1992).

For steady $(\partial/\partial t = 0)$, 2-D $(\partial/\partial z = 0)$, and fully-developed flow $(\partial/\partial x = 0)$ in an open channel as well as through an isotropic medium, the Brinkman-Forchheimer extended Darcy equation, can be readily solved using MATLAB. Specially, for the channel (or parallel-plate) case:

$$0 = -\frac{dp^*}{dx^*} + \frac{4}{Re_{D_k}}\frac{d^2u^*}{dy^{*2}} + \left\{ -\underbrace{\frac{4u^*}{Da_{H/2}Re_{D_k}}}_{Brinkman\ term} - \underbrace{\frac{C_f u^{*2}}{Da_{H/2}^{1/2}}}_{Forchheimer\ term} \right\}_{PML} \qquad (5.2.13)$$

where

$$p^* = \frac{p}{\rho U_0^2},\ u^* = \frac{u}{U_0},\ x^* = \frac{x}{H/2},\ y^* = \frac{y}{H/2},\ Da_{H/2} = \frac{\kappa}{(H/2)^2},$$

$\mathrm{Re}_{D_h} = \dfrac{\rho U_0 2H}{\mu}$, H is the channel height, and $C_F (\approx 0.55)$ is drag coefficient.

For the tubular case:

$$0 = \frac{dp^*}{dx^*} + \frac{2}{Re_D}\left(\frac{d^2u^*}{dr^2} + \frac{1}{r^*}\frac{du^*}{dr^*}\right) + \left\{ -\frac{2u^*}{Da_R Re_D} - \frac{C_F u^{*2}}{Da_R^{1/2}} \right\}_{PML} \qquad (5.2.14)$$

where $x^* = \dfrac{x}{R},\ r^* = \dfrac{r}{R},\ Da_R = \dfrac{\kappa}{R^2},\ \mathrm{Re}_D = \dfrac{\rho U_0 D}{\mu}$, R is the tube radius, and U_0 is the average velocity. The relative roughness layer thickness was calculated based on the pipe diameter, i.e., D = 2R. The terms in braces $\{\}_{PML}$ are effective only in the PML, i.e., the roughness region. Typical values of model parameters are listed in Table 5.2.2.

Table 5.2.2 Typical PML parameters.

Symbol	Meaning	Typical values/formats
α	(Volume) porosity	$0 \le \alpha \le 1.0$
κ	Permeability	$10^{-4}[cm^2] \le k < \infty$
K	Area porosity tensor	For isotropic porous media, $K_{ij} = \alpha\delta_{ij}$
Γ	Fluid diffusivity	For laminar flow, $\Gamma \equiv \mu$
R_C	Resistance constant	$R_C = f(Da, Re)$
R_F	Resistance speed factor	$R_F = f(Da, \vec{u})$
β	Resistance speed power	$\beta \approx 1.0$

The boundary conditions are no-slip velocity (u = 0) at the wall, and velocity gradient at the interface, $r^* = \xi$, between open and porous regions, i.e., $\left.\frac{du^*}{dr^*}\right|_{r^*=\xi} = \frac{\mathrm{Re}_D}{2}\frac{dp^*}{dx^*}\xi$. The interface velocity gradients should be obtained by iteration.

Figure 5.2.3 shows the Darcy number effect on the axial velocity profile. When decreasing the Darcy number, the maximum velocity at the pipe center increases, while the velocity profiles near the wall flattens. As expected, the velocity profile in the PML is a function of the Darcy number and that the effect of the Forchheimer term on the friction factor is much more significant in tubes when compared to the parallel-plate case (see Fig. 5.2.4). This can be explained by the fact that the surface area increases with the radius, attaining a maximum at the shell where the roughness layer lies. Increasing surface area causes an increase in the effects of surface phenomena, e.g., surface roughness effect in this study, while it is maintained to be constant for the parallel-plate case. Figure 5.2.5 shows comparisons between the PML model predictions and selected experimental results. Specifically, the experimental results of Mala and Li (1999), which fall into the region predicted by the PML model, indicate a strong $f^*(Re_D)$ dependence (Figure 5.2.4(a)). The experimental data sets of Guo and Li (2003) are well matched with those of the PML model (Figure 5.2.4(b)). Clearly, roughness elements of the 179.8 [μm] diameter tube have a higher Darcy number when compared to the 128.8 [μm] diameter tube. The $f^*(Re_D)$ data for the 179.8 [μm] diameter tube seem to indicate the effect of laminar-to-turbulent transition, when Re_D = 1700. They claimed that '*the form drag resulting from the roughness is one* *reason leading to the increased friction factor*'; the form drag is captured by the Forchheimer term in the PML model.

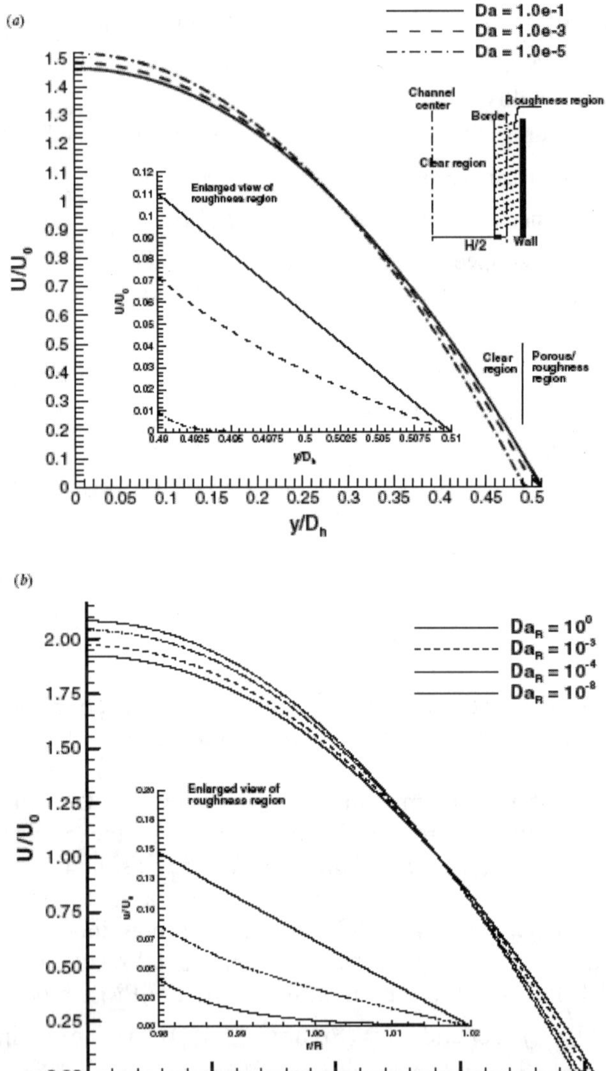

Figure 5.2.3 Effect of the Darcy number on steady 2-D fully-developed flows (Brinkman term only): (a) parallel-plate case and (b) circular tube case.

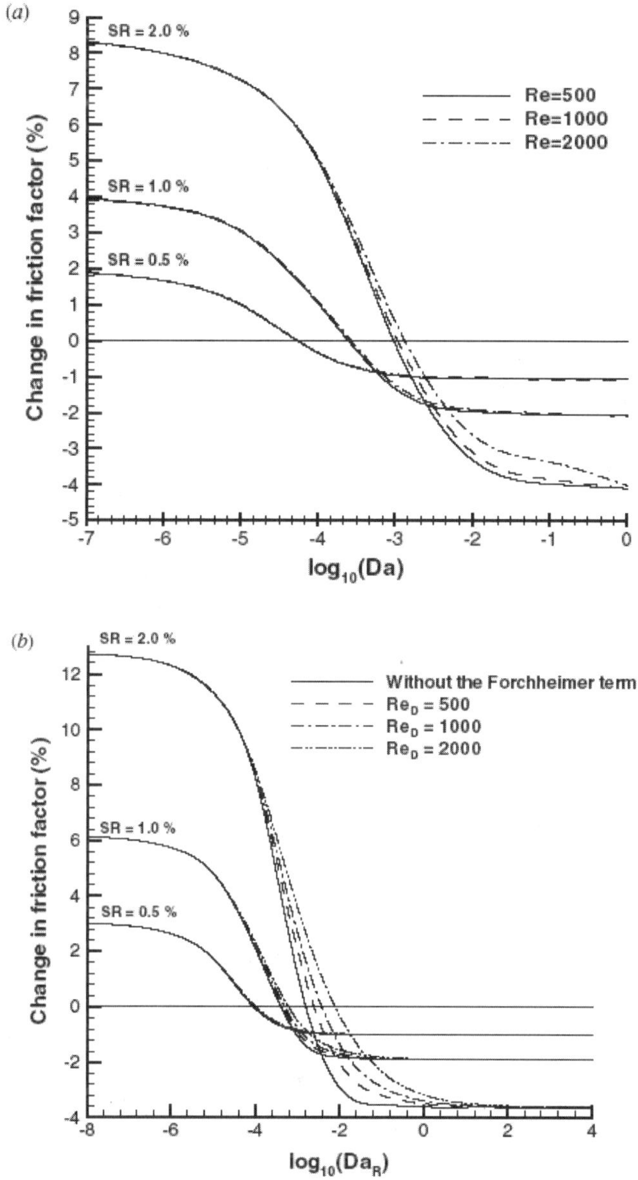

Figure 5.2.4 Effect of the Darcy number on steady 2-D fully-developed flows (Brinkman term only): (a) parallel-plate case and (b) circular tube case.

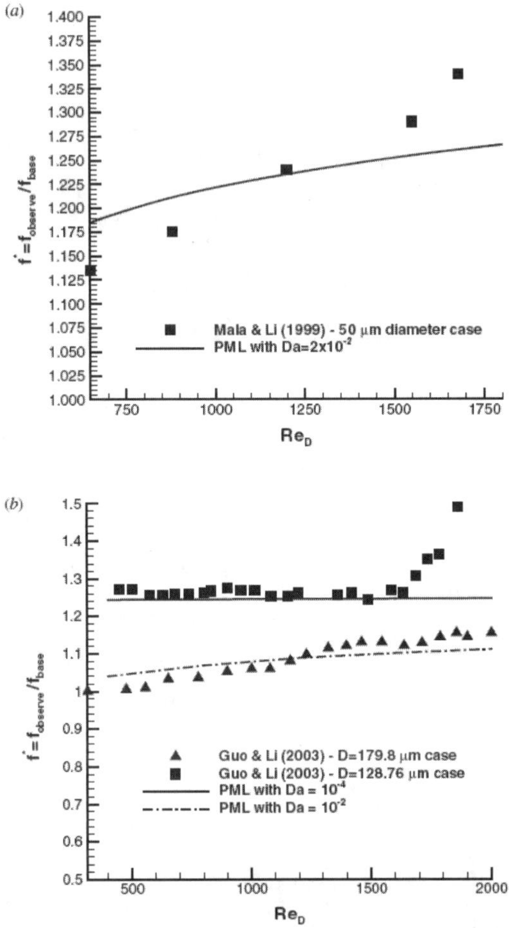

Figure 5.2.5 Comparisons of porous medium layer (PML) model predictions with experimental data: (a) Mala and Li (1999) and (b) Guo and Li (2003).

5.2.2 Controlled Nanodrug Delivery in Microchannels

In Section 5.2.1 the basics of fluid flow in microchannels were discussed with a steady 2-D liquid flow example illustrating surface roughness effects on the friction factor. In this section, the time-dependent mixing and delivery of a nanodrug solution is simulated. The goal is to analyze a multi-channel perfusion apparatus where in alternating cycles each channel delivers pre-determined *uniform* nanodrug concentrations to the channel endpoints, i.e., wells (or chambers) which contain tissue cells for testing (Fig. 5.2.6). In the next phase, the wells would then be flushed with the carrier fluid, called the purging cycle. Biochemical cell reactions to the nanodrugs (i.e., stimuli) are automatically recorded and analyzed.

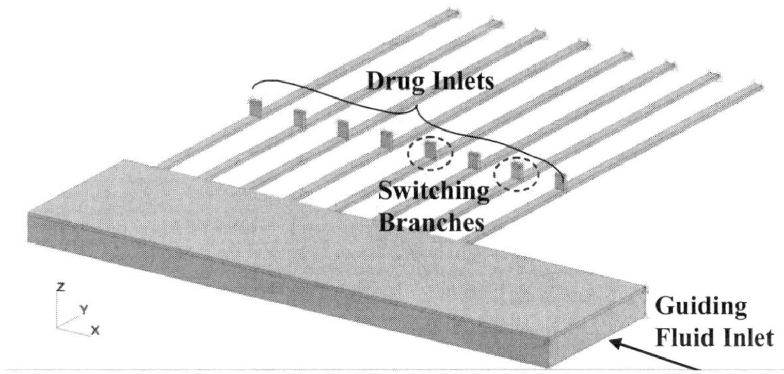

Fig. 5.2.6 Eight-channel nanodrug delivery system (Kleinstreuer & Koo, 2005).

Relevant time scales. There are three time scales to be considered for nanofluid system designs. The first one is the desired drug supply cycle time (~ 10 sec), another one is the drug supply convection time (L/U), and the third one is the radial diffusion time (a^2 / D), where L is the system length, U is the convection speed or average velocity, a is the half channel depth, and D is the nanoparticle diffusion coefficient. Considering one representative channel (see Fig. 5.2.6), the time scales for this design were selected as:

$$t_1 \text{ (cycle time scale)} = 10 \text{ [sec]}$$

$$t_2 \text{ (convection time scale)} < 1 \text{ [sec]}$$

$$t_3 \text{ (\textit{radial} diffusion time scale)} = 1 \text{ [sec]}$$

With a given system length of 2 [cm], the average velocity of each stream was set at 6 [cm/sec]. The diffusivity of drug was calculated with the Stokes-Einstein equation as:

$$D = \frac{\kappa T}{6\pi d_p} \approx \frac{1.38 \times 10^{-23} \times 300}{6 \times \pi \times 10^{-3} \times 10^{-10}} = 4 \times 10^{-9} \; [m/\text{sec}]$$

Therefore, the channel depth can be obtained from the relation (Probstein, 2003)

$$t_3 = 1\,[\text{sec}] = \frac{a^2}{D} = \frac{a^2}{4 \times 10^{-9}}$$

i.e., in this case a = 60 [μm].

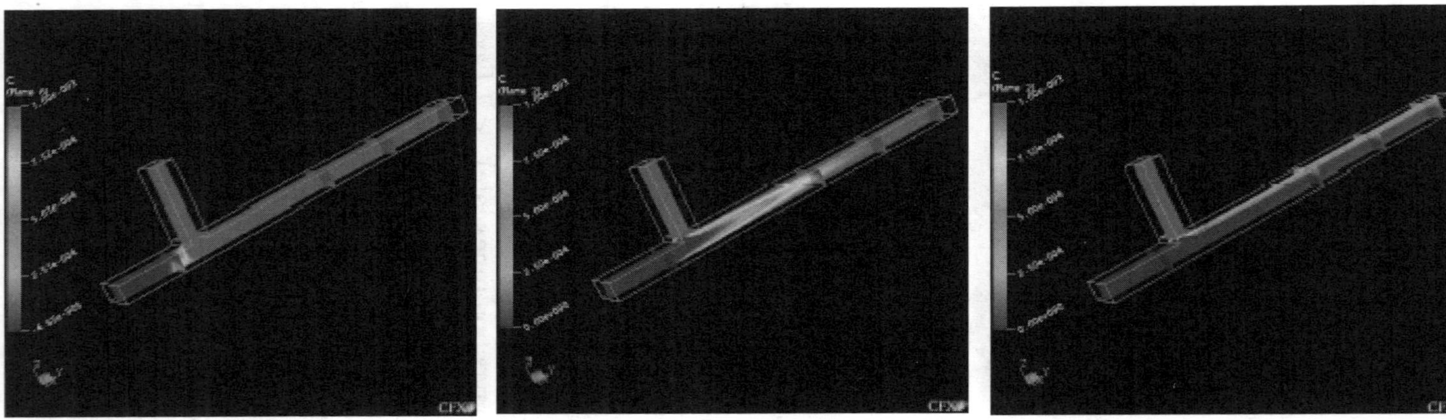

(a) (b) (c)

Figure 5.2.7 Snapshots of concentration distributions during purging process: (a) 0.001 [sec]; (b) 0.017 [sec]; (c) 0.041 [sec].

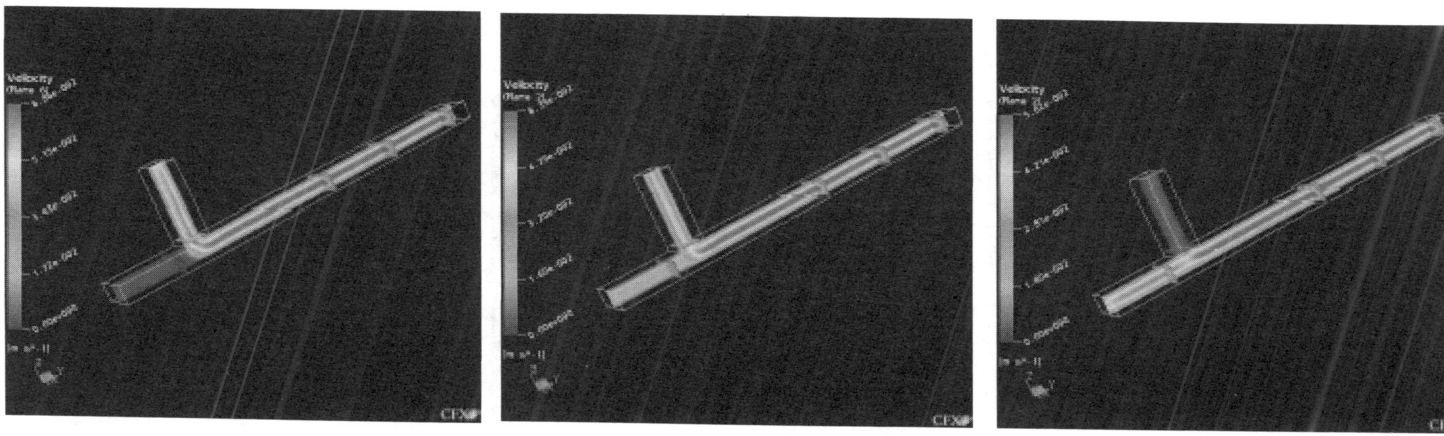

(a) (b) (c)

Figure 5.2.8 The snapshots of velocity fields during purging process: (a) 0.001 [sec]; (b) 0.003 [sec]; (c) 0.009 [sec].

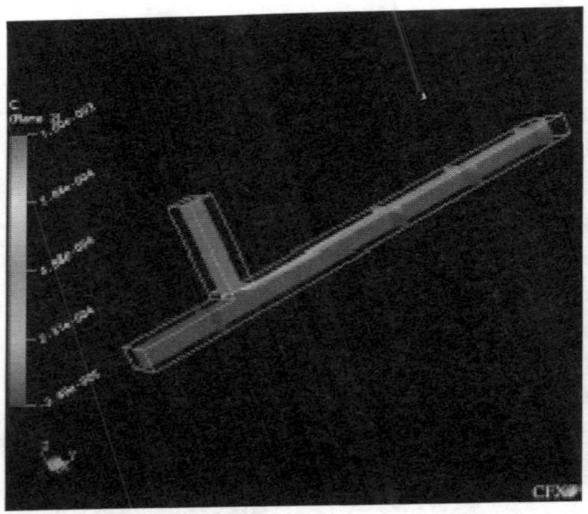

(a)

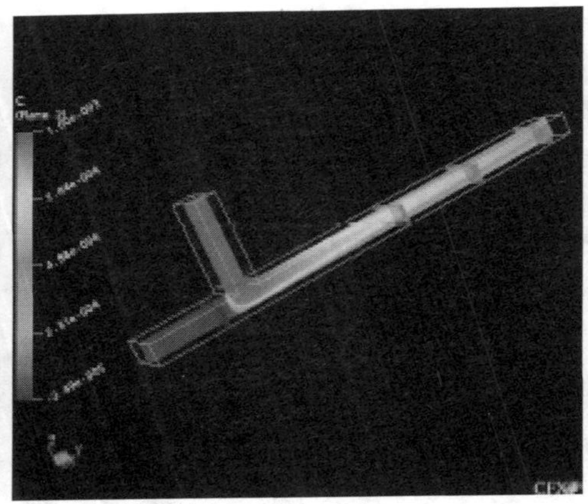

(b)

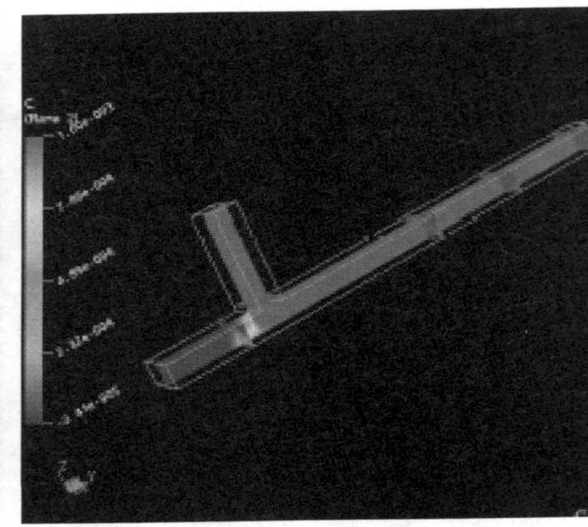

(c)

Figure 5.2.9 The snapshots of concentration distributions during drug supply process: (a) 0 [sec]; (b) 0.05 [sec]; (c) 1 [sec].

Boundary conditions. Adding the transient terms, the governing equations are presented in Sect. 5.2.1. For the current design analysis, uniform inlet velocity conditions for nano drug solution and base fluids, and ambient pressure condition for the outlet were applied. A 0.1 % drug concentration drug was assumed to enter at the drug supply inlet, and the zero flux condition was enforced at the walls. For the plenum chamber analysis, a uniform inlet velocity condition was applied to the inlet, while the ambient pressure condition was assumed for the branch-outlets. For the switching channel operation analysis, i.e., alternating between drug delivery and cleansing, the pressure boundary conditions for purging fluid and drug supply inlet were implemented. The pressure for the purging fluid was kept at 75 [Pa], while the drug supply pressure varied between 60 and 95 [Pa], depending on the operation mode.

Results and discussion. Referring to Figure 5.2.6, a plenum chamber is connected to the multi-channel inlets. It maintains a base fluid inlet velocity much lower than those in the channels owing to its large cross-sectional area, which allow the pressure drop in it to be negligible. Thus, the pressure is kept uniform throughout the chamber, which results in the same pressure gradients in all channels.

The switching branches can alter the incoming fluid to the test section by adjusting the inlet pressure of the drug solution. Specifically, setting the drug supply pressure higher than that of the purging fluid, drugs can be supplied; in contrast, the purging fluid can sweep through the test section by lowering the drug supply pressure. Figures 5.2.7-5.2.9 show snapshots of the cyclic operations. The pressure of the guiding fluid inlet side was kept to be 75 [Pa], while that of the drug inlet side varied from 60 [Pa] during the purging process to 95 [Pa] during the drug supply process. When the pressure at the drug supply inlet is lower than 60 [Pa], the contact line between the drug solution and base fluid rises in the +z-direction, whereas the drug migrates into the test section when the pressure level is higher than that.

In general, the velocity field develops much faster than the concentration field; the velocity field development was completed in less than 0.001 [sec], while it took less than 1 [sec] for the concentration to fully develop. Thus, this design ensures that the concentration field is very uniform and develops very fast.

5.3 PARTICLE DEPOSITION AND TARGETING IN HUMAN LUNG AIRWAYS

Basic information and pertinent references on the human respiratory system, including airway geometries and airflow rates, were given in Sect. 4.1.1. A special application of microparticle transport and deposition focusing on an idealized lung tumor was discussed in Sect. 4.1.2. In this section, the differences in microparticle vs. nanoparticle deposition mechanisms are analyzed and a new application of drug aerosol targeting is discussed.

In general, the breathing mode (e.g., shallow vs. rapid deep lung inhalation) and particle characteristic, especially particle size, greatly determine where in the human respiratory system and at what concentrations aerosols deposit. Particle size also influences the type of deposition mechanism. Specifically, for micro-size particles, inertial impaction is the main deposition mechanism around tubular bifurcations in the upper (large) airways while gravitional sedimentation governs their deposition in the small conducting airways and in the alveolar region (see Sect. 4.1). Lung deposition fractions of fine particles, say, $0.1 \leq d_p \leq 1\ \mu m$, are generally small. Ultrafine particles (i.e., $d_p < 100nm$) deposit mainly by diffusion and complex secondary flow structures. Figures 5.3.1a,b depict the total and local deposition fractions as adapted from Stahlhofen et al. (1989) and Kim (2000), respectively.

(a) Total lung depositions (b) Regional lung depositions

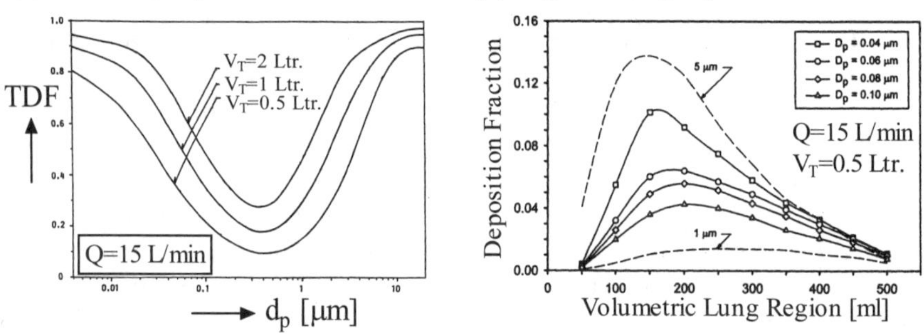

Fig. 5.3.1 Micro- and nano-particle deposition fractions: (a) TDF$\hat{=}$total lung deposition values; V_T $\hat{=}$net air volume inhaled, indicates breathing pattern; and Q$\hat{=}$mean inspiratory flow rate.

It is very difficult and cost-intensive, not to mention harmful in case of toxic aerosols, to determine *local* particle depositions by *in vivo* or *in vitro* tests. Hence, validated computational fluid-particle dynamics (CFPD) simulations can provide a non-invasive, accurate, and cost-effective means to obtain such information for both particle-size groups. Nevertheless, presently CFPD analyses are restricted to segments or regions of the respiratory system. Hence, global lung deposition models (Goo & Kim, 2003; Hofmann et al., 2002; ICRP, 1994), relying on experimental deposition correlations, algebraic and first-order rate equations, or stochastic modeling approaches, are still valuable to readily obtain averaged particle deposition data.

5.3.1 Nanoparticle and Microparticle Depositions in a Human Upper Airway Model

In a dawning age of nanotechnology, metal-oxide and carbon-based nanomaterials, $1nm \le d_p \le 100nm$, are being used in coolants to enhance heat transfer, in a variety of energy equipment, as building blocks of special composites, for biosensors, and in nanotherapeutics, to list just a few application areas.

While presently ultrafine particles in terms of mass fractions are rather low in the ambient air, they may occur in substantial concentrations where nanomaterial is being manufactured. In any case, recent studies have found that many particles in the ultrafine size range are more toxic (i.e., they may cause more severe inflammation) than large, respirable particles made of the same material (Frampton, 2001; Oberdorster, 2001, among others). For example, particles of titanium dioxide, aluminum oxide, and carbon black of less than 50nm in diameter have been shown to cause an approximately ten-fold increase in inflammation per unit mass than microparticles (Donaldson et al., 2000). These studies attributed the enhanced toxicity of nanoparticles to: (i) the greater surface area relative to the nanoparticle mass, and hence more sites to interact with cell membranes and a greater capacity to absorb and transport toxic substances such as acids; (ii) the prolonged retention time and the decreasing fraction of clearance for ultrafine particles; and (iii) the enhanced deposition of ultrafine particles in the deeper parts of the lung, including the alveolar region, where low surface tension produced by a surfactant film aids particle transfer through the liquid wall layer. However, the exact mechanisms of lung injuries by nanoparticles are still being explored. Hence, detailed comparisons between depositions of nanoparticles and microparticles in the human respiratory tract may provide more useful information for dosimetry-and-health-risk-assessments of differently sized particles.

Here we consider dilute suspension flows in a representative human upper airway model. Of interest are the local deposition patterns as well as deposition efficiencies and fractions of particles in the nano-size range of $1nm \le dp \le 150nm$ and in the micro-size range of $1\mu m \le d_{ae} \le 10\mu m$, where d_{ae} is the aerodynamic diameter. Section 5.3.1 is an updated version of the paper by Zhang et al. (2005).

5.3.2 Modeling Approach and Results

Upper airway geometry. As shown in Fig. 5.3.2, the present upper airway model consists of two parts: an oral airway model, including oral cavity, pharynx, larynx, and trachea, and a symmetric, planar, triple-bifurcation lung airway model representing generations G0 (trachea) to G3 after Weibel (1963). The dimensions of the oral airway model were adapted from a human cast as reported by Cheng et al. (1999). Specifically, the diameter variations along the present oral airway from mouth to trachea are almost the same as those for the hydraulic diameters from the cast. Variations to the actual cast only include the circular cross sections, a short mouth inlet with a diameter of 2cm, a modified soft palate, and a strong bend. The dimensions of the four-generation airway model are similar to those given by Weibel (1963) for adults with a lung volume of 3500ml. The airway conduits are assumed to be smooth and rigid. The minor effects of cartilaginous rings, which may appear especially in the upper airways, and out-of-plane bifurcations have not been considered in the present analysis. Although most bronchial bifurcations are somewhat asymmetric and nonplanar, some studies have shown that inspiratory flow in an asymmetric bifurcation exhibits the main features of the equivalent symmetric case. Nonplanar geometries only influence the flow in downstream bifurcations (Caro et al., 2002; Zhang & Kleinstreuer, 2002). For inspiration, the air and particle flow fields in the nonplanar configuration resemble those in the planar configuration, but rotated to some degree and merged with the symmetric secondary vortices (Zhang & Kleinstreuer, 2002). While the nonplanar geometry may somewhat increase microparticle deposition (Comer et al., 2000), it has only a minor effect on nanoparticle deposition (Shi et al., 2004).

Airflow equations. In order to capture the air flow structures in the laminar-to-turbulent flow regimes, i.e., $300<Re_{local}<10^4$ for the present airway configuration and inhalation rates, the low-Reynolds-number (LRN) k-ω model of Wilcox (1993) was selected and adapted. Zhang and Kleinstreuer (2003a) and Varghese & Frankel (2003) demonstrated that it is appropriate for such internal laminar-to-turbulent flows. All air flow equations, as well as initial and boundary conditions, including the adjustments of flow and particle information from the trachea of the oral airway model (see Section O-O in Fig. 5.3.2) as the inlet conditions of the bifurcating airway segment, are given in Zhang & Kleinstreuer (2003a, b).

Transport of micro-size particles. With any given ambient concentration of non-interacting spherical microparticles, a Lagrangian frame of reference for the trajectory computations of the particles can be employed. In light of the large particle-to-air density ratio, dilute particle suspensions, and negligible particle rotation, drag is the dominant point force. Hence, the particle trajectory equation can be written as (see Sect. 4.1.2):

$$\frac{d}{dt}(m_p u_i^p) = \frac{1}{8}\pi\rho\, d_p^2 C_{Dp}(u_i - u_i^p)|u_j - u_j^p| \tag{5.3.1}$$

where u_i^P and m_p are the velocity and mass of the particle, respectively; C_{DP} is the drag coefficient.

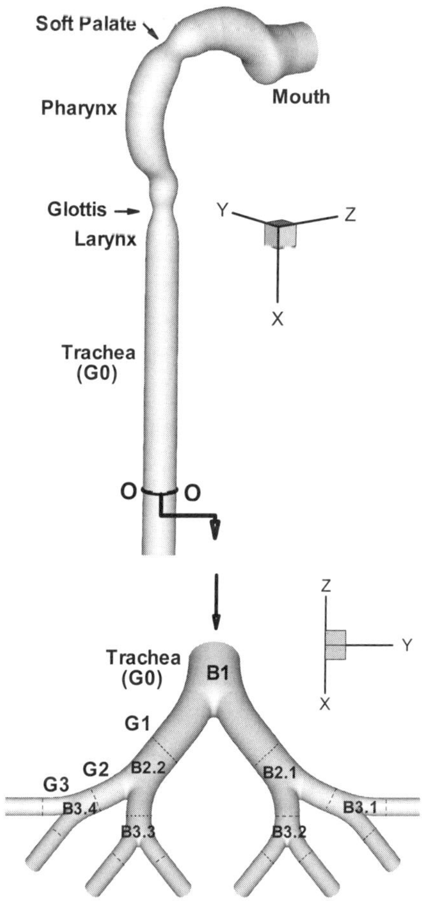

Fig. 5.3.2 3-D views of the oral airway model and bifurcation airway model (Generations G0 to G3). B1 - first bifurcation, B2.1 and B2.2 - second bifurcation, B3.1, B3.2, B3.3, and B3.4 - third bifurcation (the dashed lines indicate the segmental boundaries).

In Eq. (5.3.1), u_i is the instantaneous fluid velocity with $u_i = \bar{u}_i + u_i'$, where $\bar{u}_i$ is the time-averaged or bulk velocity of the fluid, and u_i' are its fluctuating components. Traditionally, turbulence is assumed to consist of a collection of randomly directed eddies; hence, an eddy-interaction model (EIM) is used to simulate the particle trajectories, and the fluctuating velocities u_i' are obtained by (Gosman & Ioannides, 1981; Matida et al., 2000; Schuen et al., 1983):

$$u_i' = \xi_i \left(\tfrac{2}{3}k\right)^{1/2} \qquad (5.3.2)$$

where ξ_i are random numbers with zero-mean, variance of one, and Gaussian distribution. In the EIM, each particle is allowed to interact successively with various eddies and the random numbers are maintained constant during one eddy interaction, while the corresponding turbulence intensities vary with the particle positions (MacInees & Bracco, 1992; Matida et al., 2000). The lifetime t_E and length scale l_E of the eddies which particles interact with can be determined as

$$t_E = 1.5^{\frac{1}{2}} C_\mu^{\frac{3}{4}} \frac{k}{\varepsilon} = \frac{0.2}{\omega} \text{, and } l_E = C_\mu^{\frac{3}{4}} \frac{k^{\frac{3}{2}}}{\varepsilon} = 0.164 \frac{k^{\frac{1}{2}}}{\omega} \quad (5.3.3)$$

Due to the assumption of turbulence isotropy, the fluctuating velocities normal to the wall calculated with Eq. (5.3.2) may be higher than the actual values (Kim et al., 1987), which may overpredict the particle deposition in some cases (Matida et al., 2004). As proposed by Matida et al. (2004), a near-wall correction can be used to simulate the near-wall particle trajectories, i.e., the component of fluctuating velocity normal to the wall u'_n can be expressed as

$$u'_n = f_v \xi (\tfrac{2}{3} k)^{1/2} \quad (5.3.4)$$

with

$$f_v = 1 - e^{-0.02 y^+} \quad (5.3.5)$$

where f_v is a damping function component normal to the wall considering the anisotropy of turbulence near the wall (Wang et al. 1999). Usually, Eq. (5.3.5) is used for $y^+ < 10.0$, while f_v=1 elsewhere.

At the mouth inlet, presently uniform particle distributions were prescribed. The effect of randomized non-uniform particle inlet distributions on the deposition efficiency has been discussed by Zhang & Kleinstreuer (2001). The initial particle velocities were set equal to that of the fluid, and one-way coupling was assumed between the air and particle flow fields because the maximum mass loading ratio (mass of particles/ mass of fluid) is below 10^{-4} in the present analysis. Particle positions and velocities in the cross section O-O (see Fig. 5.3.2) in the lower trachea were adjusted as the inlet particle conditions of the bifurcating airway segment.
The regional deposition of microparticles in human airways can be quantified in terms of the deposition fraction (DF) or deposition efficiency (DE) in a specific region (e.g., oral airway, first, second, and third bifurcation, etc.). They are defined as:

$$DF_{particle} = \frac{\textit{Number of deposited particles in a specific region}}{\textit{Number of partciles entering the mouth}} \quad (5.3.6)$$

$$DE_{particle} = \frac{Number\ of\ deposited\ particles\ in\ a\ specific\ region}{Number\ of\ particles\ entering\ this\ region} \tag{5.3.7}$$

The regional deposition efficiency is mainly used to develop the deposition equation for algebraic (total) lung modeling. The DEs and DFs are the same for the oral airway model in this study.

The local deposition patterns of microparticles can be quantified in terms of a deposition enhancement factor (DEF). As proposed by Balásházy et al. (1999, 2003), the particle deposition enhancement factor is defined as the ratio of local to average deposition densities, where deposition densities are computed as the number of deposited particles in a surface area divided by the size of that surface area. The mathematical expression for DEF is

$$DEF = \frac{DF_i / A_i}{\sum_{i=1}^{n} DF_i / \sum_{i=1}^{n} A_i} \tag{5.3.8a}$$

where A_i is the area of local wall cell (i), n is the number of wall cells in a specifc airway region, and DF_i is the local deposition fraction in local wall cell (i) which is given as

$$DF_i = \frac{Number\ of\ deposited\ particles\ in\ mesh\ cell\ (i)}{Number\ of\ particles\ entering\ the\ mouth} \tag{5.3.8b}$$

Clearly, if the overall and maximum DEF-values in one airway region are close to one, the distribution of deposited particles tends to be uniform. In contrast, the presence of high DEF-values indicates non-uniform deposition patterns, including "hot spots." Some "hot spots" of toxic particulate matter are related to the induction of certain lung diseases (e.g., lung cancer).

Mass transfer of nanoparticles. The convection-diffusion mass transfer equation of nanoparticles can be written as

$$\frac{\partial Y}{\partial t} + \frac{\partial}{\partial x_j}\left(u_j Y\right) = \frac{\partial}{\partial x_j}\left[\left(\tilde{D} + \frac{\nu_T}{\sigma_Y}\right)\frac{\partial Y}{\partial x_j}\right] \tag{5.3.9}$$

where Y is the mass fraction, $\sigma_Y = 0.9$ is the turbulence Prandtl number for Y, and $\tilde{D}$ is the effective aerosol diffusion coefficient which is calculated as follows (Cheng et al., 1988; Finlay, 2001):

$$\tilde{D} = (k_B T C_{slip}) / (3\pi\mu d_p) \tag{5.3.10}$$

where k_B is the Boltzmann constant (1.38×10^{-23} JK^{-1}); and C_{slip} is the Cunningham slip correction factor.

The regional deposition fraction can be determined according to Fick's law (Zhang & Kleinstreuer, 2004), i.e.,

$$DF_{region} = \sum_{i=1}^{n}[-A_i(\widetilde{D}+\frac{\nu_T}{\sigma_Y})\frac{\partial Y}{\partial n}|_i]/(Q_{in}Y_{in}) \tag{5.3.11}$$

where A_i is the area of the local wall cell (i), and n is the number of wall cells in one certain airway region, e.g., oral airway, first airway bifurcation, etc. The local deposition patterns of nanoparticles can again be quantified in terms of a deposition enhancement factor (DEF) (see Balásházy et al., 1999, 2003), which is defined as the ratio of local to average deposition densities, i.e.,

$$DEF = [(\widetilde{D}+\frac{\nu_T}{\sigma_Y})\frac{\partial Y}{\partial n}|_i]/\{\sum_{i=1}^{n}[A_i(\widetilde{D}+\frac{\nu_T}{\sigma_Y})\frac{\partial Y}{\partial n}|_i]/\sum_{i=1}^{n}A_i\} \tag{5.3.12}$$

Assuming that the airway wall is a perfect sink for aerosols upon touch, the boundary condition at the wall is Y_w=0. This assumption is reasonable for fast aerosol-wall reaction kinetics (Fan et al., 1996) and also suitable for estimating conservatively the maximum deposition of particles or toxic vapors in airways.

5.3.2.1 Numerical Method

The numerical solutions of airflow transport equations with low-Reynolds-number k-ω model were carried out with a user-enhanced commercial finite-volume based program, i.e., CFX4.4 from ANSYS, Inc. The numerical program uses a structured, multiblock, body-fitted coordinate discretization scheme. In the present simulation, the PISO algorithm with under-relaxation was employed to solve the flow equations (Issa, 1986). All variables, including velocity components, pressure, and turbulence quantities, are located at the centroids of the control volumes. An improved Rhie-Chow interpolation method was employed to obtain the velocity components, pressure and turbulence variables on the control volume faces from those at the control volume centers. A Quadratic Upwind (QUICK) differencing scheme, which is third-order accurate in space, was used to model the advective terms of the transport equations. The sets of linearized and discretized equations for all variables were solved using the Block Stone's method.

The particle transport equation (Eq. (5.3.1)) was solved with an off-line F90 code with parallelized algorithms (Longest et al., 2004). For the calculation of particle trajectories, geometry, velocity, and turbulence data at all control-volume vertices were first extracted from the CFX solution and written to arrays. A second-order improved Euler predictor-corrector method (Longest et al., 2004) was then used for the integration of the particle trajectory equation, including turbulent dispersion effects with near-wall correction (cf. Eqs. (5.3.2-5)). Particle deposition occurs when its center comes within a radius from the wall, i.e., local surface effects such as migration in mucus layers or resuspension due to clearing have been currently ignored. In the present simulations, 200,000 to 1,000,000 randomly selected, uniformly distributed particles were released at the mouth inlet. Particle deposition, including deposition efficiency and fraction, was tested to be independent of the number of particles released.

The computational mesh was generated with CFX Build4, where the near-wall region requires a very dense mesh. Specifically, the thickness of the near-wall cells was chosen to fully contain the viscous sublayers and to resolve any geometric features present there. The mesh topology was determined by refining the mesh until grid independence of the solution of flow and mass fraction fields as well as particle deposition was achieved. The final mesh features about 420,000 and 670,000 cells for the oral airway and four-generation airway model, respectively. The computations were performed on an SGI Origin 2400 workstation with 32GB RAM and multiple 450 MHz CPUs. The solution of the flow field at each time step was assumed to be converged when the dimensionless mass residual is less than 10^{-3}, i.e., (Total Mass Residual)/(Mass Flow Rate) $< 10^{-3}$. The convergence of other variables was monitored as well. The estimated maximum artificial numerical dispersion coefficient is in the order of 10^{-11} m^2/s in this study, which is still one order smaller than the physical diffusion coefficient for the largest nanoparticle considered (3.2×10^{-10} m^2/s for 150 nm particles). Typical run times for the fluid flow and mass transfer simulations on eight processors with parallel algorithm were approximately 24 - 65 hours for the oral airway model and 8 hours for the four-generation model under steady inhalation condition. Utilizing the converged flow field solution, the microparticle trajectory simulations required approximately 10 to 25 hours on four processors for each case simulated.

5.3.2.2 Model Validations

The present computational fluid-particle dynamics (CFPD) model has been validated with various experimental data sets for steady and transient laminar flows in bifurcations (Comer et al., 2000, 2001; Zhang & Kleinstreuer, 2002; Zhang et al., 2002) and for laminar, transitional, and turbulent flows in tubes with local obstructions (Kleinstreuer & Zhang, 2003; Zhang & Kleinstreuer, 2003a). Especially, the low-Reynolds-number (LRN) k-omega model has been extensively validated and has been proven to be an applicable approach to capture the velocity profiles and turbulence kinetic energy for laminar-transitional-turbulent flows in the constricted tubes of the upper airways (see Zhang & Kleinstreuer, 2003a). The current simulation of nanoparticle deposition due to diffusional transport has been validated with both analytical solutions in straight pipes and experimental data for a double-bifurcation airway model (Shi et al., 2004) as well as experimental data in an oral airway model (Zhang & Kleinstreuer, 2003b).

Simulated particle deposition fractions in the present oral airway model were compared with the observations by Cheng et al. (1999) in Fig. 5.3.3a for three inhalation rates. Following Cheng et al. (1999), the Stokes number is defined here as $St=\rho_p d_p^2\ U/9\mu D$, with ρ_p being the particle density, d_p being the particle diameter, and U being the mean velocity evaluated as (Q/A), where A is the mean cross-sectional area, and D is the minimum hydraulic diameter. A comparison of microparticle deposition in the present airway model with *in vivo* deposition data as a function of the impaction parameter is shown in Fig. 5.3.3b. Generally speaking, our computational data points agree well with the experimental data and nicely retrace the mean of the measured deposition data curve.

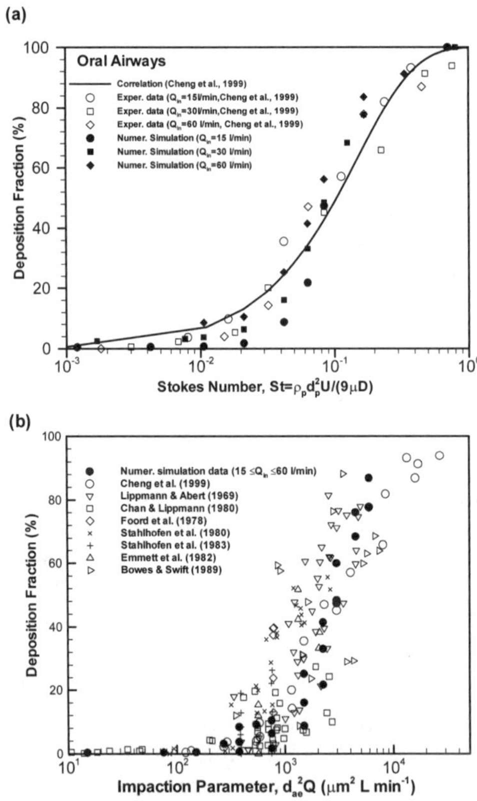

Fig.5.3.3 Comparison of the present simulated particle deposition fractions in the oral airway model with: (a) the experimental data of Cheng et al. (1999); and (b) in vitro and in vivo deposition data sets, where d_{ae} is the aerodynamic particle diameter.

In summary, the good agreements between experimental observations and theoretical predictions instill confidence that the present computer simulation model is sufficiently accurate to analyze laminar-to-turbulent fluid flow as well as mass transfer and particle deposition in three-dimensional oral and bifurcating airways.

5.3.2.3 Results and Discussion

The laminar-transitional-turbulent air flow and particle transport in the upper airway model under both steady and transient inhalation conditions have been previously analyzed (Kleinstreuer & Zhang, 2003; Zhang & Kleinstreuer, 2002, 2004). This paper will focus only on distinct features and underlying mechanisms of nanoparticle vs. microparticle depositions.

Microparticle deposition. The 3-D surface views of the local particle deposition patterns in terms of particle deposition enhancement factor DEF (see Eq. (5.3.8a)) for particles with aerodynamic diameters d_{ae}=3μm and 10μm under different inspiratory flow conditions are shown in Fig. 5.3.4. Clearly, microparticle deposition during inhalation is mainly due to impaction, secondary flow convection, and turbulent dispersion. Thus, they mainly deposit at stagnation points for axial particle motion, such as the tongue portion in the oral cavity, the outer bend of the pharynx/larynx, and the regions just upstream of the glottis and the straight tracheal tube. As shown in Fig. 5.3.4, the maximum DEF-values are in the range of 43 to 479 for d_{ae} =3 and 10μm, which vary with the flow rate. This implies that the deposition patterns of microparticles in the oral airway are highly non-uniform, and hence a small surface area, where the maximum DEF occurs, may receive hundred times higher dosages when compared to the average value for the whole airway. Such a site of massive particle deposition is usually located in the glottis region and/or at the outer bend in or just after the curved pharynx region. The contribution of turbulent dispersion on deposition seems to be stronger for small-size particles (say, d_{ae} =3μm) than for larger-size particles (say, d_{ae} =10μm) as indicated by more scattered and high DEF-values in the trachea.

Figures 5.3.5a and 5.3.5b depict the distributions of DEFs in the bifurcation airway model G0 to G3, i.e., part of the bronchial tree. As expected, for microparticles the high DEF values appear mainly around the carinal ridges due to inertial impaction, but the specific distribution of DEFs at each carina is different and varies with the inhalation flow rate as well. Some microparticles also land outside the vicinities of the cranial ridges due to secondary flows and turbulent dispersion. The strong turbulent fluctuations may occur just around the flow dividers and then decay rapidly in the straight tubular segments of the airways G0 to G3 (see Zhang & Kleinstreuer, 2004). It should be noted that no particles deposit in bifurcations B3.1 and B3.4 (see Fig. 5.3.2) in the case of Q_{in}=60 l/min and d_{ae}=10μm. In fact, no particles enter these two bifurcations because more than 80% of incoming particles deposit in the oral airway (i.e., for Q_{in}=60 l/min and d_{ae}=10μm).

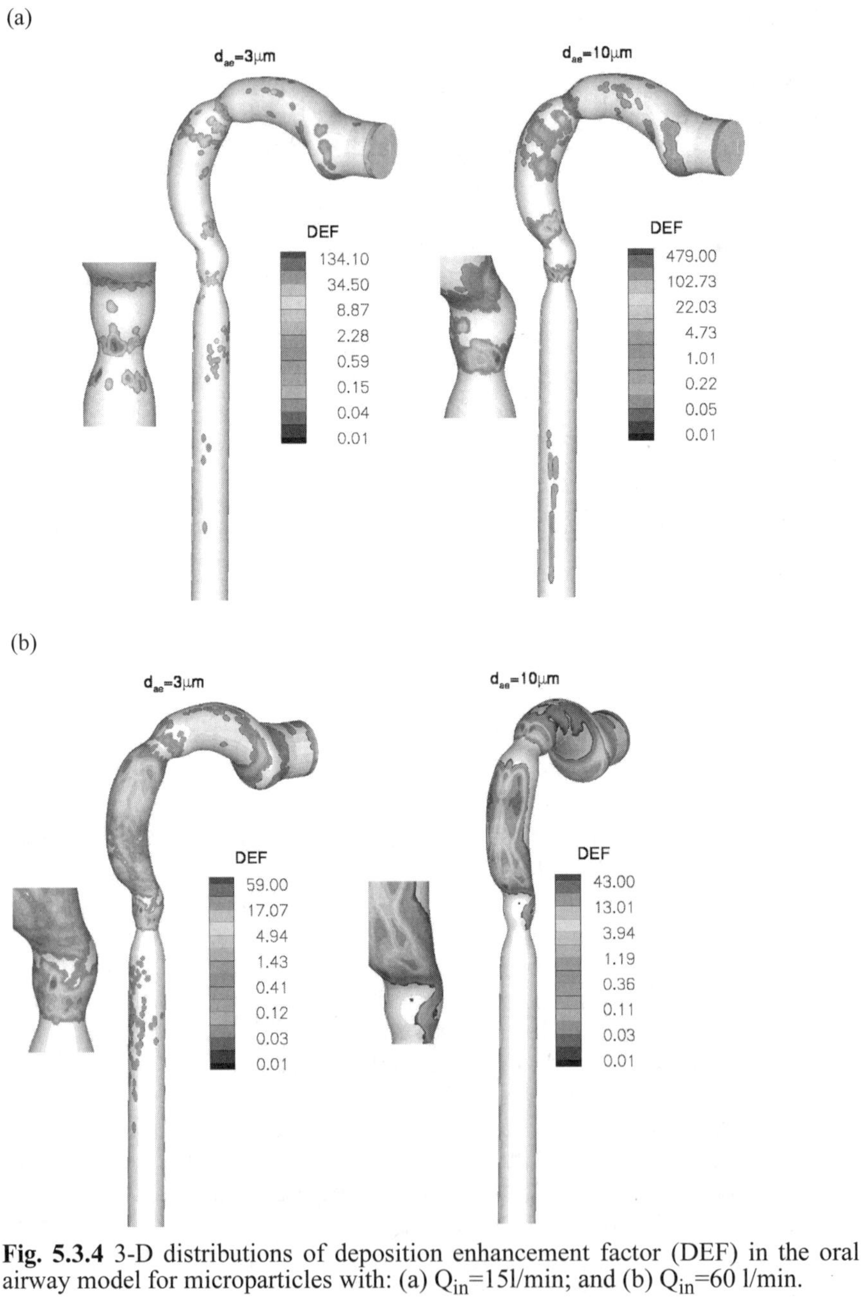

Fig. 5.3.4 3-D distributions of deposition enhancement factor (DEF) in the oral airway model for microparticles with: (a) Q_{in}=15l/min; and (b) Q_{in}=60 l/min.

(a) (b)

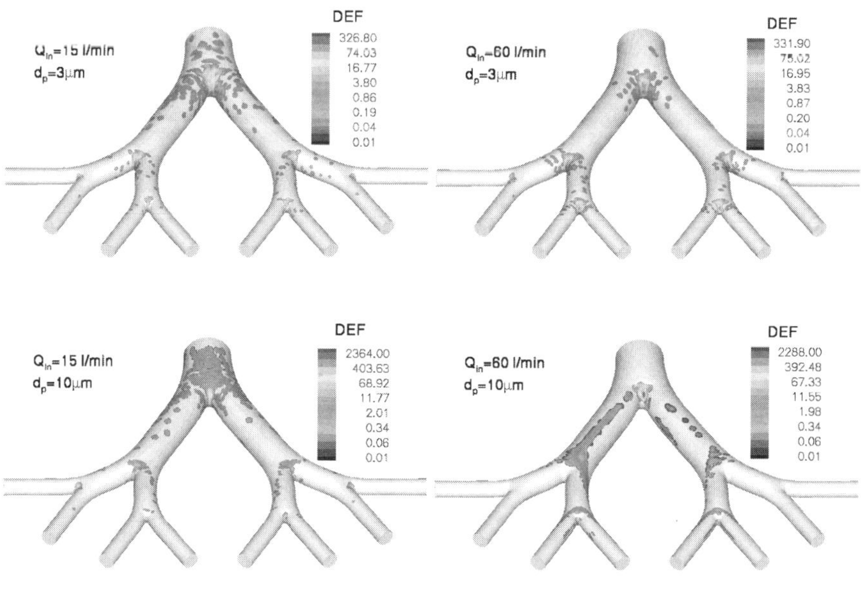

Fig. 5.3.5 3-D distributions of deposition enhancement factor (DEF) in the airways G0 to G3 for microparticles with: (a) Q_{in}=15l/min; and (b) Q_{in}=60 l/min.

Figures 5.3.6a and b show the *maximum* DEF-values in the oral airway model as a function of inspiratory flow rate and particle diameter as well as flow rate and Stokes number (St), evaluated at the trachea, i.e., $St = (\rho_p d_p^2 U)/(18\mu D)$ with D being the diameter of the trachea (G0). In the oral airway model, the maximum DEF-values range from 40 to 550 for $15 \leq Q_{in} \leq 60$ l/min and $1 \leq d_{ae} \leq 10$mm, and their variations with flow rate and particle diameter are complex. A distinct peak of DEF_{max} can be found around $St_{G0} \cong 0.04$ (i.e., Q_{in}=30 l/min, d_p=7μm and Q_{in}=15 l/min, d_p=10μm). Qualitatively, similar observations were reported by Balásházy et al. (2003) for microparticle deposition in a single-bifurcation model. With increasing Stokes number (particle size squared) the maximum number of particles deposited in a surface element may increase because of inertial impaction; however, the areas for receiving a high dose of deposited particles may also increase with Stokes number, which reduces the maximum DEF-value (Balásházy et al. 2003). Also of interest is that the maximum DEF values for the high flow rate case (Q_{in}=60 l/min) are lower than those for the low and medium flow rates. This is attributed to the strong turbulent dispersion and much broader distribution of deposited particles accompanied with the high airflow rate (see Fig. 5.3.4b).

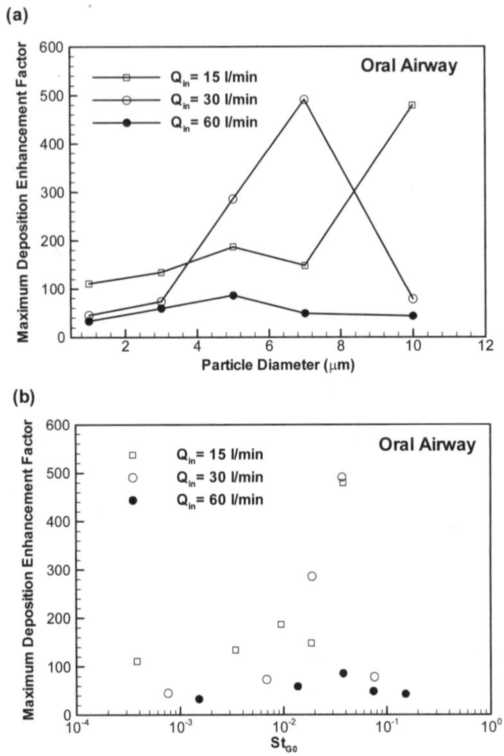

Fig. 5.3.6 Variations of the maximum deposition enhancement factor of microparticles in the oral airway model vs. flow rate and (a) particle diameter; and (b) Stokes number in the trachea.

The variations of the maximum DEFs in the airways G0 to G3 with particle size and inspiratory flow rate or Stokes number (see Figs. 5.3.7a, b) differ from those in the oral airways. Generally, the maximum DEF-values range from 200 to 600 when $St_{G0}<0.01$, but they increase rapidly with increasing St_{G0} when $0.01<St_{G0}<0.1$. The highest DEF value can be about 2400, which is much greater than that in the oral airway model. This also indicates that highly variable microparticle deposition occurs in the tracheobronchial airways when compared to the oral airways. No distinct peak of $DEF_{max}(d_p, Q_{in}=30$ l/min) can be found in the bifurcating airways. This may be attributed to the upstream flow and deposition effects, which change the uniform particle distribution at the mouth inlet to a non-uniform distribution at the inlet of the bifurcating airways so that the deposited particles accumulate more closely in very small areas of the carinal ridges. Considering the different local inlet Reynolds number, Stokes number, flow features, and particle distributions at different individual bifurcations, i.e., B1, B2.1, B2.2, B3.1–B3.4 (see Fig.5.3.2), the distributions of DEFs as well as the maximum DEF-values change greatly at each individual bifurcation. Figure 5.3.8 shows the maximum DEF-value variations vs. local inlet Stokes number for each individual bifurcation. The DEF for each individual bifurcation is calculated with Eq. (5.3.8a), where n is the number of surface elements in this bifurcation. It can be seen from Fig.5.3.8 that the maximum DEF-values range mainly from 100 to 1000, which vary with airway bifurcation due to the different local flow characteristics and particle distributions when the local Reynolds and Stokes numbers are fixed.

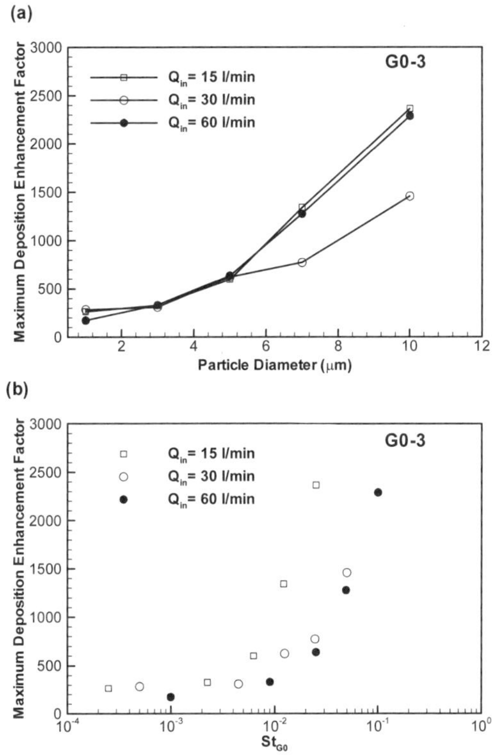

Fig.5.3.7 Variations of the maximum deposition enhancement factor of microparticles in the airways G0 to G3 vs. flow rate and (a) particle diameter; and (b) Stokes number in the trachea.

Figures 5.3.9a and b present the variations of deposition efficiencies and fractions as a function of inlet Stokes number at each individual bifurcation. In the airway bifurcation B1, both the DE and DF increase with increasing Stokes number. However, the variation of DE(St), or DF(St), is irregular in airway bifurcations B2.1, B2.2 and B3.1 to B3.4 because of the combined effects of flow developments and upstream deposition of particles entering from the mouth. Especially for relatively high Stokes numbers (say, St>0.1), the strong aerosol deposition upstream (i.e., in the oral airway and first bifurcation) will greatly reduce both DEs and DFs in the second or third bifurcation. Thus, deposition data for large micro-size particles in the tracheobronchial airways are only accurate when the oral airways with reasonable inlet conditions are included.

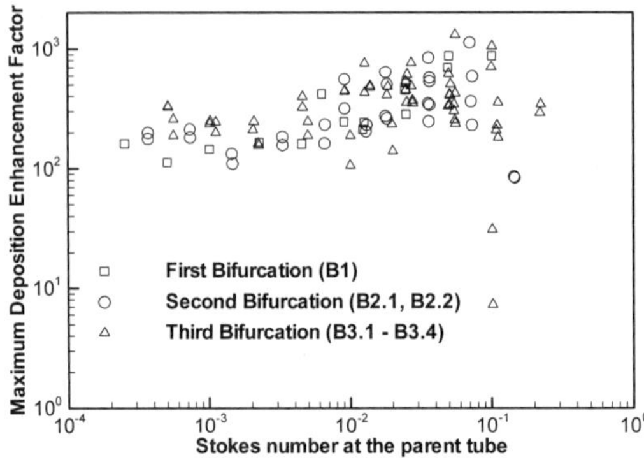

Fig.5.3.8 Plots of the maximum deposition enhancement factor of microparticles for each individual bifurcation in the bifurcation airway model against local, inlet Stokes number.

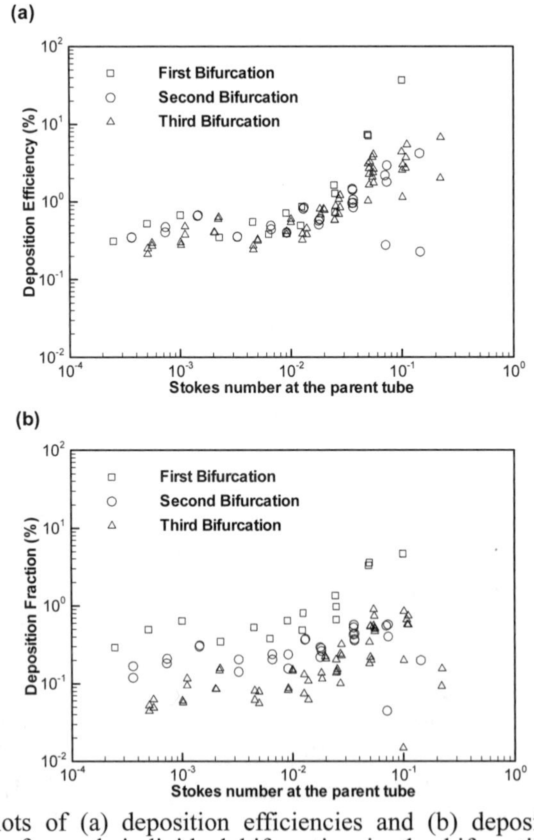

Fig. 5.3.9 Plots of (a) deposition efficiencies and (b) deposition fractions of microparticles for each individual bifurcation in the bifurcation airway model against local, inlet Stokes number.

Deposition of nanoparticles. Again, the local deposition patterns of nanoparticles can be described in terms of DEF-distributions as given in Eq. (5.3.12). For a low inhalation flow rate (Q_{in}=15 l/min), the distributions of DEF (see Fig. 5.3.10a) in the trachea are measurably different with those for higher flow rate cases (Q_{in}=60 l/min) (see Fig. 5.3.10b), because laminar flow still prevails after the throat contraction at Q_{in}=15 l/min (Kleinstreuer & Zhang, 2003). Specifically, Fig. 5.3.10b shows the distribution of DEF for different size particles in the oral airway model with an inspiratory flow rate of Q_{in}=60 l/min. For d_p =1nm particles, the maximum enhanced deposition may occur at the entrance because of a great degree of mixing, i.e., a large concentration gradient in light of the assumed plug flow and uniform particle concentration at the inlet. The deposition decreases with the development of velocity and concentration fields in the oral cavity. With further complicated variations in flow and concentration fields in the pharynx, larynx, and trachea, the deposition patterns are also inhomogeneous. The deposition of nanoparticles with 100nm in diameter becomes more uniform, i.e., the maximum DEF decreases, when compared with those particles with d_p =1nm and 10nm. This is attributed to the significantly decreasing diffusivities and more uniformly distributed concentration fields in the tubes for larger nanoparticles (see Zhang & Kleinstreuer, 2004). In other words, as the particle size increases, wall depositions decrease and the air-particle mixtures become much more uniform, featuring flat similar particle distribution profiles and hence wall gradients, which reduce the differences in local wall deposition rates. The location of maximum DEF may move from the mouth to the throat for 10nm and 100nm particles because of relatively homogenous and low deposition upstream and high near-wall gradients in particle concentrations in the areas of local tube constrictions.

An increase in inhalation flow rate reduces the particle residence time and the chance for deposition. The decrease in wall deposition may cause more uniform concentration profiles in the conduits; as a result, deposition patterns do not vary as much for higher inhalation flow rates (see Q_{in}=60 l/min in Fig. 5.3.10b) than for lower flow rates (cf. Q_{in}=15 l/min). Of interest is that the maximum DEF-values for nanoparticles are much lower than those for microparticles, i.e., the DEF_{max} for microparticles is of the order of 10^2 (see Figs. 5.3.4a, b), while DEF_{max} for nanoparticles is of the order of 1 (see Figs. 5.3.10a, b). As alluded to in the Introduction, a more uniform distribution of deposited ultrafine particles may relate to a greater toxicity effect when compared to fine particles made of the same materials. That is, not only the larger surface areas relative to the particle mass but also, more importantly, the larger surface areas with a near-uniform deposition can generate a higher probability of interaction with cell membranes and a greater capacity to absorb and transport toxic substances.

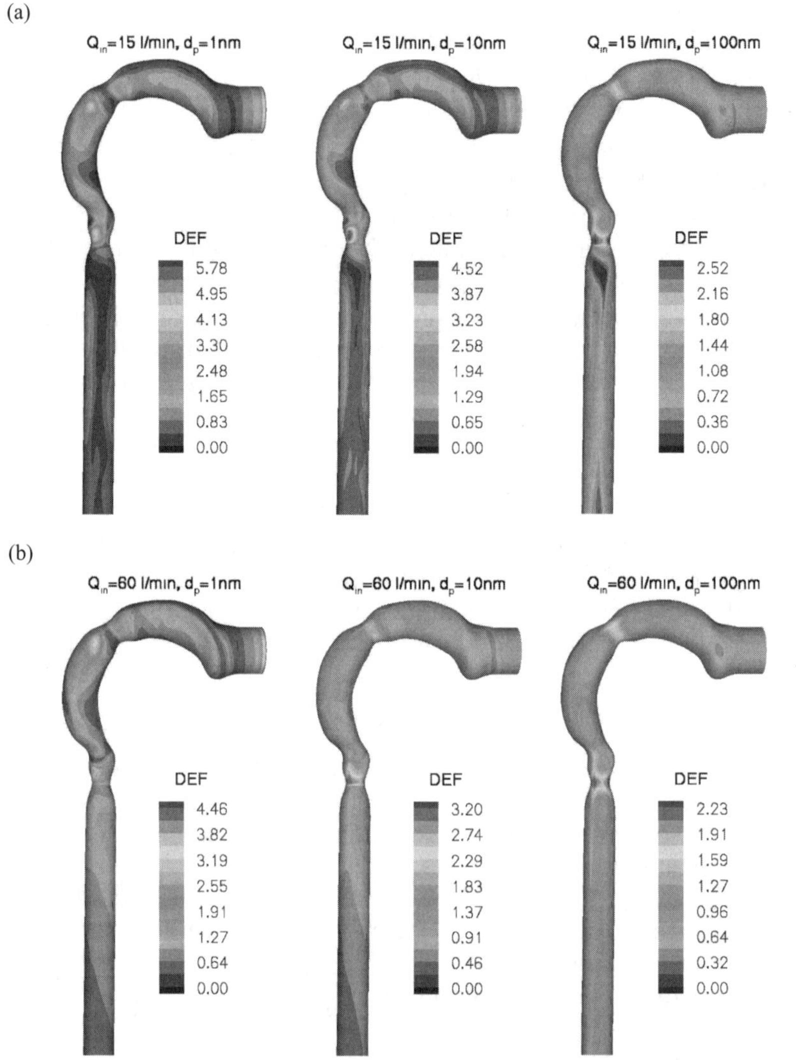

Fig.5.3.10 3-D distributions of deposition enhancement factor (DEF) of nanoparticles in the oral airway model under steady inhalation with: (a) Qin=15 l/min; and (b) Qin=60 l/min.

Turning to the bifurcation airways G0 to G3, for example with an inspiratory flow rate of Q_{in}=30 l/min, Fig. 5.3.11 depicts the DEF-distributions for 1nm≤d_p≤100nm. For both 1nm and 10nm particles, the enhanced deposition mainly occurs at the carinal ridges and the inside walls around the carinal ridges due to the complicated air flows and large particle concentration gradients in these regions. Specifically, the high concentration just upstream of the carina and zero concentration at the carinal ridge (i.e., generating high concentration gradients) lead to high diffusional depositions, as indicated by Eq. (5.3.11). This is consistent with the experimental observations of Cohen et al. (1990). For 100nm particles, the maximum DEF still occurs at the third carinal ridge; however, except at the entrance region, the DEF distribution tends to be more uniform due to the decrease in diffusion capacity. As a result, DEF_{max} for 100nm particles is smaller than those for 1nm and 10nm particles. Although the enhanced deposition sites (i.e., those with high DEF-values) in the bifurcating airways are similar for nano- and micro-size particles, the maximum DEF-values for nanoparticles are two to three orders of magnitude smaller than for microparticles.

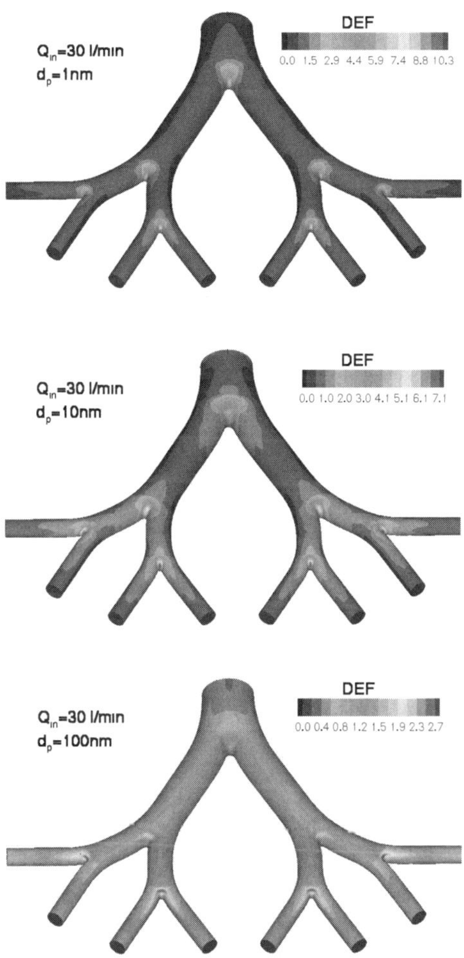

Fig.5.3.11 3-D distributions of deposition enhancement factor (DEF) of nanoparticles in the bifurcation airway model under steady inhalation with Q_{in}=30 l/min.

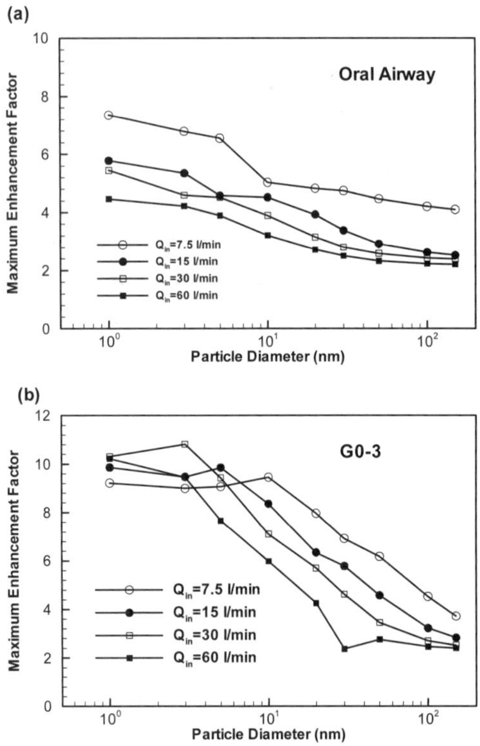

Fig.5.3.12 Variations of the maximum deposition enhancement factor of nanoparticles vs. flow rate and particle diameter in: (a) the oral airway model; and (b) the bifurcation airway model.

Figures 5.3.12a and 12b show the maximum DEF-values as a function of particle diameter and inspiratory flow rate in the oral airway model and bronchial airway model, respectively. Clearly, DEF_{max}-values indicate "hot spots," applicable to both toxic as well as therapeutic aerosols. In the oral airway, DEF_{max} may decrease with the decreasing inhalation flow rate since the deposition becomes more uniform at a higher flow rate with smaller residence times and mixing opportunities. In general, DEF_{max} also decreases with increasing particle size, but it may decrease more sharply for nanoparticles in the size range of 1 to 10nm. DEF_{max} can approach asymptotic values of 2.2 to 3.0 when the particle diameter is larger than 50nm and the inspiratory flow rate is in the range of $15 \leq Q_{in} \leq 60$ l/min. The situation for $DEF_{max}(d_p)$ in the bronchial airways becomes more complicated. When the particle diameter is smaller than 10nm, the DEF_{max} does not exhibit a monotonic variation with flow rate. The DEF_{max}-values in generations G0 to G3 vary little with particle size for very small nanoparticles and decrease sharply with particle size in the range of 10nm to 30nm. Again, the DEF_{max} can approach values around 2.4 to 3.0 when the particle diameter is larger than 100nm and the inspiratory flow rate is in the range of $15 \leq Q_{in} \leq 60$ l/min.

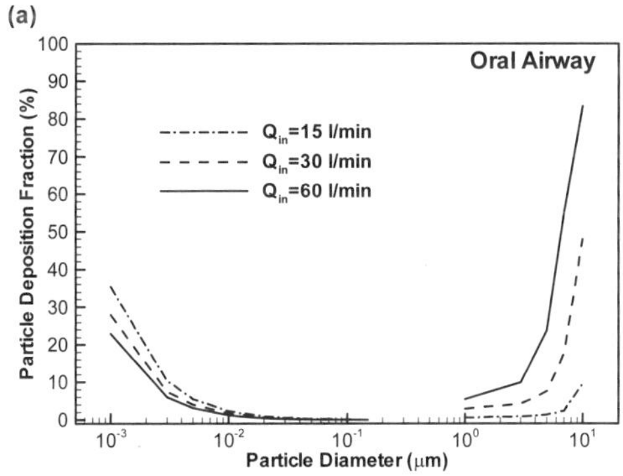

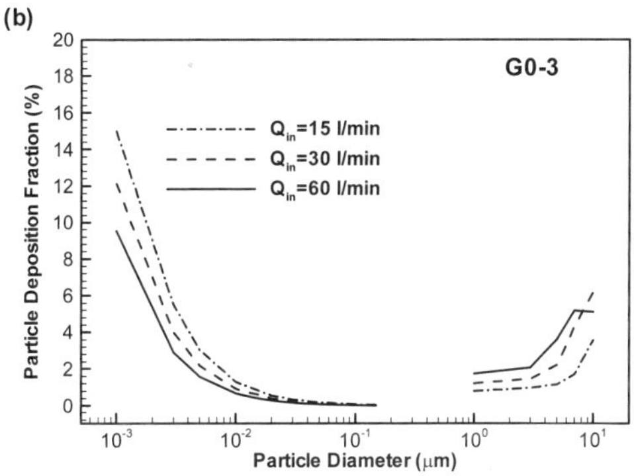

Fig.5.3.13 Variations of particle deposition fractions vs. flow rate and particle diameter in: (a) the oral airway model; and (b) the bifurcation airway model.

The variations of deposition fraction (DF) in the oral and G0-to-G3 airways vs. inhalation flow rate and particle diameter are depicted in Fig. 5.3.13. The DF is defined with Eq. (5.3.11) for nanoparticles and with Eq. (5.3.6) for microparticles. As expected, the variations of DF as a function of particle diameter are consistent with many previous experimental and theoretical studies. Specifically, with an increasing particle diameter the DFs decrease for nanoparticles because of the decrease in diffusive capacity while they may increase for microparticles due to increasing impaction. Similarly, the higher the inhalation flow rate, the lower the deposition of nanoparticles and the higher the deposition of microparticles. However, the inlet flow rate has a minor effect on nanoparticle deposition when compared to the influence of particle size. In fact, the deposition of 1nm and 10mm particles in the oral airway may be as high as 20% to 40% and 10% to 80% while the DFs of 150nm particles may be as low as 0.01% to 0.05%. The deposition fractions of very small ultrafine particles are comparable to those of large coarse particles in the upper human airways (see Fig. 5.3.13) as reported by Kim (2000) (see Fig.5.3.1b). However, it should be noted that the local deposition enhancement factors for ultrafine particles are much lower than for coarse particles. Figure 5.3.13b also shows that the DF in airways G0 to G3 may not increase with the particle size for large microparticles (say, d_{ae}>7mm) with a high inspiratory flow rate (Q_{in}=60 l/min) because of the elevated deposition in the oral airway.

5.3.2.4 Conclusions

The following conclusions can be drawn from the computational fluid-particle dynamics (CFPD) simulations:

(1) The deposition patterns of microparticles in the upper airways are highly non-uniform. The maximum deposition enhancement factor (DEF_{max}) ranges from 40 to 500 in the oral airway model and 200 to 2400 in the bronchial G0-to-G3 airway model for the parameter ranges $1 \le d_{ae} \le 10\mu m$ and $15 \le Q_{in} \le 60$ l/min. Deposition efficiency, deposition fraction, and deposition enhancement factor of microparticles in the tracheobronchial airways are dependent on the inlet flow rate and Stokes number (i.e., particle size squared) as well as the local geometry of the airway bifurcations. Some individual airway bifurcations (or generations) may receive only very small amounts of large micro-size particles due to strong upstream depositions. Hence, deposition data for large micro-size particles in the tracheobronchial airways are only accurate when the impact of the oral airways is included as well.

(2) As with micro-size particles, deposition of nanoparticles occurs at greater concentrations around the carinal ridges when compared to the straight segments in the bronchial airways; however, deposition distributions are much more uniform along the airway branches. The deposition enhancement factors vary with bifurcation, particle size, and inhalation flow rate. Specifically, the local deposition is more uniformly distributed for relatively large-size nanoparticles (say, $d_p = O(100nm)$) than for small-size nanoparticles (say, $d_p = O(1nm)$). The maximum deposition enhancement factor in the upper airways can approach asymptotic values of 2.0 to 3.0 when the nanoparticle diameter is larger than 100nm and the inspiratory flow rate is in the range of $15 \le Q_{in} \le 60$ l/min.

(3) The quite uniform distribution of deposited ultrafine particles also implies greater toxicity of ultrafine particles when compared to larger particles made of the same material. Hence, not only the greater surface area relative to the particle mass, but, more importantly, the much broader deposition area can produce more

sites to interact with cell membranes and a greater capacity to absorb and transport toxic substances.

(4) Validated CFPD simulations of particle transport in the human airways can provide local and segmental particle depositions as a function of particle size, inhalation flow rate, and local system geometry. These results are invaluable for physical insight and the analyses of toxic/therapeutic aerosol deposition impacts in the lung.

5.3.3 Micro-drug Aerosol Targeting in Lung Airways

As demonstrated in Sect.4.1.2, there is a dire need to improve targeted drug aerosol delivery. More generally, the modern trend is to administer drugs via inhalation for all kinds of diseases ranging from COPD to cancer, diabetes, and pain management. As an example, using a commercial dry-powder inhaler (DPI) (see Hickey et al., 2004) with different inhalation flow rates, Fig. 5.3.14 demonstrates that most particles deposit in the throat, rather than in the lung. Specifically, based on the aerodynamic particle size range of 2.9 to 3.7 μm, observed oropharyngeal depositions were 71% to 81%. Similar deposition data were obtained for pressurized metered dose inhalers (pMDIs). One reason, as pointed out by Finlay et al. (2001), is the initially high particle velocities generated with the DPIs and pMDIs, causing much higher depositions in the oropharnygeal region as compared to normal inhalation of aerosols, which follow the air velocity to enter the oral airway.

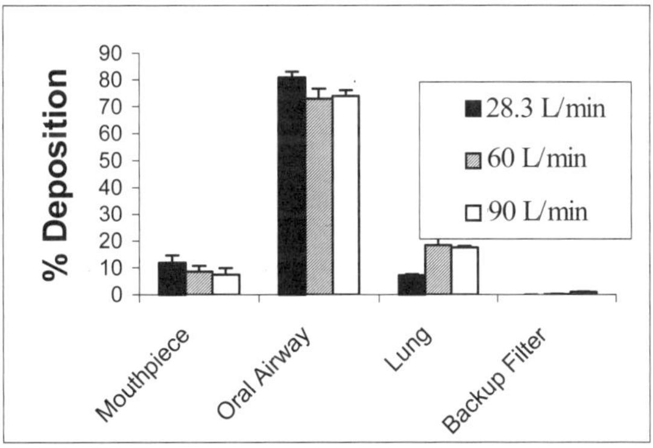

Fig.5.3.14 Deposition pattern of micro-particles from a DPI (Turbuhaler) as a function of inspiratory flow (after Cheng et al., 2003).

For optimal drug-aerosol delivery to work, a smart inhaler system is necessary. It should be a function of the fluid-particle-dynamics principles which are briefly illustrated in this section (see Kleinstreuer & Seelecke, 2005).

Clearly, appropriate particle-release locations and timing, suitable particle characteristics, and an ideal inhalation waveform will transport drug aerosols, on a case-by-case basis, to desired lung target areas. Figure 5.3.15 shows the new methodology in *virtual reality* for normal vs. controlled micro-particle releases from the mouth via a straightforward back-tracking method. Specifically, in selecting micro-particles, i.e., $5 \le d_p \le 7\mu m$, and strictly laminar flow, quasi-deterministic particle trajectories can be achieved; hence, airway landing area and particle release position at the aerosol delivery entrance correlated directly. The necessary modeling equations and CFPD program were discussed in Sect.5.3.1 plus related references.

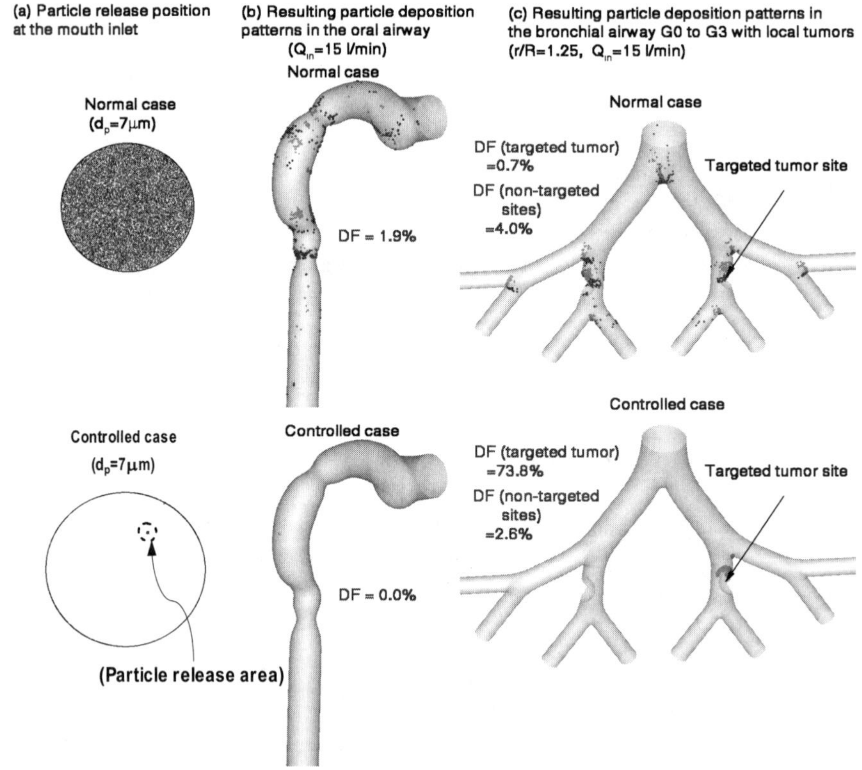

Fig. 5.3.15 Simulation-based drug delivery design for the upper airway starting from the mouth. A hemispherical tumor was placed in generation G2, comparing drug aerosol deposition fractions on the tumor surface for the normal case and controlled case.

In reality, variations in lung morphology, breathing mode, particle size, and specific lung target area for drug aerosol deposition complicate the task of achieving a controlled air-particle stream which results in optimal drug aerosol deposition. Thus, it is important to:

(a) broaden the particle release area;
(b) select the best particle characteristics; and
(c) determine an optimal inhalation waveform.

To (a): In order to accommodate different airway geometries, e.g., those of children, adults, or the elderly, and to be able to target different desirable lung areas, the mouth inlet cross section (or inhaler tube exit) is divided into eight particle release sections (see Fig. 5.3.16). The back-tracking methodology as well as trial & error runs match the optimal release segment with maximum deposition in the predetermined target region. Some well-defined particle release areas may be always excluded because aerosols from such locations deposit typically in the oral airway.

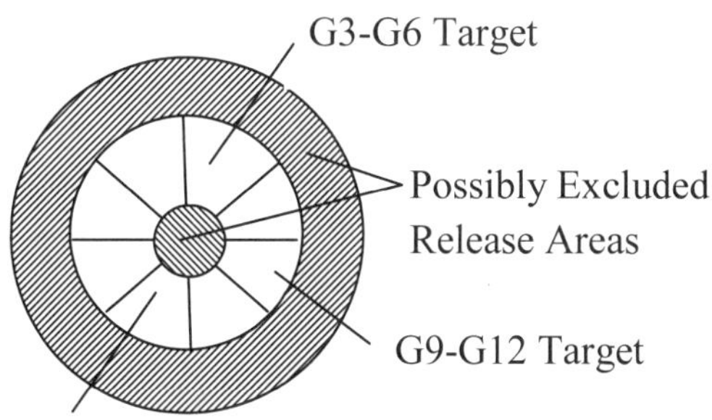

Fig. 5.3.16 Schematic of examples of inlet particle release positions in inhaler exit plane/tube.

To (b): The right effective diameter and density of micro-particles have to be determined to achieve the goal of maximum drug aerosol deposition.

To (c): Any active or passive inhalation waveform generated by a patient and/or an existing device (e.g., pMDI, DPI, or SMI (see Finlay, 2001)) has to be modified. Specifically, laminar flow, $Q_{in}<15$ L/min, and a special inhalation waveform are required to generate the highest, predetermined deposition values.

Conclusions. Drug-aerosol targeting in desired lung areas for therapeutics and pain management is an intriguing technique. While the methodology to achieve this goal is outlined in this section, it is a challenge to design and build a smart inhaler system (SIS) which produces such high DF-values (see Fig.5.3.15) for different scenarios (Kleinstreuer & Seelecke, 2005).

5.4 FLUID-STRUCTURE INTERACTIONS IN STENTED ANEURYSMS

Key physiological occurrences of coupled interactions between transient fluid flow and solid structure motion include the cardiovascular, respiratory, and synovial-joint systems. Specific examples are the pumping heart, pulsatile blood flow in viscoelastic arteries, transient airflow in the alveolar region, and elasto-hydrodynamic lubrication in the knees during high-load activities. In fluid-structure interactions, fluid flow may exert pressure on an elastic wall, i.e., a solid structure or bone material, causing small wall deformations, while soft tissue may experience finite strain and hence large deformations (see Sect.1.5 and Fig.1.5.4). As a result, the flow domain geometry changes and hence perturbs the fluid flow field. Simply put, high blood or air pressure expands a vessel, causing subsequently a decrease in fluid pressure which is followed by vessel contraction, and so on.

In general, two-way coupled multiphysics effects, which may also include thermal-mechanical and electric-thermal phenomena, have to be considered when solving certain biomechanical engineering problems. This is computationally accomplished either with direct or sequential (i.e., staggered) multi-field approaches. Direct coupled field analysis solves all of the multi-physics degrees of freedom in one solution phase. In sequential coupling, results of a single solver iteration (e.g., for the pulsatile blood flow field) are passed on as loads (e.g., pressure or stress) to the next physics field (e.g., the arterial wall or implant wall), iterating between all active physics fields until convergence criteria in the transferred loads are met. Historically, particular solvers and associated meshes have evolved separately for every physical field and hence when considering sequential coupled multi-physics analysis, the data exchange problems of dissimilar mesh interfacing had to be resolved first.

As a commercial software example, "Ansys Multiphysics" with CFX-10 (fluids) and ANSYS-10 (structure), coupled via "Workbench" software, is a powerful tool for simulating fluid-structure interaction dynamics.

5.4.1 Aneurysms and Their Possible Repairs

Aneurysms, an irreversible ballooning of weakened blood vessel segments, occur most frequently in the abdominal aorta, brain arteries or capillaries, and in the thoracic aorta, including dissecting aorta aneurysms (see Sect.3.2 and Fig.5.4.1). *It should be noted that this section is an extension of Sect.3.2 to illustrate further development steps in a complex research area.*

Causes of aneurysms are still speculative, but they may include arteriosclerosis, atheroma, congenital defects, hypertension, smoking, hereditary conditions, obesity, abnormal blood flow, etc. (see Salsac et al.,

2004). Clearly, hemodynamics, biomechanical factors, and biochemical processes play key roles in the genesis of an aneurysm.

(a) Abdominal aortic aneurysm

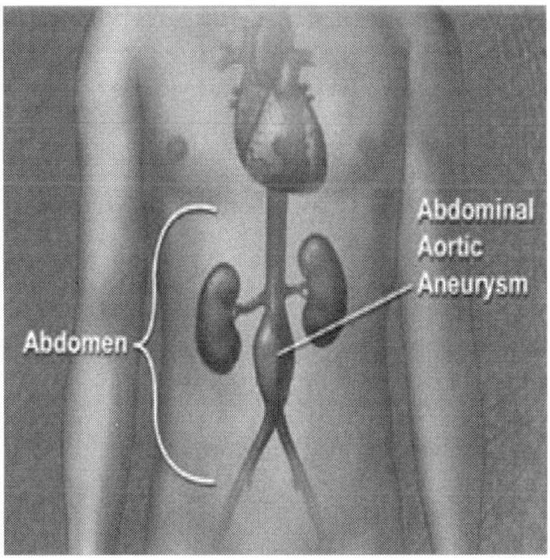

(b) Dissecting aortic aneurysm

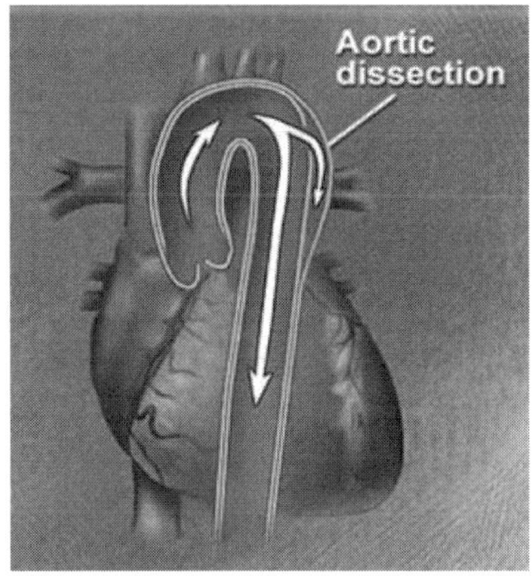

(c) Brain aneurysms

(i) Saccular Aneurysm

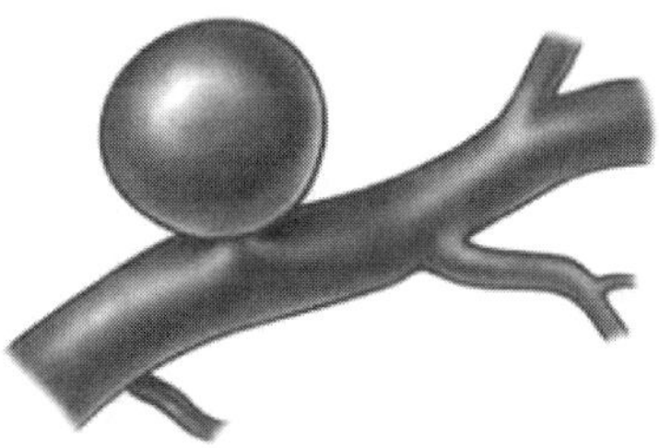

(ii) Fusiform Aneurysm

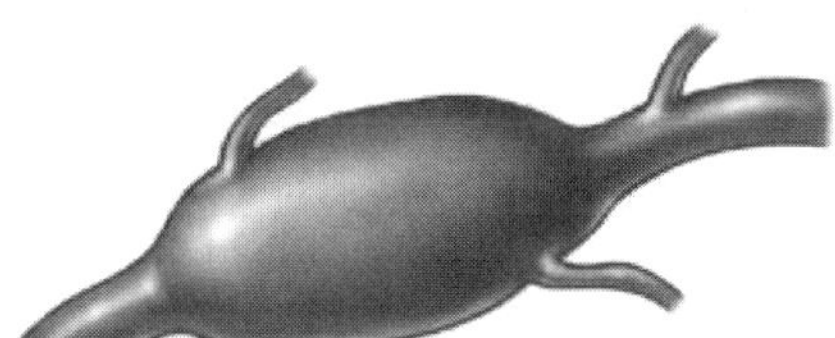

Fig.5.4.1 Frequent types of aneurysms (<u>Sources:</u> www.mayoclinic.org and eCureMe.com).

Abdominal aortic aneurysms. The etiology of aortic aneurysms is not fully understood yet. Focusing on the arterial wall mechanics, the multi-layer wall matrix components allow for the vessel function and provide integrity. These are predominantly elastin, collagen, and smooth-muscle cells. The distensible elastin is load-bearing at low pressures and responsible for the elastic recoil of the artery. Collagen is 1000 times stiffer and is load-bearing at high pressures, preventing over-dilatation and rupture of the vessel. Smooth muscle cells have the potential for contraction and relaxation, modulating the wall mechanics. The arterial wall mechanics and integrity are mainly determined by the matrix components of the wall, i.e., the collagen-to-elastin ratio is the principal determinant of wall mechanics in the aorta. An increase in collagen-to-elastin ratio may alter the wall structure, resulting in higher wall stiffness and lower tensile strength. Clinical observations show that most AAA walls become progressively stiffer as the diameter increases. This is because of biomechanical restructuring of the wall. In the normal abdominal aorta, the collagen-to-elastin ratio is about 1.58. However, the collagen-to-elastin-ratio is much higher in AAAs (Table 5.4.1).

Table 5.4.1 Composition of normal aorta and aneurysm (after He and Roach, 1994).

	Normal aorta	Aneurysm
Elastin		
Average	22.7	2.4
Maximum	32.5	6.7
Minimum	16.1	0.2
Muscle		
Average	22.6	2.2
Maximum	33.6	6.4
Minimum	15.5	0.4
Collagen and ground substances		
Average	54.8	95.5
Maximum	63	98
Minimum	48	91.4

As is shown in Fig.3.2.3, the stress-strain curve for AAAs moves to the left considerably. The elastic modulus is much higher than that in the normal aorta, while its breaking stress decreases significantly. There are three stages of AAA generation (see Fig.5.4.2). During Stage I, elastin begins degradation and more collagen is produced. The increase of collagen-to-elastin ratio is the indicator of the onset of AAA formation. In Stage II, the collagen begins degradation which is largely offset by more collagen production. However, with the accelerating collagen degradation, the remodeling ability of the AAA wall declines. Under the blood pressure load, the AAA wall begins to stretch rapidly and finally may rupture (Stage III). Clearly, elastin degradation is the key step in the development of AAAs, whereas collagen degradation is ultimately required for AAA rupture.

Even though elastin and collagen degradation are the key reason causing AAA generation, deterioration and rupture, the exact patho-physiology of elastin/collagen degradation is still unknown. Concerning larger arterial wall stiffness with age, enhanced wall stiffness is not necessarily advantageous for preventing AAA rupture, because along with the increase of wall stiffness, the wall yield stress will accordingly decrease. As a case in point, Raghavan et al. (2000) stated that Young's modulus in an AAA wall may reach 4.66 MPa, which is about three times that of a normal arterial wall while its yield stress is only 50% of the normal artery. Thus, although larger Young moduli may reduce AAA-wall stresses, the yield stress is possibly lower than the mechanical stress in the AAA wall, i.e., AAA rupture still may occur when the wall becomes stiffer (see Fig.5.4.3).

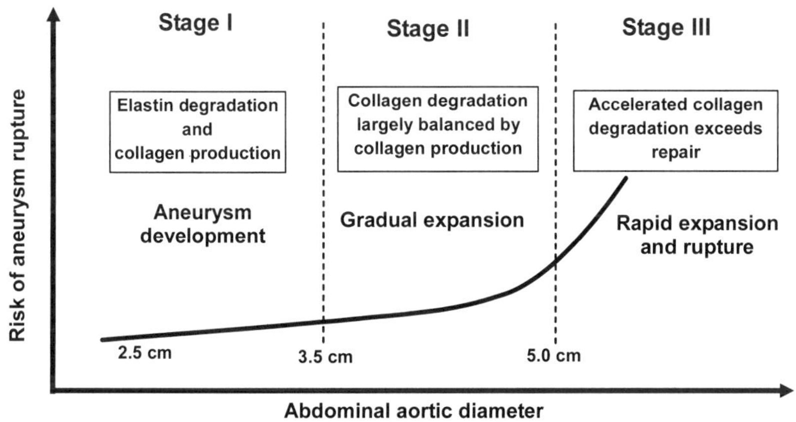

Fig.5.4.2 Three stages of AAA degeneration (after Thompson et al., 2002).

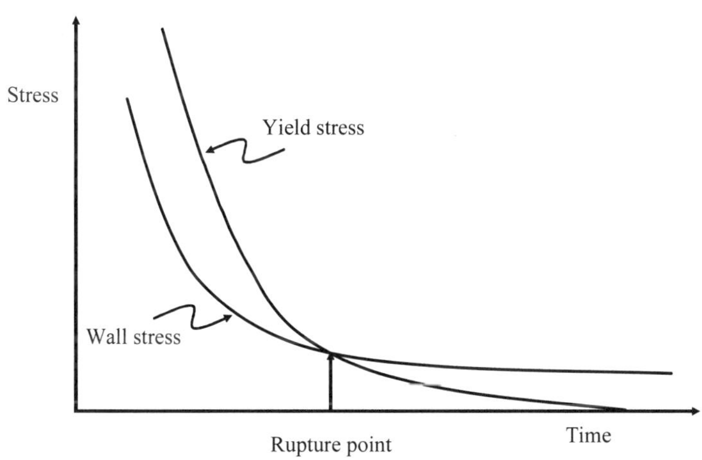

Fig.5.4.3 Effect of yield stress decreases on AAA rupture.

Open surgical repair. A major incision is made in the patient's chest and/or abdomen, the thrombus in the aneurysm is removed and the weakened portion of the aorta is replaced with a fabric tube, e.g., a PTFE or Dacron graft. Clearly, the graft is stronger than the weakened aorta, forming a new blood vessel. In general, in conventional open repair of aneurysms, patients have to deal with large incisions and possible blood transfusion. Hospitalization as well as recovery time are relatively long and surgical complications may occur (see Table 3.2.1).

Minimally invasive endovascular repair. In contrast to open aneurysm repair, a minimally invasive treatment, called endovascular aneurysm repair (EVAR), is performed remotely using a long thin tube, called a catheter. For aortic aneurysms, the surgeon makes a small incision in each upper thigh; then, guided by X-rays or other screens, passes the stent-graft (ePTFE or PET tube with an imbedded tubular metal mesh, compressed in the delivery device) through the vessel in the thigh to the aneurysm site. After it is correctly positioned in the abdominal aorta, the stent-graft is released and restores to its original shape. After balloon- or self-expansion of the stent-graft, the delivery catheter is withdrawn. The aneurysm cavity is excluded by the stent-graft and the blood pressure load is carried by the implant, preventing aneurysm rupture (Fig.5.4.4).

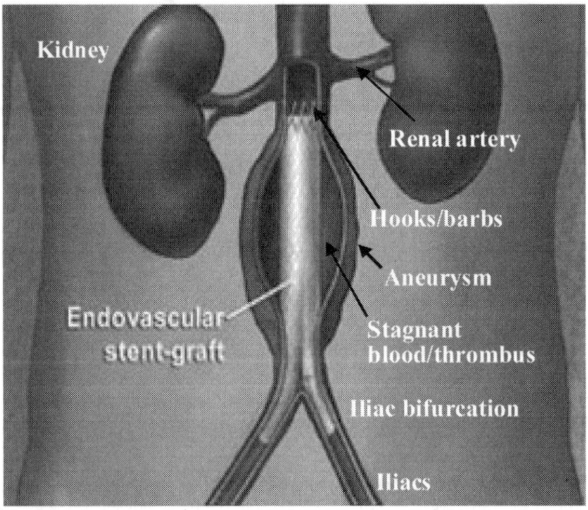

Fig.5.4.4 Endovascular repair for abdominal aortic aneurysms (Source: http://www.mayoclinic.org/aortic-aneurysm/surgery.html).

5.4.2 A Stented Abdominal Aortic Aneurysm Model

5.4.2.1 Introduction

As mentioned, for endovascular aneurysm repair (EVAR), an endovascular graft (EVG) is guided from the iliac to the affected area where the EVG expands and forms a new artificial blood vessel, shielding the aneurysm from the pulsatile blood flow (see Fig.5.4.5). The EVG, or stent-graft, is basically a cylindrical wire mesh embedded in synthetic graft material. While EVAR has shown outstanding success, especially for non-distorted abdominal aortic aneurysms (AAAs), EVG failure may occur due to blood leakage into the aneurysm cavity, which elevates the sac-pressure and may cause rupture. This may also be caused by EVG migration when the drag force exerted on the EVG exceeds the fixation force, exposing the aneurysm sac again to the pulsatile blood flow.

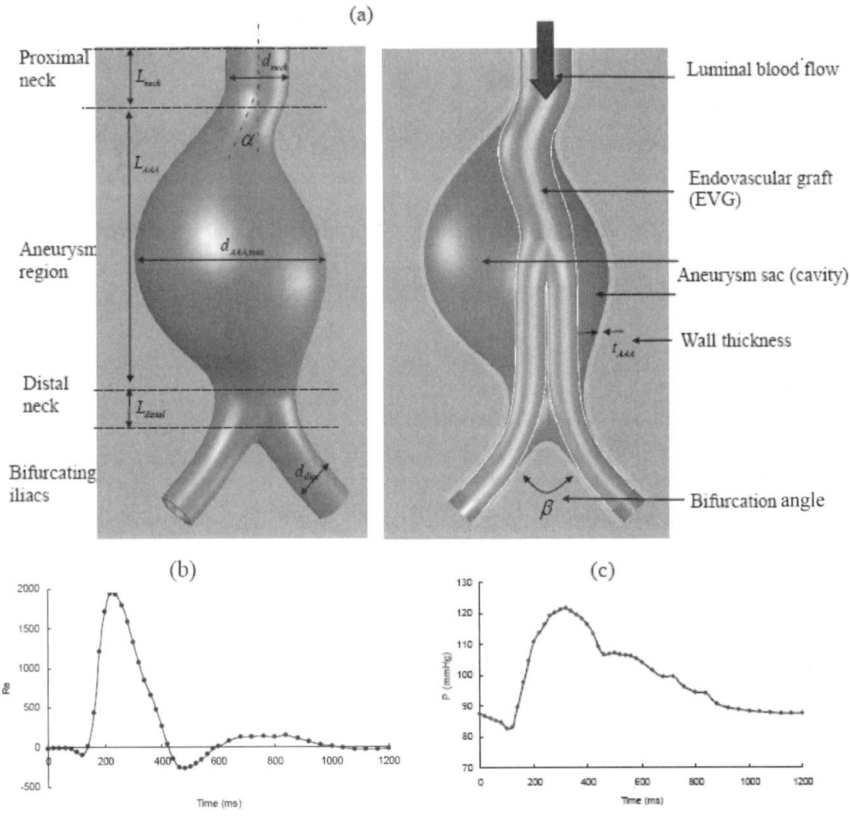

Fig.5.4.5 (a) Representative AAA with EVG; (b) Inlet velocity waveform; and (c) Outlet pressure waveform.

So far, most publications focused on AAA-wall stress or EVG-lumen flow separately. However, a stented AAA is a complex and strongly coupled system between blood flow and EVG/AAA wall. Thus, in order to evaluate blood flow patterns, wall stress distributions, sac pressure, and EVG drag forces, fluid-structure interaction (FSI) dynamics have to be considered. Hence, we follow Li & Kleisntreuer (2005) to analyze the pulsatile 3-D hemodynamics and its impact on EVG placement, EVG/AAA wall stress distributions, sac pressure generation, and EVG drag force in a representative stented AAA (see Figs.5.4.4 to 5.4.6).

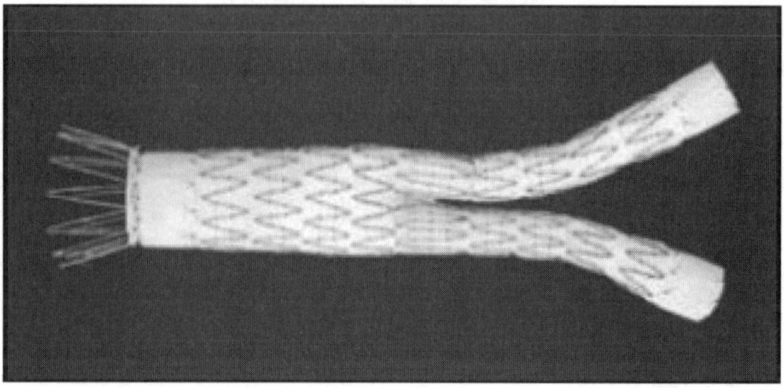

Fig.5.4.6 Actual Zenith EVG photograph (Source: Cook Incorporated, 2003).

5.4.2.2 Theory

For the reader's convenience, the governing equations, already discussed in Sect.3.2.3, are repeated here.

Flow equations. For transient three-dimensional incompressible fluid flow, the governing equations in tensor (or comma) notation, following Einstein's repeated index convention, are:

(Continuity) $$u_{i,i} = 0 \qquad (5.4.1)$$

(Momentum)

$$\rho \frac{\partial u_i}{\partial t} + \rho(u_j - \hat{u}_j)u_{i,j} = -p_{,i} + \tau_{ij,j} + \rho f_i \quad \text{in } {}^F\Omega(t) \qquad (5.4.2a)$$

(Stress tensor) $$\tau_{ij} = \eta \dot{\gamma}_{ij} \qquad (5.4.2b)$$

(Non-Newtonian fluid model) $$\eta = \frac{\eta_p}{\left[1 - \frac{1}{2}\left(\frac{k_0 + k_\infty \dot{\gamma}_r^{1/2}}{1 + \dot{\gamma}_r^{1/2}}\right) Ht\right]^2} \qquad (5.4.2c)$$

where u_i is the velocity vector, p_i is the pressure scalar, ρ is the fluid density, f_i is the body force at time t per unit mass, $\hat{u}_i$ is the wall displacement velocity at time t, ${}^F\Omega(t)$ is the moving spatial domain upon which the fluid is described, $\dot{\gamma}_{ij}$ is the shear rate tensor, η_p is the plasma viscosity, $\dot{\gamma}_r = \dot{\gamma} / \dot{\gamma}_c$ is a relative shear rate, $\dot{\gamma}_c$ is defined by a "phenomenological kinetic model" (Buchanan et al., 2000), k_0 is the lower limit Quemada viscosity constant, k_∞ is the upper limit Quemada viscosity constant, and Ht is the hemotocrit.

Structure equations. The general governing equations for structure dynamics are:

$$\text{(Momentum)} \qquad \rho a_i = \sigma_{ij,j} + \rho f_i \quad \text{in } {}^S\Omega(t) \tag{5.4.3a}$$

where a_i connects fluid flow dynamics with structure mechanics, i.e.,

$$a_i = \frac{d\hat{u}_i}{dt} \tag{5.4.3b}$$

$$\text{(Equilibrium of condition)} \quad \sigma_{ij} n_i = T_i \quad \text{on } {}^S\Gamma(t) \tag{5.4.4}$$

and

$$\text{(Constitutive)} \qquad \sigma_{ij} = D_{ijkl} \varepsilon_{kl} \quad \text{in } {}^S\Omega(t) \tag{5.4.5}$$

Here, ${}^S\Omega(t)$ is the structure domain at time t, n_i is the outward pointing normal on the wall surface ${}^S\Gamma(t)$, T_i is the surface traction vector at time t, ${}^S\Gamma(t)$ is the boundary of the structure domain, σ_{ij} is the mechanical stress tensor, D_{ijkl} is the Lagrangian elasticity tensor, and ε_{kl} is the strain tensor.

In order to analyze the stress distributions in both the endovascular graft and the arterial wall, the Von Mises stress, used as a material fracture criterion in complicated geometries, is employed. Specially,

$$\sigma_{Von\ Mises} = \frac{\sqrt{2}}{2} \sqrt{(\sigma_1 - \sigma_2)^2 + (\sigma_2 - \sigma_3)^2 + (\sigma_3 - \sigma_1)^2} \tag{5.4.6}$$

where, σ_1, σ_2, and σ_3 are the three principal stresses.

System geometry. The representative 3-D asymmetric stented AAA model is shown in Fig.5.4.5 while Fig.5.4.6 depicts a typical stent-graft (or EVG). The interacting materials include the luminal blood, EVG wall, stagnant blood in the AAA cavity, and AAA wall. The EVG is assumed to be a uniform 3-D bifurcating shell firmly attached to the proximal neck and iliac artery wall. The cavity between EVG and aneurysm wall is filled with stagnant blood, experiencing a time-dependent pressure as a result of the dynamic fluid-structure interactions. The drag force exerted on the EVG, due to net momentum change, was computed under physiologically realistic conditions to assess incipient EVG migration.

Table 5.4.2 Assumptions for blood flow and structure characteristics.

Blood flow	Structure characteristics (Artery wall, aneurysm wall, and EVG)
Incompressible	Isotropic and elastic
Non-Newtonian fluid (Quemada model)	Incompressible
Laminar blood flow	Nonlinear (large deformations)
No slip at the wall	No tissue growth on walls
Blood particle effects not considered	No residual stresses
Stagnant blood in cavity	EVG consists of equivalently uniform material

Numerical method. The underlying assumptions for simulating the coupled fluid-structure interactions are listed in Table 5.4.2.

The blood parameter values in Eq. (5.4.2c) are $\rho = 1.050 \ g/cm^3$, $k_0 = 4.58619$, $k_\infty = 1.29173$, $Ht = 40\%$, and $\eta_p = 0.014 \, dyns/cm^2$ (Buchanan et al., 2000). Table 5.4.3 lists the structure parameter values used in the present simulations. With respect to Young's moduli for the arterial wall and aneurysm, experimental data indicate that Young's modulus of an aneurysm is much higher than for a normal artery (Thubrikar et al., 2001). In this paper, the Young's modulus is assumed to be 4.66 MPa. The healthy artery section (neck) is incompressible with a Poisson ratio of 0.49, and the aneurysm wall is nearly incompressible with a Poisson ratio of 0.45 (Di Martino et al., 2001). For a bifurcated NiTi-stent interwoven with graft material (EVG), no direct experimental data is available; thus, an equivalent Young's modulus for the uniform EVG configuration was assumed to be 10 MPa (Suzuki et al., 2001).

Table 5.4.3 Parameters required in the simulation.

Parameters	Normal Artery	Aneurysm	EVG
Wall thickness	1.5 mm	1.0 mm	Equivalent: 0.2 mm
Diameter	Neck aorta (inner): 17 mm Iliac artery (inner): 11 mm	60 mm	Main body diameter 17 mm Iliac leg: 11 mm
Length	Neck aorta：30 mm Iliac artery：30 mm	80 mm	Main body: 60 mm Iliac leg: 70 mm
Young's modulus	1.2 MPa	4.66 MPa	Equivalent: 10 MPa
Poisson ratio	0.49	0.45	Equivalent: 0.27
Density	1.12 g/cm^3	1.12 g/cm^3	Equivalent: 6.0 g /cm^3

The physiologically representative inflow velocity waveform is shown in Fig.5.4.5 with a maximum Reynolds number (Re_{max}) of 1950 and average Reynolds number ($Re_{average}$) of 330. For the outlet pressure (Meter, 2000), the peak and average pressures are 122 mmHg and 98.7 mmHg, respectively. The pulse period is chosen as T=1.2 s. The inlet velocity profile is assumed to be parabolic. For the composite structures, the boundary conditions are a fixed degrees-of-freedom (DOF) at the inlet and the exit, and a free DOF on the wall. The fluid-structure interactions occur on the interfaces between luminal blood flow and EVG wall, as well as the sac's stagnant blood and aneurysm/EVG wall.

The finite element software package ANSYS for linear and nonlinear multi-physics analysis in Arbitrary Lagrangian-Eulerian (ALE) formulation has been employed to solve this fluid-structure interaction (FSI) problem. Specifically, it uses separately ANSYS FLOTRAN for the fluid domain and ANSYS Structural Solver for the solid parts. It transfers fluid forces, solid displacements, and velocities across the fluid-solid interface (see Fig.5.4.7). After the mesh-independence study, a total of 76,730 8-node fluid elements were needed for meshing the luminal blood flow region, 66,820 structure elements for the EVG and AAA wall, and 19,250 non-net-flow fluid elements for the stagnant cavity. The number of elements on the FSI interface was 7,168.

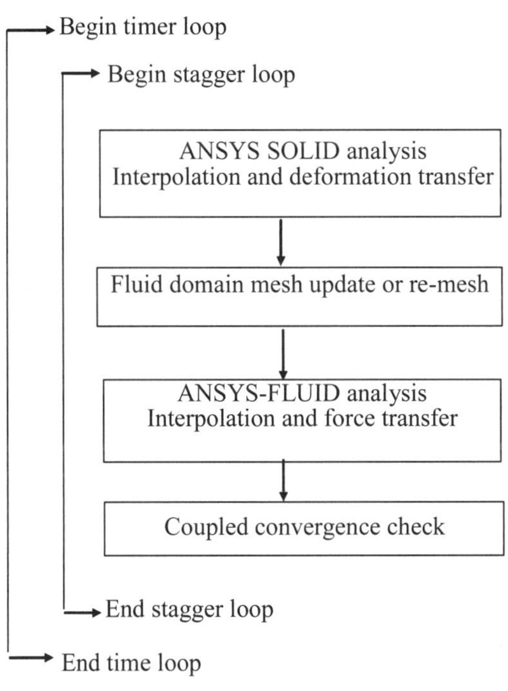

Fig.5.4.7 Flow chart of fluid-structure interaction solution algorithm.

The algorithm continues to loop through the solid and fluid analyses until convergence is reached for each time step. Convergence in the stagger loop is based on the quantities being transferred at the fluid-solid interface. A variable time step was employed, where $\Delta t_{\min} = 0.005s$ with 60 total time steps per cycle. Six cycles were required to achieve convergence for the transient analysis. Using a single processor, the total CPU time was about 50 hrs on an IBM p690 workstation.

Model validation. In order to test the accuracy of the ANSYS-ALE-FSI solver for simulating interacting flow and wall phenomena, two model validation studies were performed. Specifically, Gawenda et al. (2004) performed sac pressure experiments with in vitro stented aneurysm models (Table 5.4.4), where a sphere-like aneurysm consisted of either a 6-coat or a 12-coat latex wall. Employing the same geometry model (i.e., the sphere-like aneurysm) and assuming the same material properties for both the AAA and EVG walls under the load range of 50 to 120 mmHg, the sac pressure was simulated for the same flow conditions. Figure 5.4.8a presents the comparison between the experimental data and numerical results. It can be seen that the simulation results are in good agreement with the experimental data sets.

Table 5.4.4 Experimental parameters for stented aneurysm (Gawenda et al., 2004).

Parameters	
Aneurysm model type	Latex fusiform
EVG type	knitted polyethylene
Perfusate	Saline solution
Aneurysm Length	50 mm
Maximum aneurysm diameter	60 mm
Aneurysm volume	80 ml
Neck diameter	18 mm
Wall compliance (%/100mmHg)	6 coats: 3.45 ± 0.5; 12 coats: 1.4 ± 0.2
Pressure	50-120 mmHg
Flow rate	1.0-2.5 l/min

Flora et al. (2002) measured the wall stress distribution with in vitro latex aneurysm models consisting of different wall stiffnesses. The model and material parameters are listed in Table 5.4.5.

Adopting the same model and material properties, the AAA wall-stress at the maximum diameter was computed for similar flow conditions employing the ANSYS ALE FSI-solver. Again, the simulation results are in good agreement with the experimental data (see Fig.5.4.8b).

Table 5.4.5 Experimental parameters for aneurysm (Flora et al., 2002).

Parameters	
Aneurysm model type	Latex model
Aneurysm shape	Saccular
Aneurysm length	100 mm
Maximum aneurysm diameter	55 mm
Non-AAA artery diameter	20 mm
Aneurysm wall thickness	2 mm
Young's modulus	0.6-14 MPa (7-15 layers)
Perfusate	30% glycerin + 70% water
Pressure	80-160 mmHg

(a)

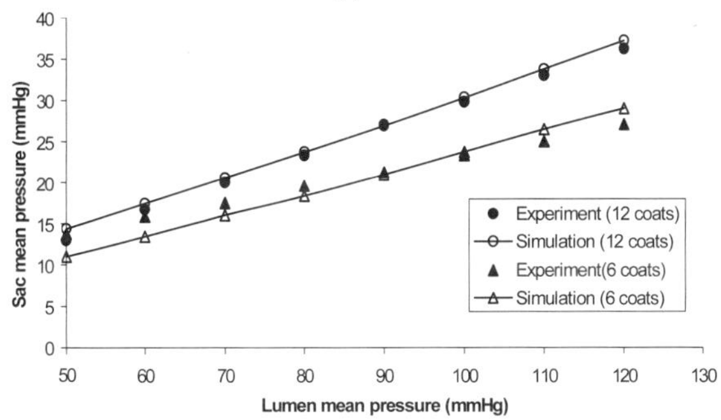

(b)

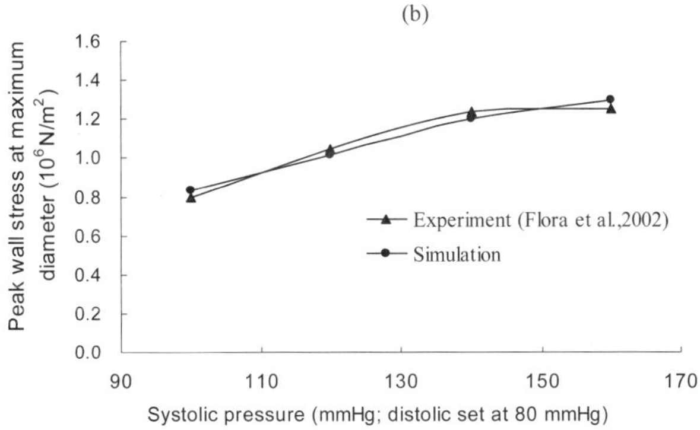

Fig.5.4.8 Comparison between simulation and experimental data: (a) Comparison of sac pressure (Experimental data from Gawenda et al., 2004); (b) Comparison of wall stress (Experimental data from Flora et al., 2002).

5.4.2.3 Results

The results are divided into three groups, i.e., the beneficial impact of EVG insertion (Figs. 5.4.9 and 5.4.10), the luminal blood velocity fields, wall stress distributions and sac-pressure levels at three selected time levels during the cardiac cycle (Figs. 5.4.11-5.4.13), and the effect of blood pressure waveforms on the transient EVG drag force, calculated from the wall shear stress distribution and the net momentum change (see Fig. 5.4.14).

EVG impact. A securely placed EVG forms a new smooth conduit for blood flow (Fig. 5.4.6) and hence protects the weakened aneurysm wall from high pressure and stress levels (Fig. 5.4.9). In fact, in case of zero leakage into the cavity, the sac pressure is reduced by a factor of 10 and the maximum wall stress decreases by a factor of 20 throughout the cardiac cycle when an EVG is inserted (see Figs. 5.4.9a and b). As a result, the largest change in maximum AAA diameter drops during systole from 2 mm to 0.17 mm (Fig. 5.4.9c). Indeed, as reported by Malina et al. (1998), EVG placement can reduce the maximum wall deformation to 0.2mm.

Clearly, these dynamic system parameter variations due to cyclic fluid-structure interactions are powered by the given blood pressure waveform (cf. Fig.5.4.6 and Fig.5.4.9, keeping in mind that the inlet pressure differs only by 5% from the outlet pressure).

A snapshot during the critical deceleration phase (t/T=0.27, peak blood pressure) shows the velocity fields as well as the pressure and wall stress distributions in both the open AAA and the stented AAA (Fig. 5.4.10). Referring to Fig. 5.4.10a, the angulated AAA neck guides the jet-like blood stream into the large cavity, forming immediately two strong vortices due to the sudden area expansion. The greatest impact of the angled blood stream occurs, as can be expected, not where the aneurysm has its maximum diameter, but well below near the iliac bifurcation, i.e., the impact area of highest net momentum change. Interestingly, the maximum wall stress and wall deflection in nonstented AAA have their highest magnitudes in different locations, i.e., the maximum wall stress (0.59 MPa) occurs on the right side above the bifurcation; while the maximum wall deformation (2.3 mm) occurs on the left side near the angled neck. Now, with an EVG in place, the blood flow is tubular and rather uniform, except for a few areas of secondary flows caused by neck and iliac angles (Fig.5.4.10b). As a result of complex fluid-structure interactions between lumen blood, EVG wall, stagnant sac, and AAA wall, the sac pressure still remains at 14.38 mmHg, i.e., 11.8% of the lumen pressure. Triggered by the EVG wall shear stress and net momentum change, the drag force acting on the EVG is at that moment almost 2 N. While the stress in the AAA wall and the pressure in the cavity are very low, the EVG carries the blood flow impact as numerically indicated with a peak EVG wall stress of 1.7 MPa at the stagnation (or bifurcation) point (see Figs. 5.4.10b and c).

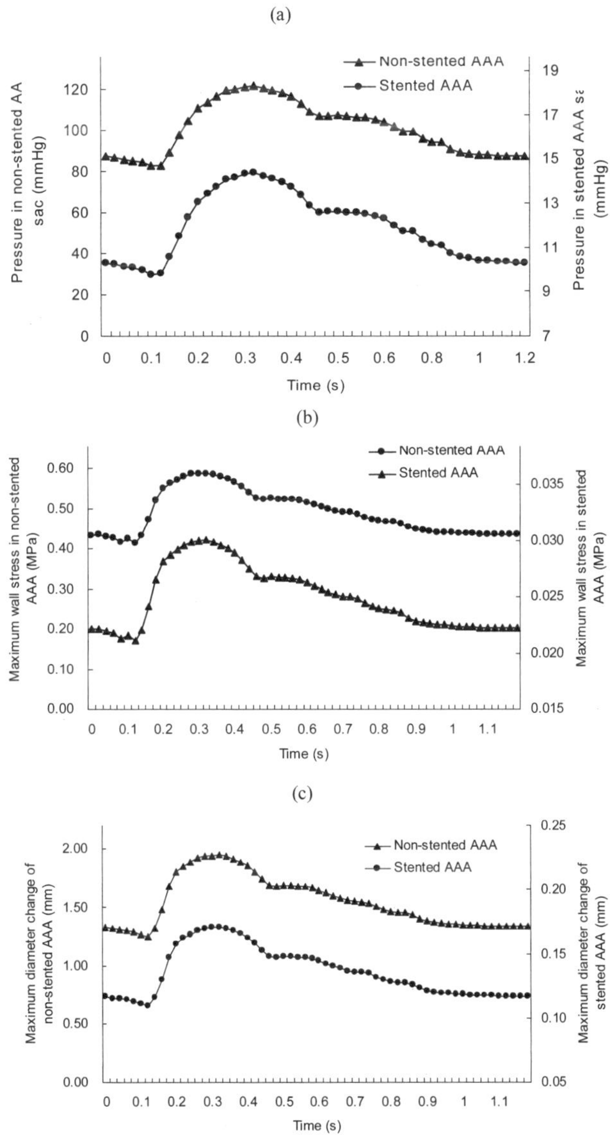

Fig. 5.4.9 EVG impacts on AAA: (a) EVG impact on sac pressure; (b) EVG impact on maximum wall stress; (c) EVG impact on $d_{AAA,max\text{-}changes}$.

Transient fluid-structure interactions. In order to illustrate the dynamics of pulsatile blood flow influencing the stented AAA parameters (see Figs. 5.4.11-13), three representative time levels were selected, i.e., reverse flow at t/T=0.1 (Re=-70), peak systole at t/T=0.2 (Re=1950), and flow deceleration at t/T=0.27 (Re=1200). In general, the sac pressure stays very

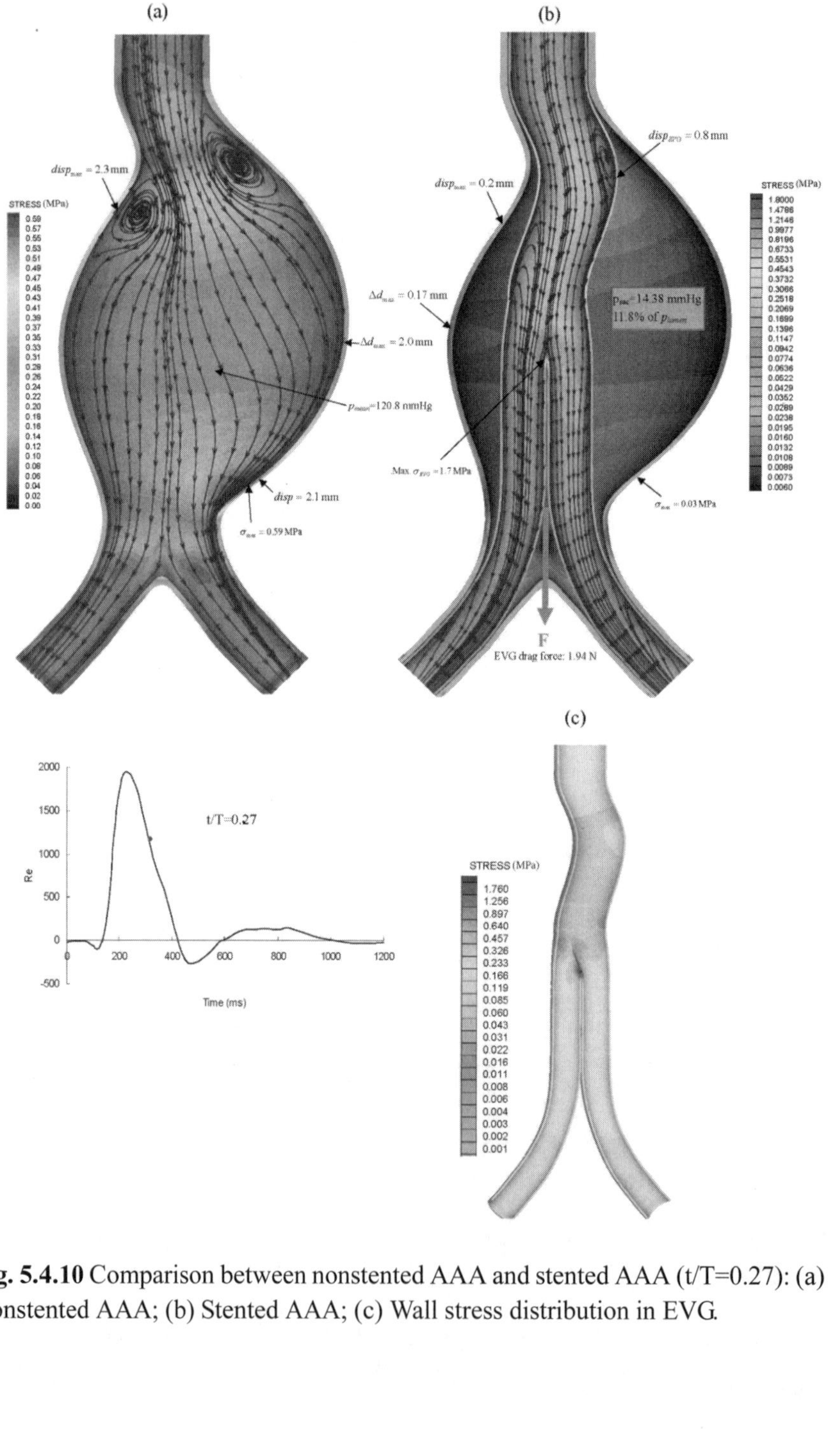

Fig. 5.4.10 Comparison between nonstented AAA and stented AAA (t/T=0.27): (a) Nonstented AAA; (b) Stented AAA; (c) Wall stress distribution in EVG.

low throughout the cycle; however, it increases slightly as the flow input waveform progresses (and time elapses) because of the blood pressure transmission from the EVG lumen via the distensible EVG wall into the aneurysm cavity. The same can be observed for the peak $\sigma_{\max}$ in the AAA wall, occurring always in the same location, which is basically an inflection point in the geometric mid-plane wall function. Naturally, the stress in the (healthy) neck tissue is the highest because of the necessary oversizing to keep the EVG anchored. The luminal blood flow meanders from the AAA neck part through the aneurysm with typically four small recirculation zones until it bifurcates into the EVG daughter tubes, i.e., the iliac legs. As expected, the maximum EVG wall stress can be found at the EVG bifurcation, nonlinearly increasing during the observation time. Although the present AAA geometry is anterior-posterior symmetric, the flow field and wall stress distributions are asymmetric, i.e., the results are fully three-dimensional (see Figs. 5.4.11-13).

Comparing Slices C-C in Figs. 5.4.11-13 more closely reveals that the location of the maximum EVG stress switches in the daughter tubes between up-flow (Fig.5.4.11) and peak down-flow (Fig.5.4.12), and back again during decelerating down-flow (Fig.5.4.13). Strong secondary flows appear before the EVG flow division (see Slices B-B in Figs. 5.4.11-13), an area which also experiences a relatively high wall stress.

EVG migration. If the actual EVG drag force starts to exceed the fixation force, the EVG will migrate or dislodge. As mentioned, the drag force is composed of the integral over the surface shear plus the net pressure, where the shear stress contribution is typically only 3% of the total. The fixation force for an EVG without barbs and hooks is the friction between the proximal EVG segment and the AAA neck which is usually oversized to supply, at least initially, solid anchoring. In addition, the EVG ends may be secured via frictional effects to the iliac lumen. In case the EVG migrates, blood can leak into the AAA cavity, leading to a dangerous pressure buildup and subsequently aneurysm rupture. As alluded to in Sect. 5.4.2.1, ideally the AAA neck should be long, cylindrical, and of healthy tissue. Obviously, AAA patients with severe hypertension (see Waveform IV in Fig.5.4.14) are especially at risk for EVG failure because in that case the drag force has a maximum and the relatively large pressure difference during the cardiac cycle may reduce the EVG fixation to a calcified, i.e., hardened, neck tissue. Specifically, the pressure descriptions of waveforms I to IV range from "normal" to "severe", where $\Delta p_{\max,normal} = 40\,\text{mmHg}$ and $\Delta p_{\max,severe} = 70\,\text{mmHg}$. For example, Mohan et al. (2002) declared that high blood pressure is one important factor to cause EVG migration, while Morris et al. (2004) found that the drag force may vary over the cardiac cycle between 3.9N and 5.5N. Thus, for an EVAR patient with severe hypertension, extra fixation should be considered.

Because of the incompressible-fluid condition, the transient drag force exhibits, with a very small time lag, basically the same trend as the given inlet pressure waveform (see Fig. 5.4.14).

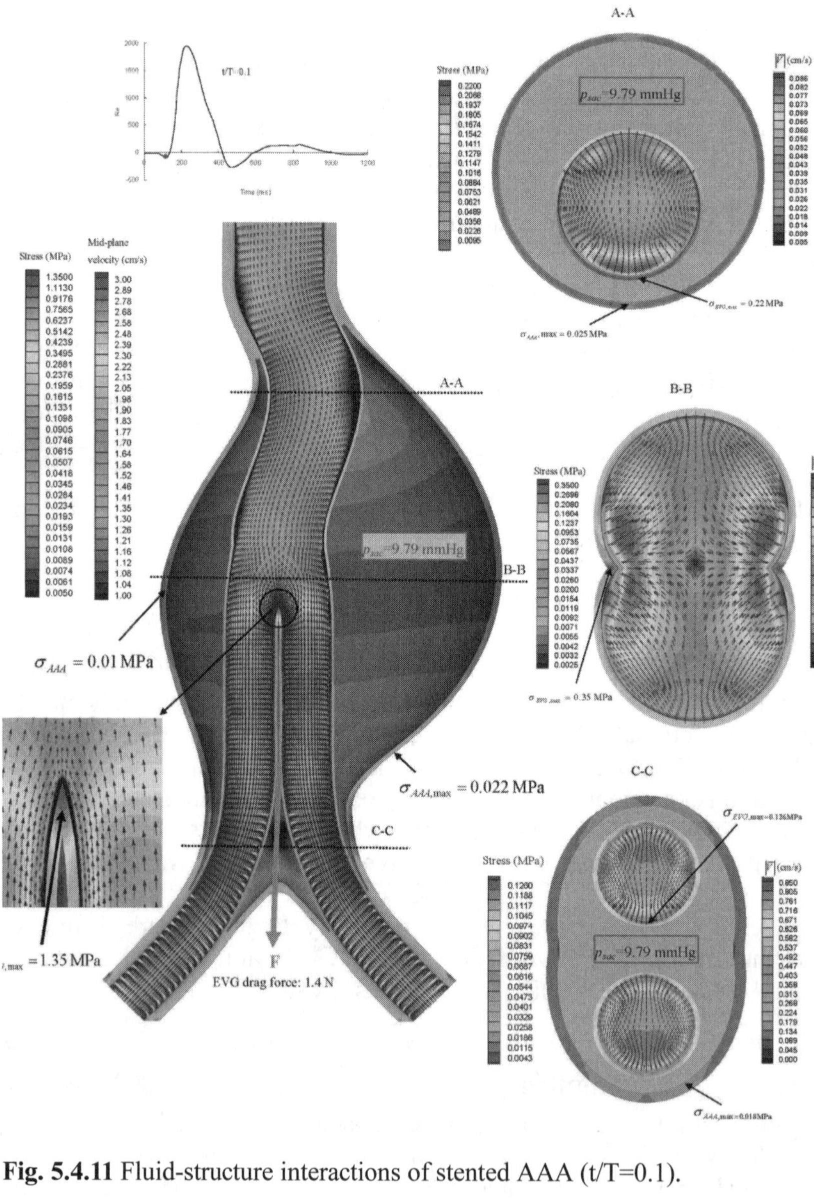

Fig. 5.4.11 Fluid-structure interactions of stented AAA (t/T=0.1).

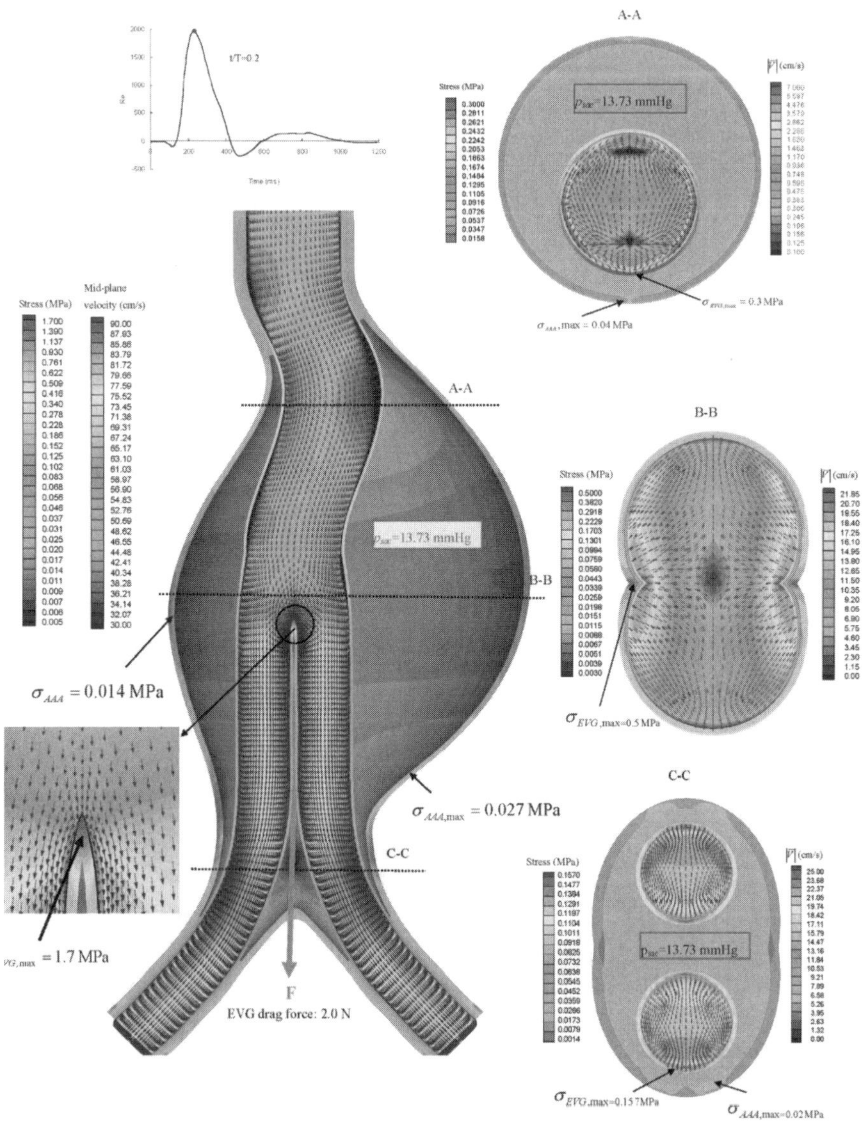

Fig. 5.4.12 Fluid-structure interactions of stented AAA (t/T=0.2).

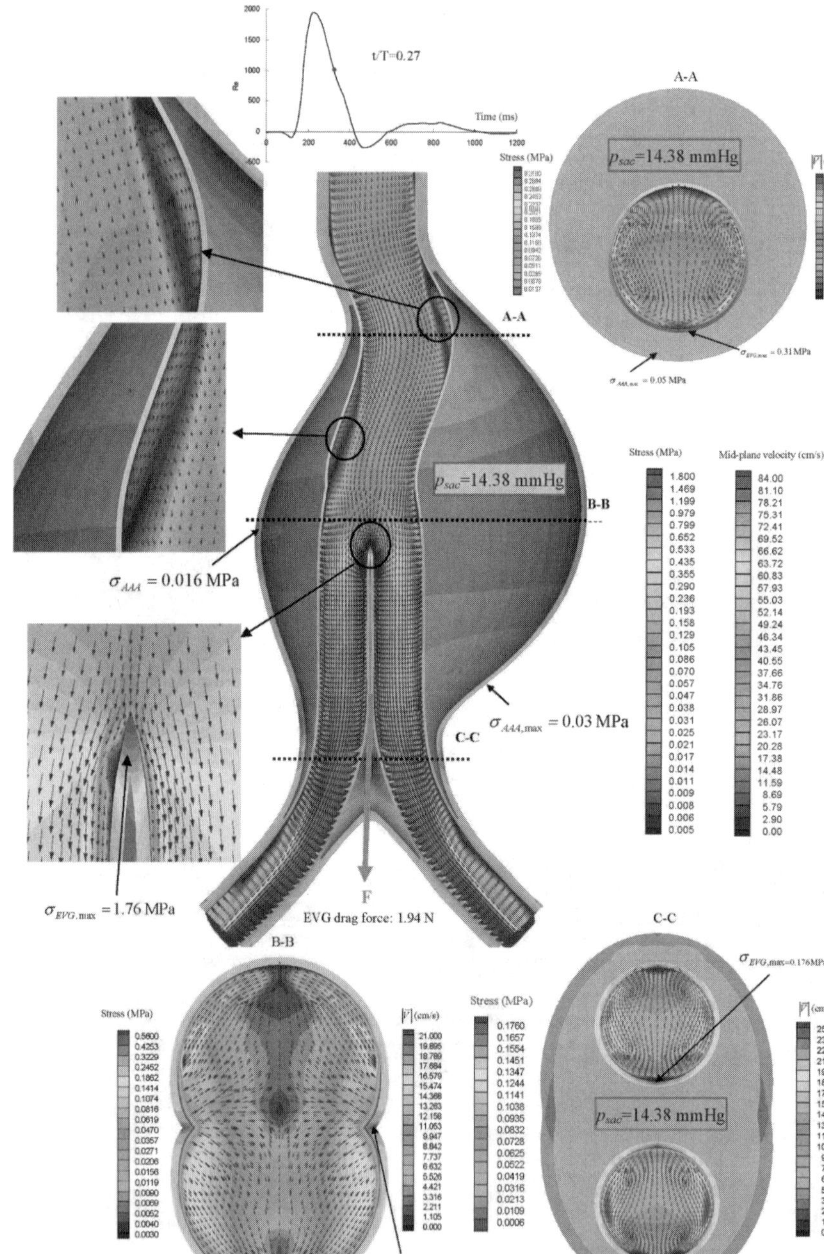

Fig. 5.4.13 Fluid-structure interactions of stented AAA (t/T=0.27).

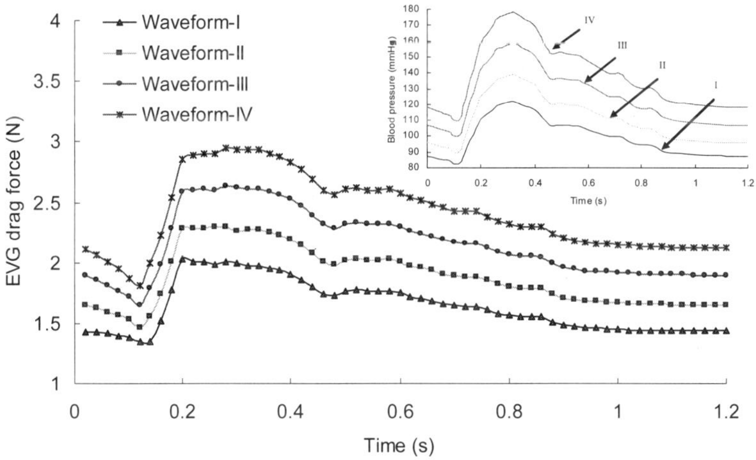

Fig. 5.4.14 Effect of blood pressure waveform on EVG drag force.

5.4.2.4 Discussion

Although minimally invasive endovascular aneurysm repair (EVAR) is very attractive, post-operative complications may occur, which are the result of excessive fluid-structure interaction dynamics. Thus, the fluid-structure interactions for a representative AAA model with and without a realistic EVG have been investigated in terms of pulsatile blood flow influencing EVG movement, which is transmitted via the stagnant blood in the cavity to the aneurysm wall. Of interest are the beneficial impact of an EVG, the blood velocity field, the highest pressure level in the aneurysm sac, the stress distributions and displacements of both of the EVG wall and the AAA wall, as well as the maximum drag force exerted on the EVG. It was readily demonstrated that a securely placed EVG shields the diseased AAA wall from the pulsatile blood pressure and hence keeps the maximum wall stress 20 times below the wall stress value in the nonstented AAA. An interesting finding is that the sac pressure is reduced significantly but is not zero after the sac is completely excluded by the EVG, i.e., no blood leakage exists. Thus, our simulation shows that even in the absence of endoleaks, the sac pressure can be generated by the complex fluid-structure interactions between luminal blood flow, EVG wall, stagnant sac blood, and aneurysm wall. The EVG/AAA wall compliance plays an important role in sac-pressure generation.

As shown in Fig.5.4.9a, the intra-sac pressure varies from 9.8 to 14.4 mmHg (4.6 mmHg pulsatility) during one cycle, while the average EVG lumen blood pressure (plumen) ranges from 82.3 to 120.3 mmHg (38 mmHg pulsatility). Sonesson et al. (2003) found that the sac pressure pulsatility varied from 0 to 6 mmHg clinically. Dias et al. (2004) indicated that the mean sac pressure pulsatility ranged from 2 to 10 mmHg for their

30 patients. Compared to the lumen pressure pulse, the pressure pulsatility in the sac is not substantial. While the AAA volume should typically shrink after successful EVAR, aneurysm enlargement might still occur because an elevated sac-pressure could cause a delayed aortic aneurysm enlargement, even after successful EVAR (Lin et al., 2003).

EVG wall deformations are insignificant due to the high material stiffness; indeed, our results show that the maximum graft deformation is less than 1mm. However, EVG wall deformation is a main factor when determining the sac pressure (Gawenda et al., 2004; Li, 2005). Furthermore, even relatively small, repetitive EVG-wall deformations may result in interactions between the metallic wire and the interwoven graft material, leading to fabric abrasion and holes (Chakfe et al., 2004). In addition, transient solid-fluid interactions can lead to material fatigue and hence device failure. Clinically, that includes metallic fracture (metal wire fracture, stent-strut and barb/hook breakage), graft fabric holes and suture breakage, and/or separation. Additionally, device corrosion has also been observed clinically. So far, there is no model which can predict such device deteriorations (Najibi et al., 2001). The reason is the serious lack of time-dependant clinical data due to short follow-ups, especially for the second generation of stent-grafts. Another reason is that most of the patients have been asymptomatic and have not as of yet needed interventions for device fatigue (Jacobs et al., 2003). In this study, we did not consider device failure cases. However, it is possible to simulate device fatigue employing a fluid-structure interaction solver for a realistic stented AAA model in order to assess such problems.

Endoleaks were not considered in this paper either. However, in cases of a loose neck attachment, graft defect, and/or minor branch backflows, blood may leak into the AAA sac after EVAR. For example, Zarins et al. (2003) indicated that endoleaks were diagnosed with CT scans in 38% of 398 patients. Endoleaks may cause an increase in sac pressure and hence higher stresses in the AAA wall. Serious endoleaks can result in EVR failure, AAA rupture, and the need for second procedures. In a preliminary study, we found that if an endoleak volume of 3% is added to the sac volume, the sac pressure and AAA wall stress can increase by 60% (Li, 2005).

Even though EVG placement may reduce significantly the sac pressure and wall stress, the hemodynamics incurs a drag force which may trigger EVG migration. It is shown that the risk of migration can be high for patients with severe hypertension (see Fig.5.4.14). EVG migration is a common problem for EVAR patients. For example, Zarins et al. (2003) reported that the migration rate was up to 8.4% for their 1119 patients within 12 months after EVAR. Serious EVG migration can cause endoleaks, EVG twisting or kinking and hence EVG failure. In this paper, the maximum EVG drag force is about 2 N because the selected EVG size is relatively small (17mm). However, our preliminary research results indicated that the EVG drag force can exceed 5 N for AAAs with a large neck angle, iliac angle, large EVG size, and aorto-uni-iliac EVG (Li, 2005). Thus, the fixation of self-expandable or balloon-expandable EVG contact may be inadequate to withstand the pulling forces caused by net momentum changes inside the endovascular graft. Means of extra fixation should be taken into account when possible. Concerning the supportive literature, Liffman et al. (2001) and Morris et al. (2004) confirmed that the drag force increases with EVG size. Mohan et al. (2002) reported that the drag force is nonlinearly increasing with iliac angle. Sternbergh et al. (2002) declared that the migration rate increases by 30% for AAAs with neck angles greater than 40 degrees.

Interestingly, compared with open surgery, the mortality of EVAR is almost the same. However, the key advantages of EVAR are reduced morbidity, shorter hospitalization, and quicker recovery; but, these benefits may be offset by the cost of the device, the need for continuous control measurements, and eventually the need for late intervention or conversion to open repair (Lester et al., 2001).

It should be noted that in this study the AAA wall was assumed to be smooth. Intraluminal thrombus and small branches were not considered. In patient-specific stented AAAs, intraluminal thrombi and small branches may affect the fluid-structure interactions, sac pressure, and hence possible EVG migration. In the range of pressure loads from 80 to 120 mmHg, linearly elastic property values are generally used (Vorp et al., 1998; Raghavan et al., 2000; Di Martino et al., 2001; Thubrikar et al., 2001). However, nonlinear wall properties may provide even more realistic results (Di Martino et al., 2001; Sternbergh et al., 2002; Finol et al., 2003; Fillinger et al., 2003). Considering the difficulty to model an actual EVG, i.e., a NiTi-wire mesh interwoven with synthetic graft material, presently the EVG is assumed to be a uniform shell made of an equivalent composite material. A detailed stent-graft mechanics analysis, assuming different static pressure loads, has been provided by Li (2005). In summary, this study provides a new technique, i.e., fully coupled fluid-structure interaction (FSI), to evaluate the impact of EVG placement, determine wall stresses and sac pressure values, calculate EVG migration forces, and provide physical insight into the biomechanics of stented AAAs.

5.5 PROJECT ASSIGNMENTS

<u>Note:</u> These PAs are open-ended, implying that any level of complexity, according to the student's ability and available resources, would be fine – as long as the work is done mostly independently and is a learning experience.

5.1 Develop a menu-driven computer program which guides the user with any biofluid dynamics problem through the steps of mathematical modeling (see Sects. 5.1.1 and 5.1.2) and computer simulation (see Sect. 5.1.3).

The input requirements include: project category (e.g., fluid flow alone or connected to heat & mass transfer or solid structure interaction), project assumptions (e.g., laminar or turbulent, transient 3-D or steady 1-D, constant coefficients, etc.), and availability of experimental data sets for model input and validation.

The expected results include: (i) the appropriate theory with necessary modeling equations, initial/boundary conditions, and input data; (ii) the best solution technique and computer software, including graphics; and (iii) a template for report writing (see page 369).

5.2 Bio-MEMS analysis (see Sect.5.2 as an example):

(a) Review three bio-MEMS applications according to different driving forces, ranging from micro-pumps and relatively high Reynolds numbers to unconventional forces creating very low Reynolds numbers.
(b) Select your favorite application with unconventional driving force and set up a mathematical model (see Sect.5.1) without considering any restriction in computer solution.
(c) Simplify 5.2b to a manageable problem which can be solved with readily available techniques.
(d) Provide a base-case solution to 5.2c and vary key system parameters to gain physical insight.

5.3 Drug targeting (see Sect.5.3 as an example):

(a) Review three methodologies of modern drug targeting, e.g., dermal transfer, in vivo release, etc.
(b) Select your favorite application and develop a mathematical model (see Sect.5.1), without any restrictions in mind.
(c) Simplify 5.3b to a manageable problem which can be solved with readily available techniques.
(d) Provide a base-case solution to 5.3c and vary key system parameters to gain physical insight.

5.4 Biomedical FSI applications (see Sects.3.1.5, 3.2.3, & 5.4 as an example):

(a) Review experimental/computational fluid-structure interactions, two applied to body implants and two applied to medical devices.

(b) Select your favorite fluid-structure system and develop a mathematical model as if there are no limits on computer resources.
(c) Reduce 5.4b to a tractable problem within given constraints of allotted time and resources.
(d) Generate a base-case solution to 5.4c and perform PSAs (parametric sensitivity analyses), provide graphical results, and suggest system improvements.

References

Balásházy, I., Hofmann, W., and Heistracher, T. (1999) Journal of Aerosol Science, 30: 185-203.

Balásházy, I., Hofmann, W., and Heistracher, T. (2003) J. Appl. Physiol., 94: 1719-1725.

Bear, J. (1972) "Dynamics of fluids in porous media," American Elsevier Publishing Company, Inc., New York, NY.

Beskok, A., and Karniadakis, G.E. (1994). Journal of Thermophysics and Heat Transfer, 8: 355–370.

Binder, K., Horbach, J., Kob, W., Paul, W., and Varnik, F. (2004) Journal of Physics: Condensed Matter, 16: S429–S453.

Bird, G.A. (1994) "Molecular Gas Dynamics and the Direct Simulation of Gas Flows," Clarendon Press, New York, NY.

Bowes, S.M., and Swift, D.L. (1989) Aerosol Sci. Technol., 11: 157-167.

Bradshaw, P., and Cebeci, T (1984) "Physical and Computational Aspects of Convective Heat Transfer," Springer-Verlag, New York, NY.

Buchanan, J.R., Kleinstreuer, C., and Comer, J.K. (2000) Computers & Fluids, 29: 695-724.

Caro, C., Schroter, R.C., Watkins, N., Sherwin, S. J., and Sauret, V. (2002) Proc. R. Soc. Lond A, 458: 791-809.

Cebeci T (2004) "Analysis of Turbulent Flows," Elsevier, Amsterdam, Oxford.

Chakfe, N., Dieval, F., Riepe, G., Mathieu, D., Zbali, I., Thaveau, F., Heintz, C., Kretz, G., and Durand, B. (2004) European Journal of Vascular and Endovascular Surgery, 27: 33–41.

Chan, T. L., and Lippmann, M. (1980) Am. Ind. Hyg. Assoc. J., 41: 399-409.

Cheng, Y.S., Yamada, Y., Yeh, H.C., and Swift, D.L. (1988) J. Aerosol Sci., 19: 741-751.

Cheng Y.S., Yazzie, D., Gao J., Muggli, D., Etter J., and Rosenthal, G.J. (2003) J. Aerosol Medicine, 16: 65-73.

Cheng, Y.S., Zhou, Y., and Chen, B. T. (1999) Aerosol Sci. Technol., 31: 286-300.

Cohen, B.S., Sussman R.G., and Lippmann, M. (1990) Aerosol Sci. and Technol., 12: 1082-1091.
Comer, J.K, Kleinstreuer, C., Hyun, S., and Kim, C. S. (2000) ASME Journal of Biomechanical Engineering, 122: 152-158.
Comer, J. K., Kleinstreuer, C., and Zhang, Z. (2001) Journal of Fluid Mechanics, 435: 25-54.
Crane, R. L., Hillstrom, K.E., and Minkoff, M. (1980) ANL-80-64.
Cuvelier, C., Segal, A., and Van Steenhoven, A. A. (1986) "Finite Element Methods and Navier-Stokes Equations," Kluwer Academic Publ., Boston, MA.
Dario, P., Carrozza, M.C., Benvenuto, A., and Menciassi, A. (2000) J. Micromech. Microeng., 10: 235-244.
Dias, N., Ivancev, K., Malina, M., Resch, T., Lindblad, B., and Sonesson, B. (2004) Journal of Vascular Surgery, 39: 1229-1235.
Di Martino, E.S., and Fumero, R. (2002) Rupture risk evaluation of abdominal aortic aneurysm by means of bi-dimensional pulsatile nonlinear structure models, Research Report.
Di Martino, E., Guadagni, G., Corno, C., Fumero, A., Spirito, R., Biglioli, P., and Redaelli, A. (2001) ASME: Bioengineering Conference, 50: 821-822.
Donaldson, K., Stone, V., Gilmour, P. S., Brown, D. M., and Macnee, W. (2000) Phil. Trans. R. Soc. Lond. A., 358: 2741-2749.
Emmett, P. C., Aitken, R. J., and Hannan, W. J. (1982) J. Aerosol Science, 13: 549-560.
Fan, B.J., Cheng, Y.S., and Yeh, H.C. (1996) Aerosol Sci. and Technol., 25: 113-120.
Ferziger, J. H., and Peric, M. (1999) "Computational Methods for Fluid Dynamics," Springer-Verlag, Berlin, Germany.
Fillinger, M.F., Marra, P.S., Raghavan, M.L., and Kennedy, E.F. (2003) Journal of Vascular Surgery, 37: 724-732.
Finlay, W.H. (2001) "The Mechanics of Inhaled Pharmaceutical Aerosols: An Introduction," Academic Press, London, UK.
Finol, E.A., Di Martino, E.S., Vorp, D.A., and Amon, C.H. (2003) Proceedings of the ASME 2003 Summer Bioengineering Conference, Key Biscayne, FL, 25-29: 75-76.
Flora, H.S., Talei, F.B., Ansdell, L., Chaloner, E.J., Sweeny, A., Grass, A., and Adiseshiah, M. (2002) J. Endovascular Therapy, 9: 665-75.
Foord, N., Black, A., and Walsh, M. (1978) J. Aerosol Science, 9: 383-390.
Frampton, M.W. (2001) Environmental Health Perspectives 109: 529-532.
Gawenda, M., Knez, P., Winter, S., Jaschke, G., Wassmer, G., Schmitz-Rixen, T., and Brunkwall, J. (2004) European Journal of Vascular and Endovascular Surgery, 27: 45-50.
Goo, J.H., and Kim, C. S. (2003) Journal of Aerosol Science, 34: 585-602.
Gosman, A. D., and Ioannides, E. (1981) J. Energy, 7: 482-490.
Guo, Z., and Li, Z. (2003) Int. J. Heat & Mass Transfer, 46: 149–159.

He, M.C., and Roach, M.R. (1994) Journal of Vascular Surgery, 20: 6-13.

Hickey, A.J. (2004) "Summary of Common Approaches to Pharmaceutical Aerosol Administration," In: Pharmaceutical Inhalation Aerosol Technology, Hickey A.J. (ed.), Marcel Dekker, Inc., New York, NY, pp.385-421.

Hirsch, C. (1990a) "Numerical Computation of Internal and External Flows: Fundamentals of Numerical Discretization," Vol. 1, John Wiley & Sons, New York, NY.

Hirsch, C. (1990b) "Numerical Computation of Internal and External Flows: Computational Methods for Inviscid and Viscous Flows," Vol. 2, John Wiley & Sons, New York, NY.

Hofmann, W., Asgharian, B., and Winkler-Heil, R. (2002) J. Aerosol Sci., 33: 219-235.

ICRP (International Commission on Radiological Protection) (1994) "Human respiratory tract model for radiological protection," Annals of the ICRP, ICRP Publication 66, Elsevier, New York.

Issa, R. I. (1986) J. Comput. Phys., 62: 40-65.

Jacobs, T.S., Won, J., Gravereaux, E.C., Faries, P.L., Morrissey, N., Teodorescu, V.J., and Hollier, L.H., Marin, M.L. (2003) Journal of Vascular Surgery, 37: 16-26.

Joseph, D. (2002) "Interrogations of direct numerical simulation of solid-liquid flow," URL: http://efluids.com/efluids/books/joseph.htm.

Kim, C. S. (2000) Respiratory Care, 45: 695-711.

Kim, J., Moin, P., and Moser, R. D. (1987) J. Fluid Mech. 177:133-166.

Kleinstreuer, C. (1997) "Engineering Fluid Dynamics," Cambridge University Press, New York, NY.

Kleinstreuer, C. (2003) "Two-Phase Flow: Theory and Applications," Taylor & Francis, New York, NY.

Kleinstreuer, C., and Koo, J. (2004) J. Fluids Engineering, 126: 1–9.

Kleinstreuer, C., and Koo, J. (2005) "Patent Application for Controlled Nanodrug Delivery in Microchannels," North Carolina State University, Raleigh, NC.

Kleinstreuer, C., and Koo, J. (2005) "Patent Application for Smart Inhaler System," North Carolina State University, Raleigh, NC.

Kleinstreuer, C., and Zhang, Z. (2003) Int. J. Multiphase Flow, 29:271-289.

Koo, J., and Kleinstreuer, C. (2003) Journal of Micromechanics and Microengineering, 13: 568–579.

Lester, J., Bosch, J., Kaufman, J., and Halpern, E. (2001) Academic Radiology, 7: 639-46.

Li, D. (2004) "Electrokinetics in microfluidics," Academic Press, New York.

Li, Z. (1997) Numerical Algorithms, 14: 269-293.

Li, Z., (2005), "Computational Analysis and Simulations of Fluid-structure Interactions for Stented Abdominal Aortic Aneurysms", PhD Dissertation, North Carolina State University.

Li, Z., and Kleinstreuer, C.(2005) Med. Eng. & Phys., 27: 369–382.

Li, Z., and Kleinstreuer, C. (2006) "Alveolar Flow Analyses using the Lattice Boltzmann Method," Physics of Fluids (in press).

Liffman, K., Lawrence, M., Semmens, B., Bui, A., Rudman, M., and Hartley, D. (2001) Journal of Endovascular Therapy, 8: 358-371.

Lin, P.H., Bush, R.L., Katzman, J.B., Zemel, G., Puente, O.A., Katzen, B.T., and Lumsden, A.B. (2003) Journal of Vascular Surgery, 38: 840-842.

Lippmann, M., and Albert, R.E. (1969) Am. Ind. Hyg. Assoc. J., 30:257-275.

Longest, P.W., Kleinstreuer, C., and Buchanan, J.R. (2004) Computers & Fluids, 33: 577-601.

Loth, E. (2000) Prog. Energy Combustion Sci., 26: 161-223.

MacInnes, J.M., and Bracco, F. V. (1992) Physics of Fluids, 4: 2809-2824.

Mala, G.M., and Li, D. (1999) Int. J. Heat and Mass Transfer, 20: 142–148.

Malina, M., Lindblad, B., and Ivancev, K. (1998) Journal of Endovascular Surgery, 5: 310-317.

Matida, E.A., Finlay, W.H., Lange, C.F., and Grgic, B. (2004) J. Aerosol Sci., 35: 1-19.

Matida, E.A., Nishino, K., and Torii, K. (2000) Int. J. Heat and Fluid Flow 21: 389-402.

Meter O. (2000) Annals of Biomedical Engineering, 28: 1281-99

Minkowycz, W. J., Torrance, K. E., Lloyd, J. R., Hartnett, J. P., Chen, T. S., and Hajisheikh, A. (1988) Int. J. Heat Mass Trans., 31 (5): 905-6.

Mohan, I.V., Harris, P.L., van Marrewijk, C.J., Laheij, R.J., and How, T.V. (2002) Journal of Endovascular Therapy, 9: 748-755.

Morris, L., Delassus, P., Walsh, M., and McGloughlin, T. (2004) Journal of Biomechanics, 37: 1087-1095.

Najibi, S., Steinberg, J., Katzen, B.T., Zemel,G., Lin, P.H., Weiss, V.J., Lumsden, A.B., and Chaikof. E.L. (2001) J. Vascular Surgery, 34: 353-356.

Nield, D.A. and Bejan, A. (1992) "Convection in Porous Media," Springer-Verlag, New York, NY.

Oberdörster, G., Ferin, J., and Lehnert, B.E. (1994) Environ. Health Perspect., 102: 173-179.

Oberdörster, G. (2001) International Archives of Occupational and Environmental Health, 74(1): 1-8.

Papautsky, I., Brazzle, J., Ameel, T., and Frazier, A.B. (1999) Sensors and Actuators, 73: 101–108.

Patankar, S. V. (1980) "Numerical Heat Transfer and Fluid Flow," Hemisphere, Washington, DC.

Peng, X.F., Peterson, G.P. and Wang, B. X. (1994) Experimental Heat Transfer, 7: 249–264.

Pepper, D. W., and Henrich, J. C. (1992) "The Finite Element Method: Basic Concepts and Applications," Hemisphere, Philadelphia, PA.

Peskin, C.S. (1977) J. Comp. Physics, 25: 220-252.

Pfhaler, J., Harley, J., Bau, H., and Zemel, J.N. (1991) Micromechanical Sensors, Actuators, and Systems, DSC, 32: 49–60.

Press, W. H., Flannery, B. P., Teukolsky, S. A., and Vetterling, W. T. (1992) "Numerical Recipes," Cambridge University Press, New York, NY.

Press, W. H., Teukolsky, S. A., Vetterling, W. T., Flannery, B. P., and Metcalf, M. (1996) "Numerical Recipes in Fortran 90," Cambridge University Press, New York, NY.

Probstein, R.F. (2003) "Physicochemical hydrodynamics," Second Edition, Wiley Interscience, Hoboken, NJ.

Qu, W., Mala, G.M., and Li, D. (2000) Int. J. Heat & Mass Transfer, 43: 353-364.

Raghavan, M. (2002) "Lecture notes on cardiovascular bio-solid mechanics section," Dept. of Biomechanical Engineering, University of Iowa.

Raghavan, M., Vorp, D., Federle, M., Makaroun, M., and Webster, M. (2000) Journal of Vascular Surgery, 31: 760-769.

Reddy, J. N., and Gartling, D. K. (1994) "The Finite Element Method in Heat Transfer and Fluid Dynamics," CRC Press, Boca Raton, FL.

Roache, P. J. (1998) "Fundamentals of Computational Fluid Dynamics," Hermosa, Albuquerque, NM.

Salsac, A.V., Sparks, S.R., and Lasheras, J.C. (2004) Annals of Vascular Surgery, 18: 14-21

Schuen, J. S., Chen, L. D., and Faeth, G. M. (1983) AICHE J., 29: 167-170.

Shah, R.K. and London, A.L. (1978) "Laminar Flow Forced Convection in Ducts," Academic Press, New York, NY.

Sharp, K. (2001) "Experimental investigation of liquid and particle laden flows in microtubes," PhD thesis, University of Illinois at Urbana-Champaign, Urbana-Champaign, IL.

Shi, H., Kleinstreuer, C., Zhang, Z., and Kim, C.S. (2004) Physics of Fluids, 16: 2199-2213.

Sonesson, B., Dias, N., Malina, M., Olofsson, P., Griffin, D., Lindbald, B., and Ivancev, K. (2003) Journal of Vascular Surgery, 37: 733-738.

Stahlhofen, W., Gebhart, J., and Heyder, J. (1980) Am. Ind. Hyg. Assoc. J., 41: 385-398.

Stahlhofen, W., Gebhart, J., Heyder, J., and Scheuch, G. (1983) J. Aerosol Sci., 14: 186-188.

Stahlhofen, W., Rudolf, G., and James, A. C. (1989) J. Aerosol Med., 2: 285-308.

Sternbergh III, W.C., Carter, G., York, J.W., Yoselevitz, M., and Money, S.R. (2002) Journal of Vascular Surgery, 35: 482-486.

Stone, H.A., and Kim, S. (2001) AIChE J., 47: 1250-1254.

Succi, S. (2001) "The Lattice Boltzmann Equation for Fluid Dynamics and Beyond," Clarendon Press, Oxford, NY.

Suzuki, K., Ishiguchi, T., Kawatsu, S., Iwai, H., Maruyama, K., and Ishigaki, T. (2001) Cardiovascular and Interventional Radiology, 24: 94-98.

Tannehill, J.C., Anderson, D.A., and Pletcher, R.H. (1997) "Computational Fluid Mechanics and Heat Transfer," Taylor & Francis, Washington, DC.

Thompson, R.W., Geraghty, P.J., and Lee, J.K. (2002) Current Problems in Surgery, 39(2).

Thubrikar, M., Al-Soudi, J., and Robicsek, F. (2001) Annals of Vascular Surgery, 15(3): 355-366.

Varghese, S.S., and Frankel, S. H. (2003) ASME Journal of Biomechanical Engineering, 125: 445-460.

Vemuri, V.R. (1978) "Modeling of Complex Systems," Academic Press, New York, NY.

Vorp, D.A., Raghavan, M., and Webster, M. (1998) J. Vascular Surgery, 27: 632-9.

Wang, Y., and James, P. W. (1999) Int. J. Multiphase Flow, 25: 551-558.

Weibel, E. R. (1963) "Morphometry of the Human Lung," Academic Press, New York, NY.

White, F.M. (1991) "Viscous Fluid Flow," Second edition, McGraw-Hill, Singapore.

White, F.M. (1999) "Fluid Mechanics," Fourth edition, McGraw-Hill, Boston, MA.

Whitesides, G.M., and Strook, A.D. (2001) Phys. Today, 54(6): 42.

Wilcox, D. C. (1993) "Turbulence Modeling for CFD," DCW Industries, Inc., LA Canada, CA.

Wilcox, D. C. (1998) "Turbulence Modeling for CFD," Second edition, DCW Industries, La Canada, CA.

Xu, B., Ooi, K.T., Wong, N. T., and Choi, W. K. (1999) Liquid flow in microchannels, In: Proc. 5th ASME/JSME Joint Thermal Engineering Conference, San Diego, CA.

Xu, B., Ooi, K.T., Wong, N.T., and Choi, W.K. (2000) International Communication of Heat and Mass Transfer, 27: 1165–1176.

Zarins, C.K., Bloch, D.A., Crabtree, T., Matsumoto, A.H., White, R.A., and Fogarty, T.J. (2003) Journal of Vascular Surgery, 38: 1264-1272.

Zhang, Z., and Kleinstreuer, C. (2001) J. Aerosol Medicine, 14: 13-29.

Zhang, Z., and Kleinstreuer, C. (2002) Physics of Fluids, 14: 862-880.

Zhang, Z., and Kleinstreuer, C. (2003a) AIAA Journal, 41: 831-840.

Zhang, Z., and Kleinstreuer, C. (2003b) Int. J. Heat and Mass Transfer, 46: 4755-4768.

Zhang, Z., and Kleinstreuer, C. (2004) J. Computational Physics, 198: 178-210.

Zhang, Z., Kleinstreuer, C., Donohue, J.F., and Kim, C.S. (2005). J. Aerosol Science, 36: 211-233.

Zhang, Z., Kleinstreuer, C., and Kim, C.S. (2002) Int. J. Multiphase Flow, 28: 1021-1046.

APPENDICES

APPENDIX A

Review of Tensor Calculus, Differential Operations, Integral Transformations, and ODE Solutions

A.1 Tensor Calculus

Here we restrict our review to tensor manipulations as needed in the text. Further information and solved examples can be found in Schey (1973), Aris (1989), and Appendix A of Bird et al. (2002).

A.1.1 Definitions

Recall: Tensors of rank n have 3^n components. For example:

- A tensor of rank "zero" is a *scalar* which has only one component, i.e., its magnitude (e.g., pressure).
- A tensor of rank "one" is a *vector* which has in general three components, i.e., 3 magnitudes and 3 directions (e.g., velocity).
- A tensor of rank "two" is usually labeled a *tensor* which has nine components, e.g., stress.

Coordinate Systems

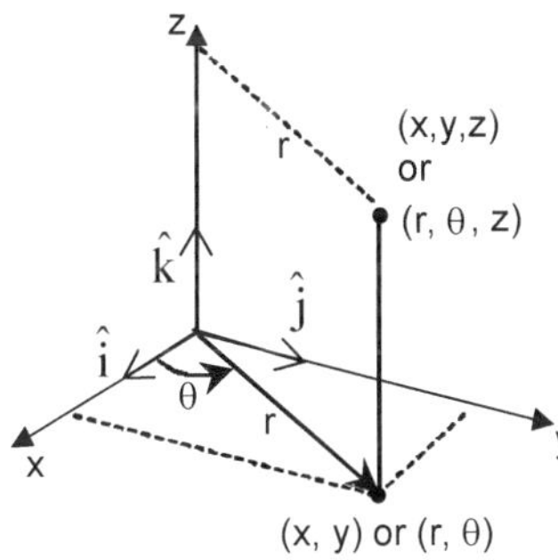

$$x = r\cos\theta$$
$$y = r\sin\theta$$
$$z = z$$

Vector Products

- Dot product

$$\vec{u}\cdot\vec{v} = \vec{v}\cdot\vec{u} = |\vec{u}||\vec{v}|\cos\alpha \rightarrow \text{scalar}$$

- Cross product

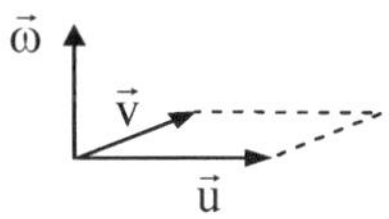

$$\vec{u}\times\vec{v} = -\vec{v}\times\vec{u} = \vec{\omega} \rightarrow \text{vector}$$

- Dyadic product:
  $\vec{u}\vec{v} = \vec{\vec{a}} \rightarrow$ tensor

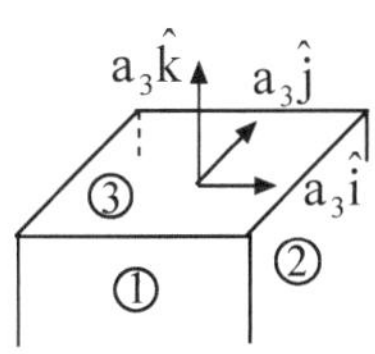

Clearly, the type of vector product may result in a scalar (see dot product) or a tensor of rank two with nine components (see dyadic product). This is further illustrated when using the del operator which has the characteristics of a vector. For example,

in rectangular coordinates: $\nabla \equiv \hat{i}\frac{\partial}{\partial x} + \hat{j}\frac{\partial}{\partial y} + \hat{k}\frac{\partial}{\partial z}$ (A.1.1a)

in cylindrical coordinates: $\nabla \equiv \hat{e}_r\frac{\partial}{\partial r} + \hat{e}_\theta\frac{1}{r}\frac{\partial}{\partial\theta} + \hat{e}_z\frac{\partial}{\partial z}$ (A.1.1b)

A.1.2 Tensor Operations

- When operating on a scalar, say, the pressure, ∇p, generates a vector, i.e., the *pressure gradient*:

$$\nabla p = (\hat{i}\frac{\partial}{\partial x} + \hat{j}\frac{\partial}{\partial y} + \hat{k}\frac{\partial}{\partial z})p = \frac{\partial p}{\partial x}\hat{i} + \frac{\partial p}{\partial y}\hat{j} + \frac{\partial p}{\partial z}\hat{k} \qquad (A.1.2)$$

- When operating on a vector, it can produce:
  a scalar in case of a dot product, e.g., $\nabla\cdot\vec{v}$;
  a vector in case of a cross product, e.g., $\nabla\times\vec{v}$;
  a tensor in case of a dyadic product, e.g., $\nabla\vec{v}$.
  Specifically,
  (a) $\nabla\cdot\vec{v} \equiv \text{div}\,\vec{v}$ is the divergence of the velocity field:

$$\nabla \cdot \vec{v} = (\hat{i}\frac{\partial}{\partial x} + \hat{j}\frac{\partial}{\partial y} + \hat{k}\frac{\partial}{\partial z}) \cdot (u\hat{i} + v\hat{j} + w\hat{k})$$

$$\begin{aligned} &= (\hat{i}\cdot\hat{i})\frac{\partial u}{\partial x} + (\hat{i}\cdot\hat{j})\frac{\partial v}{\partial x} + (\hat{i}\cdot\hat{k})\frac{\partial w}{\partial x} + \\ &(\hat{j}\cdot\hat{i})\frac{\partial u}{\partial y} + (\hat{j}\cdot\hat{j})\frac{\partial v}{\partial y} + (\hat{j}\cdot\hat{k})\frac{\partial w}{\partial y} + \\ &(\hat{k}\cdot\hat{i})\frac{\partial u}{\partial z} + (\hat{k}\cdot\hat{j})\frac{\partial v}{\partial z} + (\hat{k}\cdot\hat{k})\frac{\partial w}{\partial z} = \frac{\partial u}{\partial x} + \frac{\partial v}{\partial y} + \frac{\partial w}{\partial z} \end{aligned} \qquad \text{(A.1.3)}$$

because $\hat{i}\cdot\hat{i} = |i||i|\cos\alpha$; $\alpha = 0 \Rightarrow \hat{i}\cdot\hat{i} = 1$, while $\hat{i}\cdot\hat{j} = 0$ <α = 90°>

In general, $\hat{\delta}_i \cdot \hat{\delta}_j = \delta_{ij} := \begin{cases} 1 & \text{if } i = j \\ 0 & \text{if } i \neq j \end{cases}$, known as the *Kronecker delta*

(b) $\nabla \times \vec{v} \equiv \text{curl}\,\vec{v}$ is the curl (or rotation) of the velocity field:

$$\nabla \times \vec{v} = (\hat{i}\frac{\partial}{\partial x} + \hat{j}\frac{\partial}{\partial y} + \hat{k}\frac{\partial}{\partial z}) \times (u\hat{i} + v\hat{j} + w\hat{k})$$

$$\begin{aligned} &= (\hat{i}\times\hat{i})\frac{\partial u}{\partial x} + (\hat{i}\times\hat{j})\frac{\partial v}{\partial x} + (\hat{i}\times\hat{k})\frac{\partial w}{\partial x} + \\ &(\hat{j}\times\hat{i})\frac{\partial u}{\partial y} \ldots etc. \end{aligned} \qquad \text{(A.1.4)}$$

Recalling the results for the cross products between unit (or base) vectors:

$$\begin{array}{lll} \hat{i}\times\hat{j} = \hat{k} & \hat{j}\times\hat{i} = -\hat{k} & \hat{i}\times\hat{i} = 0 \\ \hat{j}\times\hat{k} = \hat{i} & \hat{k}\times\hat{j} = -\hat{i} & \hat{j}\times\hat{j} = 0 \\ \hat{k}\times\hat{i} = \hat{j} & \hat{i}\times\hat{k} = -\hat{j} & \hat{k}\times\hat{k} = 0 \end{array}$$

we obtain:

$$\nabla \times \vec{v} = \hat{i}\left(\frac{\partial w}{\partial y} - \frac{\partial v}{\partial z}\right) + \hat{j}\left(\frac{\partial u}{\partial z} - \frac{\partial w}{\partial x}\right) + \hat{k}\left(\frac{\partial v}{\partial x} - \frac{\partial u}{\partial y}\right)$$

or

$$\nabla \times \vec{v} - \text{curl}\vec{v} - \begin{vmatrix} \hat{i} & \hat{j} & \hat{k} \\ \dfrac{\partial}{\partial x} & \dfrac{\partial}{\partial y} & \dfrac{\partial}{\partial z} \\ u & v & w \end{vmatrix} \equiv \vec{\zeta} \quad \text{(vorticity vector)} \quad (A.1.5)$$

(c) $\nabla\vec{v} \equiv \text{grad}\vec{v}$ is the dyadic product, or gradient, of the velocity field:

$$\nabla\vec{v} = (\hat{i}\frac{\partial}{\partial x} + \hat{j}\frac{\partial}{\partial y} + \hat{k}\frac{\partial}{\partial z})(u\hat{i} + v\hat{j} + w\hat{k})$$

$$= (\hat{i}\hat{i})\frac{\partial u}{\partial x} + (\hat{i}\hat{j})\frac{\partial v}{\partial x} + (\hat{i}\hat{k})\frac{\partial w}{\partial x} + \quad (A.1.6)$$

$$(\hat{j}\hat{i})\frac{\partial u}{\partial y} + \dots etc.$$

Now, the unit vector dyadic product $\hat{\delta}_i\hat{\delta}_j$ indicates the location (i.e., $\hat{\delta}_i$ is the unit normal to the particular surface) and direction (i.e., $\hat{\delta}_j$ gives the direction) of a tensor. Thus,

$$\nabla\vec{v} = \text{grad}\vec{v} = \begin{vmatrix} \dfrac{\partial u}{\partial x} & \dfrac{\partial v}{\partial x} & \dfrac{\partial w}{\partial x} \\ \dfrac{\partial u}{\partial y} & \dfrac{\partial v}{\partial y} & \dfrac{\partial w}{\partial y} \\ \dfrac{\partial u}{\partial z} & \dfrac{\partial v}{\partial z} & \dfrac{\partial w}{\partial z} \end{vmatrix} \sim \vec{\vec{\tau}} \qquad \text{<stress tensor>} \qquad (A.1.7)$$

Notes:

- The use of $\nabla \cdot \vec{v}$, $\nabla \times \vec{v}$, and $\nabla\vec{v}$ plus illustrations were introduced in Chapter I.
- The dot product reduces the rank of a tensor, e.g., $\nabla \cdot \vec{v} \rightarrow$ scalar and $\nabla \cdot \vec{\vec{\tau}} \rightarrow$ vector.
- The dyadic (or gradient) product increases the rank, e.g., $\nabla p \rightarrow$ vector and $\nabla\vec{v} \rightarrow$ tensor.
- The divergence of a scalar gradient is $\nabla \cdot \nabla s = \nabla^2 s = \Sigma \dfrac{\partial^2 s}{\partial x^2}$, where ∇^2 *is the Laplacian operator* producing a scalar field, e.g.,

$$\nabla^2 \equiv \frac{\partial^2}{\partial x^2} + \frac{\partial^2}{\partial y^2} + \frac{\partial^2}{\partial z^2} \tag{A.1.8}$$

- The transpose of a second-order tensor, $\vec{\vec{a}}$, with components a_{ij} is denoted by $\vec{\vec{a}}^{tr}$ and is defined by

$$\left[a^{tr}\right]_{ij} = a_{ji} \tag{A.1.9a}$$

For example,

$$\vec{\vec{a}} \equiv \vec{v}\vec{w} = \begin{pmatrix} v_1 w_1 & v_1 w_2 & v_1 w_3 \\ v_2 w_1 & v_2 w_2 & v_2 w_3 \\ v_3 w_1 & v_3 w_2 & v_3 w_3 \end{pmatrix} \tag{A.1.9b}$$

whereas

$$\vec{\vec{a}}^{tr} \equiv (\vec{v}\vec{w})^{tr} = \begin{pmatrix} v_1 w_1 & v_2 w_1 & v_3 w_1 \\ v_1 w_2 & v_2 w_2 & v_3 w_2 \\ v_1 w_3 & v_2 w_3 & v_3 w_3 \end{pmatrix} \tag{A.1.9c}$$

Sample Problem Solutions

To illustrate a few tensor manipulations, the following sample problems are solved. Given the components of a symmetric tensor $\vec{\vec{\tau}}$, i.e., $\tau_{ij} = \tau_{ji}$:

$$\tau_{xx} = 3, \quad \tau_{xy} = 2, \quad \tau_{xz} = -1,$$
$$\tau_{yy} = 2, \quad \tau_{yz} = 1,$$
$$\tau_{zz} = 0$$

and the components of a vector $\vec{v}$, e.g., $v_x = 5, v_y = 3, v_z = 7$, evaluate:

(a) $\vec{\vec{\tau}} \cdot \vec{v}$; (b) $\vec{v} \cdot \vec{\vec{\tau}}$; (c) $\vec{v}\vec{v}$; and (d) $\vec{\vec{\tau}} \cdot \vec{\vec{\delta}}$

where $\vec{\vec{\delta}}$ is the unit tensor, i.e.,

$$\delta_{ij} = \begin{pmatrix} 1 & & \phi \\ & 1 & \\ \phi & & 1 \end{pmatrix}$$

<u>*Solution:*</u> A good preliminary exercise is to write down the vectors and tensors in component form.

(a) $$\vec{\vec{\tau}}\cdot\vec{v} = \Sigma_i\vec{\delta}_i\left\{\Sigma_j\tau_{ij}v_j\right\} = \left(\vec{\delta}_1\vec{\delta}_2\vec{\delta}_3\right)\begin{pmatrix}3 & 2 & -1\\ 2 & 2 & 1\\ -1 & 1 & 0\end{pmatrix}\begin{pmatrix}5\\3\\7\end{pmatrix}$$

$$= 14\vec{\delta}_1 + 23\vec{\delta}_2 - 2\vec{\delta}_3 \qquad \text{<a vector>}$$

where $\vec{\delta}_i \triangleq$ unit vector in the i-direction, with i=1, 2, 3.

(b) $\vec{v}\cdot\vec{\vec{\tau}} = \vec{\vec{\tau}}\cdot\vec{v}$ since $\vec{\vec{\tau}}$ is symmetric

(c) $$\vec{v}\vec{v} = \Sigma_i\Sigma_j\vec{\delta}_i\vec{\delta}_j v_i v_j = 25\vec{\delta}_1\vec{\delta}_1 + 15\vec{\delta}_1\vec{\delta}_2 + 35\vec{\delta}_1\vec{\delta}_3 + 15\vec{\delta}_2\vec{\delta}_1 + 9\vec{\delta}_2\vec{\delta}_2 + 21\vec{\delta}_2\vec{\delta}_3 + 35\vec{\delta}_3\vec{\delta}_1 + 21\vec{\delta}_3\vec{\delta}_2 + 49\vec{\delta}_3\vec{\delta}_3$$

(d) $$\vec{\vec{\tau}}\cdot\vec{\vec{\delta}} = \Sigma_i\Sigma_l\vec{\delta}_i\vec{\delta}_l\left(\Sigma_j\tau_{ij}\delta_{jl}\right) = \vec{\vec{\tau}} = 3\vec{\delta}_1\vec{\delta}_1 + 2\vec{\delta}_1\vec{\delta}_2 - 1\vec{\delta}_1\vec{\delta}_3 + 2\vec{\delta}_2\vec{\delta}_1 + 2\vec{\delta}_2\vec{\delta}_2 + 1\vec{\delta}_2\vec{\delta}_3 - 1\vec{\delta}_3\vec{\delta}_1 + 1\vec{\delta}_3\vec{\delta}_2 + 0\vec{\delta}_3\vec{\delta}_3$$

A.1.3 Tensor Identities

$$\nabla rs = r\nabla s + s\nabla r \qquad \text{(A.1.10)}$$

$$(\nabla\cdot s\vec{v}) = (\nabla s\cdot\vec{v}) + s(\nabla\cdot\vec{v}) \qquad \text{(A.1.11)}$$

$$(\nabla\cdot[\vec{v}\times\vec{w}]) = (\vec{w}\cdot[\nabla\times\vec{v}]) - (\vec{v}\cdot[\nabla\times\vec{w}]) \qquad \text{(A.1.12)}$$

$$[\nabla\times s\vec{v}] = [\nabla s\times\vec{v}] + s[\nabla\times\vec{v}] \qquad \text{(A.1.13)}$$

$$[\nabla\cdot\nabla\vec{v}] = \nabla(\nabla\cdot\vec{v}) - [\nabla\times[\nabla\times\vec{v}]] \qquad \text{(A.1.14)}$$

$$[\vec{v}\cdot\nabla\vec{v}] = \tfrac{1}{2}\nabla(\vec{v}\cdot\vec{v}) - [\vec{v}\times[\nabla\times\vec{v}]] \qquad \text{(A.1.15)}$$

$$[\nabla\cdot\vec{v}\vec{w}] = [\vec{v}\cdot\nabla\vec{w}] + \vec{w}(\nabla\cdot\vec{v}) \qquad \text{(A.1.16)}$$

$$[\nabla\cdot s\vec{\vec{\delta}}] = \nabla s \qquad \text{(A.1.17)}$$

$$[\nabla\cdot s\vec{\vec{\tau}}] = [\nabla s\cdot\vec{\vec{\tau}}] + s[\nabla\cdot\vec{\vec{\tau}}] \qquad \text{(A.1.18)}$$

$$\nabla(\vec{v}\cdot\vec{w}) = [(\nabla\vec{v})\cdot\vec{w}] + [(\nabla\vec{w})\cdot\vec{v}] \tag{A.1.19}$$

A.2 Differentiation

A.2.1 Differential Time Operators

In order to understand and solve fluid mechanics problems, the basic skills in linear algebra, as well as differentiating and integrating functions, graphing and analyzing functions, and curve fitting, are definitely *prerequisites*. If a review is necessary, the reader may want to consult M. Spiegel (1971), SCHAUM's Outline series, or M. D. Greenberg (1998), Prentice-Hall, among many other texts.

The different notations and the physical meanings of various time derivatives (i.e., differential operators) are presented as follows:

- Partial time derivative: $\partial\#/\partial t \triangleq$ Changes in variable "#" with time observed from a fixed position in space, i.e., stationary observer.
- Substantial or material time derivate: $D\#/Dt \triangleq$ Changes of variable "#" with time following the fluid/material element in motion. The Stokes, or material derivative, is defined as:

$$\frac{D\#}{Dt} \equiv \frac{\partial\#}{\partial t} + (\vec{\mathbf{v}}\cdot\nabla)\# \tag{A.2.1}$$

In Eq. (A.2.1) the Lagrangian time rate of change is expressed in terms of Eulerian derivatives. For example, c being a species concentration $[M/L^3]$, the material time derivative is

$$\frac{Dc}{Dt} \equiv \frac{\partial c}{\partial t} + (\vec{\mathbf{v}}\cdot\nabla)c \tag{A.2.2a}$$

In rectangular coordinates

$$\frac{Dc}{Dt} = \frac{\partial c}{\partial t} + u\frac{\partial c}{\partial x} + \mathrm{v}\frac{\partial c}{\partial y} + w\frac{\partial c}{\partial z} \tag{A2.2b}$$

while in tensor notation

$$\frac{Dc}{Dt} = \frac{\partial c}{\partial t} + \mathrm{v}_k\frac{\partial c}{\partial x_k}; k = 1, 2, 3 \tag{A.2.2c}$$

where $\frac{\partial c}{\partial t} \triangleq$ local time derivative (i.e., accumulation of species c) and

$\mathrm{v}_k\frac{\partial c}{\partial x_k} \triangleq$ convective derivatives (i.e., mass transfer by convection).

Note: Repeated indices imply summation of these terms, here $k = 1, 2, 3$ <Einstein convention>.

- Total time derivative: $d\#/dt \triangleq$ Changes of # with respect to time observed from a point *moving differently* than the flow field. For example:

$$\frac{dc}{dt} = \frac{\partial c}{\partial t} + \frac{dx}{dt}\frac{\partial c}{\partial x} + \frac{dy}{dt}\frac{\partial c}{\partial y} + \frac{dz}{dt}\frac{\partial c}{\partial z} \qquad \text{(A.2.3)}$$

where $dx/dt = u$, $dy/dt = v$, and $dz/dt = w$ are the velocity components of the moving observer.

A.2.2 The Chain Rule

Dependent variables describing transport phenomena, such as fluid velocity, pressure, and species concentration, are often a function of more than one independent variable. For example, the fluid velocity is a function of three spatial coordinates, say, x, y, and z and, if the flow is unsteady, time t. Thus, $\vec{v} = \vec{v}(x, y, z, t)$. The total differential is defined as

$$d\vec{v} = \frac{\partial \vec{v}}{\partial x}dx + \frac{\partial \vec{v}}{\partial y}dy + \frac{\partial \vec{v}}{\partial z}dz + \frac{\partial \vec{v}}{\partial t}dt \qquad \text{(A.2.4)}$$

If the spatial coordinates x, y, and z are also functions of time, then the total (particle) time derivative is:

$$\frac{d\vec{v}}{dt} = \frac{\partial \vec{v}}{\partial x}\frac{dx}{dt} + \frac{\partial \vec{v}}{\partial y}\frac{dy}{dt} + \frac{\partial \vec{v}}{\partial z}\frac{dz}{dt} + \frac{\partial \vec{v}}{\partial t} \qquad \text{(A.2.5)}$$

Such differentiation can be extended to the calculation of fluid acceleration and mass transport where the local quantities change with time. For a scalar, say the pressure, we have

$$dp = \frac{\partial p}{\partial x}dx + \frac{\partial p}{\partial y}dy + \frac{\partial p}{\partial z}dz \qquad \text{(A.2.6)}$$

A.2.3 Truncated TAYLOR Series Expansions and Binomial Theorem

In order to approximate a function, say, y(x), around some point $x = x_0$, we employ two or three terms of the Taylor series. For one independent variable,

$$y(x) = y\Big|_{x=x_0} + \frac{dy}{dx}\Big|_{x=x_0}(x - x_0) + \frac{1}{2}\frac{d^2y}{dx^2}\Big|_{x=x_0}(x - x_0)^2 + \frac{1}{6}\frac{d^3y}{dx^3}\Big|_{x=x_0}(x - x_0)^3 + \ldots \tag{A.2.7}$$

Clearly, the first two terms provide a straight-line fit and the first three a parabolic fit of y(x) at $x = x_0$.

If we want to estimate the value of y a very small distance away from the known y(x), i.e., what is $y(x+\varepsilon)$ where $\varepsilon << 1$, we can write with Eq. (A.2.7):

$$y(x+\varepsilon) \approx y(x) + \frac{dy}{dx}\Big|_{x=\varepsilon}\varepsilon \tag{A.2.8}$$

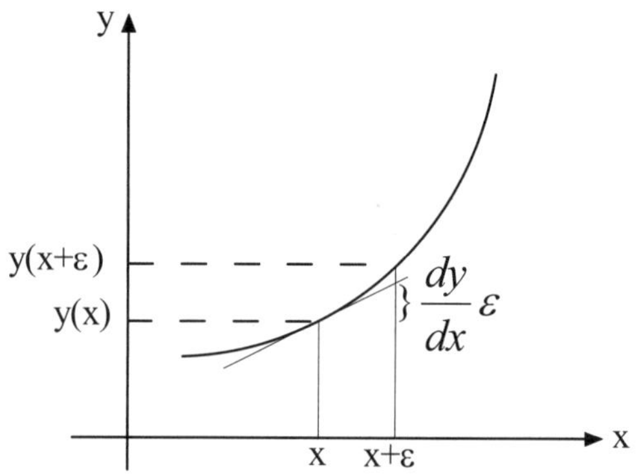

As the graph indicates, the step size ε and local curvature of y(x) determine the accuracy of Eq. (A.2.8). Equation (A.2.8) is employed in Sect. 1.3 to derive equations in differential form. For functions of two variables, e.g., $f(x, y)$, we write

$$f(x+\varepsilon, y+\delta) \approx f(x,y) + \frac{\partial f}{\partial x}\Big|_{x=\varepsilon}\varepsilon + \frac{\partial f}{\partial y}\Big|_{y=\delta}\delta \tag{A.2.9}$$

When dealing with rational *fractional* functions, it is often advantageous to express them in terms of partial fractions and then expand them, using the *binomial theorem*. For example, the expansion

$$(\mathrm{c}+\varepsilon)^{\mathrm{n}} = \mathrm{c}^{\mathrm{n}} + \mathrm{nc}^{\mathrm{n}-1}\varepsilon + \frac{\mathrm{n}(\mathrm{n}-1)}{2!}\mathrm{c}^{\mathrm{n}-2}\varepsilon^2 + \frac{\mathrm{n}(\mathrm{n}-1)(\mathrm{n}-2)}{3!}\mathrm{c}^{\mathrm{n}-3}\varepsilon^3 + \cdots \tag{A.2.10}$$

is valid for all values of n if $|\varepsilon| < |c|$. If $|c| < |\varepsilon|$, the expansion is valid only if n is a nonnegative integer.

A.3 Integral Transformations

A.3.1 The Divergence Theorem

As established by Gauss, the divergence theorem states that the integration over the dot product of a vector field, $\vec{v}$, with a closed regular area element, $d\vec{A}$, is equal to the integration of the divergence of $\vec{v}$, i.e., $\nabla \cdot \vec{v}$, over the interior volume, $\forall$:

$$\iint_A \vec{v} \cdot d\vec{A} = \iiint_\forall (\nabla \cdot \vec{v}) d\forall \tag{A.3.1}$$

Equation (A.3.1) is being used in Sect.1.3 when converting all surface integrals in the Reynolds Transport Theorem into volume integrals in order to express the conservation laws of mass, momentum, and energy in *differential form*.

<u>Sample Problem:</u> Given $\vec{v} = 4xz\hat{i} - y^2\hat{j} + yz\hat{k}$, i.e., $u = 4xz$, $v = -y^2$, and $w = yz$, in a unit cube, i.e., $0 \le x \le 1$, $0 \le y \le 1$, and $0 \le z \le 1$, show that Eq. (A.3.1) holds.

<u>Solution:</u>

(a) Six-sided surface integral: $\iint_A \vec{v} \cdot d\vec{A} = \int_0^1\int_0^1 \vec{v} \cdot \vec{n} dA$

where

$\hat{n} = \hat{i}$ at x = 1 and $\hat{n} = -\hat{i}$ at x = 0 with dA= dydz

$\hat{n} = \hat{j}$ at y = 1 and $\hat{n} = -\hat{j}$ at y = 0 with dA= dxdz

and

$\hat{n} = \hat{k}$ at z = 1 and $\hat{n} = -\hat{k}$ at z = 0 with dA= dxdy.

Thus,

$$\int_0^1\int_0^1 \vec{v} \cdot \hat{i} dA = \int_0^1\int_0^1 4xz\Big|_{x=1} dydz = 4y\Big|_0^1 \left(\frac{1}{2}z^2\Big|_0^1\right) = 2$$

Similarly,

$$\int_0^1\int_0^1 \vec{v} \cdot \hat{j} dA = -1; \qquad \int_0^1\int_0^1 \vec{v} \cdot \hat{k} dA = \frac{1}{2}; \qquad \text{and}$$

the other three negative surface integrals are zero. Hence,

$$\iint_A \vec{v} \cdot \hat{n} dA = 2 - 1 + \frac{1}{2} + 0 + 0 + 0 = \frac{3}{2}$$

(b) Volume integral: $\iiint_{\forall} \nabla \cdot \vec{v} d\forall = \int_0^1\int_0^1\int_0^1 \left(\frac{\partial u}{\partial x} + \frac{\partial v}{\partial y} + \frac{\partial w}{\partial z} \right) dxdydz$

$$\int_0^1\int_0^1\int_0^1 (4z - 2y + y) dxdydz = \left| \frac{4z^2}{2} - \frac{z}{2} \right|_0^1 = \frac{3}{2}$$

<u>*Comment:*</u> It is evident that such integral operations over *variable* vector fields result in scalars (i.e., numbers); this implies that the Reynolds Transport Theorem generates scalar quantities, i.e., numerical values for flow rates, averaged velocities, forces, wall stresses, pressure drops, etc.

A.3.2 Leibniz Rule

A switch in the order of operation is justified with Leibniz Rule: If $F(t) = \int_a^b f(x,t)dx$ and a, b are constants, then

$$\frac{dF}{dt} \equiv \frac{d}{dt}\left[\int_a^b f(x,t)dx \right] = \int_a^b \frac{\partial f}{\partial t} dx \tag{A.3.2}$$

Equation (A.3.2) is occasionally applied when dealing with the transient term <volume integral> in the RTT. In general, Leibniz Rule reads

$$\frac{d}{dt}\int_{a(t)}^{b(t)} f(x,t)dx = \int_a^b \frac{\partial f}{\partial t} dx + b' f[b(t),t] - a' f[a(t),t]$$

A.3.3 Error Function

Numerous natural phenomena follow exponentional functions. For example,

$$erf(x) = \frac{2}{\sqrt{\pi}} \int_0^x e^{-\xi^2} d\xi \tag{A.3.3a}$$

where

$$erf(0) = 0 \text{ and } erf(\infty) = 1 \tag{A.3.3b,c}$$

The integrand $\exp(-\xi^2)$ is a normal probability distribution. Thus, Eq. (A.3.3) is a solution part of processes governed by Gaussian-type distributions. The "complementary error function" is defined as

$$1 - erf(x) \equiv erfc(x) \tag{A.3.4}$$

A.3.4 Integral Methods

Two solution techniques dealing with integral equations are briefly discussed. The first method starts with the integration of a given set of partial differential equations that describe a given flow system, known as the integral method. The second approach starts with balance equations in integral form, i.e., the Reynolds Transport Theorem, which assures the conservation laws for a control volume.

Von Karman integral method. In contrast to separation of variables and similarity theory, the integral method is an approximation method. The Von Karman integral method is the most famous member of the family of integral relations, which in turn is a special case of the method of weighted residuals (MWR). Specifically, a transport equation in normal form can be written as [cf. Eq. (1.7)]

$$L(\phi) \equiv \frac{\partial \phi}{\partial t} + \nabla \cdot (\vec{v}\phi) - \nu\nabla^2\phi - S = 0 \tag{A.3.5}$$

where L(•) is a (nonlinear) operator, ϕ is a dependent variable, and S represents sink/source terms. Now, the unknown ϕ-function is replaced by an *approximate* expression, i.e., a "profile" or functional $\tilde{\phi}$ that satisfies the boundary conditions, but contains a number of unknown coefficients or parameters. As can be expected,

$$L(\tilde{\phi}) \neq 0, \text{ i.e., } L(\tilde{\phi}) \equiv R \tag{A.3.6a,b}$$

where R is the residual. In requiring that

$$\int_\Omega W R d\Omega = 0 \tag{A.3.6c}$$

we force the weighted residual over the computational domain Ω to be zero and thereby determine the unknown coefficients or parameters in the assumed $\tilde{\phi}$-function. The type of weighing function W determines the special case of the MWR, e.g., integral method, collocation method, Galerkin finite element method, control volume method, etc. (cf. Finlayson, 1978).

The Von Karman method is best applicable to laminar/turbulent similar or nonsimilar *boundary-layer type* flows for which appropriate velocity, concentration, and temperature profiles are known, i.e., thin and thick wall shear layers as well as plumes, jets, and wakes. Solutions of such problems yield global or integral system parameters, such as flow rates, fluxes, forces, boundary-layer thicknesses, shape factors, drag coefficients, etc.

In general, a two-dimensional partial differential equation is integrated in one direction, typically normal to the main flow, and thereby transformed into an ordinary differential equation, which is then solved analytically or numerically. Implementation of the integral method rests on two general characteristics of boundary-layer type problems: (i) the boundary conditions for a particular system simplify the integration process significantly so that a simpler differential equation is obtained, and (ii) all extra unknown functions, or parameters, remaining in the governing differential equation are approximated on physical grounds or by empirical relationships. Thus, closure is gained using, for example, the entrainment concept for plumes, jets, and wakes or by expressing velocity and temperature profiles with power expansions for high Reynolds number flows past submerged bodies.

A.4 Ordinary Differential Equations

For most applications in biofluids, the key differential equations are the equation of continuity, i.e., conservation of fluid mass, and the Navier-Stokes equations, i.e., conservation of linear momentum for constant fluid properties, as well as the scalar transport equations for species mass and heat transfer. They reflect the conservation laws in terms of differential balances for fluid mass, momentum, species concentration, and energy.

If the dependent variable, say, the velocity, is a function of more than one independent variable (e.g., x, y, z, t or r, θ, z, t) then the describing equation is a partial differential equation (PDE); otherwise, it is an ordinary differential equation (ODE). Clearly, solving PDEs requires usually elaborate transformations or numerical algorithms (see Özisik, 1993 and Hoffman, 2001). For that reason and to gain direct physical insight, *simplified base-case problems* are discussed (see Chapter I) where the continuity equation is fully satisfied and the Navier-Stokes equations are reduced to ODEs. Exact solutions for ODEs are listed in Polyamin & Zaitsev (1995). Numerical ODE solutions may be obtained with commercial software such as Matlab and Mathcad, which for their underlying finite difference approximations rely on selected terms of the Taylor Series (see Sect.A.2.3).

After developing a mathematical model describing approximately the fluid flow problem at hand, the resulting ODE (or system of coupled ODEs) has to be classified. One has to determine if the ODE, say, for y(x) is:

- linear or nonlinear (e.g., y^2, yy', $\sqrt{y''}$, etc.);
- with constant coefficients or not;
- homogeneous or inhomogeneous;
- of first, second, or n-th order;
- an initial value problem (IVP) or a boundary value problem (BVP).

The last two types of ODEs can be solved numerically with the Runge-Kutta method (IVPs) or a shooting method (BVPs) as available from www.netlib.org/odepack.

Fortunately, most introductory fluid flow problems are governed by ODEs of the form

$$\frac{d^n}{dy^n}\left[f(y)\right]=g(y) \tag{A.4.1}$$

which can be solved by direct integration, subject to n boundary conditions (BCs).

Typically, n=2 and $g(y)\equiv K$, a constant, so that Eq. (A.4.1) can be rewritten as

$$\frac{d}{dy}\left[\frac{df}{dy}\right]=K \tag{A.4.2a}$$

or

$$\mathrm{d}\left[\frac{\mathrm{df}}{\mathrm{dy}}\right]=\mathrm{Kdy} \tag{A.4.2b}$$

Hence,

$$\frac{df}{dy}=Ky+C_1 \tag{A.4.2c}$$

and integrating again yields

$$f(y)=\frac{K}{2}y^2+C_1y+C_2 \tag{A.4.3}$$

where the integration constants are determined with two given boundary conditions for f(y).

In cylindrical coordinates, we may encounter an ODE somewhat similar to Eq. (A.4.1) of the form

$$\frac{1}{r}\frac{df(r)}{dr}+\frac{d^2f(r)}{dr^2}=g(r) \tag{A.4.4a}$$

which can be rewritten for direct integration as:

$$\frac{1}{r}\frac{d}{dr}\left(r\frac{df(r)}{dr}\right)=g(r) \tag{A.4.4b}$$

For example, with $g(r)\equiv K$=constant and $f\equiv \mathrm{v}$, say the fully-developed axial velocity in a tube of radius R, we have

$$d\left(r\frac{d\mathrm{v}(r)}{dr}\right)=Krdr \tag{A.4.5a}$$

so that after integration:

$$\frac{d\mathrm{v}(r)}{dr}=\frac{K}{2}r+\frac{C_1}{r} \tag{A.4.5b}$$

Integrating again yields

$$\mathrm{v}(r)=\frac{K}{4}r^2+C_1\ln r+C_2 \tag{A.4.6}$$

The differences between the solutions of (A.4.3) and (A.4.6) as well as the impact of different BCs on Eq. (A.4.6) were discussed in Sec. 1.3.

Another interesting case where term contraction allows for direct integration is as an ODE of the form

$$\frac{df(x)}{dx}+\frac{n}{x}f(x)\equiv\frac{1}{x^n}\frac{d}{dx}\left[x^n f(x)\right]=g(x); \qquad n=1{,}2 \tag{A.4.7a}$$

which yields after the first integration:

$$x^n f(x)=\int x^n g(x)dx+C_1 \tag{A.4.7b}$$

The term $\int x^n g(x)dx$ could be solved via integration by parts, i.e.,

$$\int_a^b udv=(uv)\Big|_a^b-\int_a^b vdu \tag{A.4.8}$$

Numerous natural processes can be described by a linear, inhomogeneous 2nd-order ODE with constant coefficients, i.e.,

$$f''+Af'+Bf=F(x) \tag{A.4.9}$$

where F(x) is a prescribed (forcing) function. Typically,

$$f(x)=f_{\text{homog.}}+f_{\text{inhomog.}} \tag{A.4.10a}$$

In general, $f_{\text{homog.}}$ admits exponential solutions, e.g.,

$$f(x)\sim e^{\lambda x} \tag{A.4.10b}$$

where λ can be obtained from the quadratic equation

$$\lambda^2+A\lambda+B=0 \tag{A.4.10c}$$

so that

$$f(x)=C_1e^{\lambda_1 x}+C_2e^{\lambda_2 x} \tag{A.4.10d}$$

Table A.4.1 summarizes typical ODEs describing transport phenomena, where f and g are functions of x and the quantities a, b, and c are real constants.

Table A4.1 Typical ODEs and their general solutions.

Equation	General Solution
$\frac{dy}{dx} = \frac{f(x)}{g(y)}$	$\int g dy = \int f dx + C_1$
$\frac{dy}{dx} + f(x)y = g(x)$	$y = e^{-\int f dx}(\int e^{\int f dx} g dx + C_1)$
$\frac{d^2 y}{dx^2} + a^2 y = 0$	$y = C_1 \cos ax + C_2 \sin ax$
$\frac{d^2 y}{dx^2} - a^2 y = 0$	$y = C_1 \cosh ax + C_2 \sinh ax$ or $y = C_3 e^{+ax} + C_4 e^{-ax}$
$\frac{1}{x}\frac{d}{dx}\left(x^2 \frac{dy}{dx}\right) + a^2 y = 0$	$y = \frac{C_1}{x}\cos ax + \frac{C_2}{x}\sin ax$

APPENDIX B

Single-Phase Field Equations

Equation of Motion: $\frac{\partial}{\partial t}(\rho\vec{\mathbf{v}})+\nabla\cdot(\rho\vec{\mathbf{v}}\vec{\mathbf{v}})=-\nabla p+\nabla\cdot\vec{\vec{\tau}}+\rho\vec{\vec{\tau}}$

Shear stress tensor: $\vec{\vec{\tau}}=2\mu\vec{\vec{\varepsilon}}$ (Newtonian fluids)

$\mu\rightarrow\eta=\eta(\dot{\gamma})$ for non-Newtonian fluids, e.g., $\eta=K\dot{\gamma}^{n-1}$

Mass Conservation Equations:

Fluid mass (continuity): $\frac{\partial\rho}{\partial t}+\nabla\cdot(\rho\vec{\mathbf{v}})=0$

Species mass (convection-diffusion equation): $\frac{\partial c}{\partial t}+\nabla\cdot(c\vec{\mathbf{v}})=D\nabla^2 c+S_c$

Navier-Stokes Equation: $\partial\vec{v}/\partial t+(\nabla\cdot\vec{v})\vec{v}=(-1/\rho)\nabla p+(\mu/\rho)\nabla^2\vec{v}+\vec{g}$; *Continuity:* $\nabla\cdot\vec{v}=0$ (for incompressible flow)

Inviscid (frictionless): Euler Eqn.: $D\vec{v}/Dt=(-1/\rho)\nabla p+\vec{g}$; $<\mu\equiv 0>$

Negligible inertia: Stokes Eqn.: $\nabla p\approx\mu\nabla^2\vec{v}$ (Re<1.0)

Expanded N-S Equations (Constant Fluid Props):

Rectangular Coordinates: Continuity Equation: $\partial u/\partial x+\partial v/\partial y+\partial w/\partial z=0$

x-momentum: $$\rho\left(\frac{\partial u}{\partial t}+u\frac{\partial u}{\partial x}+v\frac{\partial u}{\partial y}+w\frac{\partial u}{\partial z}\right)=F_x-\frac{\partial p}{\partial x}+\mu\left(\frac{\partial^2 u}{\partial x^2}+\frac{\partial^2 u}{\partial y^2}+\frac{\partial^2 u}{\partial z^2}\right)$$

y-momentum: $$\rho\left(\frac{\partial v}{\partial t}+u\frac{\partial v}{\partial x}+v\frac{\partial v}{\partial y}+w\frac{\partial v}{\partial z}\right)=F_y-\frac{\partial p}{\partial y}+\mu\left(\frac{\partial^2 v}{\partial x^2}+\frac{\partial^2 v}{\partial y^2}+\frac{\partial^2 v}{\partial z^2}\right)$$

z-momentum: $$\rho\left(\frac{\partial w}{\partial t}+u\frac{\partial w}{\partial x}+v\frac{\partial w}{\partial y}+w\frac{\partial w}{\partial z}\right)=F_z-\frac{\partial p}{\partial z}+\mu\left(\frac{\partial^2 w}{\partial x^2}+\frac{\partial^2 w}{\partial y^2}+\frac{\partial^2 w}{\partial z^2}\right);\quad F_z=\rho g_z$$

Acceleration	Shear Stresses:	Vorticity:
$a_x=\frac{Du}{Dt}=\frac{\partial u}{\partial t}+u\frac{\partial u}{\partial x}+v\frac{\partial u}{\partial y}+w\frac{\partial u}{\partial z}$	$\tau_{xy}=\mu\left(\frac{\partial v}{\partial x}+\frac{\partial u}{\partial y}\right)$	$\omega_x=\left(\frac{\partial w}{\partial y}-\frac{\partial v}{\partial z}\right)$
$a_y=\frac{Dv}{Dt}=\frac{\partial v}{\partial t}+u\frac{\partial v}{\partial x}+v\frac{\partial v}{\partial y}+w\frac{\partial v}{\partial z}$	$\tau_{yz}=\mu\left(\frac{\partial w}{\partial y}+\frac{\partial v}{\partial z}\right)$	$\omega_y=\left(\frac{\partial u}{\partial z}-\frac{\partial w}{\partial x}\right)$
$a_z=\frac{Dw}{Dt}=\frac{\partial w}{\partial t}+u\frac{\partial w}{\partial x}+v\frac{\partial w}{\partial y}+w\frac{\partial w}{\partial z}$	$\tau_{zx}=\mu\left(\frac{\partial u}{\partial z}+\frac{\partial w}{\partial x}\right)$	$\omega_z=\left(\frac{\partial v}{\partial x}-\frac{\partial u}{\partial y}\right)$

Cylindrical Coordinates: Continuity: $\frac{1}{r}\frac{\partial}{\partial r}(rv_r)+\frac{\partial v_z}{\partial z}+\frac{1}{r}\frac{\partial}{\partial\theta}(v_\theta)=0$

r-Momentum:

$$\rho\left(\frac{\partial v_r}{\partial t}+v_r\frac{\partial v_r}{\partial r}+\frac{v_\theta}{r}\frac{\partial v_r}{\partial \theta}-\frac{v_\theta^{\,2}}{r}+v_z\frac{\partial v_r}{\partial z}\right)=\rho g_r-\frac{\partial p}{\partial r}$$
$$+\mu\left[\frac{\partial}{\partial r}\left(\frac{1}{r}\frac{\partial}{\partial r}[rv_r]\right)+\frac{1}{r^2}\frac{\partial^2 v_r}{\partial \theta^2}-\frac{2}{r^2}\frac{\partial v_\theta}{\partial \theta}+\frac{\partial^2 v_r}{\partial z^2}\right]$$

θ-Momentum:

$$\rho\left(\frac{\partial v_\theta}{\partial t}+v_r\frac{\partial v_\theta}{\partial r}+\frac{v_\theta}{r}\frac{\partial v_\theta}{\partial \theta}+\frac{v_r v_\theta}{r}+v_z\frac{\partial v_\theta}{\partial z}\right)=\rho g_\theta-\frac{1}{r}\frac{\partial p}{\partial \theta}$$
$$+\mu\left[\frac{\partial}{\partial r}\left(\frac{1}{r}\frac{\partial}{\partial r}[rv_\theta]\right)+\frac{1}{r^2}\frac{\partial^2 v_\theta}{\partial \theta^2}-\frac{2}{r^2}\frac{\partial v_r}{\partial \theta}+\frac{\partial^2 v_\theta}{\partial z^2}\right]$$

z-Momentum:

$$\rho\left(\frac{\partial v_z}{\partial t}+V_r\frac{\partial v_z}{\partial r}+\frac{v_\theta}{r}\frac{\partial v_z}{\partial \theta}+v_z\frac{\partial v_z}{\partial z}\right)=+\rho g_z-\frac{\partial p}{\partial z}+\mu\left[\frac{1}{r}\frac{\partial}{\partial r}\left(r\frac{\partial v_z}{\partial r}\right)+\frac{1}{r^2}\frac{\partial^2 v_z}{\partial \theta^2}+\frac{\partial^2 v_z}{\partial z^2}\right]$$

Acceleration:

$$a_r=\frac{\partial v_r}{\partial t}+v_r\frac{\partial v_r}{\partial r}+\frac{v_\theta}{r}\frac{\partial v_r}{\partial \theta}-\frac{v_\theta^{\,2}}{r}+v_z\frac{\partial v_r}{\partial z}$$
$$a_\theta=\frac{\partial v_\theta}{\partial t}+v_r\frac{\partial v_\theta}{\partial r}+\frac{v_\theta}{r}\frac{\partial v_\theta}{\partial \theta}+\frac{v_r v_\theta}{r}+v_z\frac{\partial v_\theta}{\partial z}$$
$$a_z=\frac{\partial v_z}{\partial t}+v_r\frac{\partial v_z}{\partial r}+\frac{v_\theta}{r}\frac{\partial v_z}{\partial \theta}+v_z\frac{\partial v_z}{\partial z}$$

Shear Stresses:

$$\tau_{r\theta}=\mu\left[r\frac{\partial}{\partial r}\left(\frac{v_\theta}{r}\right)+\frac{1}{r}\frac{\partial v_r}{\partial \theta}\right]$$
$$\tau_{\theta z}=\mu\left[\frac{\partial v_\theta}{\partial z}+\frac{1}{r}\frac{\partial v_z}{\partial \theta}\right]$$
$$\tau_{rz}=\mu\left[\frac{\partial v_r}{\partial z}+\frac{\partial v_z}{\partial r}\right],$$

Vorticity:

$$\omega_r=\left[\frac{1}{r}\frac{\partial v_z}{\partial \theta}-\frac{\partial v_\theta}{\partial z}\right]$$
$$\omega_\theta=\left[\frac{1}{r}\frac{\partial v_z}{\partial \theta}-\frac{\partial v_\theta}{\partial z}\right]$$
$$\omega_z=\left[\frac{1}{r}\frac{\partial (rv_\theta)}{\partial r}-\frac{1}{r}\frac{\partial v_r}{\partial \theta}\right]$$

APPENDIX C

Suitable CFD Solvers

Clearly, the complexity of a given biofluids problem determines the type of numerical solution method and computer platform to be employed. For example, quasi-homogeneous mixture flows described by ODEs (or linear PDEs) can be simulated and graphed using math software (e.g., MathCad, MATLAB, Mathematica, etc.) on a PC, uncoupled fluid-particle flows can be solved with CFD software (e.g., Fluent, CFX, etc.) on Unix workstations (or NT-based PCs), and coupled multiphysics problems can be solved with two-field solvers (e.g., commercial software CFD-TWOPHASE, ANSYS10-CFX10, etc.) on multiprocessor computers, including terascale supercomputers.

C.1 Basic ODE and PDE Solvers

There are various tools available for solving basic problems in the engineering sciences, ranging from scientific calculators, programming languages and spreadsheets, to symbolic/numeric PC software such as Maple, MathCad, MATLAB, FEMLAB, Mathematica, Macsyma, etc. While MathCad, MATLAB, and FEMLAB stress numerical problem solving, the others feature symbolic math programs. Needless to say, as new versions are being issued annually, the differences in capability and performance are getting smaller.

The focus in this section is on MATLAB 7 (MATrix LABoratory, Version 7) and FEMLAB 3.1 for MS Windows. MATLAB is a computer programming language devoted to processing data in the form of matrices where numeric/symbolic computations and visualizations with a diverse family of built-in functions are integrated into a flexible computer environment (Fig. C.1a). The basic building block is the matrix and hence array-heavy problems, for example, in electrical and structural engineering, are best tackled with MATLAB. Supporting engineering texts of increasing depth include Pratrap (2002), and Highham & Highham (2005).

Algebraic systems, ordinary differential equations, and *linear* two-dimensional partial differential equations can be directly solved and the solutions graphed, where special tasks can be addressed with add-on software modules called *toolboxes*, which include linear PDE solvers, Symbolic Computations, Control Systems, Statistics, Image Processing, etc.

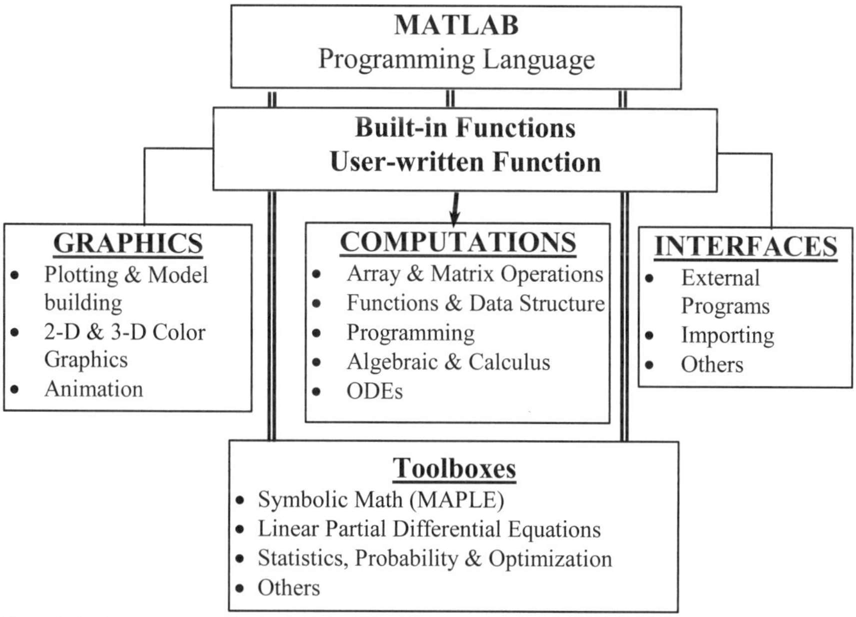

Fig. C.1a Features of MATLAB.

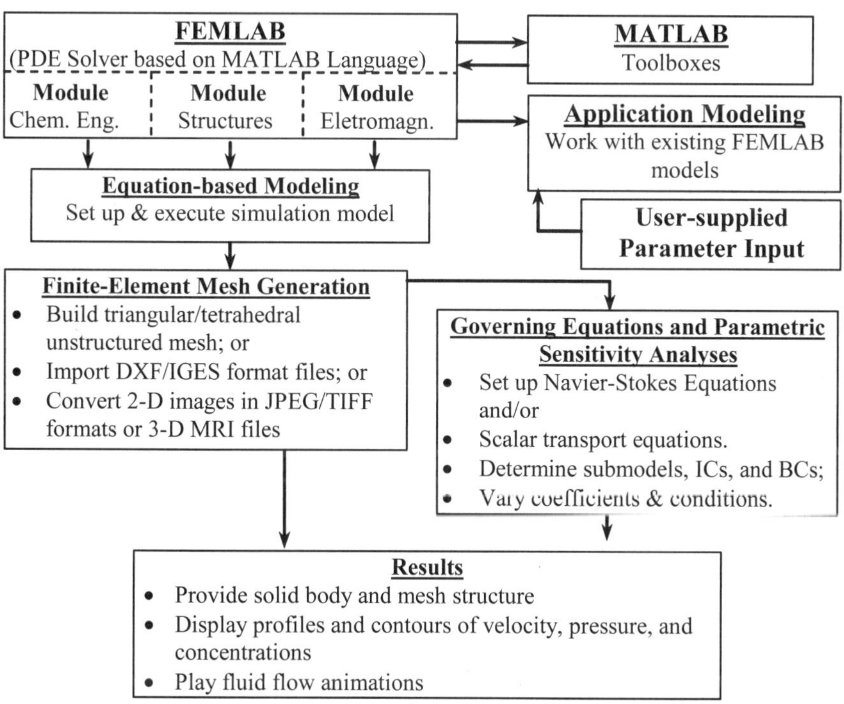

Fig. C.1b Fluid flow modeling with FEMLAB.

FEMLAB, using the MATLAB language, is a *nonlinear* PDE solver for two- and three-dimensional engineering science problems as encountered in transport phenomena, structural analyses, and electromagnetism (cf. Fig. C.1b). The user can either work with existing FEMLAB models in changing equation coefficients, boundary conditions, system parameters, etc., or set up a mathematical model for a specific problem solution. The modeling capabilities and sequential steps are given in Fig. C.1b. Employing Graphic User Interface (GUI) the advantage is that, within limits, FEMLAB is easy to use, i.e., menu-driven operations and interactions with MS Windows, MATLAB, etc.

C.2 Advanced PDE Solvers

Table C.2.1 lists popular commercial software packages for solving mass, momentum, and heat transfer problems. While the CFD software CFX4 and an associated particle tracking code was discussed by Kleinstreuer (2003), here highlights are given for FLEXPDE5 and ANSYS-CFX10.

Table C.2.1 Some Commercial CFD Codes.

CFD Code	Mesh and Solver	Description
CFD-ACE (CFD Research Corp., USA)	• Structured and unstructured mesh • Finite volume method	• Transient 3-D turbulent flow simulation • Turbulence modeling ($k-\varepsilon$ and eddy viscosity model for two-phase flow) • Lagrangian particle tracking • Two-fluid model using VOF method* • User-friendly interfaces
CFD 2000 (Adaptive Research Corp., USA)	• Structured mesh • Moving grid • Finite volume method	• Transient 3-D turbulent flow simulation • Turbulence modeling (state-of-the-art turbulence models) • Lagrangian particle tracking with particle merging, breakup, evaporating, and ablation • Two-fluid model using VOF method • User-friendly interfaces
CFX (CFX-5, Ansys, Inc.)	• Unstructured mesh (CFX-5) • Moving grid • Finite volume method	• Transient 3-D flow simulation • Turbulence modeling (state-of-the-art turbulence models) • Lagrangian particle tracking • User-friendly interfaces
Flex PDE (PDE Solutions Inc.)	• Unstructured finite element mesh • Finite element method	• Scripted finite element model builder and numerical solver • Solver for first or second order partial differential equations in one, two, or three-dimensional Cartesian geometry, or in axi-symmetric two-dimensional geometry • Any number of simultaneous equations (steady state or time dependent, linear or nonlinear) can be solved

Fluent (Fluent, USA) FIDAP	• Unstructured mesh • Finite volume method • Finite element method	• Transient 3-D flow simulation • Turbulence modeling (state-of-the-art turbulence models) • Lagrangian particle tracking • Two-fluid model using VOF method • Fluid-structure interactions
ANSYS-CFX10 (Ansys, Inc.)	• Unstructured mesh • Moving grid • Finite volume and finite element methods	• Updated version of CFX5 and combines with ANSYS10 (structure analysis software) • Coupling Fluid-structure interactions • Transient particle tracking
Star-CD (CD Adapco, UK)	• Unstructured mesh • Finite volume method	• Transient 3-D flow simulation • Turbulence modeling (state-of-the-art turbulence models) • Lagrangian particle tracking • Two-fluid model using VOF-method • User-friendly interfaces

[*] VOF, Volume-of-fluid (cf. resolved-volume method; Loth, 2000).

FlexPDE5. FlexPDE is a "scripted finite element model builder and numerical solver." It is different from the traditional commercial CFD codes (e.g., CFX), which has the pre-defined equation list (e.g., momentum equations, scalar equations) and standard equation terms (e.g., convection, diffusion, and source terms). The choice of partial differential equations in FlexPDE is totally up to the user. FlexPDE addresses the mathematical basis of different fields such as heat and fluid flow, stress analysis, chemical reactions by treating the equations rather than the application. However, the import of complicated geometry files generated by other mesh generators is difficult and the solution convergence and robustness for many complicated problems need to be improved. In general, FlexPDE is easy to use for solving systems of first or second-order partial differential equations in 1D, 2D, or 3D, which can be easily defined in Cartesian or axisymmetric 2D geometries.

The FlexPDE interface provides a full text (problem description script), editing facility, and a graphical domain preview. The script, a readable text file, consists of a number of sections, each identified by a header. The results generated by FlexPDE can also be reported in many formats, including simple tables, full finite element mesh data, or TecPlot compatible files. The student version of the FlexPDE can be downloaded from http://www.pdesolutions.com. More details for using FlexPDE as well as the well-documented "User Guide," "Reference Manual," and sample scripts for various applications can also be found on that website.

Similarly, Fluent Inc., at http://www.fluent.com, invites CFD people to try out studentFluent, which is a mesh-size restricted version of Fluent without the mesh generator Gambit.

ANSYS-CFX10. ANSYS-CFX10 combines the finite-volume based CFD code CFX with the commonly used structural analysis software ANSYS. The CFX part of this software is updated from CFX5 and it introduces both new transient physical models (such as Transient Particle Tracking), as well as algorithmic and transient efficiency improvements (adaptive time stepping and extrapolate initial solutions). ANSYS is a finite element software package for linear and nonlinear multi-physics analysis in a static or dynamic environment. It can handle even the most complex assemblies, especially those involving nonlinear contact and is the ideal choice for determining stresses, temperatures, displacements, and contact pressure distributions on all component and assembly designs. Therefore, ANSYS-CFX10 is an applicable solver for various Fluid Structure Interaction (FSI) applications (see Sect.5.4). In ANSYS-CFX10, the CFX fluid flow and ANSYS FE analysis software packages are applied simultaneously to a single coupled simulation. Inter-process communication (IPC) is used to exchange data dynamically between CFX and ANSYS solvers to obtain a coupled steady or transient fluid-solid simulation. CFX and ANSYS solvers communicate via proprietary two-way coupling. Each solver can run on the same or separate computers, permitting mixed hardware types, CFX parallel processing, and communication across networks, if needed.

APPENDIX D
Physical Properties

Table D.1 Physical Properties of Gases at Atmospheric Pressure.

T, K	ρ, $\frac{kg}{m^3}$	c_p, $\frac{kJ}{kg \cdot °C}$	μ, $\frac{kg}{m \cdot s}$	ν, $\frac{m^2}{s} \times 10^6$	k, $\frac{W}{m \cdot K}$	α, $\frac{m^2}{s} \times 10^4$	Pr
Air							
100	3.60.10	1.0266	0.6924×10-5	1.923	0.009246	0.02501	0.770
150	2.3675	1.0099	1.0283	4.343	0.013735	0.05745	0.753
200	1.7684	1.0061	1.3289	7.490	0.01809	0.10165	0.739
250	1.4128	1.0053	1.488	10.53	0.02227	0.13161	0.722
300	1.1774	1.0057	1.983	16.84	0.02624	0.22160	0.708
350	0.9980	1.0090	2.075	20.76	0.03003	0.2983	0.697
400	0.8826	1.0140	2.286	25.90	0.03365	0.3760	0.689
450	0.7833	1.0207	2.484	31.71	0.03707	0.4222	0.683
500	0.7048	1.0295	2.671	37.90	0.04038	0.5564	0.680
550	0.6423	1.0392	2.848	44.34	0.04360	0.6532	0.680
600	0.5879	1.0551	3.018	51.34	0.04659	0.7512	0.680
650	0.5430	1.0635	3.177	58.51	0.04953	0.8578	0.682
700	0.5030	1.0752	3.332	66.25	0.05230	0.9672	0.684
750	0.4709	1.0856	3.481	73.91	0.05509	1.0774	0.686
800	0.4405	1.0978	3.625	82.29	0.05779	1.1951	0.689
850	0.4149	1.1095	3.765	90.75	0.06028	1.3097	0.692
900	0.3925	1.1212	3.899	99.3	0.06279	1.4271	0.696
950	0.3716	1.1321	4.023	108.2	0.06525	1.5510	0.699
1000	0.3524	1.1417	4.152	117.8	0.06752	1.6779	0.702
1100	0.3204	1.160	4.44	138.6	0.0732	1.969	0.704
1200	0.2947	1.179	4.69	159.1	0.0782	2.251	0.707
1300	0.2707	1.197	4.93	182.1	0.0837	2.583	0.705
1400	0.2515	1.214	5.17	205.5	0.0891	2.920	0.705
1500	0.2355	1.230	5.40	229.1	0.0946	3.262	0.705
1600	0.2211	1.248	5.63	254.5	0.100	3.609	0.705
1700	0.2082	1.267	5.85	280.5	0.105	3.977	0.705
Steam (H_2O) vapor)							
380	0.5863	2.060	12.71×10^{-6}	21.6	0.0246	0.2036	1.060
400	0.5542	2.014	13.44	24.2	0.0261	0.2338	1.040
450	0.4902	1.980	15.25	31.1	0.0299	0.307	1.010
500	0.4405	1.985	17.04	38.6	0.0339	0.387	0.996
550	0.4005	1.997	18.81	47.0	0.0379	0.475	0.991
600	0.3652	2.026	20.67	56.6	0.0422	0.573	0.986
650	0.3380	2.056	22.47	64.4	0.0464	0.666	0.995
700	0.3140	2.085	24.26	77.2	0.0505	0.772	1.000
750	0.2931	2.119	26.04	88.8	0.0549	0.883	1.005
800	0.2739	2.152	27.86	102.0	0.0592	1.001	1.010
850	0.2579	2.186	29.69	115.2	0.0637	1.130	1.019

Table D.2 Physical Properties of Saturated Liquids.

T,°C	ρ, $\frac{kg}{m^3}$	c_p, $\frac{kJ}{kg \cdot ° C}$	ν, $\frac{m^2}{s} \times 10^6$	k, $\frac{W}{m \cdot K}$	α, $\frac{m^2}{s} \times 10^4$	Pr
Water H_2O						
0	1,002.28	4.2178	1.788×10^{-6}	0.552	1.308	13.6
20	1,000.52	4.1818	1.006	0.597	1.430	7.02
40	994.59	4.1784	0.658	0.628	1.512	4.34

Table D.3 Relative Density of Common Aerosol Materials (Solids).
(Multiply values by 1000 for density in kg/m^3)

Solids	Relative Density	Solids	Relative Density
Aluminim	2.7	Natural fibers	1–1.6
Aluminum oxide	4.0	Plastics	1–1.6
Aluminum sulfate	1.8	Pollens	0.45–1.05
Asbestos	2.0–2.8	Quartz	2.6
Coal	1.2–1.8	Sodium chloride	2.2
Glass	2.4–2.8	Starch	2.1
Ice	0.92	Monocyte (10–14 μm)	1.0–1.05
Iron	7.9	Red blood cell (5 μm)	1.0
Lead	11.3	White blood cell (2–3 μm)	1.0

Table D.4 Physical Properties of Biological Materials (adapted from Chato, 1985).

Description	**Thermal Conductivity W/mK**	**Thermal Diffusivity $\times 10^7$ m^2/s**
Solid materials [b]		
Dry bone	0.22	
Fat	0.094-0.37	
Heart	0.48-0.59	1.4-1.5
Kidney	0.49-0.63	1.3-1.8
Liver	0.42-0.57	1.1-2.0
Lung		2.4-2.8
Muscle	0.34-0.68	1.8
Skin	0.21-0.41	0.82-1.2
Skin [b]	0.48-2.8	0.4-1.6
Spleen	0.45-0.60	1.3-1.6
Tumors, General range	0.47-0.58	
Biological fluids		
Blood (at room to body temperature)	0.57-0.12H [c]	
Blood (-10°C)	1.6	8.7
Blood (-20°C)	1.7	10.4
Blood (-40°C)	1.9	13.6
Blood (-60°C)	2.1	16.9
Blood (-80°C)	2.4	20.4
Blood (-100°C)	2.7	23.7
Blood plasma (at room to body temperature)	0.57-0.60	
Humor, aqueous and vitreous	0.58-0.59	
Urine	0.56	

[a] Unless noted otherwise all are human materials in vitro from room to body temperatures.
[b] It is the material *in vivo* (values depend on blood perfusion).
[c] H is the hematocrit fraction.

References

Aris, R. (1989) "Vectors, Tensors, and the Basic Equations of Fluid Mechanics," Dover, New York.

Bird, R. B., Stewart, W. E., and Lightfoot, E. N. (2002) "Transport Phenomena," John Wiley, New York, NY.

Chato, J.C. (1985) Selected thermophysical properties of biological materials, In: Heat Transfer in Medicine and Biology, A. Shitzer and R.C. Eberhart (eds.), Plenum Press, New York.

Greenberg, M.D. (1998) "Advanced Engineering Mathematics," Prentice Hall, Upper Saddle River, NJ.

Highham, D.J., and Highham, N.J. (2005) "Matlab Guide," 2nd Edition, Society for Industrial and Applied Mathematics, Philadelphia.

Kleinstreuer, C. (2003) "Two-Phase Flow: Theory and Applications," Taylor & Francis, New York, NY.

Loth, E. (2000) Prog. Energy Combustion Sci., 26: 161-223.

Pratrap, R. (2002) "Getting Started with Matlab," Oxford University Press.

Schey, H. M. (1973) "Div, Grad, Curl, and All That," W.W. Norton & Co.

Spiegel, M.R. (1971) "Schaum's Outline of Theory and Problems of Advanced Mathematics for Engineers and Scientists," McGraw-Hill, New York, NY.

INDEX

Note: Most biomedical terms and some BME correlations used in the text are listed and defined in the Glossary, pages xi-xxiii. They are not repeated in this index.

A

D

E

F

G

H

I

J

K

L

M

N

O

P

Q

R

T

RELATED TITLES

Reliable Design of Medical Devices, Second Edition
Richard Fries
ISBN: 0824723759

The Medical Device R&D Handbook
Theodore R. Kucklick
ISBN: 0849327172

Biological Performance of Materials: Fundamentals of Biocompatibility, Fourth Edition
Jonathan Black
ISBN: 0849339596

An Introduction to Biomaterials
Scott A. Guelcher and Jeffrey O. Hollinger
ISBN: 0849322820

Biomedical Signal and Image Processing
Kayvan Najarian and Robert Splinter
ISBN: 0849320992